Beam Deflections and Slopes (continued)

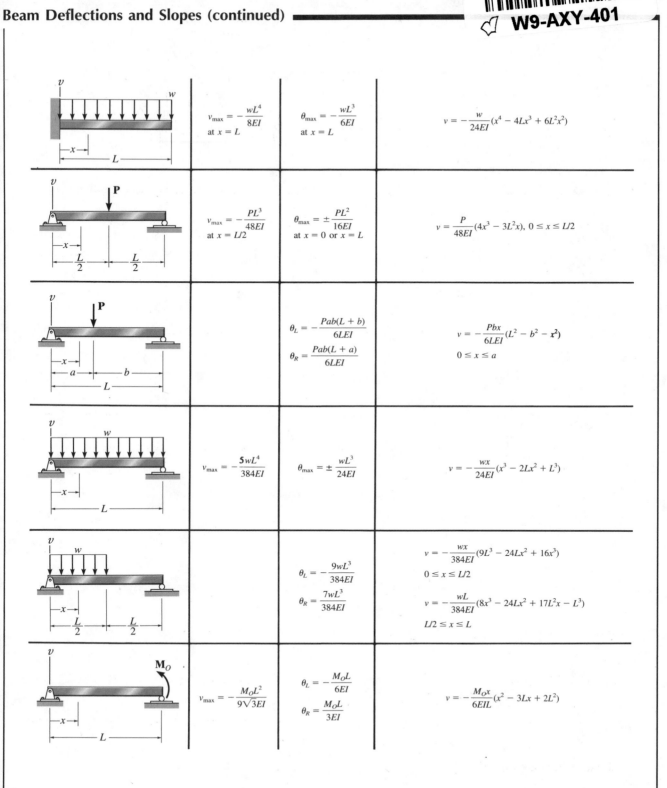

	$v_{max} = -\dfrac{wL^4}{8EI}$ at $x = L$	$\theta_{max} = -\dfrac{wL^3}{6EI}$ at $x = L$	$v = -\dfrac{w}{24EI}(x^4 - 4Lx^3 + 6L^2x^2)$
	$v_{max} = -\dfrac{PL^3}{48EI}$ at $x = L/2$	$\theta_{max} = \pm\dfrac{PL^2}{16EI}$ at $x = 0$ or $x = L$	$v = \dfrac{P}{48EI}(4x^3 - 3L^2x),\ 0 \le x \le L/2$
		$\theta_L = -\dfrac{Pab(L + b)}{6LEI}$ $\theta_R = \dfrac{Pab(L + a)}{6LEI}$	$v = -\dfrac{Pbx}{6LEI}(L^2 - b^2 - x^2)$ $0 \le x \le a$
	$v_{max} = -\dfrac{5wL^4}{384EI}$	$\theta_{max} = \pm\dfrac{wL^3}{24EI}$	$v = -\dfrac{wx}{24EI}(x^3 - 2Lx^2 + L^3)$
		$\theta_L = -\dfrac{9wL^3}{384EI}$ $\theta_R = \dfrac{7wL^3}{384EI}$	$v = -\dfrac{wx}{384EI}(9L^3 - 24Lx^2 + 16x^3)$ $0 \le x \le L/2$ $v = -\dfrac{wL}{384EI}(8x^3 - 24Lx^2 + 17L^2x - L^3)$ $L/2 \le x \le L$
	$v_{max} = -\dfrac{M_OL^2}{9\sqrt{3}EI}$	$\theta_L = -\dfrac{M_OL}{6EI}$ $\theta_R = \dfrac{M_OL}{3EI}$	$v = -\dfrac{M_Ox}{6EIL}(x^2 - 3Lx + 2L^2)$

Structural Analysis

Russell C. Hibbeler

Structural Analysis

THIRD EDITION

Prentice Hall, Upper Saddle River, New Jersey 07458

Hibbeler, R. C.
    Structual analysis / Russell C. Hibbeler.—3rd ed.
        p.   cm.
    Includes index.
    ISBN 0-02-354041-9
    1. Structural analysis (Engineering)   I. Title.
TA645.H47 1994
624.1'71—dc20   93-41397
                                                                                    CIP

Editor: William Stenquist
Production Supervisor: Margaret Comaskey
Production Managers: Paul Smolenski/Nicholas Sklitsis
Text Designer: Robert Freese
Cover Designer: Robert Vega
Cover Art (Photo): John V.A.F. Neal/INTERNATIONAL STOCK
Photo Researcher: Barbara Scott
Illustrations: Susan Sibille

©1995 by Prentice-Hall, Inc.
A Simon & Schuster Company
Upper Saddle River, New Jersey 07458

Printed in the United States of America

10 9 8 7 6 5 4

ISBN 0-02-354041-9

Prentice-Hall International (UK) Limited, *London*
Prentice-Hall of Australia Pty. Limited, *Sydney*
Prentice-Hall Canada Inc., *Toronto*
Prentice-Hall Hispoanoamericana, S.A., *Mexico*
Prentice-Hall of India Private Limited, *New Delhi*
Prentice-Hall of Japan, Inc., *Tokyo*
Simon & Schuster Asia Pte. Ltd., *Singapore*
Editora Prentice-Hall do Brazil, Ltda., *Rio de Janeiro*

Preface

This book is intended to provide the student with a clear and thorough presentation of the theory and application of structural analysis as it applies to trusses, beams, and frames. Emphasis is placed on developing the student's ability to both model and analyze a structure. In this *third edition*, the presentation and arrangement of some of the topics have been improved, and the examples and problem assignments have been expanded. In particular, a chapter on cables and arches has been added, and the contents of a previous chapter on a general explanation of displacement methods have been incorporated into the chapter on the slope-deflection method. Also, a topic using the method of integration to determine the slope and deflection of a beam has been included. Additional photographs and improved art throughout the book have been used to enhance the understanding of the material, and particular attention has been given to numerical accuracy and clarity in both defining and developing the concepts.

Organization and Approach

The contents of each chapter are arranged into sections with specific topics categorized by title headings. Discussions relevant to a particular theory are brief yet thorough. In most cases this is followed by a "procedure for analysis" guide, which provides the student with a summary of the important concepts and a systematic approach for applying the theory. The example problems are solved using this outlined method in order to clarify its numerical application.

The book is divided into three parts. The classicial methods of analysis are covered in Chapters 2–7 for statically determinate structures and Chapters 8–12 for statically indeterminate structures. The third part of the book, Chapters 13–15, presents matrix methods for structural analysis. During recent years there has been a growing emphasis on using computers to analyze structures by matrix analysis. These developments are most welcome, because they relieve the engineer of the often lengthy calculations required when large or complicated structures are analyzed using classical methods. Although matrix methods are more efficient for a structural analysis, it is the author's opinion that students taking a first course in this subject should also be well versed in the classicial methods. Practice in applying these methods will develop a deeper understanding of the basic engineering sciences of statics and mechanics of materials. Also, problem-solving skills are further developed when the various techniques are thought out and applied in a clear and orderly way. By experience, one can better grasp the way loads are transmitted through structures and obtain a more complete understanding of the way structures deform under load. Finally, the classicial methods provide a means of checking computer results rather than simply relying on the generated output.

Problems

Most of the problems in the book depict realistic situations encountered in practice. It is hoped that this realism will both stimulate the student's interest in structural analysis and develop the skill to reduce any such problem from its physical description to a model or symbolic representation to which the appropriate theory can be applied. Both SI and FPS units are used throughout the book. The problems are presented at the end of each chapter and are arranged to cover the material in sequential order; moreover, for any topic they are arranged in approximate order of increasing difficulty. The intent here has been to develop problems that test the student's ability to apply the theory, keeping in mind that those problems requiring tedious calculations can be relegated to computer analysis. The answers to all but every fourth problem (indicated by an asterisk) are listed in the back of the book. All the solutions have been reviewed by others, and the answers have been carefully checked for numerical accuracy.

Contents

This book consists of 15 chapters. Chapter 1 provides a brief discussion of the various types of structural forms and loads. The analysis of statically determinate structures is covered in the next six chapters. Chapter 2 discusses the determination of forces at a structure's supports and connections. The analysis of various types of statically determinate trusses is given in Chapter 3, and shear and bending-moment functions and diagrams for beams and frames are presented in Chapter 4. In Chapter 5, the analysis of simple cable and arch systems is presented, and in Chapter 6 influence lines for beams, girders, and trusses are discussed. Finally, in Chapter 7 several common techniques for the approximate analysis of statically indeterminate structures are considered.

In the second part of the book, the analysis of statically indeterminate structures is covered in five chapters. Both geometrical and energy methods for computing deflections are discussed in Chapter 8. Chapter 9 covers the analysis of statically indeterminate structures using the force method of analysis, in addition to a discussion of influence lines for beams. Then the displacement methods consisting of the slope-deflection method in Chapter 10 and moment distribution in Chapter 11 are discussed. Finally, beams and frames having nonprismatic members are considered in Chapter 12.

The third part of the book treats the analysis of structures using matrix methods. Chapter 13 presents a brief discussion of matrix algebra. Chapters 14 and 15 develop the analysis of trusses and frames, respectively, using the global stiffness method. This method has wide acceptance for use on a computer.

Acknowledgments

Many of my colleagues in the teaching profession and my students have made constructive criticisms that have helped in the development of this revision, and I would like to hereby acknowledge all of their valuable suggestions and comments. In particular I would like to thank the reviewers contracted by my editor: Ahmed M. Abdel-Ghaffar, Princeton University; Ahmet Emin Aktan, Louisana State University; Richard A. Behr, University of Missouri at Rolla; William L. Bingham, North Carolina State University; Michael C. Constantinou, State University of New York at Buffalo; Mike Gingrich, Olivet Nazarene University; Phillip L. Gould, Washington University in St. Louis; Harry D. Knostman, Kansas State University; Daryl Logan, Rose-Hulman Institute of Technology; Seroj Mackertich, Pennsylvania State University at Harrisburg; David F. Mazurek, U.S. Coast Guard Academy; B. B. Muvdi, Bradley University; John F. Ritter, Youngstown State University; Masoud Sanayei, Tufts University; Scott D. Schiff, Clemson University; Jack W. Schwalbe, Florida Institute of Technology; Charles T. Stephens, Oregon Institute of Technology; Leon R. L. Wang, Old Dominion University; R. Warren, Point

Park College; and Aspasia Zerva and Tolga Ergunay, Drexel University. Also, many thanks go to Susan Sibille for producing the many fine illustrations used throughout the book and Barry Nolan for his help in producing the latest version of the STRAN computer program. Lastly, I would like to acknowledge the help I received from my wife Conny, who has always been very supportive and helpful in preparing the manuscript for publication.

R. C. H.

Contents

Contents

4 Internal Loadings Developed in Structural Members

5 Cables and Arches

6 Influence Lines for Statically Determinate Structures

7   Approximate Analysis of Statically Indeterminate Structures

8   Deflections

9 Analysis of Statically Indeterminate Structures by the Force Method

10 Displacement Method of Analysis: Slope-Deflection Equations

11 Displacement Method of Analysis: Moment Distribution

12 Analysis of Beams and Frames Consisting of Nonprismatic Members

13 Matrix Algebra for Structural Analysis

14 Truss Analysis Using the Stiffness Method

15   Beam and Plane Frame Analysis Using the Stiffness Method

Structural Analysis

The warehouse loading, as shown here, can be quite large. To properly design the building, it is necessary to estimate the load and to specify where it will be placed within the building. *(Photo courtesy of Portland Cement Association)*

1

Types of Structures and Loads

In this chapter we will explain the various phases necessary to produce a structure. Also, the basic types of structures, and their elements and supports, will be defined. Finally, a brief explanation is given of the types of loads that must be considered for an appropriate structural analysis and design.

1.1 Introduction

A *structure* consists of a series of connected parts used to support a load. Notable examples include buildings, bridges, towers, tanks, and dams. The process of creating any of these structures requires planning, analysis, design, and construction. A brief discussion of each of these phases follows in order to illustrate what must be considered in practice.

Planning

When creating a structure to serve a specified function for public use, consideration must first be given to selecting a structural form that is safe, esthetic, and economical. This is usually the most difficult and yet the most important phase of structural engineering. Often it requires several independent studies of different solutions before final judgment can be made as to which form (arch, truss, frame, etc.) is the most appropriate. When this has been decided, the loadings, materials, arrangement of the members, and their overall dimensions are then specified. Obviously, the skill necessary to execute these planning activities normally comes when one has had several years of experience in both the art and science of engineering.

Analysis

To analyze a structure properly, certain idealizations must be made as to how the members are supported and connected together. Once this is determined and the loadings have been specified, the forces in the members and their displacements can be found using the theory of structural mechanics, which is the subject matter of this text.

Design

After the internal loadings of a member have been specified, the size of the member can then be found so that it meets the criteria for strength, stability, and deflection as noted in acceptable codes and specifications. Furthermore, the connections between members may be designed for strength and the dimensions detailed so that all the parts fit together properly.

Construction

This final phase requires ordering the various components of the structure and planning the activities that involve the actual erection of the structure. In this regard, all phases of construction must be inspected to make sure that they are in accordance with the specified design drawings.

1.2   Classification of Structures

It is important for a structural engineer to recognize the various types of elements composing a structure and to be able to classify structures as to their form and function. We will introduce some of these aspects now and expand on them at appropriate points throughout the text.

Structural Elements

Some of the more common elements from which structures are composed are as follows.

Tie Rods. Structural members subjected to a *tensile force* are often referred to as *tie rods* or *bracing struts*. Due to the nature of this load, these members are rather slender, and are chosen from rods, bars, angles, channels, and so on, Fig. 1–1.

Beams. Beams are usually straight horizontal members used primarily to carry vertical loads, Fig. 1–2a. Quite often they are classified according to the way they are supported, as indicated in Fig. 1–2b. In particular, when the cross section varies the beam is referred to as a

rod   bar

angle   channel

typical cross sections

tie rod

Fig. 1–1

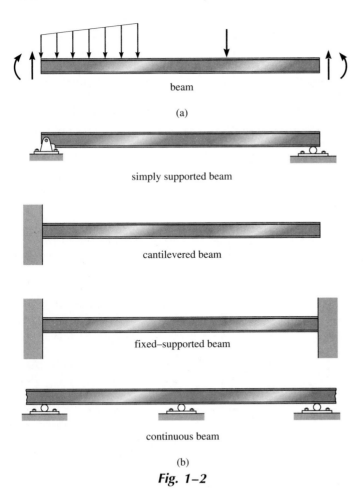

beam

(a)

simply supported beam

cantilevered beam

fixed–supported beam

continuous beam

(b)

Fig. 1–2

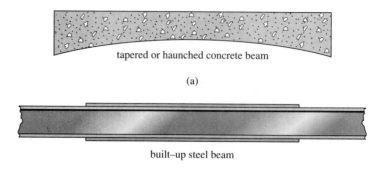

tapered or haunched concrete beam

(a)

built–up steel beam

Fig. 1–3

(b)

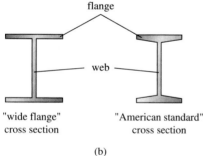

(a)

Fig. 1–4

flange

web

"wide flange"
cross section

"American standard"
cross section

(b)

tapered or haunched beam, Fig. 1–3a. Beam cross sections are also "built up" by adding plates to their top and bottom, Fig. 1–3b.

Beams are most often designed only to resist bending moment, Fig. 1–2a; however, if they are short and carry large loads, the internal shear force may become quite large and this force may govern their design.

When the material used for its construction is a metal such as steel or aluminum, a beam's cross section is most efficient when it is shaped as shown in Fig. 1–4a. Here the forces developed in the top and bottom *flanges* of the beam form the necessary couple used to resist the applied moment **M,** whereas the *web* is effective in resisting the applied shear **V.** This cross section is commonly referred to as a "wide flange," Fig. 1–4b, and it is normally fabricated as a single unit in a rolling mill in lengths up to 75 ft. If shorter lengths are needed, the "American Standard" cross section, having tapered flanges, is sometimes selected, Fig. 1–4b. When the beam is required to have a very large span and the loads applied are rather large, the cross section may take the form of a *plate girder*. This member is fabricated by using a large plate for the web

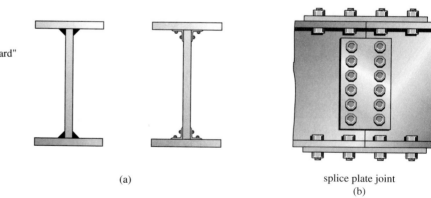

(a)

splice plate joint
(b)

Fig. 1–5

plate girder

and welding or bolting plates to its ends for flanges, Fig. 1–5*a*. The girder is often transported to the field in segments, and the segments are designed to be spliced or joined together at points where the girder carries small internal moments, Fig. 1–5*b*.

Concrete beams generally have rectangular cross sections since it is easy to construct this form directly in the field. Because concrete is rather weak in resisting tension, steel "reinforcing rods" are cast into the beam within regions of the cross section subjected to tension, Fig. 1–6*a*. Precast concrete beams or girders are fabricated at a shop or yard in the same manner, Fig. 1–6*b*, and then transported to the job site.

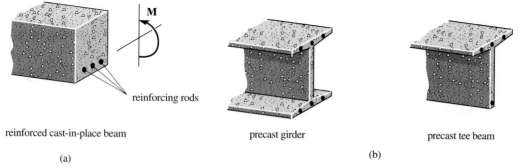

reinforcing rods

reinforced cast-in-place beam

precast girder

precast tee beam

(a)

(b)

Fig. 1–6

Beams made from timber may be sawn from a solid piece of wood or laminated. *Laminated* beams are constructed from solid sections of wood, which are fastened together using high-strength glues. Typical examples of beam cross sections made from sawn and laminated timber are shown in Fig. 1–7.

sawn timber beam

laminated wood beam

Fig. 1–7

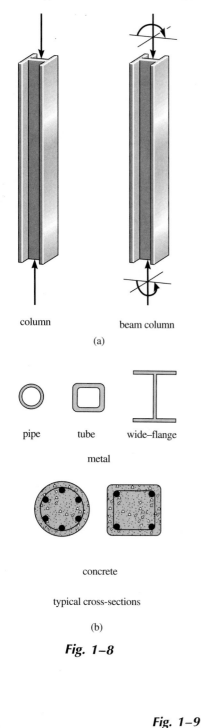

column   beam column

(a)

pipe   tube   wide–flange

metal

concrete

typical cross-sections

(b)

Fig. 1–8

Columns. Members that are generally vertical and resist axial compressive loads are referred to as *columns*, Fig. 1–8*a*. Tubes and wide-flange cross sections are often used for metal columns, and circular and square cross sections with reinforcing rods are used for those made of concrete, Fig. 1–8*b*. Occasionally, columns are subjected to both an axial load and a bending moment as shown in Fig. 1–8*a*. These members are referred to as *beam columns*.

Types of Structures

The combination of structural elements and the materials from which they are composed is referred to as a *structural system*. Each system is constructed of one or more of four basic types of structures. Ranked in order of complexity of their force analysis, they are as follows.

Trusses. When the span of a structure is required to be large and its depth is not an important criterion for design, a truss may be selected. *Trusses* consist of tension ties and slender column elements, usually arranged in triangular fashion. *Planar trusses* are composed of members that lie in the same plane and are frequently used for bridge and roof support, whereas *space trusses* have members extending in three dimensions and are suitable for derricks and towers.

Due to the geometric arrangement of its members, loads that cause bending of the truss, Fig. 1–9, are converted into tensile or compressive forces in the members, and because of this, one of the primary advantages of a truss, compared to a beam, is that it uses less material to support a given load. Also, a truss is constructed from relatively *light-weight elements*, which can be arranged in various ways to support an imposed load. Most often it is economically feasible to use a truss to cover spans ranging from 30 ft to 400 ft, although they have been used on occasion for spans greater than 400 ft.

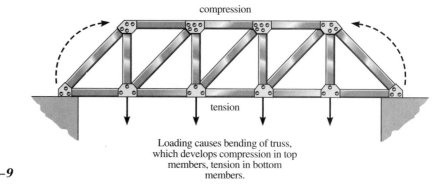

Loading causes bending of truss, which develops compression in top members, tension in bottom members.

Fig. 1–9

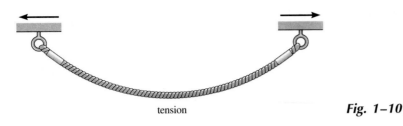

tension

Fig. 1–10

Cables and Arches.   Two other forms of structures used to span long distances are the cable and the arch. *Cables* are usually flexible and carry their loads in tension. Unlike tension ties, however, the external load is not applied along the axis of the cable, and consequently the cable takes a form that has a defined sag, Fig. 1–10. Cables are commonly used to support bridges and building roofs. When used for these purposes, the cable has an advantage over the beam and the truss, especially for spans that are greater than 150 ft. Because they are always in tension, cables will not become unstable and suddenly collapse, as may happen with beams or trusses. Furthermore, the truss will require added costs for construction and increased depth as the span increases. Use of cables, on the other hand, is limited only by their weight and methods of anchorage.

The arch achieves its strength in compression, since it has a reverse curvature to that of the cable, Fig. 1–11. The arch, however, must be rigid in order to maintain its shape, and this results in secondary loadings involving shear and moment, which must be considered in its design. Arches are frequently used in bridge structures, dome roofs, and for openings in masonry walls.

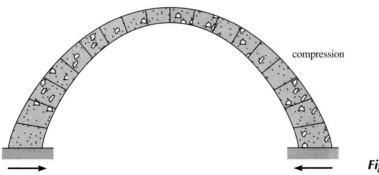

compression

Fig. 1–11

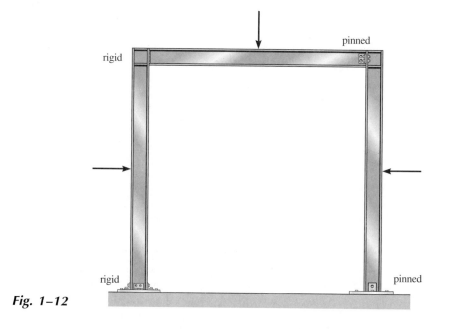

Fig. 1–12

Frames. Frames are often used in buildings and are composed of beams and columns that are either pin- or fixed-connected, Fig. 1–12. Like trusses, frames extend in two or three dimensions. The loading on a frame causes bending of the members, and due to the rigid joint connections, this structure is generally "indeterminate" from a standpoint of analysis. The strength of the frame is derived from the moment interactions between the beams and the columns at the rigid joints, and as a result, the economical benefits of using a frame depend on the efficiency gained in using reduced beam sizes versus increasing the necessary size of the columns due to the "beam-column" action caused by bending at the joints.

Surface Structures. A *surface structure* is made from a material having a very small thickness compared to its other dimensions. Sometimes this material is very flexible and can take the form of a tent or air-inflated structure. In both cases the material acts as a membrane that is subjected to pure tension.

Surface structures may also be made of rigid materials such as reinforced concrete. As such they may be shaped as folded plates, cylinders, or hyperbolic paraboloids, and are referred to as *thin plates* or *shells*. These structures act like cables or arches since they support loads primarily in tension or compression, with very little bending. In spite of

The ribs and ring girders form a framework used to support the domed roof of an ice-skating rink.

this, plate or shell structures are generally very difficult to analyze, due to the three-dimensional geometry of their surface. Such an analysis is beyond the scope of this text and is instead covered in texts devoted entirely to this subject.

1.3 Loads

The design loading for a structure is often specified in codes. In general, the structural engineer works with two types of codes: general building codes and design codes. *General building codes* specify the requirements of governmental bodies for minimum design loads on structures and minimum standards for construction. *Design codes* provide detailed technical standards and are used to establish the requirements for the actual structural design. Table 1–1 lists some of the important codes used in practice. It should be realized, however, that codes provide only a general guide for design. The ultimate responsibility for the design lies with the structural engineer.

The actual design of a structure begins with those elements that are subjected to the primary loads the structure is intended to carry and proceeds in sequence to the various supporting members until the foundation is reached. Thus, a building floor slab would be designed

Table 1–1   Codes

General Building Codes

American National Standard Building Code, American National Standards Institute (ANSI)

Basic Building Code, Building Officials and Code Administrators International

Standard Building Code, Southern Building Code Congress

Uniform Building Code, International Conference of Building Officials

Design Codes

Aluminum Construction Manual, Aluminum Association

Building Code Requirements for Reinforced Concrete, American Concrete Institute (ACI)

Manual of Steel Construction, American Institute of Steel Construction (AISC)

PCI Design Handbook, Prestressed Concrete Institute (PCI)

Standard Specifications for Highway Bridges, American Association of State Highway and Transportation Officials (AASHTO)

Timber Construction Manual, American Institute of Timber Construction (AITC)

Manual for Railway Engineering, American Railway Engineering Association (AREA)

first, followed by the supporting beams, columns, and last, the foundation footings. In order to design a structure, it is therefore first necessary to specify the loads that act on it. A structure is generally subjected to several types of loads. A brief discussion of these loadings will now be presented to illustrate how one must consider their effects in practice.

Dead Loads

Dead loads consist of the weights of the various structural members and the weights of any objects that are permanently attached to the structure. Hence, for a building, the dead loads include the weights of the columns, beams, and girders; the floor slab; roofing; walls; windows; plumbing; electrical fixtures; and other miscellaneous attachments.

In some cases a structural dead load can be estimated satisfactorily from simple formulas based on the weights and sizes of similar structures. Through experience one can also derive a "feeling" for the magnitude of these loadings. Ordinarily, though, once the materials and sizes of the various components of the structure are determined, their weight can be found from tables that list their density. A portion of one such table, taken from the American National Standard Building Code, is given in Table 1–2. Calculation of dead loads based on use of tabulated data as in Table 1–2 is rather straightforward, as shown in Example 1–1 appearing on the following page.

Table 1–2   Minimum Design Dead Loads*

Walls	*psf*
4-in. clay brick, high absorption	34
4-in. sand-lime brick	38
4-in. concrete brick, heavy aggregate	46
4-in. concrete brick, light aggregate	33
8-in. clay brick, high absorption	69
8-in. clay brick, medium absorption	79
8-in. clay brick, low absorption	89
8-in. sand-lime brick	74
8-in. concrete brick, heavy aggregate	89
8-in. concrete brick, light aggregate	68
8-in. concrete block, heavy aggregate	55
12-in. concrete block, heavy aggregate	85
8-in. concrete block, light aggregate	35
12-in. concrete block, light aggregate	55
Partitions	
2-in. solid plaster	20
4-in. solid plaster	32
Wood studs 2×4, unplastered	4
Wood studs 2×4, plastered one side	12
Wood studs 2×4, plastered two sides	20
Concrete Slabs	
Concrete, reinforced-stone, per inch of thickness	$12\frac{1}{2}$
Concrete, plain stone, per inch of thickness	12
Concrete reinforced, lightweight, per inch of thickness	9
Concrete, plain, lightweight, per inch of thickness	$8\frac{1}{2}$
Cinder concrete, per inch	9
Ceilings	
Plaster on tile or concrete	5
Suspended metal lath and gypsum plaster	10
Asphalt shingles	2
Fiberboard, $\frac{1}{2}$-in.	0.75

*Reproduced with permission from American National Standard Minimum Design Loads for Buildings and Other Structures, ANSI A58.1-1982. Copies of this standard may be purchased from ANSI at 1430 Broadway, New York, N.Y. 10018.

Example 1–1

The floor beam in Fig. 1–13 is used to support a reinforced light-weight concrete slab having a width of 6 ft and thickness of 4 in. The slab serves as a ceiling for the floor below and therefore its bottom is coated with plaster. Furthermore, an 8-ft-high, 12-in. light-aggregate concrete block wall bears directly on the top flange of the beam. Determine the loading on the beam measured per foot of linear length.

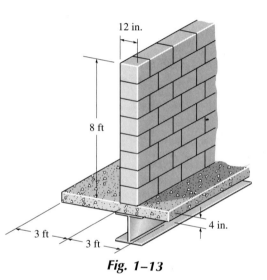

Fig. 1–13

Solution

Using the data in Table 1–2, we have

Concrete slab:	$[9 \text{ lb}/(\text{ft}^2 \cdot \text{in.})](4 \text{ in.})(6 \text{ ft}) =$	216 lb/ft
Plaster ceiling:	$(5 \text{ lb}/\text{ft}^2)(6 \text{ ft}) =$	30 lb/ft
Block wall:	$(55 \text{ lb}/\text{ft}^2)(8 \text{ ft}) =$	440 lb/ft
Total load		686 lb/ft **Ans.**

Live Loads

Live loads can vary both in their magnitude and location. They may be caused by the weights of objects temporarily placed on a structure, moving vehicles, or natural forces. The minimum live loads specified in codes are determined from studying the history of their effects on existing structures. Usually, these loads include an allowance for some protection against excessive deflection or sudden overload. In Chapter 5 we will develop techniques for specifying the proper location of live loads on the structure so that they cause the greatest stress or deflection of the members. Various types of live loads will now be discussed.

Building Loads.   The floors of buildings are assumed to be subjected to *uniform live loads*, which depend on the purpose for which the building is designed. These loadings are generally tabulated in local, state, or national codes. A representative example of such *maximum uniform loadings*, taken from the American National Standard Building Code, is shown in Table 1–3. These values are determined from a history of loading various buildings. They include some protection against the possibility of overload due to construction loads, and serviceability requirements. In addition to uniform loads, some codes specify *minimum concentrated live loads*, caused by hand carts, automobiles, etc., which must also be applied anywhere to the floor system. For example, both uniform and concentrated live loads must be considered in the design of an automobile parking deck.

Table 1–3   Minimum Live Loads*

Occupancy or Use	Live Load (psf)	Occupancy or Use	Live Load (psf)
Assembly areas and theaters		Office buildings	
Fixed seats	60	Lobbies	100
Movable seats	100	Offices	50
Dance halls and ballrooms	100	Residential	
Garages (passenger cars only)	50	Dwellings (one- and two-family)	40
Storage warehouse		Hotels and multifamily houses	40
Light	125	School classrooms	40
Heavy	250		

*Reproduced with permission from American National Standard Minimum Design Loads for Buildings and Other Structures, ANSI A58.1-1982, p. 10.

For some types of structures many codes will allow a reduction in the uniform live load for a *floor* since it is unlikely that the prescribed live load will occur simultaneously throughout the entire structure at any one time. For example, ANSI A58.1-1982 allows a reduction of live load on a member having an *influence area* of 400 ft² or more. This reduced live load is calculated using the following equation:

$$L = L_o\left(0.25 + \frac{15}{\sqrt{A_I}}\right) \tag{1-1}$$

where
L = reduced design live load per square foot of area supported by the member

L_o = unreduced design live load per square foot of area supported by the member (see Table 1–3)

A_I = influence area in square feet equal to four times the tributary or effective load-carrying floor area for a column, and two times the tributary or effective load-carrying floor area for a beam*

The reduced live load defined by Eq. 1–1 is limited to not less than 50% of L_o for members supporting one floor, or not less than 40% of L_o for members supporting more than one floor. No reduction is allowed for structures used for public assembly, garages, or roofs. The following example illustrates its application.

Bridge Loads.   Design live loadings for highway bridges are specified in the code of the American Association of State Highway and Transportation Officials (AASHTO), whereas railroad bridge design follows the specifications of the American Railway Engineering Association (AREA). Both of these codes give wheel loadings and spacing for different types of trucks and trains. For design, a series of such loadings are placed back to back within critical regions of the bridge, and the maximum live-load stress in the members is calculated. Furthermore, since vehicles are in constant motion, any bouncing that occurs results in an impact of their weights on the bridge. To account for this the AASHTO and AREA codes give empirical formulas used to determine the *impact fraction*, which specifies the percentage by which the maximum live load should be increased. Specific examples of such formulas and vehicle loadings are discussed further in Sec. 5.6.

---

*Specific examples of the determination of tributary areas for beams and columns are given in Sec. 2.1.

Example 1–2

A two-story office building has interior columns that are spaced 22 ft apart in two perpendicular directions. If the (flat) roof loading is 20 lb/ft^2, determine the reduced live load supported by a typical interior column located at ground level.

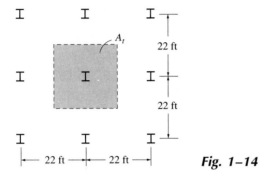

Fig. 1–14

Solution

As shown in Fig. 1–14, each interior column has a tributary area or effective loaded area of $A_t = (22 \text{ ft})(22 \text{ ft}) = 484 \text{ ft}^2$. A ground-floor column therefore supports a roof live load of

$$F_R = (20 \text{ lb/ft}^2)(484 \text{ ft}^2) = 9680 \text{ lb} = 9.68 \text{ k*}$$

This load is not reduced, since it is not a floor load. For the second story, the live load is taken from Table 1–3: $L_o = 50$ lb/ft^2. Since $A_I = 4A_T = 4(484 \text{ ft}^2) = 1936 \text{ ft}^2$ and 1936 ft^2 > 400 ft^2, the live load can be reduced using Eq. 1–1. Thus,

$$L = 50\left(0.25 + \frac{15}{\sqrt{1936}}\right) = 29.55 \text{ lb/ft}^2$$

The load reduction here is (29.55/50)100% = 59.1% > 50%. O.K. Therefore

$$F_F = (29.55 \text{ lb/ft}^2)(484 \text{ ft}^2) = 14,300 \text{ lb} = 14.3 \text{ k}$$

The total live load supported by the ground-floor column is thus

$$F = F_R + F_F = 9.68 \text{ k} + 14.3 \text{ k} = 24.0 \text{ k} \qquad \textbf{Ans.}$$

*The unit k stands for "kip," which symbolizes kilopounds. Hence, 1 k = 1000 lb.

Wind Loads.   When structures block the flow of wind, the wind's kinetic energy is converted into potential energy of pressure, which causes a wind loading. The effect of wind on a structure depends upon the density and velocity of the air, the angle of incidence of the wind, the shape and stiffness of the structure, and the roughness of its surface. For design purposes, wind loadings can be treated using either a static or a dynamic approach.

For the *static approach*, the fluctuating pressure caused by a constantly blowing wind is approximated by a *mean pressure* that acts on the windward and leeward sides of the structure. This pressure q is defined by its kinetic energy, $q = \frac{1}{2}\rho v^2$, where ρ is the density of air. If we take the specific weight of air to be $\gamma = 0.0765$ lb/ft^3 and specify the wind velocity v in miles per hour, then

$$q = \frac{1}{2}\left(\frac{0.0765 \text{ lb/ft}^3}{(32.2 \text{ ft/s}^2)(3600 \text{ s/h})^2}\right)(v \text{ mi/h})^2(5280 \text{ ft/mi})^2$$

$$q = 0.00256v^2 \tag{1-2}$$

Here q is measured in pounds per square foot and acts on a flat surface that is perpendicular to the wind's velocity. A 100-mph wind is often used for design of many low-rise structures in the United States. However, more exact values of velocity, which depend on the structure's geographic location and its elevation from the ground, can be obtained from *wind-zone maps* generally given in codes.* These maps are compiled by the U.S. Weather Bureau and generally represent the largest wind velocity at a specified ground elevation during a 50-year recurrence period. Ground elevation is important here since the velocity of the wind increases with elevation. Consequently, the higher the structure, the more severe wind loadings become.

Once the mean pressure q of the wind has been calculated, its magnitude is multiplied by various coefficients to obtain the design static pressure p that is applied to the structure. In the ANSI A58.1-1982 code, for example, there are coefficients that account for the importance of the structure, the possibility of wind gust, and the pressure difference outside and inside the structure. A choice of these factors depends on the height and orientation of the structure, its vibrational characteristics, and the number of openings in the structure. In the ASCE report, geometric or shape coefficients are also used. For example, for a building with *vertical sides*, the shape factor for the building's

*See Wind Forces on Structures, *Trans. ASCE*, Vol. 126, Part 2, 1961; also, *Building Code Requirements for Minimum Design Loads in Buildings and Other Structures*, ANSI A58.1-1982.

windward side is 0.8, and for the leeward (opposite) side it is 0.5. Hence, in a 100-mph wind the pressure (or pushing) acting on the windward side is $p = 0.8(0.00256)(100)^2 = 20.5$ psf, and the suction (or pulling) on the leeward side is $p = 0.5(0.00256)(100)^2 = 12.8$ psf.

For high-rise buildings or those having a shape or location that makes them wind-sensitive, it is recommended that a *dynamic approach* be used to determine the wind loadings. This requires wind-tunnel tests to be performed on a scale model of the building and those surrounding it, in order to simulate the natural environment. The pressure effects of the wind on the building are determined from pressure transducers attached to the model. Also, if the model has the same stiffness characteristics as the building, then the dynamic deflections of the building can be determined.

Snow Loads.   In some parts of the country, roof loading due to snow can be quite severe, and therefore protection against possible failure is of primary concern. Design loadings typically depend on the building's general shape and roof geometry, wind exposure, and location. Like wind, snow loads are generally determined from a zone map reporting 50-year recurrence intervals of an extreme snow depth. For example, in some western states and in the Northeast, 45 lb/ft^2 is commonly used for design. No single code can cover all the implications of this type of loading. Instead, the engineer must use judgment regarding the possibility of additional effects caused by rain, snow drifting or movement, and whether the building to be designed is to be heated.

Earthquake Loads.   Earthquakes produce loadings on a structure through the interaction of the ground motion and the response characteristics of the structure. These loadings result from the structure's distortion caused by the ground's motion and the lateral resistance of the structure. Their magnitude depends on the amount and type of ground accelerations and the mass and stiffness of the structure. In order to provide some insight as to the nature of earthquake loads, consider the simple structural model shown in Fig. 1–15. This model may represent a single-story building, where the block is the "lumped" mass of the roof, and the column is the lumped stiffness of all the building's columns. During an earthquake the ground vibrates both horizontally and vertically. The vertical motion is slight, and is usually neglected in design. As a consequence of the horizontal accelerations, shear forces in the column try to put the block in sequential motion with the ground. If the column is *stiff* and the block has a *small* mass, the period of vibration of the block will be *short* and the block will accelerate with the same motion as the ground and undergo only slight relative displacements. For a structure this is beneficial, since less stress is devel-

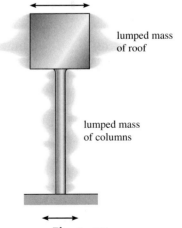

lumped mass of roof

lumped mass of columns

Fig. 1–15

oped in the members. On the other hand, if the column is very flexible and the block has a large mass, then earthquake-induced motion will cause small accelerations of the block and large relative displacements.

In practice the effects of a structure's acceleration, velocity, and displacement can be determined and represented as an *earthquake response spectrum.** Once this graph is established, the earthquake loadings can be calculated using a *dynamic analysis* based on the theory of structural dynamics. This analysis is often quite elaborate and requires the use of a computer. Although this may be the case, such an analysis becomes mandatory if the structure is large.

Some codes require that specific attention be given to earthquake design, especially in areas of the country where strong earthquakes predominate. Also, these loads should be seriously considered when designing high-rise buildings or nuclear power plants. In order to assess the importance of earthquake design consideration, one can check a seismic map published in ANSI A58.1-1982. This map divides the country into five zones, each zone specifying the amount of probable risk. The scale ranges from 0 (low risk, such as parts of Texas) to 4 (very high risk, such as along the west coast of California).

For small structures, a *static analysis* for earthquake design may be satisfactory. This method approximates the dynamic loads by a set of externally applied *static forces* that are applied laterally to the structure. One such formula for doing this is reported in ANSI A58.1-1982. It is used to determine the "base shear" V in the columns and can be written as

$$V = ZIKCSW \qquad (1-3)$$

Here the terms represent factors that are tabulated in the ANSI code and depend on the earthquake zone (Z), the importance of the building regarding occupancy (I), its framework configuration (K), its vibration characteristics (C), the type of supporting soil (S), and the weight of the structure (W). With each new publication of the ANSI code, values of these coefficients are updated as more accurate data about earthquake response become available.

Hydrostatic and Soil Pressure. When structures are used to retain water, soil, or granular materials, the pressure developed by those loadings becomes an important criterion for their design. Examples of such types of structures include tanks, dams, ships, bulkheads, and retaining

*See J. A. Blume et al., *Design of Multi-Story Reinforced Concrete Buildings for Earthquake Motions*, Portland Cement Association, Chicago, 1961.

walls. Here the laws of hydrostatics and soil mechanics are applied to define the intensity of the loadings on the structure.

Other Natural Loads. Several other types of live loads may also have to be considered in the design of a structure, depending on its location or use. These include the effect of blast, temperature changes, and differential settlement of the foundation. Like the other effects mentioned above, design codes usually specify the limitations of these loadings and which combinations of them should be applied simultaneously so as to cause the maximum, yet realistic live-load effects on the structure.

REFERENCES

American Institute of Steel Construction, Inc. *Manual of Steel Construction*, 8th ed., AISC, Inc., New York, 1980.

American National Standard Building Code Requirements for Minimum Design Loads in Buildings and Other Structures, A58.1-1972, Am. Nat. Standards Inst., New York.

American Society of Civil Engineers, "Locomotive Loadings for Railroad Bridges," *Transactions*, ASCE, Vol. 86, 1923.

American Society of Civil Engineers, "Wind Bracing in Steel Buildings," *Transactions*, ASCE, Vol. 105, 1940, 1713–1739.

American Society of Civil Engineers, "Wind Forces on Structures," *Transactions*, ASCE, Structural Division, 84, Part I, 1958, and 126, Part II, 1961.

PCI Design Handbook, Prestressed Concrete Institute, Chicago, 1971.

Specifications for Steel Railway Bridges, American Railway Engineering Association, Chicago, 1965.

Standard Specifications for Highway Bridges, American Association of State Highway and Transportation Officials, 12th ed., Washington, D.C., 1973.

Uniform Building Code, International Conference of Building Officials, Whittier, Ca., 1976.

PROBLEMS

1–1. A building wall consists of 12-in. light-aggregate concrete block and 2-in. solid plaster on both sides. If the wall is 8 ft high, determine the load in pounds per foot length of wall that it exerts on the floor.

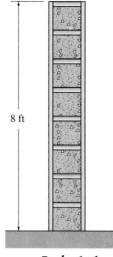

8 ft

Prob. 1–1

1–2. The beam supports the roof made from asphalt shingles and wood sheathing boards. If the boards have a thickness of $1\frac{1}{2}$ in. and a specific weight of 50 lb/ft^3, and the roof's angle of slope is 30°, determine the dead load of the roofing—per square foot—that is supported in the x and y directions by the purlins.

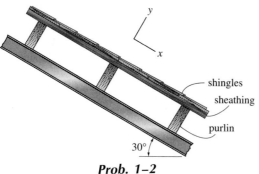

y

x

shingles
sheathing
purlin

30°

Prob. 1–2

1–3. The second floor of a light manufacturing building is constructed from a 4-in.-thick reinforced-stone concrete slab with an added 3-in. cinder concrete fill as shown. If the suspended ceiling of the first floor consists of metal lath and gypsum plaster, determine the dead load for design in pounds per square foot of floor area.

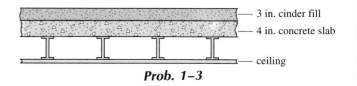

3 in. cinder fill
4 in. concrete slab
ceiling

Prob. 1–3

*1–4. The precast floor beam is made from concrete having a specific weight of 150 lb/ft^3. If it is to be used for a floor in an office of an office building, calculate its dead and live loadings per foot length of beam.

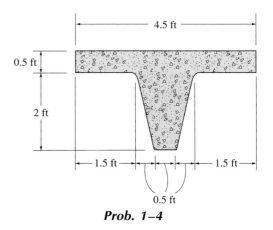

4.5 ft

0.5 ft

2 ft

1.5 ft

1.5 ft

0.5 ft

Prob. 1–4

1–5. The *T*-beam used in a heavy storage warehouse is made of concrete having a specific weight of 125 lb/ft³. Determine the dead load per foot length of beam, and the live load on the top of the beam per foot length of beam. Neglect the weight of the steel reinforcement.

1–7. The second floor of a light manufacturing building is constructed from a 5-in.-thick reinforced-stone concrete slab with an added 4-in. cinder concrete fill as shown. If the suspended ceiling of the first floor consists of metal lath and gypsum plaster, determine the dead load for design in pounds per square foot of floor area.

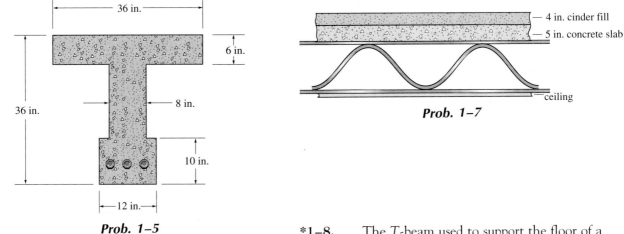

Prob. 1–7

Prob. 1–5

*1–8. The *T*-beam used to support the floor of a light storage warehouse is made from concrete having a specific weight of 150 lb/ft³. Determine the dead load and live load per foot length of beam. Neglect the weight of the steel reinforcement.

1–6. The floor of a heavy storage warehouse building is made of 6-in.-thick reinforced lightweight concrete. If the floor is a slab having a length of 15 ft and width of 10 ft, determine the resultant force caused by the dead load and that caused by the live load.

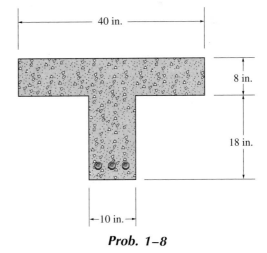

Prob. 1–8

21

1–9.   The face of the sign is subjected to a 125-mph wind. Determine the resultant force of the wind and specify the x and y coordinates of its location on the sign. Use a shape factor of 0.8.

1–10.   Determine the pressure p acting on the face of the sign if it is subjected to a 75-mph wind. Use a shape factor of 0.8. If the sign has a width of 12 ft and a height of 7 ft, as indicated, what is the resultant force of this pressure? Specify the x and y coordinates on the face of the sign where this force acts?

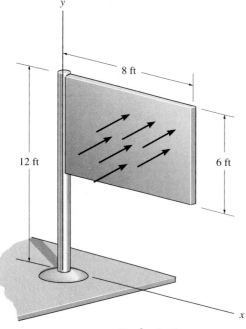

Prob. 1–9

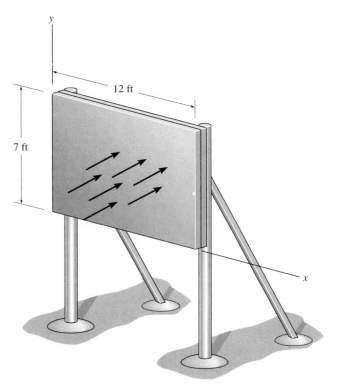

Prob. 1–10

1–11. Wind blows directly on the front of the building with a speed of 100 mph. If the shape factor for the wall *ABCDE* on the windward side is 0.8, determine the resultant force pushing on the windward-side wall.

*1–12. Solve Prob. 1–11 if the wind is blowing on the side *DEFG*. Use a shape factor of 0.8.

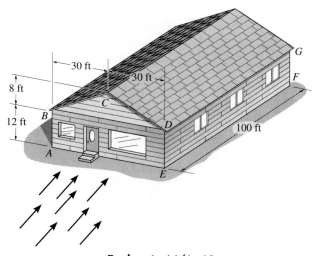

Probs. 1–11/1–12

1–13. A two-story school has interior columns that are spaced 15 ft apart in two perpendicular directions. If the loading on the flat roof is estimated to be 20 lb/ft², determine the reduced live load supported by a typical interior column at (a) the ground-floor level, and (b) the second-floor level.

1–14. A three-story hotel has interior columns that are spaced 20 ft apart in two perpendicular directions. If the loading on the flat roof is estimated to be 30 lb/ft², determine the live load supported by a typical interior column at (a) the ground-floor level, and (b) the second-floor level.

1–15. A four-story office building has interior columns spaced 30 ft apart in two perpendicular directions. If the flat-roof loading is estimated to be 30 lb/ft², determine the reduced live load supported by a typical interior column located at ground level.

Oftentimes the elements of a structure, like the beams and girders of this building frame, are connected together in a manner whereby the analysis is statically determinate.

Analysis of Statically Determinate Structures

The most common form of structure that the engineer will have to analyze is one that lies in a plane and is subjected to a force system that lies in the same plane. Hence, in this chapter attention will be directed to analyzing this type of structure. We begin by discussing the importance of choosing an appropriate analytical model for a structure so that the forces in the structure may be determined with reasonable accuracy. Then the criteria necessary for structural stability are discussed. Finally, the analysis of statically determinate, planar, pin-connected structures is presented.

2.1   Idealized Structure

In the real sense an exact analysis of a structure can never be carried out since estimates always have to be made of the loadings and the strength of the materials composing the structure. Furthermore, points of application for the loadings must also be estimated. It is important, therefore, that the structural engineer develop the ability to model or idealize a structure so that he or she can perform a practical force analysis of the members. In this section we will develop the basic techniques necessary to do this.

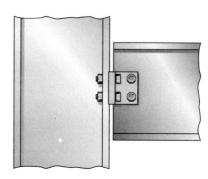

typical "pin-supported" connection (metal)

(a)

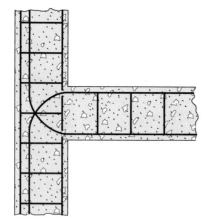

typical "roller-supported" connection (concrete)

(a)

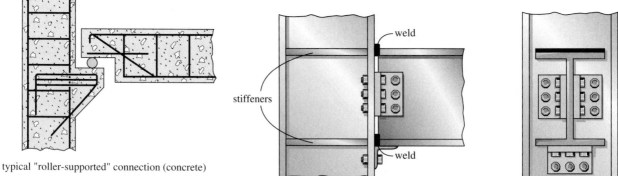

typical "fixed-supported" connection (metal)

(b)

Fig. 2–1

typical "fixed-supported" connection (concrete)

(b)

Fig. 2–2

Support Connections

Structural members are joined together in various ways depending on the intent of the designer. The two types of joints most often specified are the pin connection and the fixed joint. A pin-connected joint allows some freedom for slight rotation, whereas the fixed joint allows no relative rotation between the connected members and is consequently more expensive to fabricate. Examples of these joints, fashioned in metal and concrete, are shown in Figs. 2–1 and 2–2, respectively. For most timber structures, the members are assumed to be pin-connected, since bolting or nailing them to their supports will not sufficiently restrain them from rotating with respect to each other.

Idealized models used in structural analysis that represent pinned and fixed supports and pin-connected and fixed-connected joints are

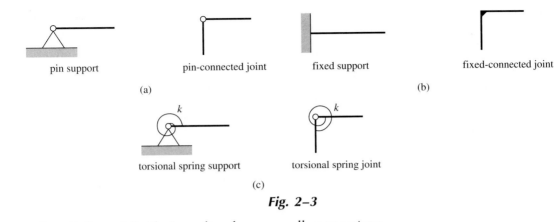

pin support pin-connected joint fixed support fixed-connected joint

(a) (b)

k k

torsional spring support torsional spring joint

(c)

Fig. 2–3

shown in Figs. 2–3*a* and 2–3*b*. In reality, however, all connections exhibit some stiffness toward joint rotations, owing to friction and material behavior. In this case a more appropriate model for a support or joint might be that shown in Fig. 2–3*c*. If the torsional spring constant $k = 0$, the joint is a pin, and if $k \to \infty$, the joint is fixed.

When selecting a particular model for each support or joint, the engineer must be aware of how the assumptions will affect the actual performance of the member and whether the assumptions are reasonable for the structural design. For example, consider the beam shown in Fig. 2–4*a*, which is used to support a concentrated load **P.** The angle connection at support *A* is like that in Fig. 2–1*a* and can therefore be idealized as a typical pin support. Furthermore, the support at *B* provides an approximate point of smooth contact and so it can be idealized as a roller. The beam's thickness can be neglected since it is small in comparison to its length, and therefore the idealized model of the beam is as shown in Fig. 2–4*b*. The analysis of the loadings in this beam should give results that closely approximate the loadings in the actual beam. To show that the model is appropriate, consider a specific case of a beam made of steel with $P = 8$ k and $L = 20$ ft. One of the

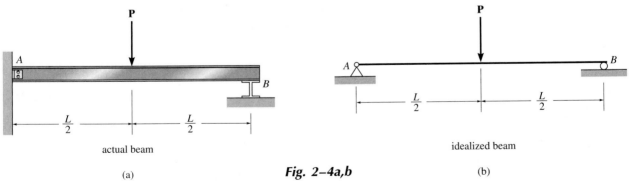

actual beam idealized beam

(a) (b)

Fig. 2–4a,b

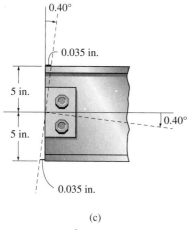

(c)

Fig. 2–4c

major simplifications made here was assuming the support at A to be a pin. Design of the beam using standard code procedures* indicates that a W 11 × 22 would be adequate for supporting the load. Using one of the deflection methods of Chapter 8, the rotation at the "pin" support can be calculated as $\theta = 0.0070$ rad $= 0.40°$. From Fig. 2–4c, such a rotation only moves the top or bottom flange a distance of $\Delta = \theta$ $r = (0.0070$ rad$)(5$ in.$) = 0.035$ in.! This *small amount* would certainly be accommodated by the connection fabricated as shown in Fig. 2–1a, and therefore the pin serves as an appropriate model.

Other types of connections most commonly encountered on planar structures are given in Table 2–1. It is important to be able to recognize the symbols for these connections and the kinds of reactions they exert on their attached members. This can easily be done by noting how the connection *prevents* any degree of freedom or displacement of the member. In particular, the support will develop a *force* on the

Table 2–1 Supports for Coplanar Structures

Type of Connection	Idealized Symbol	Reaction	Number of Unknowns
(1) light cable weightless link			One unknown. The reaction is a force that acts in the direction of the cable or link.
(2) rollers rocker			One unknown. The reaction is a force that acts perpendicular to the surface at the point of contact. Symbolically, the force can have either sense of direction (i.e., up or down).
(3) smooth contacting surface			One unknown. The reaction is a force that acts perpendicular to the surface at the point of contact.

*Using the *Manual of Steel Construction*, American Institute of Steel Construction, with steel having a yield point of $F_y = 36$ ksi.

Table 2-1   Supports for Coplanar Structures (continued)

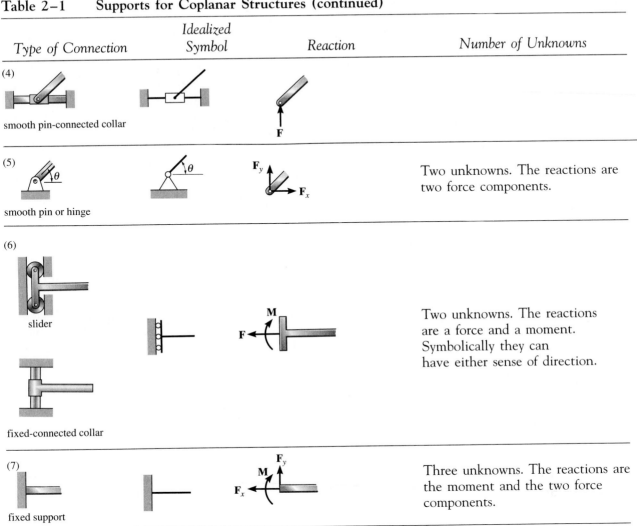

Type of Connection	Idealized Symbol	Reaction	Number of Unknowns
(4) smooth pin-connected collar		F	
(5) smooth pin or hinge	θ	F_y, F_x	Two unknowns. The reactions are two force components.
(6) slider / fixed-connected collar		M, F	Two unknowns. The reactions are a force and a moment. Symbolically they can have either sense of direction.
(7) fixed support		F_y, M, F_x	Three unknowns. The reactions are the moment and the two force components.

member if it *prevents translation* of the member, and it will develop a *moment* if it *prevents rotation* of the member. For example, a member in contact with a smooth surface (3) is prevented from translating only in one direction, which is perpendicular or normal to the surface. Hence, the surface exerts only a *normal force* F on the member in this direction. The magnitude of this force represents *one unknown*. Also note that the member is free to rotate on the surface, so that a moment cannot be developed by the surface on the member. As another example, the fixed support (7) prevents *both* translation and rotation of a member at the point of connection. Therefore, this type of support

29

exerts two force components and a moment on the member. The "curl" of the moment lies in the plane of the page, since rotation is prevented in that plane. Hence, there are *three unknowns* at a fixed support.

In reality, all structural supports actually exert *distributed surface loads* on their contacting members. The concentrated forces and moments shown in Table 2–1 represent the *resultants* of these load distributions. This representation is, of course, an idealization; however, it is used here since the surface area over which the distributed load acts is considerably *smaller* than the *total* surface area of the connecting members.

Idealized Structure

Having stated the various ways in which the connections on a structure can be idealized, we are now ready to discuss some of the techniques used by structural engineers to represent various structural systems by idealized models.

As a first example, consider the jib crane and trolley in Fig. 2–5a. For the structural analysis we can neglect the thickness of the two main members and will assume that the joint at B is fabricated to be rigid. Furthermore, the support connection at A can be modeled as a fixed support and the details of the trolley excluded. Thus, the members of the idealized structure are represented by two connected lines, and the load on the hook is represented by a single concentrated force **F**, Fig. 2–5b. This idealized structure shown here as a *line drawing* can now be used as a means for applying the principles of structural analysis, which will eventually lead to the design of its two main members.

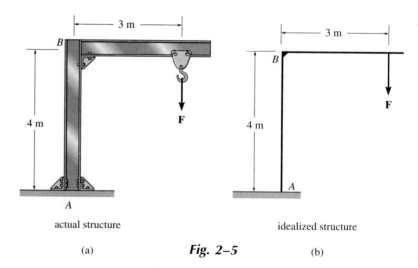

actual structure   idealized structure

(a)   ***Fig. 2–5***   (b)

A typical rocker support used for a bridge girder.

This bearing support is made from a material having a low resistance to friction; therefore the support can be modeled as a roller.

Rollers and associated bearing pads are used to support the prestressed concrete girders of a highway bridge.

A typical welded, rigid knee-joint connection in a steel frame. Notice how the splice plates are used to form a rigid column connection.

Observe the thermal expansion of the deck on this highway bridge, due to the tipping of the rocker. Also note the use of stiffeners, or vertical plates, over the rocker and pin supports. The stiffeners are necessary to prevent localized buckling of the girder flange, or web, which can be caused by the large reactions at the support.

31

Beams and girders are often used to support building floors. In particular, a *girder* is the main load carrying element of the floor, whereas, the smaller elements having a shorter span and connected to the girders are called *beams*. Often the loads that are applied to a beam or girder are transmitted to it by the floor that is supported by the beam or girder. Again, it is important to be able to appropriately idealize the system as a series of models, which can be used to determine, to a close approximation, the forces acting in the members. Consider, for example, the framing used to support a typical floor slab in a building, Fig. 2–6a. Here the slab is supported by *floor joists* located at 6-ft intervals and these in turn are supported by the two side girders *AB* and *CD*. For analysis it is reasonable to assume that the joists are pin- and/or roller-connected to the girders and that the girders are pin- and/or roller-connected to the columns. The top view of the structural framing plan for this system is shown in Fig. 2–6b. In this "graphic" scheme, notice that the "lines" representing the joists do not touch the girders and the lines for the girders do not touch the columns. This symbolizes pin- and/or roller-supported connections. On the other hand, if the framing plan is intended to represent fixed-connected members, such as those that are welded instead of simply bolted connections, then the lines for the beams or girders would touch the columns as in Fig. 2–7. Similarly,

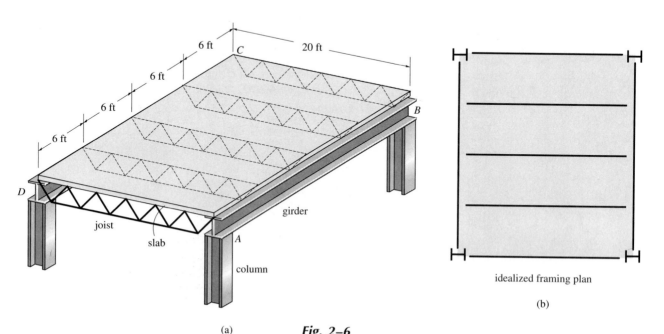

(a)

idealized framing plan

(b)

Fig. 2–6

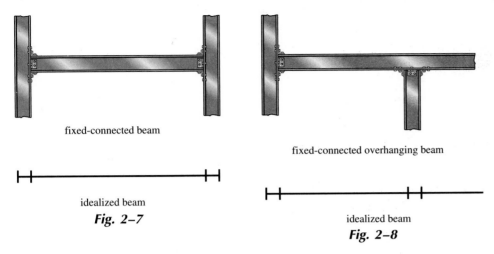

fixed-connected beam

fixed-connected overhanging beam

idealized beam

Fig. 2–7

idealized beam

Fig. 2–8

a fixed-connected overhanging beam would be represented in top view as shown in Fig. 2–8. If reinforced concrete construction is used, the beams and girders are represented by double lines. These systems are generally all fixed-connected and therefore the members are drawn to the supports. For example, the structural graphic for the cast-in-place reinforced concrete system in Fig. 2–9a is shown in top view in Fig. 2–9b. The lines for the beams are dashed because they are below the slab.

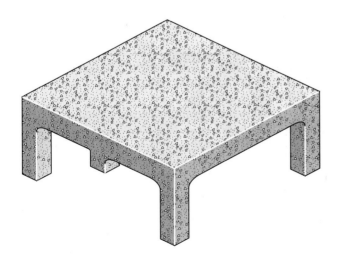

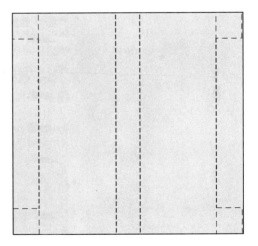

idealized framing plan

(a)

(b)

Fig. 2–9

33

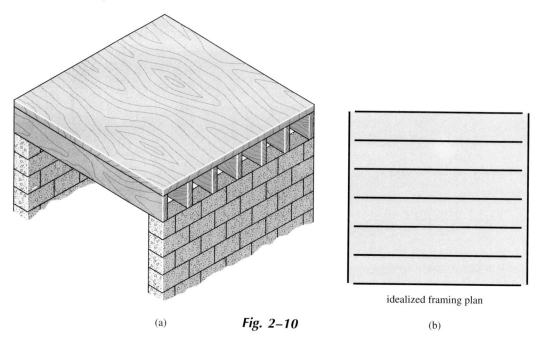

(a)   ***Fig. 2–10***   (b)

idealized framing plan

Structural graphics and idealizations for timber structures are similar to those made of metal. For example, the structural system shown in Fig. 2–10*a* represents beam-wall construction, whereby the roof deck is supported by wood joists, which deliver the load to a masonry wall. The joists can be assumed to be simply-supported on the wall, so that the structural framing plan would be like that shown in Fig. 2–10*b*.

Tributary Loadings

When flat surfaces such as walls, floors, or roofs are supported by a structural frame, it is necessary to determine how the load on these surfaces is transmitted to the various structural elements used for their support. There are generally two ways in which this can be done. The choice depends on the geometry of the structural system, the material from which it is made, and the method of its construction.

One-Way System. A slab or deck that is supported as shown in Fig. 2–10 is said to deliver its load to the supporting members by one-way action, and the slab is often referred to as a *one-way slab*. To illustrate the method of load transmission, consider the framing system shown in Fig. 2–11*a* where the beams *AB*, *CD*, and *EF* rest on the girders *AE* and *BF*. If a uniform load of 100 lb/ft^2 is placed on the slab, then the

The structural framework of this building consists of concrete floor joists, which were formed on site using metal pans. These joists are simply-supported on the girders, which in turn are simply-supported on the columns.

An example of one-way slab construction of a steel-frame building with a poured concrete floor on a corrugated metal deck. The load on the floor is transmitted to the beams, not to the girders.

The floor load in this building is first transmitted by one-way action to the bar joists, which in turn are supported by the side, or spandrel, girders. The girders then transmit the load to the columns. Masonry block is used for the walls, but the walls are not intended as structural support for the girders in the design.

center beam *CD* is assumed to support the load acting on the *tributary area* shown shaded on the structural framing plan in Fig. 2–11b. Member *CD* is therefore subjected to a *linear* distribution of load of $(100 \text{ lb/ft}^2)(5 \text{ ft}) = 500 \text{ lb/ft}$, shown on the idealized beam in Fig. 2–11c. The reactions on this beam (2500 lb) would then be applied to the center of the girders *AE* (and *BF*), shown idealized in Fig. 2–11d. Using this same concept, do you see how the remaining portion of the slab loading is transmitted to the ends of the girder as 1250 lb?

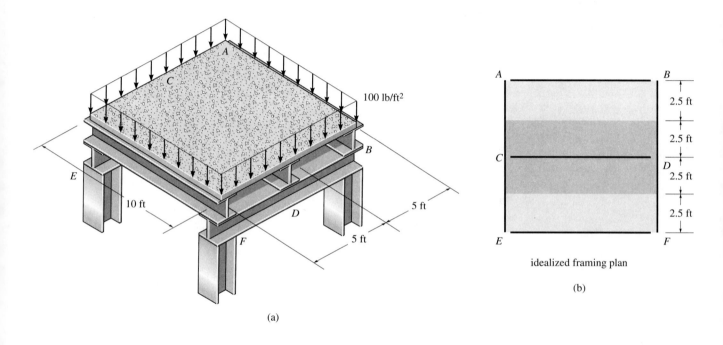

(a)

idealized framing plan

(b)

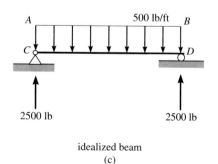

idealized beam

(c)

Fig. 2–11

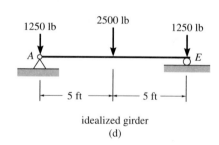

idealized girder

(d)

For some floor systems the beam and girders are connected to the columns at the *same elevation*, as in Fig. 2–12a. If this is the case, the slab can in some cases also be considered a "one-way slab." For example, if the slab is reinforced concrete with reinforcement in *only one direction*, or the concrete is poured on a corrugated metal deck, as in Fig. 2–12b, then one-way action of load transmission can be assumed. On the other hand, if the slab is flat on top and bottom and is reinforced in *two directions*, then consideration must be given to the *possibility* of the load being transmitted to the supporting members from either one or two directions. For example, consider the slab and framing plan in Fig. 2–12c. As a general rule, *if $L_2 \geq L_1$ and if the span ratio $(L_2/L_1) \geq 2$, the slab will behave as a one-way slab*, since as L_1 becomes smaller, the beams AB, CD, and EF provide the greater stiffness to carry the load.

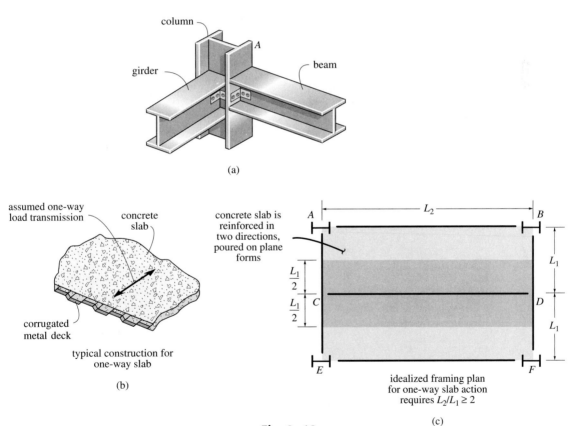

(a)

(b)

(c)

Fig. 2–12

Two-Way System.   If the support ratio in Fig. 2–12c is $(L_2/L_1) < 2$, the load is delivered to the supporting beams and girders in two directions, and when this is the case the slab is referred to as a *two-way slab*. To show one method of treating this case, consider the square reinforced concrete slab in Fig. 2–13a, which is supported by the four

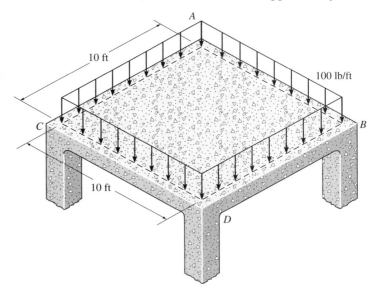

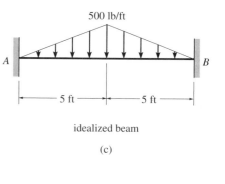

(a)

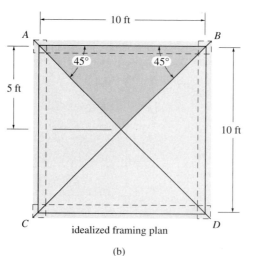

idealized framing plan

(b)

idealized beam

(c)

Fig. 2–13

10-ft-long edge beams. Here $L_2/L_1 = 1$. Due to two-way slab action, the assumed *tributary area* for beam AB is shown shaded in Fig. 2–13b. This area is determined by constructing diagonal $45°$ lines as shown. Hence if a uniform load of 100 lb/ft^2 is applied to the slab, a peak intensity of (100 lb/ft^2)(5 ft) = 500 lb/ft will be applied to the center of beam AB, resulting in a *triangular* load distribution shown in Fig. 2–13c. For other geometries that cause two-way action, a similar procedure can be used. For example, in the case of $L_2/L_1 = 1.5$, Fig. 2–14a, it is necessary to construct $45°$ lines that intersect as shown. A 100-lb/ft^2 loading placed on the slab will then produce *trapezoidal* and *triangular* distributed loads on members AB and AC as shown in Fig. 2–14b and 2–14c, respectively.

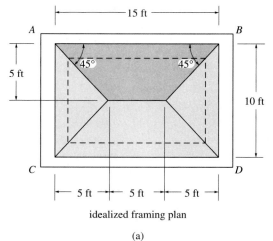

idealized framing plan

(a)

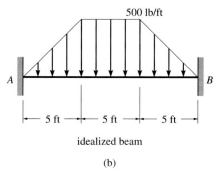

idealized beam

(b)

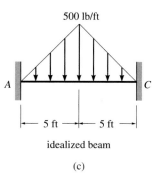

idealized beam

(c)

Fig. 2–14

The ability to reduce an actual structure to an idealized form, as shown by these examples, can only be gained by experience. To provide practice at doing this, the example problems and the problems for solution throughout this book are presented in somewhat realistic form, and the associated problem statements aid in explaining how the connections and supports can be modeled by those listed in Table 2–1. In engineering practice, if it becomes doubtful as to how to model a structure or transfer the loads to the members, it is best to consider *several* idealized structures and loadings and then design the actual structure so that it can resist the loadings in all the idealized models.

2.2   Principle of Superposition

The principle of superposition forms the basis for much of the theory of structural analysis. It may be stated as follows: *The total displacement or stress at a point in a structure subjected to several loadings can be determined by adding together the displacements or stresses caused by each of the loads acting separately.* For this statement to be valid it is necessary that a *linear relationship* exist among the loads, stresses, and displacements.

Two requirements must be imposed for the principle of superposition to apply:

1. The structural material must behave in a linear-elastic manner, so that Hooke's law is valid, and therefore the load will be proportional to displacement.

2. The geometry of the structure must not undergo significant change when the loads are applied. Large displacements will significantly change the position and orientation of the loads. An example would be a column subjected to a buckling load.

Throughout this text, these two requirements will be satisfied. Here only linear-elastic material behavior occurs; and the displacements produced by the loads will not significantly change the directions of applied loadings nor the dimensions used to compute the moments of forces.

2.3 Equations of Equilibrium _____

It may be recalled from statics that a structure or one of its members is in equilibrium when it maintains a balance of force and moment. In general this requires that the force and moment equations of equilibrium be satisfied along three independent axes, namely,

$$\Sigma F_x = 0 \qquad \Sigma F_y = 0 \qquad \Sigma F_z = 0$$
$$\Sigma M_x = 0 \qquad \Sigma M_y = 0 \qquad \Sigma M_z = 0 \qquad (2\text{--}1)$$

The principal load-carrying portions of most structures, however, lie in a single plane, and since the loads are also coplanar, the above requirements for equilibrium reduce to

$$\Sigma F_x = 0$$
$$\Sigma F_y = 0 \qquad (2\text{--}2)$$
$$\Sigma M_O = 0$$

Here ΣF_x and ΣF_y represent, respectively, the algebraic sums of the x and y components of all the forces acting on the structure or one of its members, and ΣM_O represents the algebraic sum of the moments of these force components about an axis perpendicular to the x–y plane (z axis) and passing through point O.

Whenever these equations are applied, *it is first necessary to draw a free-body diagram of the structure or its members.* If a member is selected, it must be *isolated* from its supports and surroundings and its outlined shape drawn. All the forces and couple moments must be shown that act *on the member.* In this regard, the types of reactions at the supports can be determined using Table 2–1. Also, recall that forces common to two members act with equal magnitudes but opposite directions on the respective free-body diagrams of the members.

41

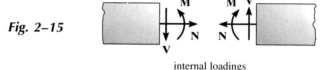

internal loadings

Fig. 2–15

If the *internal loadings* at a specified point in a member are to be determined, the *method of sections* must be used. This requires that a "cut" or section be made perpendicular to the axis of the member at the point where the internal loading is to be determined. A free-body diagram of either segment of the "cut" member is isolated and the internal loads are then determined from the equations of equilibrium applied to the segment. In general, the internal loadings acting at the cut section of the member will consist of a normal force **N,** shear force **V,** and bending moment **M,** as shown in Fig. 2–15.

We will cover the principles of statics that are used to apply the equilibrium equations to determine the external reactions on structures in Sec. 2–5. Internal loadings in structural members will be discussed in Chapter 4.

2.4 Determinacy and Stability

Before the force analysis of a structure is begun, it is necessary to determine the determinacy and stability of the structure.

Determinacy

The equilibrium equations provide both the *necessary and sufficient* conditions for equilibrium. When all the forces in a structure can be determined strictly from these equations, the structure is referred to as *statically determinate*. Structures having more unknown forces than available equilibrium equations are called *statically indeterminate*. As a general rule, a structure can be identified as being either statically determinate or statically indeterminate by drawing free-body diagrams of all its members, or selective parts of its members, and then comparing the total number of unknown reactive force and moment components with the total number of available equilibrium equations.* For a

---

*Drawing the free-body diagrams is not strictly necessary, since a "mental count" of the number of unknowns can also be made and compared with the number of equilibrium equations.

coplanar structure there are at most *three* equilibrium equations for each part, so that if there is a total of n parts and r force and moment reaction components, , we have

$$
\begin{aligned}
r &= 3n, \text{ statically determinate} \\
r &> 3n, \text{ statically indeterminate}
\end{aligned}
\qquad (2\text{--}3)
$$

In particular, if a structure is *statically indeterminate*, the additional equations needed to solve for the unknown reactions are obtained by relating the applied loads and reactions to the displacement or slope at different points on the structure. These equations, which are referred to as *compatibility equations*, must be equal in number to the *degree of indeterminacy* of the structure. Compatibility equations involve the geometric and physical properties of the structure and will be discussed further in Chapter 8.

We will now consider some examples to show how to classify the determinacy of a structure. The first example considers beams; the second example, pin-connected structures; and in the third we will discuss frame structures. Classification of trusses will be considered in Chapter 3.

The structural system for the floor of this building can be modeled as a series of pin-connected floor beams and girders which are statically determinate.

Example 2–1

Classify each of the beams shown in Fig. 2–16a through 2–16e as statically determinate or statically indeterminate. If statically indeterminate, report the number of degrees of indeterminacy. The beams are subjected to external loadings that are assumed to be known and can act anywhere on the beams.

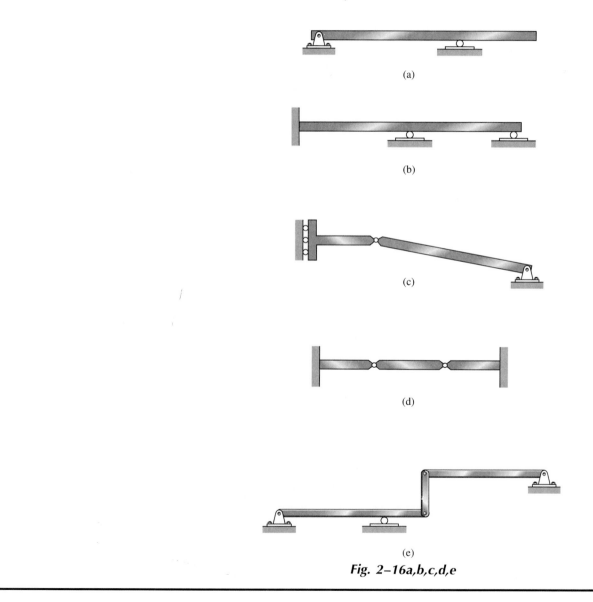

(a)

(b)

(c)

(d)

(e)

Fig. 2–16a,b,c,d,e

Solution

Compound beams, that is, those in Fig. 2–16c through 2–16e, which are composed of pin-connected members or members connected by short links, must be disassembled. Note that in these cases, the unknown reactive forces must be shown in equal but opposite pairs. The free-body diagrams of each member are shown in Fig. 2–16f through 2–16j. Using $r = 3n$ or $r > 3n$, the resulting classifications are therefore

Fig. 2–16f, $r = 3$, $n = 1, 3 = 3(1)$ Statically determinate
Fig. 2–16g, $r = 5$, $n = 1, 5 > 3(1)$ Statically indeterminate to the second degree
Fig. 2–16h, $r = 6$, $n = 2, 6 = 3(2)$ Statically determinate
Fig. 2–16i, $r = 10$, $n = 3, 10 > 3(3)$ Statically indeterminate to the first degree
Fig. 2–16j, $r = 9$, $n = 3, 9 = 3(3)$ Statically determinate

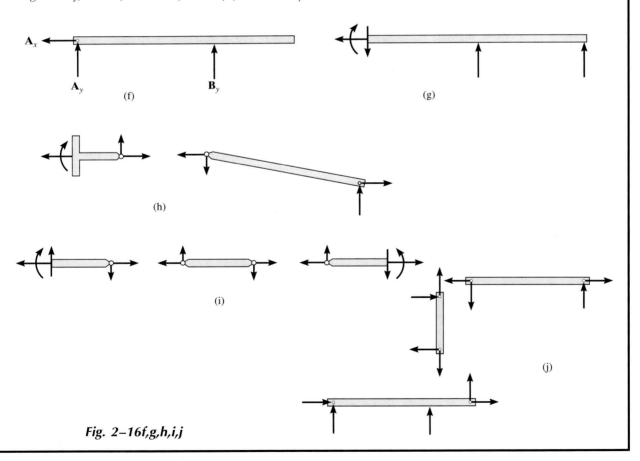

Fig. 2–16f,g,h,i,j

Example 2–2

Classify each of the pin-connected structures shown in Fig. 2–17a through 2–17d as statically determinate or statically indeterminate. If statically indeterminate, report the number of degrees of indeterminacy. The structures are subjected to arbitrary external loadings that are assumed to be known and can act anywhere on the structures.

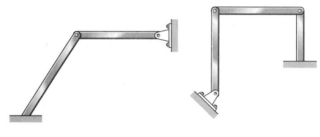

(a)

(b)

(c)

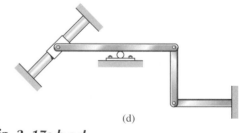
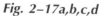

(d)

Fig. 2–17a,b,c,d

Solution

Classification of pin-connected structures is similar to that of beams. The free-body diagrams of the members are shown in Fig. 2–17e through 2–17h. Applying $r = 3n$ or $r > 3n$, the resulting classifications are as follows

Fig. 2–17e, $r = 7$, $n = 2$, $7 > 6$,   Statically indeterminate to the first degree

Fig. 2–17f, $r = 9$, $n = 3$, $9 = 9$,   Statically determinate

Fig. 2–17g, $r = 10$, $n = 2$, $10 > 6$, Statically indeterminate to the fourth degree

Fig. 2–17h, $r = 9$, $n = 3$, $9 = 9$,   Statically determinate

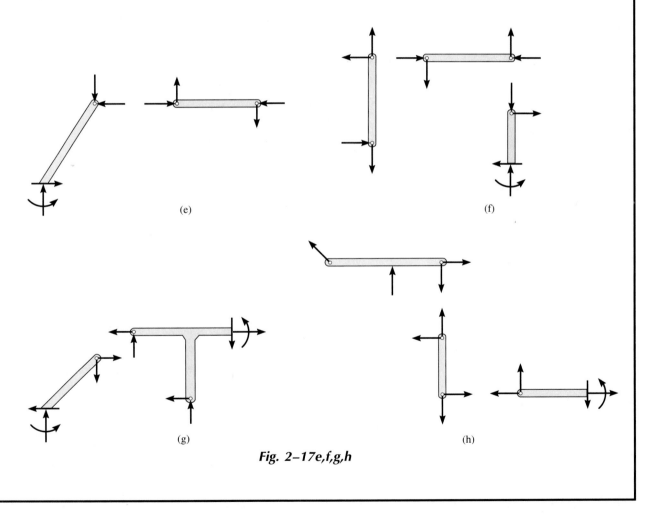

(e)

(f)

(g)

(h)

Fig. 2–17e,f,g,h

Example 2–3

Classify each of the frames shown in Fig. 2–18a through 2–18c as statically determinate or statically indeterminate. If statically indeterminate, report the number of degrees of indeterminacy. The frames are subjected to external loadings that are assumed to be known and can act anywhere on the frames.

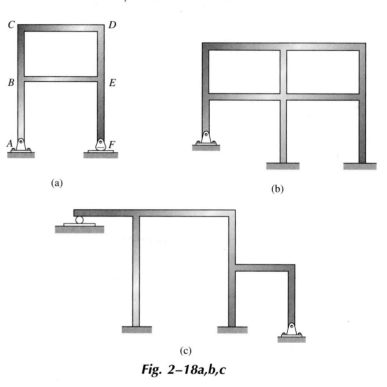

(a)

(b)

(c)

Fig. 2–18a,b,c

Solution

Unlike the beams and pin-connected structures of the previous examples, *these frame structures consist of members that are connected together by rigid joints and form closed loops.* For example, in Fig. 2–18a members *BCDE* form a closed loop. In order to classify these structures, it is necessary to use the method of sections and "cut" the loop apart. The free-body diagrams of the sectioned parts are drawn and the frame can then be classified. Notice that only *one section* through the loop is required, since once the unknowns at the section are determined, the internal forces at any point in the mem-

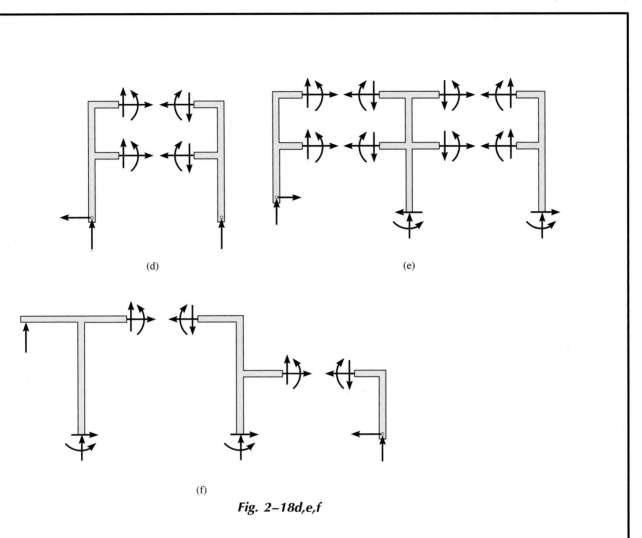

(d)   (e)

(f)

Fig. 2–18d,e,f

bers can then be computed using the method of sections and the
equations of equilibrium. With reference to Fig. 2–18d through
2–18f, the resulting classifications are as follows:

Fig. 2–18d, $r = 9$, $n = 2$, $9 > 6$,   Statically indeterminate to the third degree
Fig. 2–18e, $r = 20$, $n = 3$, $20 > 9$, Statically indeterminate to the eleventh degree
Fig. 2–18f, $r = 15$, $n = 3$, $15 > 9$, Statically indeterminate to the sixth degree

This frame has no closed loops.

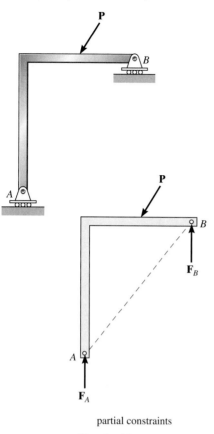

partial constraints

Fig. 2–19

Stability

To ensure the equilibrium of a structure or its members, it is not only necessary to satisfy the equations of equilibrium, but the members must also be properly held or constrained by their supports. Two situations may occur where the conditions for proper constraint have not been met.

Partial Constraints. In some cases a structure or one of its members may have *fewer* reactive forces than equations of equilibrium that must be satisfied. The structure then becomes only *partially constrained*. For example, consider the member shown in Fig. 2–19 with its corresponding free-body diagram. Here the equation $\Sigma F_x = 0$ will not be satisfied for the loading conditions and therefore the member will be unstable.

Improper Constraints. In some cases there may be as many unknown forces as there are equations of equilibrium; however, *instability* or movement of a structure or its members can develop because of *improper constraining* by the supports. This can occur if all the *support reactions are concurrent* at a point. An example of this is shown in Fig. 2–20. From the free-body diagram of the beam it is seen that the summation of moments about point O will *not* be equal to zero ($Pd \neq 0$); thus rotation about point O will take place.

Another way in which improper constraining leads to instability occurs when the *reactive forces* are all *parallel*. An example of this case is shown in Fig. 2–21. Here when an inclined force **P** is applied, the summation of forces in the horizontal direction will not equal zero.

In general, then, a structure will be geometrically unstable—that is, it will move slightly or collapse—if there are fewer reactive forces than equations of

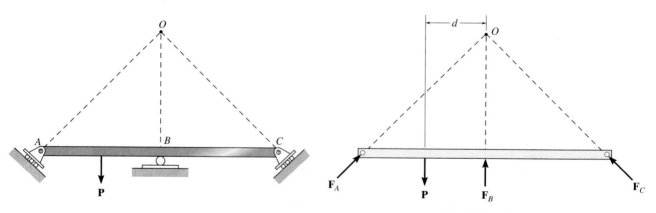

concurrent reactions

Fig. 2–20

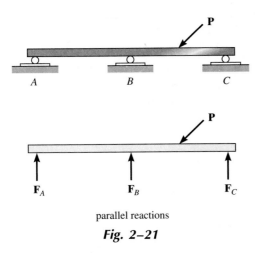

parallel reactions

Fig. 2–21

equilibrium; or if there are enough reactions, instability will occur if the lines of action of the reactive forces intersect a common point or are parallel to one another. If the structure consists of several members or components, local instability of one or several of these members can generally be determined *by inspection.* If the members form a collapsible mechanism, the structure will be unstable. We will now formulize these statements for a *coplanar structure* having n members or components with r unknown reactions. Since three equilibrium equations are available for each member or component, we have

$$
\begin{array}{ll}
r < 3n & \text{unstable} \\
r \geq 3n & \text{unstable if member reactions are} \\
& \text{concurrent or parallel or some of the} \\
& \text{components form a collapsible mechanism}
\end{array}
\qquad (2\text{–}4)
$$

If the structure is unstable, *it does not matter* if it is statically determinate or indeterminate. In all cases such types of structures must be avoided in practice.

The following examples illustrate how structures or their members can be classified as stable or unstable. Structures in the form of a truss will be discussed in Chapter 3.

Example 2–4

Classify each of the structures in Fig. 2–22a through 2–22f as stable or unstable. The structures are subjected to arbitrary external loads that are assumed to be known.

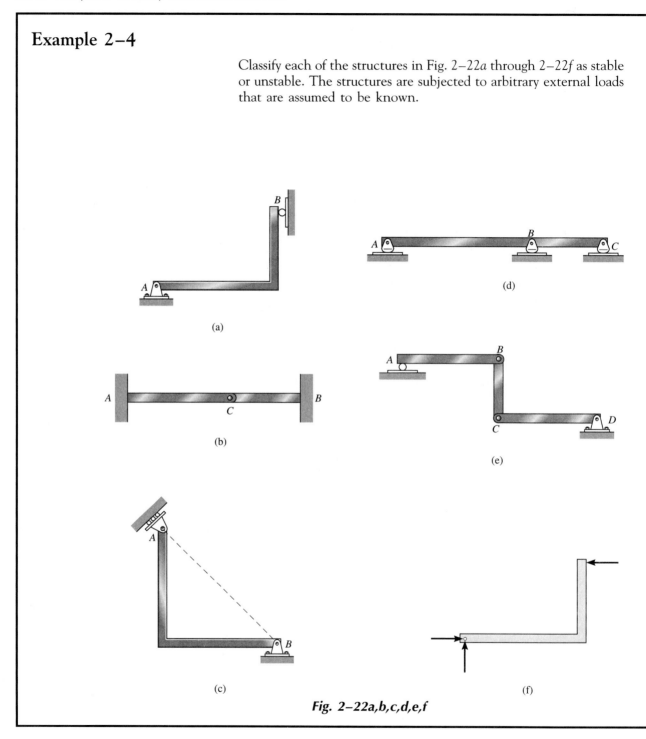

(a)

(b)

(c)

(d)

(e)

(f)

Fig. 2–22a,b,c,d,e,f

Solution

The structures are classified as follows:

Fig. 2–22a.   The member is *stable* since the reactions are nonconcurrent or nonparallel, Fig. 2–22f. It is also statically determinate.

Fig. 2–22b.   The compound beam is *stable*, Fig. 2–22g. It is also indeterminate to the second degree.

Fig. 2–22c.   The member is *unstable* since the three reactions are concurrent at B, Fig. 2–22h.

Fig. 2–22d.   The beam is *unstable* since the three reactions are all parallel, Fig. 2–22i.

Fig. 2–22e.   The structure is *unstable* since $r = 7$, $n = 3$, Fig. 2–22j, so that, by Eq. 2–4, $7 < 9$. Also, this can be seen by inspection, Fig. 2–22e, since AB can move horizontally without restraint.

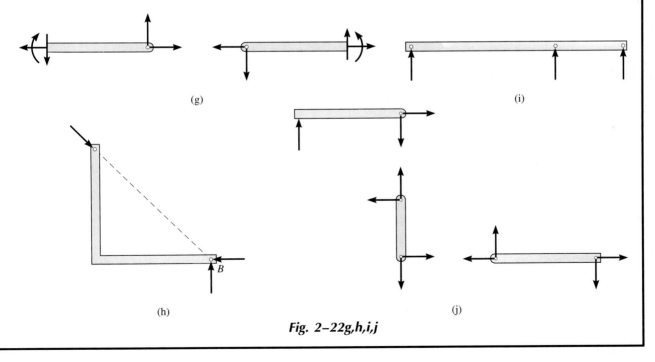

Fig. 2–22g,h,i,j

53

2.5 Application of the Equations of Equilibrium _____

Occasionally, the members of a structure are connected together in such a way that the joints can be assumed as pins. Building frames and trusses are typical examples that are often constructed in this manner. Provided that a pin-connected structure is properly constrained and contains no more supports or members than are necessary to prevent collapse, the forces acting at the joints and supports can be determined by applying the three equations of equilibrium ($\Sigma F_x = 0$, $\Sigma F_y = 0$, $\Sigma M_O = 0$) to each member. Understandably, once the forces at the joints are obtained, the size of the members, connections, and supports can then be determined on the basis of design code restrictions.

To illustrate the method of force analysis, consider the three-member frame shown in Fig. 2–23a, which is subjected to loads $\mathbf{P}_1$ and $\mathbf{P}_2$. The free-body diagrams of each member are shown in Fig. 2–23b. In total there are nine unknowns; however, nine equations of

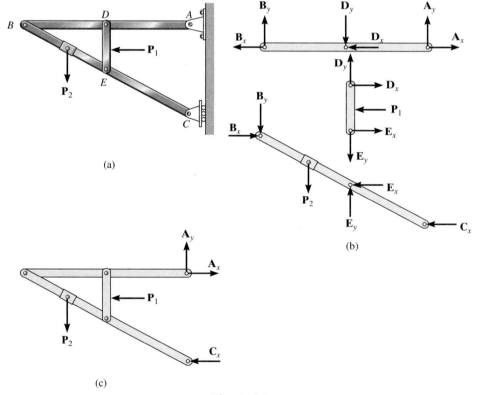

(a)

(b)

(c)

Fig. 2–23

equilibrium can be written, three for each member, so the problem is *statically determinate*. For the actual solution it is *also* possible, and sometimes convenient, to consider a portion of the frame or its entirety when applying some of these nine equations. For example, a free-body diagram of the entire frame is shown in Fig. 2–23c. One could determine the three reactions A_x, A_y, and C_x on this "rigid" pin-connected system, then analyze *any* two of its members, Fig. 2–23b, to obtain the other six unknowns. Furthermore, the answers can be checked in part by applying the three equations of equilibrium to the remaining "third" member. To summarize, then, this problem can be solved by writing *at most* nine equilibrium equations using free-body diagrams of any members and/or combinations of connected members. Any more than nine equations written would *not* be unique from the original nine and would only serve to check the results.

An example of a simply supported highway girder bridge. Note the use of roller supports at the left pier.

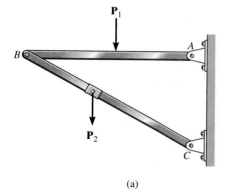

(a)

Consider now the two-member frame shown in Fig. 2–24a. Here the free-body diagrams of the members reveal six unknowns, Fig. 2–24b; however, six equilibrium equations, three for each member, can be written, so again the problem is statically determinate. As in the previous case, a free-body diagram of the entire frame can also be used for part of the analysis, Fig. 2–24c. Although, as shown, the frame has a tendency to collapse without its supports, by rotating about the pin at B, this will not happen since the force system acting on it must still hold it in equilibrium. Hence, if so desired, all six unknowns can be determined by applying the three equilibrium equations to the entire frame, Fig. 2–24c, and also to either one of its members.

The above two examples illustrate that if a structure is properly supported and contains no more supports or members than are necessary to prevent collapse the frame becomes statically determinate, and so, the unknown forces at the supports and connections can be determined from the equations of equilibrium applied to each member. Also, if the structure remains *rigid* (noncollapsible) when the supports are removed (Fig. 2–23c), all three support reactions can be determined by applying the three equilibrium equations to the entire structure. However, if the structure appears to be nonrigid (collapsible) after removing the supports (Fig. 2–24c), it must be dismembered and equilibrium of the individual members must be considered in order to obtain enough equations to determine *all* the support reactions.

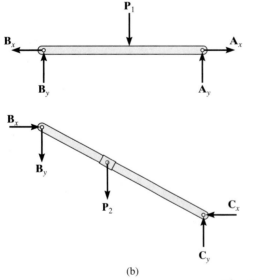

(b)

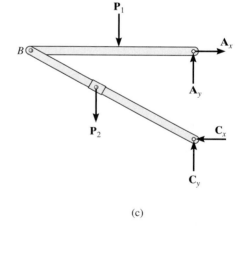

(c)

Fig. 2–24

Procedure for Analysis

The following procedure provides a method for determining the *joint reactions* for structures composed of pin-connected members.

Free-Body Diagrams.   Disassemble the structure and draw a free-body diagram of each member. Also, it may be convenient to supplement a member free-body diagram with a free-body diagram of the *entire structure*. Some or all of the support reactions can then be determined using this diagram.

Recall that reactive forces common to two members act with equal magnitudes but opposite directions on the respective free-body diagrams of the members. Specifically, the unknown reactions acting at the joints should be represented by their rectangular components. Also, all two-force members should be identified. These members, regardless of their shape, have no external loads on them, and therefore their free-body diagrams are represented with equal but opposite collinear forces acting on their ends. In many cases it is possible to tell by inspection the proper arrowhead sense of direction of an unknown force or couple moment; however, if this seems difficult, the directional sense can be assumed.

Equations of Equilibrium.   Count the total number of unknowns to make sure that an equivalent number of equilibrium equations can be written for solution. Except for two-force members, recall that in general three equilibrium equations can be written for each member. Many times, the solution for the unknowns will be straightforward if the moment equation $\Sigma M_O = 0$ is applied about a point (O) that lies at the intersection of the lines of action of as many unknown forces as possible. Also, when applying the force equations $\Sigma F_x = 0$ and $\Sigma F_y = 0$, orient the x and y axes along lines that will provide the simplest reduction of the forces into their x and y components. If the solution of the equilibrium equations yields a *negative* magnitude for an unknown force or couple moment, it indicates that its arrowhead sense of direction is *opposite* to that which was assumed on the free-body diagram.

The following example problems illustrate this procedure numerically.

Example 2–5

Determine the reactions on the beam shown in Fig. 2–25a.

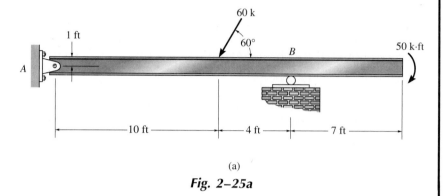

(a)

Fig. 2–25a

Solution

Free-Body Diagram.   As shown in Fig. 2–25b, the 60-k force is resolved into x and y components. Furthermore, the 7-ft dimension line is not needed since a couple moment is a *free vector* and can therefore act anywhere on the beam for the purpose of computing the external reactions.

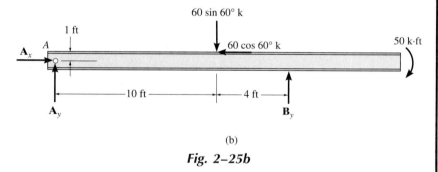

(b)

Fig. 2–25b

Equations of Equilibrium.   Applying Eqs. 2–2 in a sequence, using previously calculated results, we have

$$\xrightarrow{+}\Sigma F_x = 0; \quad A_x - 60\cos 60° = 0 \qquad\qquad A_x = 30.0\,\text{k} \quad \textbf{Ans.}$$
$$\curvearrowright+\Sigma M_A = 0; \quad 60\sin 60°(10) - 60\cos 60°(1) - B_y(14) + 50 = 0$$
$$B_y = 38.5\,\text{k} \quad \textbf{Ans.}$$
$$+\uparrow\Sigma F_y = 0; \quad -60\sin 60° + 38.5 + A_y = 0 \quad A_y = 13.4\,\text{k} \quad \textbf{Ans.}$$

Example 2–6

Determine the reactions on the beam in Fig. 2–26a.

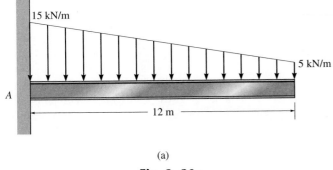

(a)

Fig. 2–26a

Solution

Free-Body Diagram.   As shown in Fig. 2–26b, the trapezoidal distributed loading is segmented into a triangular and uniform load. The *areas* under the triangle and rectangle represent the *resultant* forces. These forces act through the centroid of their corresponding areas.

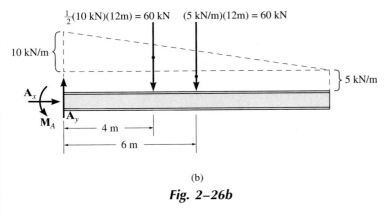

(b)

Fig. 2–26b

Equations of Equilibrium

$$\xrightarrow{+}\Sigma F_x = 0; \quad A_x = 0 \hspace{5.5cm} \textbf{\textit{Ans.}}$$
$$+\uparrow\Sigma F_y = 0; \quad A_y - 60 - 60 = 0 \quad\quad A_y = 120\ \text{kN} \quad\quad \textbf{\textit{Ans.}}$$
$$\curvearrowleft+\Sigma M_A = 0; \quad 60(4) + 60(6) - M_A = 0 \quad M_A = 600\ \text{kN}\cdot\text{m} \quad \textbf{\textit{Ans.}}$$

Example 2–7

Determine the reactions on the beam in Fig. 2–27a. Assume the support at B is a roller (smooth surface) and A is a pin.

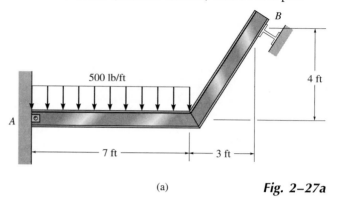

(a)   **Fig. 2–27a**

Solution

Free-Body Diagram. As shown in Fig. 2–27b, the support ("roller") at B exerts a *normal force* on the beam at its point of contact. The line of action of this force is defined by the 3–4–5 triangle.

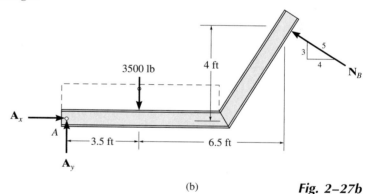

(b)   **Fig. 2–27b**

Equations of Equilibrium. Resolving N_B into x and y components and summing moments about A yields a direct solution for N_B. Why? Using this result, we can then obtain A_x and A_y.

$$\zeta + \Sigma M_A = 0; \quad 3500(3.5) - (\tfrac{4}{5})N_B(4) - (\tfrac{3}{5})N_B(10) = 0$$
$$N_B = 1332 \text{ lb} \quad \textbf{Ans.}$$
$$\xrightarrow{+} \Sigma F_x = 0; \quad A_x - \tfrac{4}{5}(1331.5) = 0 \qquad A_x = 1065 \text{ lb} \quad \textbf{Ans.}$$
$$+\uparrow \Sigma F_y = 0; \quad A_y - 3500 + \tfrac{3}{5}(1331.5) = 0 \quad A_y = 2701 \text{ lb} \quad \textbf{Ans.}$$

Example 2–8

The compound beam in Fig. 2–28a is fixed at A. Determine the reactions at A, B, and C. Assume that the connection at B is a pin and C is a roller.

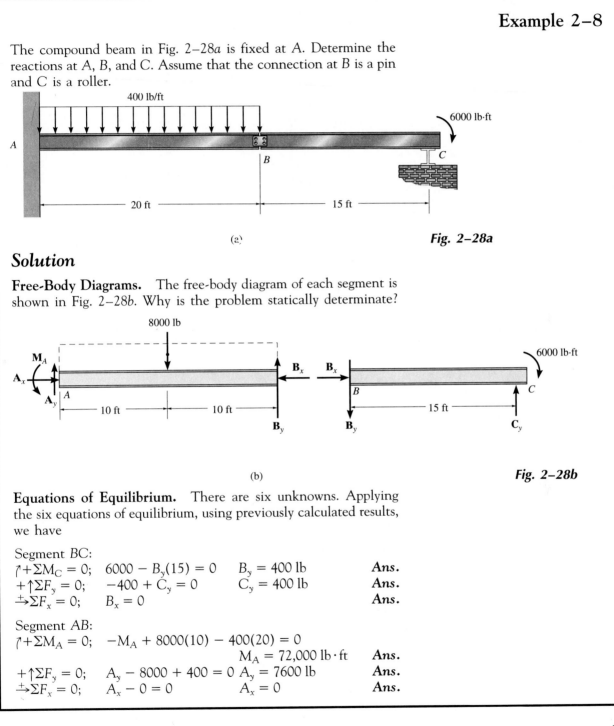

(a)

Fig. 2–28a

Solution

Free-Body Diagrams. The free-body diagram of each segment is shown in Fig. 2–28b. Why is the problem statically determinate?

(b)

Fig. 2–28b

Equations of Equilibrium. There are six unknowns. Applying the six equations of equilibrium, using previously calculated results, we have

Segment BC:

$\zeta + \Sigma M_C = 0;$ $6000 - B_y(15) = 0$ $B_y = 400 \text{ lb}$ ***Ans.***

$+\uparrow \Sigma F_y = 0;$ $-400 + C_y = 0$ $C_y = 400 \text{ lb}$ ***Ans.***

$\xrightarrow{+} \Sigma F_x = 0;$ $B_x = 0$ ***Ans.***

Segment AB:

$\zeta + \Sigma M_A = 0;$ $-M_A + 8000(10) - 400(20) = 0$

$M_A = 72,000 \text{ lb} \cdot \text{ft}$ ***Ans.***

$+\uparrow \Sigma F_y = 0;$ $A_y - 8000 + 400 = 0$ $A_y = 7600 \text{ lb}$ ***Ans.***

$\xrightarrow{+} \Sigma F_x = 0;$ $A_x - 0 = 0$ $A_x = 0$ ***Ans.***

Example 2–9

The "three-hinged arched" is subjected to the two concentrated forces shown in Fig. 2–29a. Determine the horizontal and vertical components of reaction at the pins A, B, and C.

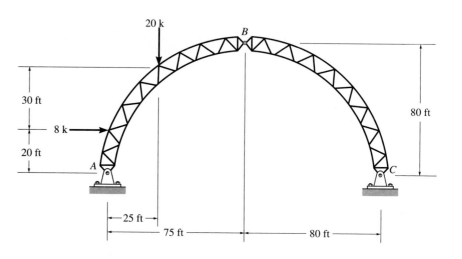

(a)

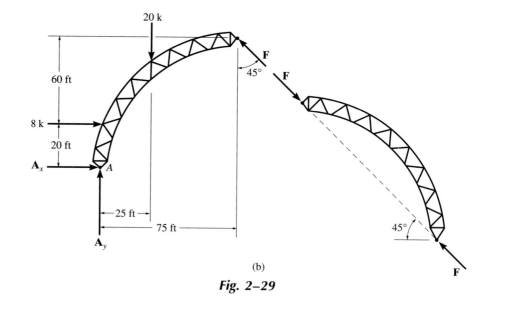

(b)

Fig. 2–29

Solution

Free-Body Diagrams.   Member BC is a two-force member, so the resultant forces acting at B and C are equal, opposite, and collinear, as shown in Fig. 2–29b. Realize that this greatly simplifies the analysis, since there are only three unknowns in this problem.*

Equations of Equilibrium.   Applying the three equations of equilibrium to member AB, we have

$$\uparrow + \Sigma M_A = 0; \quad 8(20) + 20(25) - (F \cos 45°)(75)$$
$$- (F \sin 45°)(80) = 0 \qquad F = 6.02 \text{ k}$$

Thus, although not shown in Fig. 2–29b,

$$B_x = C_x = 6.02(\sin 45°) = 4.26 \text{ k} \qquad \textbf{Ans.}$$
$$B_y = C_y = 6.02(\cos 45°) = 4.26 \text{ k} \qquad \textbf{Ans.}$$

Also,

$$\xrightarrow{+} \Sigma F_x = 0; \quad A_x + 8 - 6.02(\sin 45°) = 0 \quad A_x = -3.74 \text{ k} \qquad \textbf{Ans.}$$
$$+\uparrow \Sigma F_y = 0; \quad A_y - 20 + 6.02(\cos 45°) = 0 \quad A_y = 15.7 \text{ k} \qquad \textbf{Ans.}$$

---

*Try working the problem by placing two components $\mathbf{B}_x$, $\mathbf{B}_y$ at B and $\mathbf{C}_x$, $\mathbf{C}_y$ at C.

Example 2–10

The side of the building in Fig. 2–30a is subjected to a wind loading that creates a uniform *normal* pressure of 15 kPa on the windward side and a suction pressure of 5 kPa on the leeward side. Determine the horizontal and vertical components of reaction at the pin connections A, B, and C of the supporting gable arch.

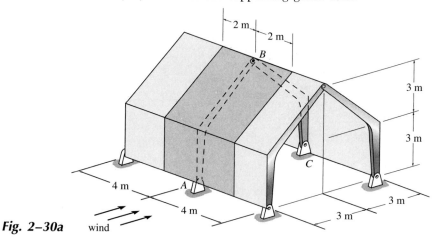

Fig. 2–30a   wind

Solution

Since the loading is evenly distributed, the central gable arch supports a loading acting on the walls and roof of the shaded tributary area. This represents a uniform distributed load of $(15 \text{ kN/m}^2)(4 \text{ m}) = 60 \text{ kN/m}$ on the windward side and $(5 \text{ kN/m}^2)(4 \text{ m}) = 20 \text{ kN/m}$ on the suction side, Fig. 2–30b.

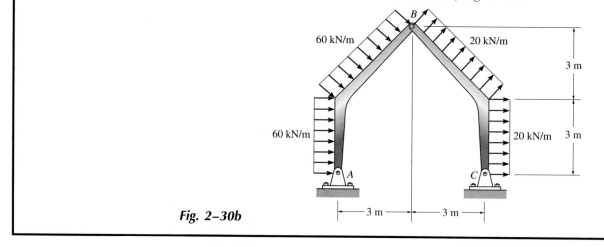

Fig. 2–30b

Free-Body Diagrams.   Simplifying the distributed loadings, the free-body diagrams of the entire frame and each of its parts are shown in Fig. 2–30c.

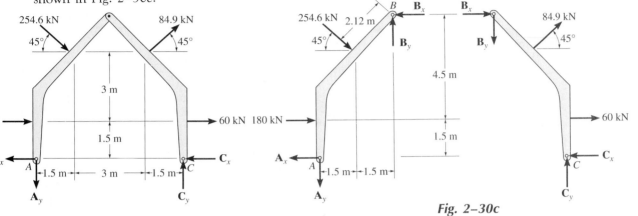

Fig. 2–30c

Equations of Equilibrium.   Simultaneous solution of equations is avoided by applying the equilibrium equations in the following sequence using previously computed results.*

Entire Frame:

$\curvearrowright + \Sigma M_A = 0$; $(180 + 60)(1.5) + (254.6 + 84.9)\cos 45°(4.5)$
$\qquad + (254.6 \sin 45°)(1.5) - (84.9 \sin 45°)(4.5) - C_y(6) = 0$
$\qquad\qquad\qquad C_y = 240.0 \text{ kN}$ **Ans.**

$+\uparrow \Sigma F_y = 0$; $-A_y - 254.6 \sin 45° + 84.9 \sin 45° + 240.0 = 0$
$\qquad\qquad\qquad A_y = 120.0 \text{ kN}$ **Ans.**

Member AB:

$\curvearrowright + \Sigma M_B = 0$; $A_x(6) - 120.0(3) - 180(4.5) - 254.6(2.12) = 0$
$\qquad\qquad\qquad A_x = 285.0 \text{ kN}$ **Ans.**

$\xrightarrow{+} \Sigma F_x = 0$; $-285.0 + 180 + 254.6 \cos 45° - B_x = 0$
$\qquad\qquad\qquad B_x = 75.0 \text{ kN}$ **Ans.**

$+\uparrow \Sigma F_y = 0$; $-120 - 254.6 \sin 45° + B_y = 0$
$\qquad\qquad\qquad B_y = 300.0 \text{ kN}$ **Ans.**

Member CB:

$\xrightarrow{+} \Sigma F_x = 0$; $-C_x + 60 + 84.9 \cos 45° + 75.0 = 0$
$\qquad\qquad\qquad C_x = 195.0 \text{ kN}$ **Ans.**

*The problem can also be solved by applying the six equations of equilibrium only to the two members. If this is done, it is best to first sum moments about point A on member AB, then point C on member CB. By doing this one obtains two equations to be solved simultaneously for B_x and B_y.

Example 2–11

Determine the horizontal and vertical components of reaction at the pins A, B, and C of the two-member frame shown in Fig. 2–31a.

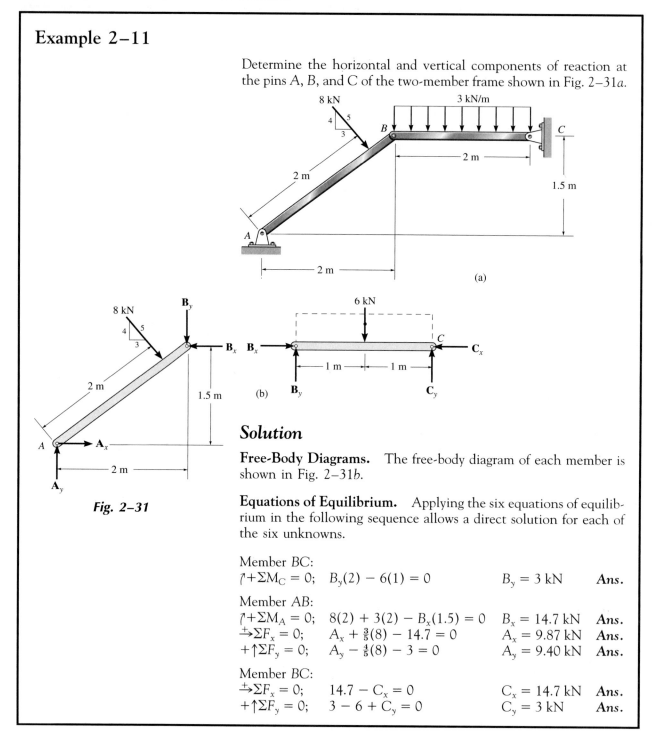

Fig. 2–31

Solution

Free-Body Diagrams. The free-body diagram of each member is shown in Fig. 2–31b.

Equations of Equilibrium. Applying the six equations of equilibrium in the following sequence allows a direct solution for each of the six unknowns.

Member BC:
$$\curvearrowleft +\Sigma M_C = 0; \quad B_y(2) - 6(1) = 0 \qquad\qquad B_y = 3 \text{ kN} \qquad \textbf{Ans.}$$

Member AB:
$$\curvearrowleft +\Sigma M_A = 0; \quad 8(2) + 3(2) - B_x(1.5) = 0 \quad B_x = 14.7 \text{ kN} \quad \textbf{Ans.}$$
$$\xrightarrow{+}\Sigma F_x = 0; \quad A_x + \tfrac{3}{5}(8) - 14.7 = 0 \qquad A_x = 9.87 \text{ kN} \quad \textbf{Ans.}$$
$$+\uparrow\Sigma F_y = 0; \quad A_y - \tfrac{4}{5}(8) - 3 = 0 \qquad A_y = 9.40 \text{ kN} \quad \textbf{Ans.}$$

Member BC:
$$\xrightarrow{+}\Sigma F_x = 0; \quad 14.7 - C_x = 0 \qquad\qquad C_x = 14.7 \text{ kN} \quad \textbf{Ans.}$$
$$+\uparrow\Sigma F_y = 0; \quad 3 - 6 + C_y = 0 \qquad\qquad C_y = 3 \text{ kN} \qquad \textbf{Ans.}$$

PROBLEMS

2–1. The roof deck of the single story building is subjected to a dead plus live load of 95 lb/ft². If the purlins are spaced 5 ft and the bents are spaced 20 ft apart, determine the distributed loading that acts along the purlin *DI*, and the loadings that act on the bent at *A*, *B*, *C*, *D*, *E*, *F*, *G*, and *H*.

2–2. The roof deck consists of precast lightweight reinforced concrete that is 3 in. thick and rests on the flanges of *T*-bars that are spaced 2.5 ft apart on centers. The subpurlins are clipped to the top of the main purlins that are spaced 6 ft apart. Determine the distributed loading that the deck exerts on the subpurlin *AB* and the concentrated loading on the main purlin *CD* located along the edge of the deck. Assume the subpurlins are simply-supported on the ends of each main purlin. *Hint:* See Table 1–2.

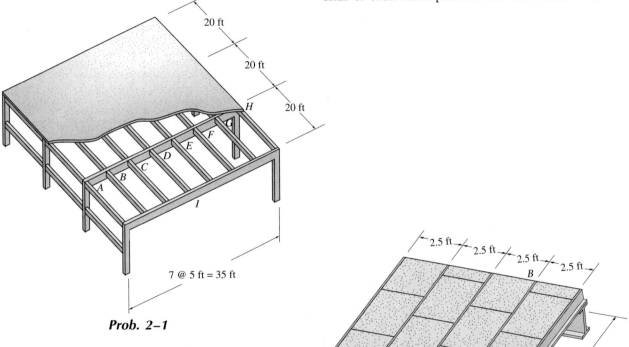

Prob. 2–1

Prob. 2–2

67

2–3. The frame is used to support the wood deck in a residential dwelling. Sketch the loading that acts along members *BG* and *ABCD*. Set $b = 10$ ft, $a = 5$ ft. *Hint:* See Table 1–3.

*2–4. Solve Prob. 2–3 if $b = 8$ ft, $a = 8$ ft.

2–5. Solve Prob. 2–3 if $b = 15$ ft, $a = 8$ ft.

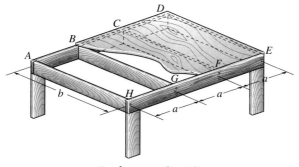

Probs. 2–3/2–4/2–5

2–6. The steel framework is used to support the 4-in. reinforced lightweight concrete slab that carries a uniform live loading of 500 lb/ft². Sketch the loading that acts along members BE' and *FD*. Set $b = 10$ ft, $a = 6$ ft. *Hint:* See Table 1–2.

2–7. Solve Prob. 2–6, with $b = 12$ ft, $a = 4$ ft.

Probs. 2–6/2–7

*2–8. Classify each of the structures as statically determinate, statically indeterminate, or unstable. If indeterminate, specify the degree of indeterminacy.

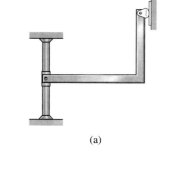

(a)

(b)

(c)

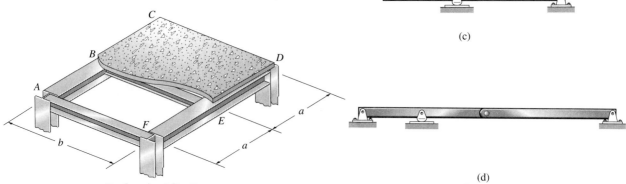

(d)

Prob. 2–8

2–9. Classify each of the structures as statically determinate, statically indeterminate, or unstable. If indeterminate, specify the degree of indeterminacy.

2–10. Classify each of the structures as statically determinate, statically indeterminate, or unstable. If indeterminate, specify the degree of indeterminacy. The supports or connections are to be assumed as stated.

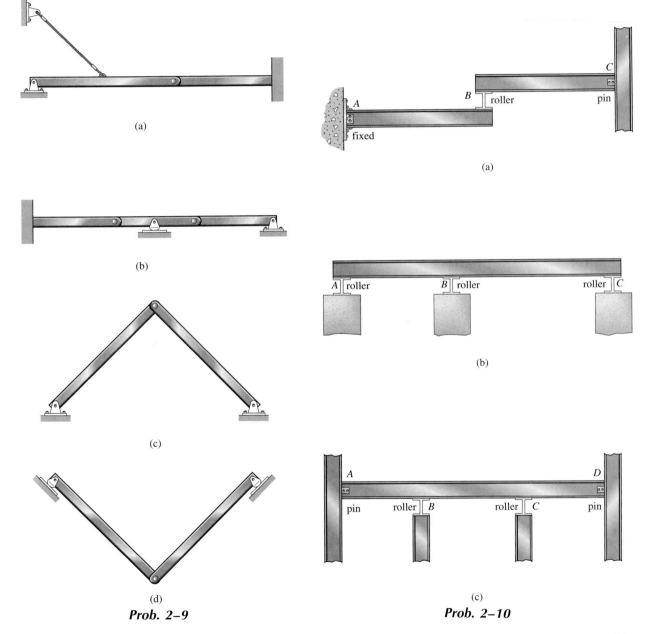

(a)

(b)

(c)

(d)

Prob. 2–9

(a)

(b)

(c)

Prob. 2–10

2–11.   Classify each of the structures as statically determinate, statically indeterminate, or unstable. If indeterminate, specify the degree of indeterminacy. The supports or connections are to be assumed as stated.

*2–12.   Classify each of the structures as statically determinate, statically indeterminate, or unstable. If indeterminate, specify the degree of indeterminacy.

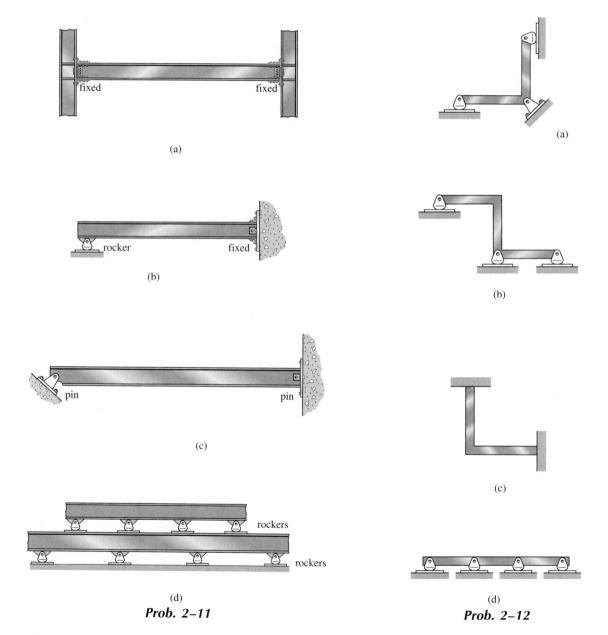

(a)

(b)

(c)

(d)

Prob. 2–11

(a)

(b)

(c)

(d)

Prob. 2–12

2–13. Classify each of the structures as statically determinate, statically indeterminate, or unstable. If indeterminate, specify the degree of indeterminacy. The supports or connections are to be assumed as stated.

2–14. Classify each of the structures as statically determinate, statically indeterminate, or unstable. If indeterminate, specify the degree of indeterminacy. The supports or connections are to be assumed as stated.

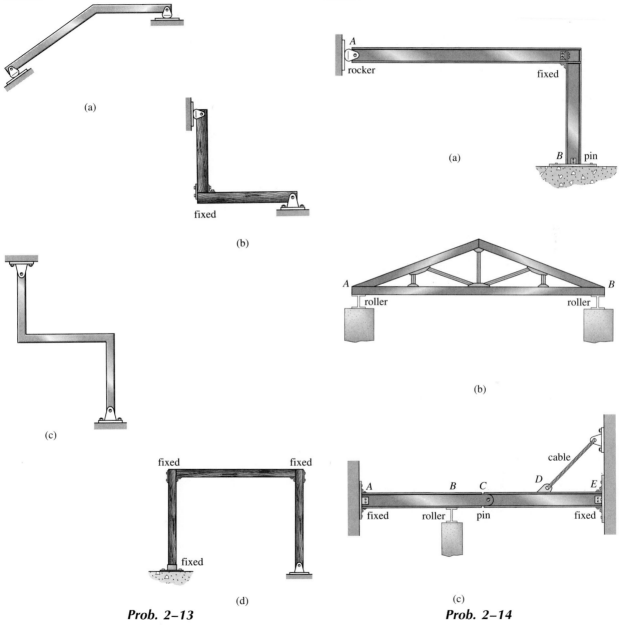

(a)

fixed

(b)

(c)

fixed   fixed

fixed

(d)

Prob. 2–13

rocker   fixed

(a)

roller   roller

(b)

cable

fixed   roller   pin   fixed

(c)

Prob. 2–14

2–15. Classify each of the frames as statically determinate, statically indeterminate, or unstable. If indeterminate, specify the degree of indeterminacy. All internal joints are fixed-connected.

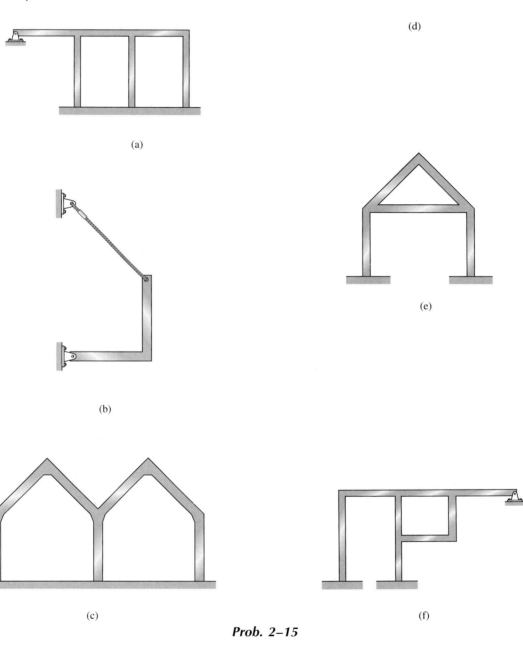

(d)

(a)

(b)

(e)

(c)

(f)

Prob. 2–15

*2–16. Classify each of the frames as statically determinate, statically indeterminate, or unstable. If indeterminate, specify the degree of indeterminacy. All internal joints are fixed-connected.

2–17. Determine the reactions on the beam.

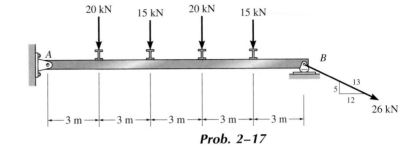

Prob. 2–17

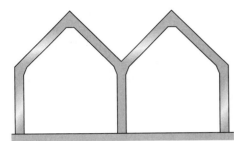

(a)

(b)

2–18. Determine the reactions on the beam. The support at B can be assumed as a roller.

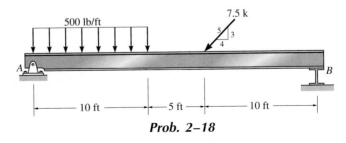

Prob. 2–18

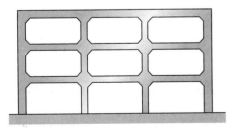

(c)

Prob. 2–16

73

2–19. Determine the reactions on the beam. The support at B can be assumed to be a roller.

2–21. Determine the reactions on the beam.

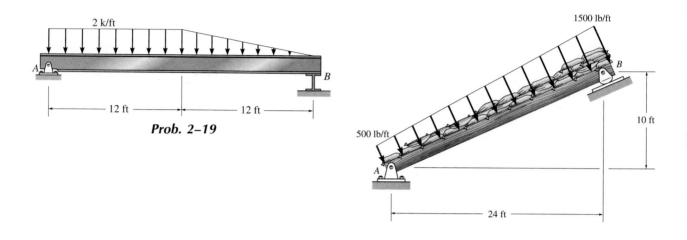

Prob. 2–19

Prob. 2–21

*2–20. Determine the support reactions on the beam.

2–22. Determine the reactions on the beam.

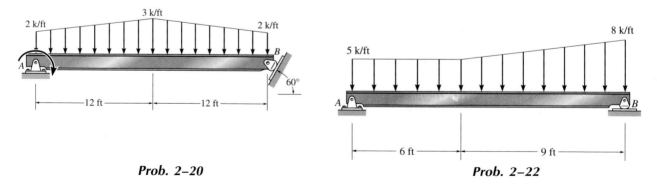

Prob. 2–20

Prob. 2–22

2–23. Determine the vertical reactions at the supports A and B. Assume A is a roller and B is a pin.

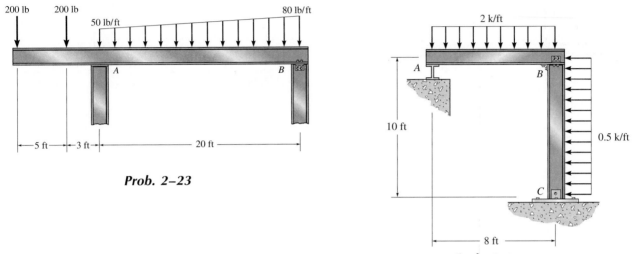

Prob. 2–23

2–25. Determine the reactions at the supports A and B. Assume the support at A is a roller, B is a fixed-connected joint and C is a pin.

Prob. 2–25

*2–24. Determine the reactions at the supports A, C, and D.

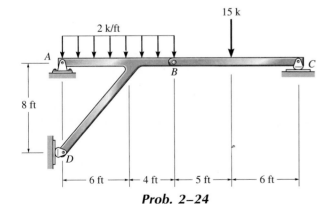

Prob. 2–24

2–26. Determine the reactions at the smooth support A and pin support B. The connection at C is fixed.

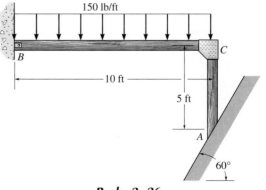

Prob. 2–26

75

2–27. Determine the reactions at the supports A, C, and E. Assume A and C are rollers, E is fixed, and B and D are pins.

2–29. Determine the reactions at the supports A and B of the frame. Assume that the support at A is a roller.

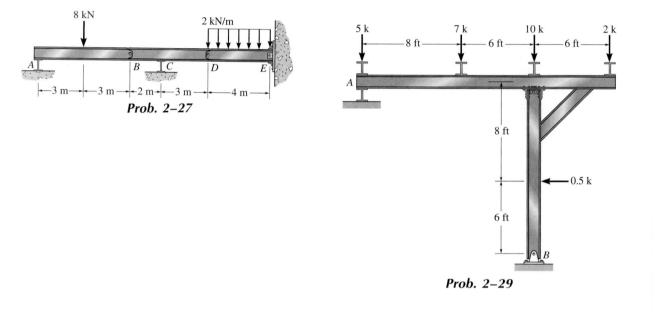

Prob. 2–27

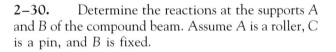

Prob. 2–29

*2–28. The construction features of a cantilever truss bridge are shown in the figure. Here it can be seen that the center truss CD is suspended by the cantilever arms ABC and DEF. C and D are pins. Determine the vertical reactions at the supports A, B, E, and F if a 15-kip load is applied to the center truss.

2–30. Determine the reactions at the supports A and B of the compound beam. Assume A is a roller, C is a pin, and B is fixed.

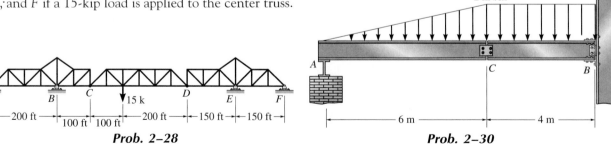

Prob. 2–28

Prob. 2–30

2–31. Determine the reactions at the supports A and B.

2–33. Determine the reactions at the supports A and C in terms of the load P.

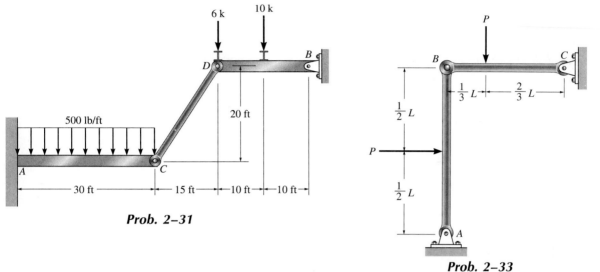

Prob. 2–31

Prob. 2–33

2–34. Determine the horizontal and vertical components of reaction at the supports A, B, and C. Assume the frame is pin-connected at A, B, D, E, and F, and there is a fixed-connected joint at C.

*2–32. Determine the reactions at A, B, and E. Assume A and B are roller-supported.

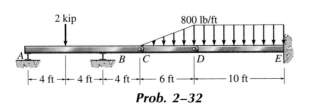

Prob. 2–32

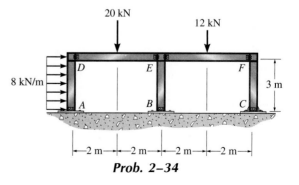

Prob. 2–34

2–35. Determine the reactions at the supports A and C. The frame is pin-connected at A, B, and C.

2–37. Determine the reactions at the supports A and D. Assume A is *fixed* and B, and C, and D are pins.

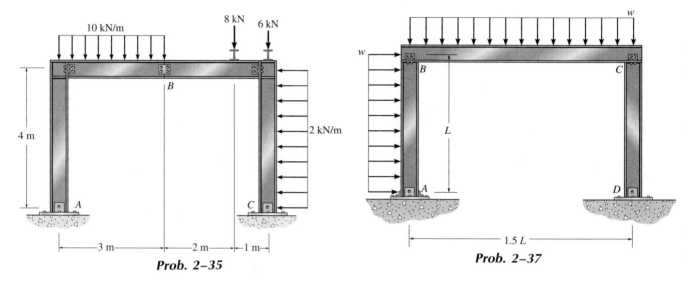

Prob. 2–35

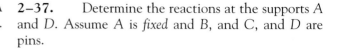

Prob. 2–37

*2–36. Determine the horizontal and vertical components of force at the connections A, B, and C. Assume each of these connections is a pin.

2–38. Determine the reactions at the truss supports A and B. The distributed loading is caused by wind pressure.

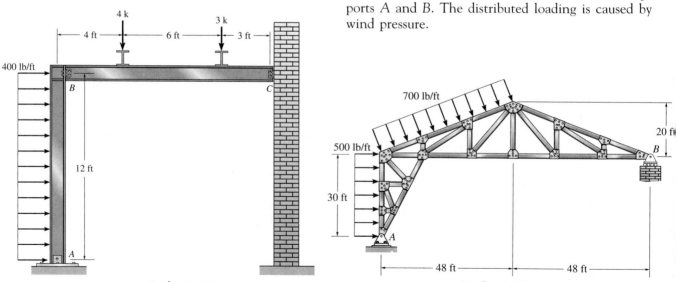

Prob. 2–36

Prob. 2–38

2–39. Determine the horizontal and vertical re-actions at the connections A and C of the gable frame. Assume that A, B, and C are pin connections. The purlin loads such as D and E are applied perpendicular to the center line of each girder.

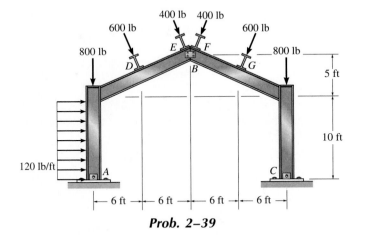

Prob. 2–39

*2–40. Determine the horizontal and vertical components of reaction acting at the pins A and C.

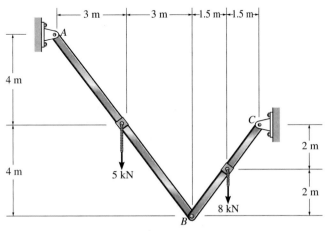

Prob. 2–40

This bridge consists of two parallel Warren trusses (with verticals). The forces in its members can be readily analyzed using either the method of joints or the method of sections. *(Photo courtesy Bethlehem Steel)*

Analysis of Statically Determinate Trusses

In this chapter we will develop the procedures for analyzing statically determinate trusses using the method of joints and the method of sections. First, however, the determinacy and stability of a truss will be discussed. Then the analysis of three forms of planar trusses will be considered: simple, compound, and complex. Finally, at the end of the chapter we will consider the analysis of a space truss.

3.1 Common Types of Trusses

A *truss* is a structure composed of slender members joined together at their end points. The members commonly used in construction consist of wooden struts, metal bars, angles, or channels. The joint connections are usually formed by bolting or welding the ends of the members to a common plate, called a *gusset plate*, as shown in Fig. 3–1, or by simply passing a large bolt or pin through each of the members. Planar trusses lie in a single plane and are often used to support roofs and bridges.

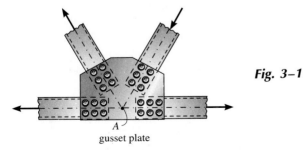

A

gusset plate

Fig. 3–1

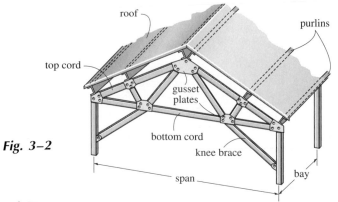

Fig. 3–2

Roof Trusses

Roof trusses are often used as part of an industrial building frame, such as the one shown in Fig. 3–2. Here, the roof load is transmitted to the truss at the joints by means of a series of *purlins*. The roof truss along with its supporting columns is termed a *bent*. Ordinarily, roof trusses are supported either by columns of wood, steel, or reinforced concrete, or by masonry walls. To keep the bent rigid, and thereby capable of resisting horizontal wind forces, knee braces are sometimes used at the supporting columns. When the span is large, often a rocker or roller is used to support one end. This type of support allows freedom for expansion or contraction due to temperature or application of loads. The space between adjacent bents is called a *bay*. Bays are economically spaced at about 15 ft (4.5 m) for spans around 60 ft (18 m) and about 20 ft (6 m) for spans of 100 ft (30 m). Bays are often tied together using diagonal bracing in order to maintain rigidity of the building's structure.

Truss forms used to support roofs are selected on the basis of the span, slope, and the kind of roof material. Some of the more common types used are shown in Fig. 3–3. In particular, the scissors truss, Fig. 3–3*a* can be used for short spans that require overhead clearance. The Howe and Pratt trusses, Fig. 3–3*b* and 3–3*c*, are used for roofs of moderate span, about 60 ft (18 m) to 100 ft (30 m). If larger spans are required to support the roof, the fan truss or Fink truss may be used, Figs. 3–3*d* and 3–3*e*. These trusses may be built with a cambered bottom cord such as that shown in Fig. 3–3*f*. If a flat roof or nearly flat roof is to be selected, the Warren truss, Fig. 3–3*g*, is often used. Also, the Howe and Pratt trusses may be modified for flat roofs. Sawtooth trusses, Fig. 3–3*h*, are often used where column spacing is not objectionable and uniform lighting is important. A textile mill would be an example. The bowstring truss, Fig. 3–3*i*, is sometimes selected for garages and small airplane hangars; and the arched truss, Fig. 3–3*j*, although relatively expensive, can be used for high rises and long spans such as field houses, gymnasiums, and so on.

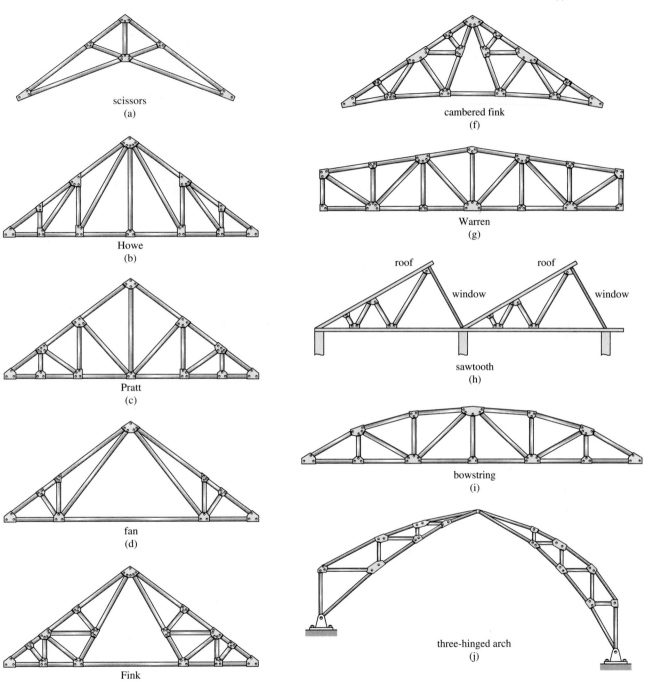

scissors
(a)

Howe
(b)

Pratt
(c)

fan
(d)

Fink
(e)

cambered fink
(f)

Warren
(g)

roof

roof

window

window

sawtooth
(h)

bowstring
(i)

three-hinged arch
(j)

Fig. 3–3

83

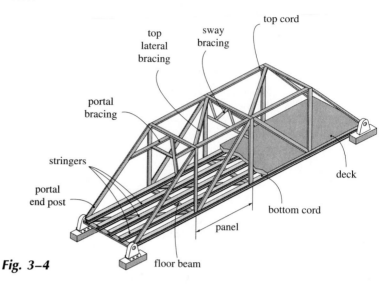

Fig. 3–4

Bridge Trusses

The main structural elements of a typical bridge truss are shown in Fig. 3–4. Here it is seen that a load on the *deck* is first transmitted to *stringers,* then to *floor beams,* and finally to the joints of the two supporting side trusses. The top and bottom cords of these side trusses are connected by top and bottom *lateral bracing,* which serves to resist the lateral forces caused by wind and the sidesway caused by moving vehicles on the bridge. Additional stability is developed by the *portal* and *sway bracing.* As in the case of many long-span roof trusses, a roller is provided at one end of a bridge truss to allow for thermal expansion.

A few of the typical forms of bridge trusses currently used for single spans are shown in Fig. 3–5. In particular, the Pratt, Howe, and Warren trusses are normally used for spans up to 180 ft (55 m) to 200 ft (61 m) in length. The most common form is the Warren truss with verticals, Fig. 3–5c. For larger spans, a truss with a polygonal upper cord, such as the Parker truss, Fig. 3–5d, is used for some savings in material. The Warren truss with verticals can also be fabricated in this manner for spans up to 300 ft (91 m). The greatest economy of material is obtained if the diagonals have a slope between 45° and 60° with the horizontal. If this rule is maintained, then for spans greater than 300 ft (91 m), the depth of the truss must increase and consequently the panels will get longer. This results in a heavy deck system and, to keep the weight of the deck within tolerable limits, *subdivided* trusses have been developed. Typical examples include the Baltimore and subdivided Warren trusses, Fig. 3–5e and 3–5f. Finally, the K-truss shown in Fig. 3–5g can also be used in place of a subdivided truss, since it accomplishes the same purpose.

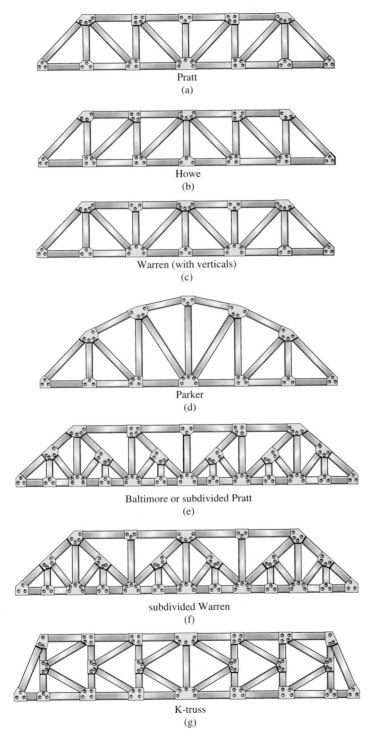

Pratt
(a)

Howe
(b)

Warren (with verticals)
(c)

Parker
(d)

Baltimore or subdivided Pratt
(e)

subdivided Warren
(f)

K-truss
(g)

Fig. 3–5

Assumptions for Design

To design both the members and the connections of a truss, it is first necessary to determine the *force* developed in each member when the truss is subjected to a given loading. In this regard, two important assumptions will be made in order to idealize the truss.

1. *The members are joined together by smooth pins.* In cases where bolted or welded joint connections are used, this assumption is generally satisfactory provided the center lines of the joining members are concurrent at a point, as in the case of point A in Fig. 3–1. It should be realized, however, that the actual connections do give some *rigidity* to the joint and this in turn introduces bending of the connected members when the truss is subjected to a load. The bending stress developed in the members is called *secondary stress*, whereas the stress in the members of the idealized truss, having pin-connected joints, is called *primary stress*. Most often a secondary stress analysis of a truss is seldom performed, although for some types of truss geometries these stresses may be large.*

2. *All loadings are applied at the joints.* In most situations, such as for bridge and roof trusses, this assumption is true. Frequently in the force analysis, the weight of the members is neglected, since the force supported by the members is large in comparison with their weight. If the weight is to be included in the analysis, it is generally satisfactory to apply it as a vertical force, half of its magnitude applied at each end of the member.

Because of these two assumptions, *each truss member acts as an axial force member*, and therefore the forces acting at the ends of the member must be directed along the axis of the member. If the force tends to *elongate* the member, it is a *tensile force (T)*, Fig. 3–6a; whereas if the force tends to *shorten* the member, it is a *compressive force (C)*, Fig. 3–6b. In the actual design of a truss it is important to state whether the nature of the force is tensile or compressive. Most often, compression members must be made *thicker* than tension members, because of the buckling or sudden instability that may occur in compression members.

T   C

T   C

(a)   (b)

Fig. 3–6

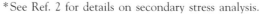

*See Ref. 2 for details on secondary stress analysis.

3.2   Classification of Coplanar Trusses _____

Before beginning the force analysis of a truss, it is important to classify the truss as simple, compound, or complex, and then to be able to specify its determinacy and stability.

Simple Truss

To prevent collapse, the framework of a truss must be rigid. Obviously, the four-bar frame $ABCD$ in Fig. 3–7 will collapse unless a diagonal, such as AC, is added for support. The simplest framework that is rigid

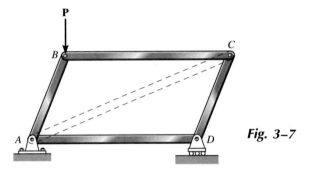

Fig. 3–7

or stable is a *triangle*. Consequently, a *simple truss* is constructed by starting with a basic triangular element, such as ABC in Fig. 3–8, and connecting two members (AD and BD) to form an additional element. Thus it is seen that as each additional element of two members is placed on the truss, the number of joints is increased by one.

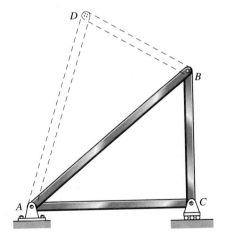

Fig. 3–8

87

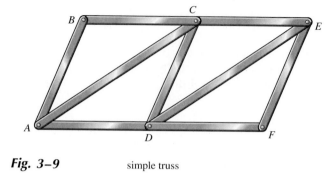

Fig. 3–9   simple truss

An example of a simple truss is shown in Fig. 3–9, where the basic "stable" triangular element is *ABC*, from which the remainder of the joints, *D*, *E*, and *F*, are established in alphabetical sequence. For this method of construction, however, it is important to realize that simple trusses *do not* have to consist entirely of triangles. An example is shown in Fig. 3–10, where starting with triangle *ABC*, bars *CD* and *AD* are added to form joint *D*. Finally, bars *BE* and *DE* are added to form joint *E*.

Fig. 3–10

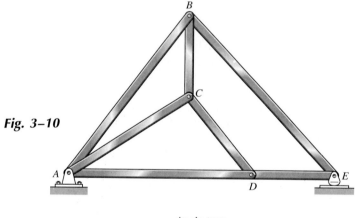

simple truss

Compound Truss

A *compound truss* is formed by connecting two or more simple trusses together. Quite often this type of truss is used to support loads acting over a *large span*, since it is cheaper to construct a compound truss than to use a single simple truss. (Refer to the discussion on using subdivided bridge trusses in Sec. 3.1.)

There are three ways in which simple trusses are joined together to form a compound truss.

Type 1: The trusses may be connected by a common joint and bar. An example is given in Fig. 3–11a, where the shaded truss *ABC* is connected to the shaded truss *CDE* in this manner.

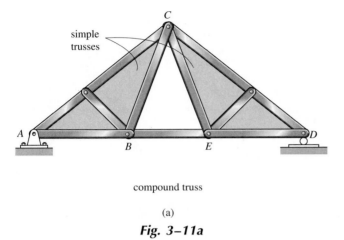

simple trusses

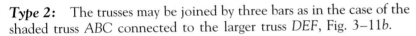

compound truss

(a)

Fig. 3–11a

Type 2: The trusses may be joined by three bars as in the case of the shaded truss *ABC* connected to the larger truss *DEF*, Fig. 3–11b.

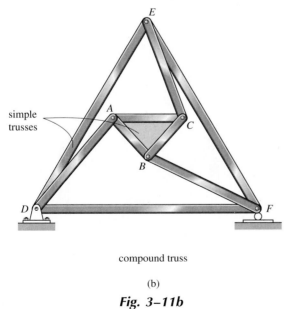

simple trusses

compound truss

(b)

Fig. 3–11b

Type 3: The trusses may be joined where bars of a large simple truss, called the *main truss*, have been *substituted* by simple trusses, called *secondary trusses*. An example is shown in Fig. 3–11c, where the dashed members of the main truss *ABCD* have been *replaced* by the secondary shaded trusses. If this truss carried roof loads, the use of the secondary trusses might be more economical, since the dashed members may be subjected to excessive bending, whereas the secondary trusses can better transfer the load.

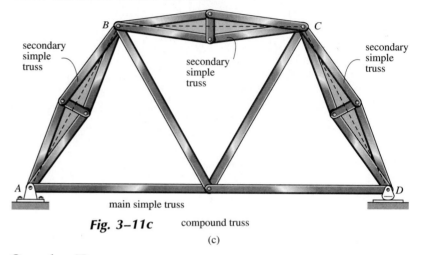

secondary simple truss

secondary simple truss

secondary simple truss

main simple truss

Fig. 3–11c   compound truss

(c)

Complex Truss

A *complex truss* is one that cannot be classified as being either simple or compound. The truss in Fig. 3–12 is an example.

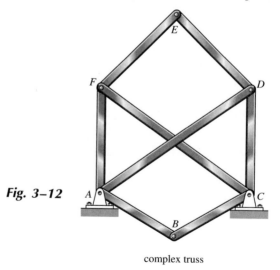

Fig. 3–12

complex truss

The members of this truss bridge must be designed to support both the intended bridge loading and the dead-weight loading imposed on it during its construction. *(Courtesy of Bethlehem Steel Corporation)*

Determinacy

For any problem in truss analysis, it should be realized that the total number of *unknowns* includes the forces in b number of bars of the truss and the total number r of external support reactions. Since the truss members are all straight axial force members lying in the same plane, the force system acting at each joint is *coplanar and concurrent*. Consequently, rotational or moment equilibrium is automatically satisfied at the joint (or pin), and it is only necessary to satisfy $\Sigma F_x = 0$ and $\Sigma F_y = 0$ to ensure translational or force equilibrium. Therefore, only two equations of equilibrium can be written for each joint, and since there are j number of joints, the total number of equations available for solution is $2j$. By simply comparing the total number of unknowns $(b + r)$ with the total number of available equilibrium equations, it is therefore possible to specify the determinacy for either a simple, compound, or complex truss. We have

$$
\begin{aligned}
b + r = 2j \qquad &\text{statically determinate} \\
b + r > 2j \qquad &\text{statically indeterminate}
\end{aligned}
\qquad (3\text{--}1)
$$

In particular, the *degree of indeterminacy* is specified by the difference in the numbers $(b + r) - 2j$.

Stability

If $b + r < 2j$, a truss will be *unstable*, that is, it will collapse, since there will be an insufficient number of bars or reactions to constrain all the joints. A truss can also be unstable if it is statically determinate or statically indeterminate. In this case the stability will have to be determined either by inspection or by a force analysis.

External Stability. As stated in Sec. 2.4, a *structure (or truss) is externally unstable if its reactions are concurrent or parallel*. For example, the two trusses in Fig. 3–13 are externally unstable since the support reactions have lines of action that are either concurrent or parallel.

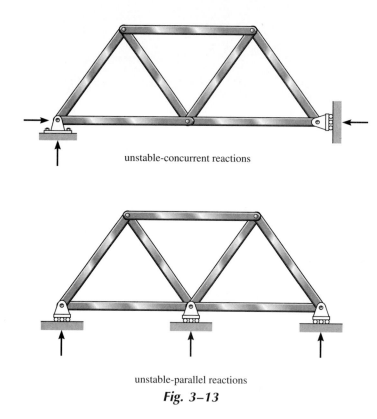

unstable-concurrent reactions

unstable-parallel reactions

Fig. 3–13

Internal Stability. The internal stability of a truss can often be checked by careful inspection of the arrangement of its members. If it can be determined that each joint is held fixed so that it cannot move in a "rigid body" sense with respect to the other joints, then the truss will be stable. Notice that *a simple truss will always be internally stable*, since the nature of its construction requires starting from a basic triangular element and adding successive "elements," each containing two additional members and a joint. The truss in Fig. 3–14 exemplifies this construction, where, starting with the shaded triangle element *ABC*, the successive joints *D, E, F, G, H* have been added.

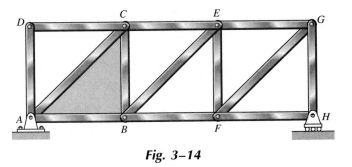

Fig. 3–14

If a truss is not constructed so that it holds its joints in a fixed position, it will be unstable or have a "critical form." An obvious example of this is shown in Fig. 3–15, where it can be seen that since no restraint or fixity is provided between joints *C* and *F* or *B* and *E*, the truss will collapse under load.

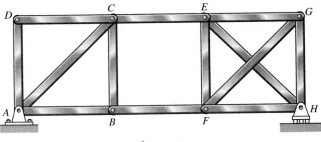

Fig. 3–15

93

To determine the internal stability of a compound truss, it is necessary to identify the way in which the simple trusses are connected together. For example, the compound truss in Fig. 3–16 is unstable since the inner simple truss *ABC* is connected to the outer simple truss *DEF* using three bars, *AD*, *BE*, and *CF*, which are *concurrent* at point *O*. Thus an external load can be applied to joint *A*, *B*, or *C* and cause the truss *ABC* to rotate slightly.

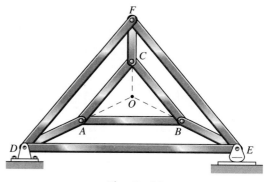

Fig. 3–16

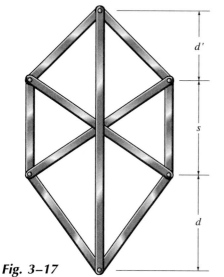

Fig. 3–17

If a truss is identified as complex, it may not be possible to tell by inspection if it is stable. For example, it can be shown by the analysis discussed in Sec. 3.7 that the complex truss in Fig. 3–17 is unstable or has a "critical form" only if the dimension $d = d'$. If $d \neq d'$ it is stable. The best way to classify the stability of a complex truss, however, is simply to perform a force analysis of the members. If the analysis yields a *unique solution* for the forces in the members, the truss is *stable*.*

In general, the stability of any form of truss, be it simple, compound, or complex, may be checked by using a computer to solve the $2j$ simultaneous equations written for all the joints of the truss. If Cramer's rule is used for this solution, then if the coefficients of the $2j$ equations have a determinant $D \neq 0$, the force analysis will be unique and the truss will be stable. However, if $D = 0$ the equations will have an infinite number of solutions and the truss will be unstable or have a critical form.

---

*The instability of complex trusses can also be identified using the zero-load test. See the second reference at the end of the chapter.

If a computer analysis is not performed, the methods discussed above can be used to check the stability of the truss. To summarize, if the truss has b bars, r external reactions, and j joints, then if

$$b + r < 2j \quad \text{unstable}$$
$$b + r \geq 2j \quad \text{unstable if truss support reactions}$$

are concurrent or parallel or if
some of the components of the
truss form a collapsible mechanism

$(3-2)$

Bear in mind, however, that *if a truss is unstable, it does not matter whether it is statically determinate or indeterminate*. Obviously, the use of an unstable truss is to be avoided in practice.

The following example illustrates how trusses are classified as statically determinate or indeterminate, stable or unstable.

Many types of electrical transmission towers are made from connected planar trusses that can be analyzed using the methods of this chapter.
(Myron Wood/Photo Researchers, Inc.)

Example 3–1

Classify each of the trusses in Fig. 3–18 as stable, unstable, statically determinate, or statically indeterminate. Also, if the truss is statically determinate and stable, specify the type of truss, i.e., simple, compound, or complex.

Solution

Fig. 3–18a.   *Externally stable*, since the reactions are not concurrent or parallel. Since $b = 19$, $r = 3$, $j = 11$, then $b + r = 2j$ or $22 = 22$. Therefore, the truss is *statically determinate*. By inspection the truss is *internally stable* and is classified as a *compound truss* (type 1).

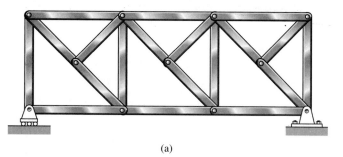

(a)

Fig. 3–18b.   *Externally stable*. Since $b = 15$, $r = 4$, $j = 9$, then $b + r > 2j$ or $19 > 18$. The truss is *statically indeterminate* to the first degree. By inspection the truss is *internally stable* and is classified as a *simple* truss.

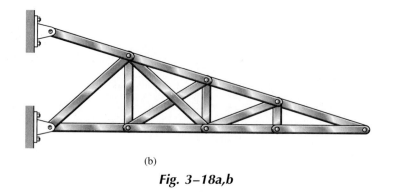

(b)

Fig. 3–18a,b

Fig. 3–18c. *Externally stable.* Since $b = 9$, $r = 3$, $j = 6$, then $b + r = 2j$ or $12 = 12$. The truss is *statically determinate.* By inspection the truss is *internally stable* and can be classified as a *compound truss* (type 2).

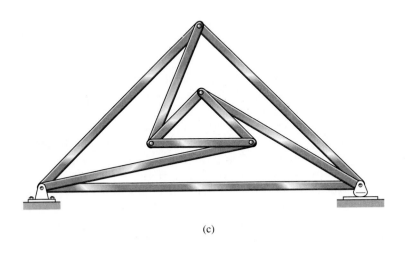

(c)

Fig. 3–18d. *Externally stable.* Since $b = 12$, $r = 3$, $j = 8$, then $b + r < 2j$ or $15 < 16$. The truss is *internally unstable.*

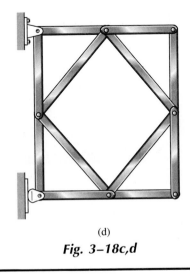

(d)

Fig. 3–18c,d

3.3   The Method of Joints

If a truss is in equilibrium, then each of its joints must also be in equilibrium. Hence, the method of joints consists of satisfying the equilibrium conditions $\Sigma F_x = 0$ and $\Sigma F_y = 0$ for the forces exerted *on the pin* at each joint of the truss.

When using the method of joints, it is necessary to draw the joint's free-body diagram before applying the equilibrium equations. In this regard, recall that the *line of action* of each member force acting on the joint is *specified* from the geometry of the truss, since the force in a member passes along the axis of the member. As an example, consider joint B of the truss in Fig. 3–19*a*. From the free-body diagram, Fig. 3–19*b*, the only unknowns are the *magnitudes* of the forces in members BA and BC. As shown, $\mathbf{F}_{BA}$ is "pulling" on the pin, which indicates that member BA is in *tension*, whereas $\mathbf{F}_{BC}$ is "pushing" on the pin, and consequently member BC is in *compression*. These effects are clearly demonstrated by using the method of sections and isolating the joint with small segments of the member connected to the pin, Fig. 3–19*c*. Notice that pushing or pulling on these small segments indicates the effect of the member being either in compression or tension.

In all cases, the joint analysis should start at a joint having at least one known force and at most two unknown forces, as in Fig. 3–19*b*. In

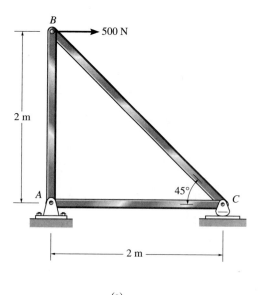

(a)

Fig. 3–19a

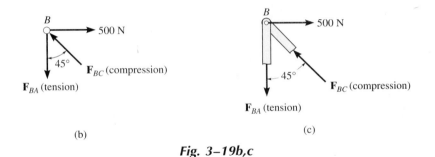

B
$\longrightarrow$ 500 N

45°

$\mathbf{F}_{BC}$ (compression)

$\mathbf{F}_{BA}$ (tension)

(b)

B
$\longrightarrow$ 500 N

45°

$\mathbf{F}_{BC}$ (compression)

$\mathbf{F}_{BA}$ (tension)

(c)

Fig. 3–19b,c

this way, application of $\Sigma F_x = 0$ and $\Sigma F_y = 0$ yields two algebraic equations that can be solved for the two unknowns. The correct arrowhead sense of direction of an unknown member force can, in many cases, be determined "by inspection." For example, $\mathbf{F}_{BC}$ in Fig. 3–19*b* must push on the pin (compression) since its horizontal component $F_{BC} \sin 45°$ must balance the 500-N force ($\Sigma F_x = 0$). Likewise, $\mathbf{F}_{BA}$ is a tensile force since it balances the vertical component $F_{BC} \cos 45°$ ($\Sigma F_y = 0$). In more complicated cases, the sense of direction of an unknown member force can be *assumed*; then, after applying the equilibrium equations, the assumed directional sense can be verified from the numerical results. A *positive* answer indicates that the sense of direction is *correct*, whereas a *negative* answer indicates that the sense of direction shown on the free-body diagram must be *reversed*.

Example of a Pratt truss used to support the roof of a metal building.

Procedure for Analysis

The following procedure provides a means for analyzing a truss using the method of joints.

First determine the reactions at the supports by considering the equilibrium of the entire truss. Then draw the free-body diagram of a joint having at least one known force and at most two unknown forces. (If this joint is at one of the supports, it may be necessary to know the external reactions at the support.) By inspection, attempt to show the unknown forces acting on the joint with the correct sense of direction. The x and y axes should be oriented such that the forces on the free-body diagram can be easily resolved into their x and y components, and then apply the two force equilibrium equations $\Sigma F_x = 0$ and $\Sigma F_y = 0$. It should also be noted that if one of the axes is oriented along one of the unknown forces, the other unknown can be determined by summing forces along the other axis. Solve for the two unknown member forces and verify their correct directional sense.

Continue to analyze each of the other joints, where again it is necessary to choose a joint having at most two unknowns and at least one known force. Realize that once the force in a member is found from the analysis of a joint at one of its ends, the result can be used to analyze the forces acting on the joint at its other end. Strict adherence to the principle of action, equal but opposite force reaction, must, of course, be observed. Remember, a member in *compression* "pushes" on the joint and a member in *tension* "pulls" on the joint.

Once the force analysis of the truss has been completed, the size of the members and their connections can then be determined from an appropriate design code.

Example 3–2

Determine the force in each member of the roof truss shown in Fig. 3–20a. State whether the members are in tension or compression. The reactions at the supports are given.

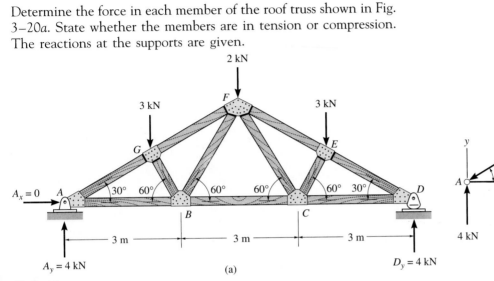

(a)

Solution

Only the forces in half the members have to be determined, since the truss is symmetric with respect to *both* loading and geometry.

Joint A, Fig. 3–20b. We can start the analysis at joint A. Why? The free-body diagram is shown in Fig. 3–20b.

$+\uparrow\Sigma F_y = 0;$ $4 - F_{AG}\sin 30° = 0$ $F_{AG} = 8\text{ kN (C)}$ **Ans.**
$\Rightarrow\Sigma F_x = 0;$ $F_{AB} - 8\cos 30° = 0$ $F_{AB} = 6.93\text{ kN (T)}$ **Ans.**

Joint G, Fig. 3–20c. In this case note how the orientation of the x, y axes avoids simultaneous solution of equations.

$+\nwarrow\Sigma F_y = 0;$ $F_{GB} - 3\cos 30° = 0$ $F_{GB} = 2.60\text{ kN (C)}$ **Ans.**
$+\nearrow\Sigma F_x = 0;$ $8 - 3\sin 30° - F_{GF} = 0$ $F_{GF} = 6.50\text{ kN (C)}$ **Ans.**

Joint B, Fig. 3–20d

$+\uparrow\Sigma F_y = 0;$ $F_{BF}\sin 60° - 2.60\sin 60° = 0$
$$F_{BF} = 2.60\text{ kN (T)} \quad \textbf{Ans.}$$
$\Rightarrow\Sigma F_x = 0;$ $F_{BC} + 2.60\cos 60° + 2.60\cos 60° - 6.93 = 0$
$$F_{BC} = 4.33\text{ kN (T)} \quad \textbf{Ans.}$$

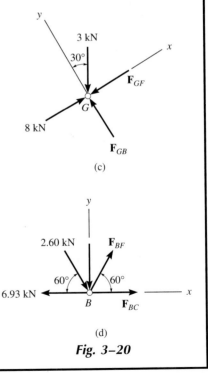

(b)

(c)

(d)

Fig. 3–20

101

Example 3–3

Determine the force in each member of the scissors truss shown in Fig. 3–21a. State whether the members are in tension or compression. The reactions at the supports are given.

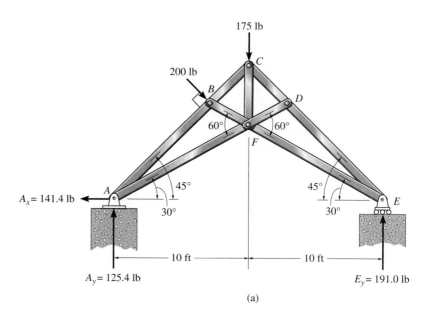

(a)

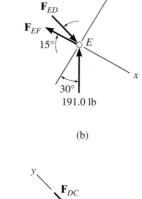

(b)

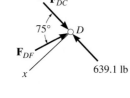

(c)

Fig. 3–21a,b,c

Solution

The truss will be analyzed in the following sequence:

Joint E, Fig. 3–21b. Note that simultaneous solution of equations is avoided by the x, y axes orientation.

$$+\nearrow \Sigma F_y = 0; \quad 191.0 \cos 30° - F_{ED} \sin 15° = 0$$
$$F_{ED} = 639.1 \text{ lb (C)} \qquad \textbf{Ans.}$$
$$+\searrow \Sigma F_x = 0; \quad 639.1 \cos 15° - F_{EF} - 191.0 \sin 30° = 0$$
$$F_{EF} = 521.8 \text{ lb (T)} \qquad \textbf{Ans.}$$

Joint D, Fig. 3–21c

$$+\swarrow \Sigma F_x = 0; \quad -F_{DF} \sin 75° = 0 \qquad F_{DF} = 0 \qquad \textbf{Ans.}$$
$$+\nwarrow \Sigma F_y = 0; \quad -F_{DC} + 639.1 = 0 \qquad F_{DC} = 639.1 \text{ lb (C)} \quad \textbf{Ans.}$$

Joint C, Fig. 3–21d

$\xrightarrow{+} \Sigma F_x = 0; \quad F_{CB} \sin 45° - 639.1 \sin 45° = 0$
$$F_{CB} = 639.1 \text{ lb (C)} \qquad \textbf{\textit{Ans.}}$$
$+\uparrow \Sigma F_y = 0; \quad -F_{CF} - 175 + 2(639.1) \cos 45° = 0$
$$F_{CF} = 728.8 \text{ lb (T)} \qquad \textbf{\textit{Ans.}}$$

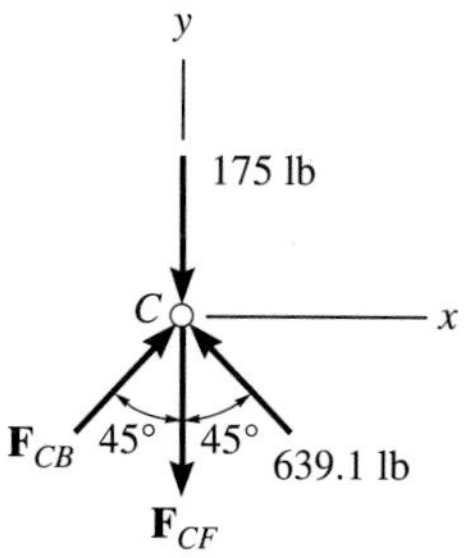

(d)

Joint B, Fig. 3–21e

$+\nwarrow \Sigma F_y = 0; \quad F_{BF} \sin 75° - 200 = 0$
$$F_{BF} = 207.1 \text{ lb (C)} \qquad \textbf{\textit{Ans.}}$$
$+\swarrow \Sigma F_x = 0; \quad 639.1 + 207.1 \cos 75° - F_{BA} = 0$
$$F_{BA} = 692.7 \text{ lb (C)} \qquad \textbf{\textit{Ans.}}$$

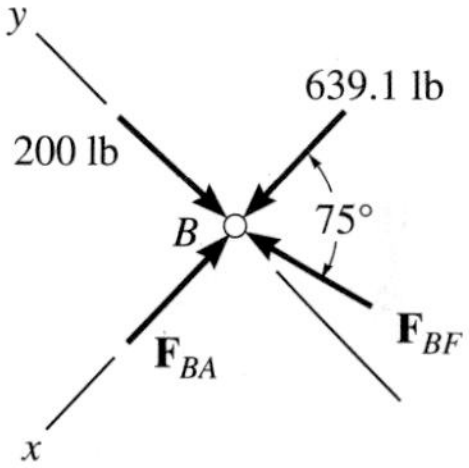

(e)

Joint A, Fig. 3–21f

$\xrightarrow{+} \Sigma F_x = 0; \quad F_{AF} \cos 30° - 692.7 \cos 45° - 141.4 = 0$
$$F_{AF} = 728.9 \text{ lb (T)} \qquad \textbf{\textit{Ans.}}$$
$+\uparrow \Sigma F_y = 0; \quad 125.4 - 692.7 \sin 45° + 728.9 \sin 30° \equiv 0 \quad \text{check}$

A further check of the calculations can be made by analyzing joint F.

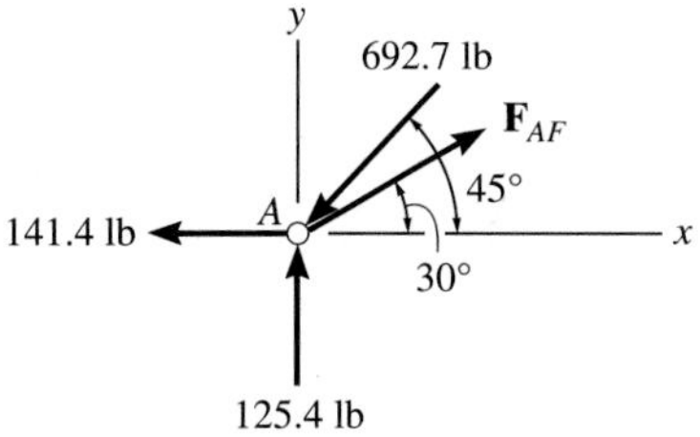

(f)

Fig. 3–21d,e,f

3.4 Zero-Force Members

Truss analysis using the method of joints is greatly simplified if one is able to first determine those members that support *no loading*. These *zero-force members* may be necessary for the stability of the truss during construction and to provide support if the applied loading is changed.

The zero-force members of a truss can generally be determined by inspection of the joints. For example, consider the truss in Fig. 3–22a. The two members at joint C are connected together at a right angle *and* there is no external load on the joint. The free-body diagram of joint C, Fig. 3–22b, indicates that the force in each member must be zero in order to maintain equilibrium. Furthermore, as in the case of joint A, Fig. 3–22c, this must be true regardless of the angle, say θ, between the members.

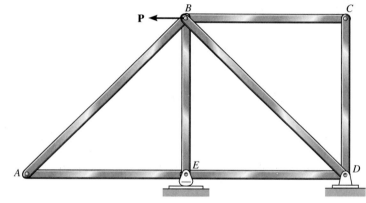

(a)

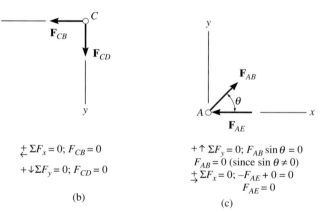

$$\underset{\leftarrow}{+}\ \Sigma F_x = 0; \ F_{CB} = 0$$

$$+\downarrow \Sigma F_y = 0; \ F_{CD} = 0$$

(b)

$$+\uparrow \Sigma F_y = 0; \ F_{AB}\sin\theta = 0$$

$$F_{AB} = 0 \ (\text{since } \sin\theta \neq 0)$$

$$\underset{\rightarrow}{+}\ \Sigma F_x = 0; \ -F_{AE} + 0 = 0$$

$$F_{AE} = 0$$

(c)

Fig. 3–22

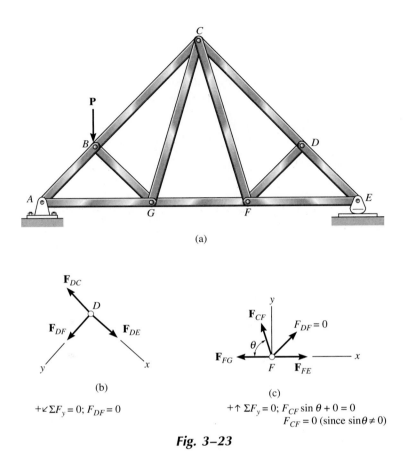

(a)

(b)

$+\swarrow\Sigma F_y = 0; \; F_{DF} = 0$

(c)

$+\uparrow \Sigma F_y = 0; \; F_{CF} \sin\theta + 0 = 0$
$F_{CF} = 0 \; (\text{since } \sin\theta \neq 0)$

Fig. 3–23

   Zero-force members also occur at joints having a geometry as joint D in Fig. 3–23a. Here *no external load acts on the joint*, so that a force summation in the y direction, Fig. 3–23b, which is perpendicular to the two collinear members, requires that $F_{DF} = 0$. Using this result, CF is also a zero-force member, as indicated by the force analysis of joint F, Fig. 3–23c. Particular attention should be directed to these conditions of joint geometry and loading, since the analysis of a truss can be considerably simplified by *first* spotting the zero-force members.

Example 3–4

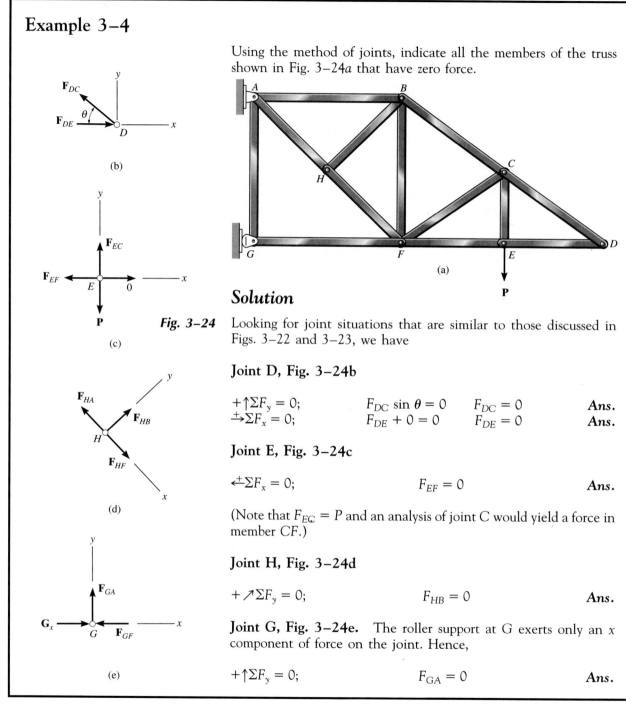

Using the method of joints, indicate all the members of the truss shown in Fig. 3–24a that have zero force.

Fig. 3–24

Solution

Looking for joint situations that are similar to those discussed in Figs. 3–22 and 3–23, we have

Joint D, Fig. 3–24b

$+\uparrow\Sigma F_y = 0;$	$F_{DC} \sin \theta = 0$	$F_{DC} = 0$	**Ans.**
$\xrightarrow{+}\Sigma F_x = 0;$	$F_{DE} + 0 = 0$	$F_{DE} = 0$	**Ans.**

Joint E, Fig. 3–24c

$$\xleftarrow{+}\Sigma F_x = 0; \qquad\qquad F_{EF} = 0 \qquad\qquad \textbf{Ans.}$$

(Note that $F_{EC} = P$ and an analysis of joint C would yield a force in member CF.)

Joint H, Fig. 3–24d

$$+\nearrow\Sigma F_y = 0; \qquad\qquad F_{HB} = 0 \qquad\qquad \textbf{Ans.}$$

Joint G, Fig. 3–24e. The roller support at G exerts only an x component of force on the joint. Hence,

$$+\uparrow\Sigma F_y = 0; \qquad\qquad F_{GA} = 0 \qquad\qquad \textbf{Ans.}$$

3.5   The Method of Sections

If the forces in only a few members of a truss are to be found, the method of sections generally provides the most direct means of obtaining these forces. The *method of sections* consists of passing an *imaginary section* through the truss, thus cutting it into two parts. Provided the entire truss is in equilibrium, each of the two parts must also be in equilibrium; and as a result, the three equations of equilibrium may be applied to either one of these two parts to determine the member forces at the "cut section."

When the method of sections is used to determine the force in a particular member, a decision must be made as to how to "cut" or section the truss. Since only *three* independent equilibrium equations ($\Sigma F_x = 0$, $\Sigma F_y = 0$, $\Sigma M_O = 0$) can be applied to the isolated portion of the truss, try to select a section that, in general, passes through not more than *three* members in which the forces are unknown. For example, consider the truss in Fig. 3–25a. If the force in member GC is to be determined, section *aa* will be appropriate. The free-body diagrams of the two parts are shown in Fig. 3–25b and 3–25c. In particular, note that the line of action of each force in a sectioned member is specified from the *geometry* of the truss, since the force in a member passes along the axis of the member. Also, the member forces acting on one part of the truss are equal but opposite to those acting on the other part— Newton's third law. As shown, members assumed to be in *tension* (BC and GC) are subjected to a "pull," whereas the member in *compression* (GF) is subjected to a "push."

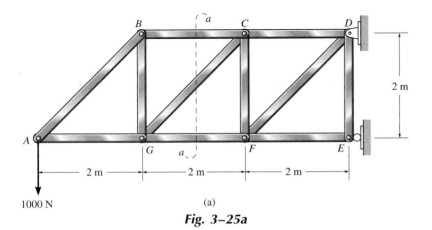

(a)

Fig. 3–25a

107

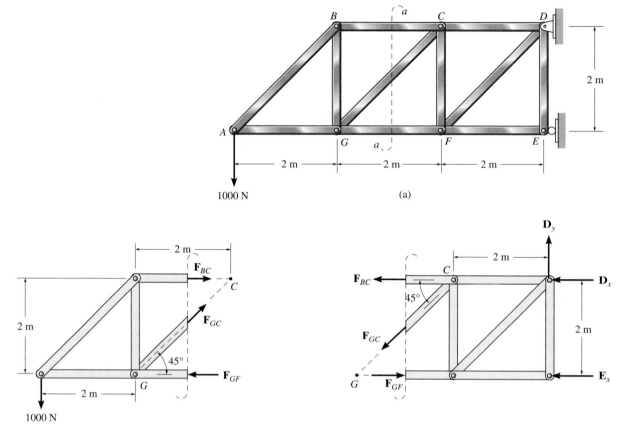

Fig. 3–25

The three unknown member forces $\mathbf{F}_{BC}$, $\mathbf{F}_{GC}$, and $\mathbf{F}_{GF}$ can be obtained by applying the three equilibrium equations to the free-body diagram in Fig. 3–25b. If, however, the free-body diagram in Fig. 3–25c is considered, the three support reactions $\mathbf{D}_x$, $\mathbf{D}_y$, and $\mathbf{E}_x$ will have to be determined *first*. Why? (This, of course, is done in the usual manner by considering a free-body diagram of the *entire truss*.) When applying the equilibrium equations, consider ways of writing the equations so as to yield a *direct solution* for each of the unknowns, rather than having to solve simultaneous equations. For example, summing moments about C in Fig. 3–25b would yield a direct solution for $\mathbf{F}_{GF}$ since $\mathbf{F}_{BC}$ and $\mathbf{F}_{GC}$ create zero moment about C. Likewise, $\mathbf{F}_{BC}$ can be obtained directly by summing moments about G. Finally, $\mathbf{F}_{GC}$ can be found directly from a force summation in the vertical direction, since $\mathbf{F}_{GF}$ and $\mathbf{F}_{BC}$ have no vertical components.

The correct arrowhead sense of direction of an unknown member force can in many cases be determined by inspection. For example, $\mathbf{F}_{BC}$ is a tensile force as represented in Fig. 3–25b, since moment equilibrium about G requires that $\mathbf{F}_{BC}$ create a moment opposite to that of the 1000-N force. Also, $\mathbf{F}_{GC}$ is tensile since its vertical component must balance the 1000-N force acting downward. In more complicated cases, the sense of an unknown member force may be *assumed*. If the solution yields a *negative* magnitude, it indicates that its sense of direction is *opposite* to that shown on the free-body diagram.

Procedure for Analysis

The following procedure provides a means for applying the method of sections to determine the forces in the members of a truss.

Free-Body Diagram   Make a decision as to how to "cut" or section the truss through the members where forces are to be determined. Before isolating the appropriate section, it may first be necessary to determine the truss's *external* reactions, so that the three equilibrium equations are used *only* to solve for member forces at the cut section. Draw the free-body diagram of that part of the sectioned truss which has the least number of forces on it. By inspection, attempt to show the unknown member forces acting in the correct sense of direction.

Equations of Equilibrium   Try to apply the three equations of equilibrium such that simultaneous solution of equations is avoided. In this regard, moments should be summed about a point that lies at the intersection of the lines of action of two unknown forces; in this way, the third unknown force is determined directly from the moment equation. If two of the unknown forces are *parallel*, forces may be summed *perpendicular* to the direction of these unknowns to determine *directly* the third unknown force.

The following examples illustrate these concepts numerically.

Example 3–5

Determine the force in members CF and GC of the roof truss shown in Fig. 3–26a. State whether the members are in tension or compression. The reactions at the supports have been calculated.

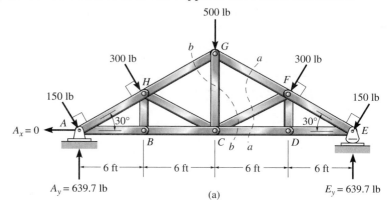

(a)

Solution

Member CF

Free-Body Diagram.   The force in member CF can be obtained by considering the section *aa* in Fig. 3–26a. The free-body diagram of the right part of this section is shown in Fig. 3–26b.

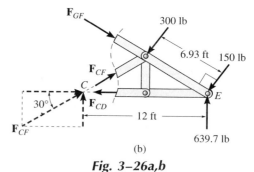

(b)

Fig. 3–26a,b

Equations of Equilibrium.   A direct solution for $\mathbf{F}_{CF}$ can be obtained by applying $\Sigma M_E = 0$. Why? For simplification, $\mathbf{F}_{CF}$ is extended to point C (principle of transmissibility), Fig. 3–26b. Thus,

$$\zeta + \Sigma M_E = 0; \qquad F_{CF} \sin 30°(12) - 300(6.93) = 0$$
$$F_{CF} = 346.4 \text{ lb (C)} \qquad \qquad \textbf{Ans.}$$

Member GC

Free-Body Diagram.   The force in GC can be obtained by using section *bb* in Fig. 3–26a. This section cuts four members; however, the force in CF has been calculated. The free-body diagram of the left portion of the section is shown in Fig. 3–26c.

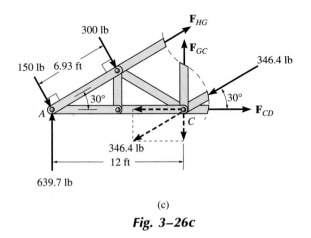

(c)

Fig. 3–26c

Equations of Equilibrium.   Moments will be summed about point A in order to eliminate the unknowns $\mathbf{F}_{HG}$ and $\mathbf{F}_{CD}$. Again extending $\mathbf{F}_{CF}$ (346.4 lb) to point C, we have

$$\zeta+\Sigma M_A = 0; \quad 300(6.93) - F_{GC}(12) + 346.4 \sin 30°(12) = 0$$
$$F_{GC} = 346.4 \text{ lb (T)} \qquad \textbf{Ans.}$$

Example 3–6

Determine the force in members GF and GD of the truss shown in Fig. 3–27a. State whether the members are in tension or compression. The reactions at the supports have been calculated.

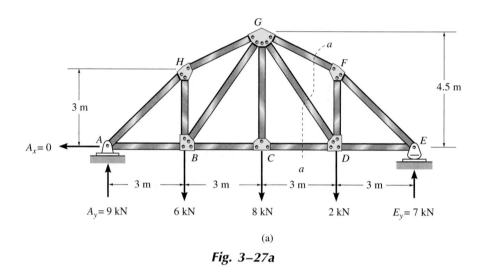

(a)

Fig. 3–27a

Solution

Free-Body Diagram. Section aa in Fig. 3–27a will be considered. Why? The free-body diagram to the right of this section is shown in Fig. 3–27b. The distance EO can be determined by proportional triangles or realizing that member GF drops vertically $4.5 - 3 = 1.5$ m in 3 m, Fig. 3–27a. Hence to drop 4.5 m from G the distance from C to O must be 9 m. Also, the angles that $\mathbf{F}_{GD}$ and $\mathbf{F}_{GF}$ make with the horizontal are $\tan^{-1}(4.5/3) = 56.3°$ and $\tan^{-1}(4.5/9) = 26.6°$, respectively.

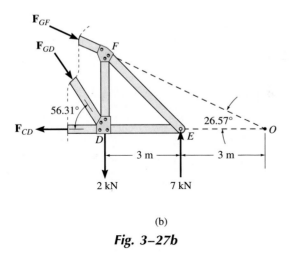

(b)

Fig. 3–27b

Equations of Equilibrium.   The force in GF can be determined directly by applying $\Sigma M_D = 0$. Why? For the calculation use the principle of transmissibility and apply $\mathbf{F}_{GF}$ at point O. Thus,

$$\zeta+\Sigma M_D = 0; \qquad F_{GF} \sin 26.6°(6) - 7(3) = 0$$
$$F_{GF} = 7.83 \text{ kN (C)} \qquad \text{\textbf{Ans.}}$$

The force in GD is determined directly by applying $\Sigma M_O = 0$. For simplicity use the principle of transmissibility and apply $\mathbf{F}_{GD}$ at D. Hence,

$$\zeta+\Sigma M_O = 0; \qquad 7(3) - 2(6) - F_{GD} \sin 56.3°(6) = 0$$
$$F_{GD} = 1.80 \text{ kN (C)} \qquad \text{\textbf{Ans.}}$$

Example 3–7

Determine the force in members BC and MC of the K-truss shown in Fig. 3–28a. State whether the members are in tension or compression. The reactions at the supports have been calculated.

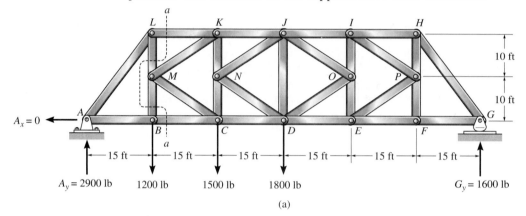

(a)

Solution

Free-Body Diagram. Although section aa shown in Fig. 3–28a cuts through four members, it is possible to solve for the force in member BC using this section. The free-body diagram of the left portion of the truss is shown in Fig. 3–28b.

Fig. 3–28a,b

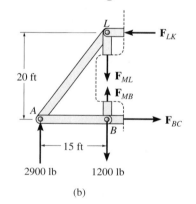

(b)

Equations of Equilibrium. Summing moments about point L eliminates *three* of the unknowns, so that

$$\zeta + \Sigma M_L = 0; \qquad 2900(15) - F_{BC}(20) = 0$$
$$F_{BC} = 2175 \text{ lb (T)} \qquad \textbf{Ans.}$$

Free-Body Diagrams. The force in MC can be obtained indirectly by first obtaining the force in MB from vertical force equilibrium of joint B, Fig. 3–28c, i.e., $F_{MB} = 1200$ lb (T). Then from the free-body diagram in Fig. 3–28b.

$$+\uparrow\Sigma F_y = 0; \quad 2900 - 1200 + 1200 - F_{ML} = 0$$
$$F_{ML} = 2900 \text{ lb (T)}$$

Using these results, the free-body diagram of joint M is shown in Fig. 3–28d.

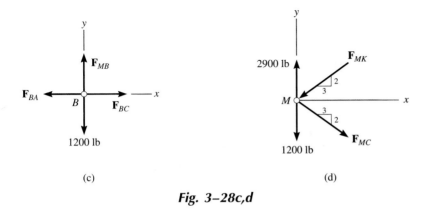

(c)

(d)

Fig. 3–28c,d

Equations of Equilibrium

$$\xrightarrow{+}\Sigma F_x = 0; \qquad \left(\frac{3}{\sqrt{13}}\right)F_{MC} - \left(\frac{3}{\sqrt{13}}\right)F_{MK} = 0$$

$$+\uparrow\Sigma F_y = 0; \quad 2900 - 1200 - \left(\frac{2}{\sqrt{13}}\right)F_{MC} - \left(\frac{2}{\sqrt{13}}\right)F_{MK} = 0$$

$$F_{MK} = 1532 \text{ lb (C)} \qquad F_{MC} = 1532 \text{ lb (T)} \qquad \textbf{Ans.}$$

Sometimes, as in this example, application of both the method of sections and the method of joints leads to the most direct solution to the problem.

It is also possible to solve for the force in MC by using the result for $\mathbf{F}_{BC}$. In this regard, pass a vertical section through LK, MK, MC, and BC, Fig. 3–28a. Isolate the left section and apply $\Sigma M_K = 0$.

3.6 Compound Trusses

In Sec. 3.2 it was stated that compound trusses are formed by connecting two or more simple trusses together either by bars or by joints. Occasionally this type of truss is best analyzed by applying *both* the method of joints and the method of sections. It is often convenient to first recognize the type of construction as listed in Sec. 3.2 and then perform the analysis using the following procedure.

Procedure for Analysis

Type 1 Determine the external reactions on the truss and then, using the method of sections, cut the truss through the bar connecting the two simple trusses so that this bar force may be obtained when one of the sectioned parts is isolated as a free body. Once this force is obtained, proceed to analyze the simple trusses using the method of joints. (See Example 3–8.)

Type 2 Determine the external reactions on the truss. Use the method of sections and cut each of the three bars that connect the two simple trusses together. From a free-body diagram of one of the sectioned parts, determine each of the three bar forces. Proceed to analyze each simple truss using the method of joints. (See Example 3–9.)

Type 3 Although many of these types of trusses can be analyzed using the method of sections combined with the method of joints, we will instead use a more general method. Remove the secondary trusses and replace them by dashed members so as to construct the main truss. The loads that the secondary trusses exert on the main truss are also placed on the main truss at the joints where the secondary trusses are connected to the main truss. Determine the forces in the dashed members of the main truss using the method of joints or the method of sections. These forces are then applied to the joints of the secondary trusses and then, using the method of joints, the bar forces in the secondary trusses can be obtained. (See Example 3–10.)

Example 3–8

Indicate how to analyze the compound truss shown in Fig. 3–29a. The reactions at the supports have been calculated.

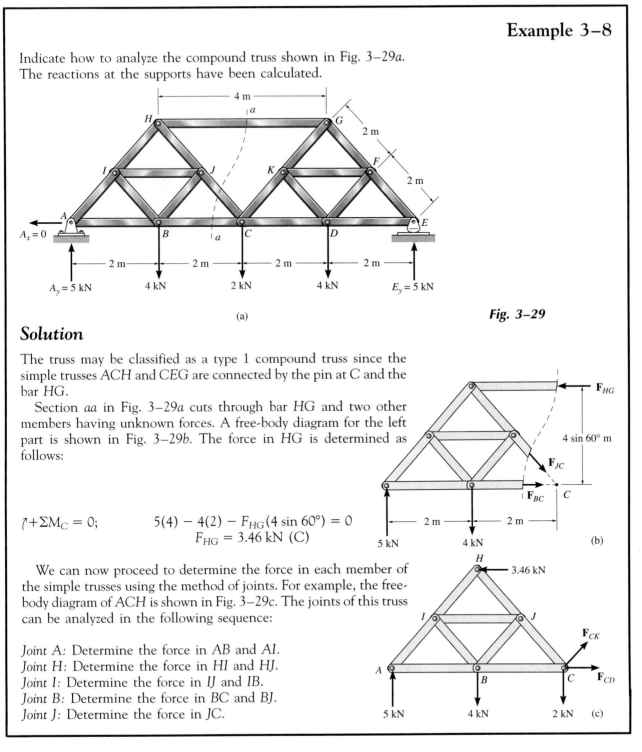

(a)

Fig. 3–29

Solution

The truss may be classified as a type 1 compound truss since the simple trusses ACH and CEG are connected by the pin at C and the bar HG.

Section aa in Fig. 3–29a cuts through bar HG and two other members having unknown forces. A free-body diagram for the left part is shown in Fig. 3–29b. The force in HG is determined as follows:

$$\zeta+\Sigma M_C = 0; \qquad 5(4) - 4(2) - F_{HG}(4\sin 60°) = 0$$
$$F_{HG} = 3.46 \text{ kN (C)}$$

We can now proceed to determine the force in each member of the simple trusses using the method of joints. For example, the free-body diagram of ACH is shown in Fig. 3–29c. The joints of this truss can be analyzed in the following sequence:

Joint A: Determine the force in AB and AI.
Joint H: Determine the force in HI and HJ.
Joint I: Determine the force in IJ and IB.
Joint B: Determine the force in BC and BJ.
Joint J: Determine the force in JC.

(b)

(c)

Example 3–9

Indicate how to analyze the compound truss shown in Fig. 3–30a. The reactions at the supports have been calculated.

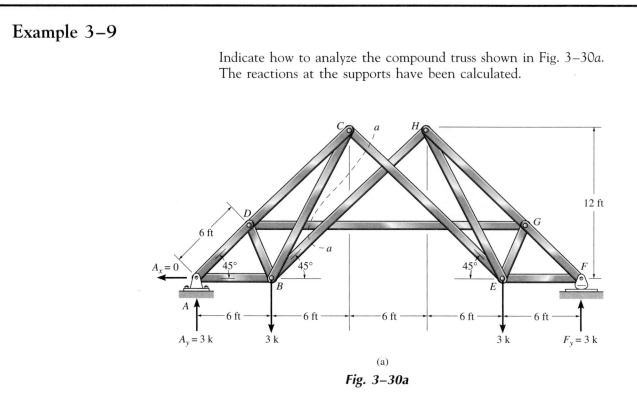

(a)

Fig. 3–30a

Solution

The truss may be classified as a type 2 compound truss since the simple trusses $ABCD$ and $FEHG$ are connected by three nonparallel or nonconcurrent bars, namely, CE, BH, and DG.

Using section aa in Fig. 3–30a we can determine the force in each connecting bar. The free-body diagram of the left part of this section is shown in Fig. 3–30b. Hence,

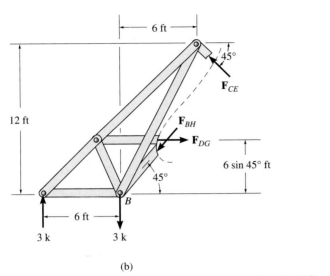

(b)

(c)

Fig. 3–30b,c

$\zeta+\Sigma M_B = 0;\quad 3(6) + F_{DG}(6 \sin 45°) - F_{CE} \cos 45°(12)$
$$- F_{CE} \sin 45°(6) = 0 \qquad (1)$$
$+\uparrow\Sigma F_y = 0;\quad 3 - 3 - F_{BH} \sin 45° + F_{CE} \sin 45° = 0 \qquad (2)$
$\xrightarrow{+}\Sigma F_x = 0;\quad -F_{BH} \cos 45° + F_{DG} - F_{CE} \cos 45° = 0 \qquad (3)$

From Eq. (2), $F_{BH} = F_{CE}$; then solving Eqs. (1) and (3) simultaneously yields

$$F_{BH} = F_{CE} = 2.68 \text{ k (C)} \qquad F_{DG} = 3.78 \text{ k (T)}$$

Analysis of each connected simple truss can now be performed using the method of joints. For example, from Fig. 3–30c, this can be done in the following sequence.

Joint A: Determine the force in *AB* and *AD*.
Joint D: Determine the force in *DC* and *DB*.
Joint C: Determine the force in *CB*.

119

Example 3–10

Indicate how to analyze the symmetrical compound truss shown in Fig. 3–31a. The reactions at the supports have been calculated.

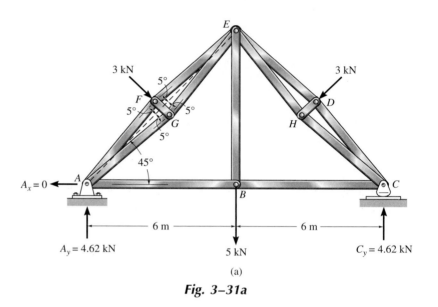

(a)

Fig. 3–31a

Solution

The truss can be classified as a type 3 compound truss, where the secondary trusses are *AGEF* and *CHED* and the main truss is *ABCE*.

By removing each secondary truss, Fig. 3–31b, one of the two reactions, specifically 1.5 kN, at each pin connection *A* and *E* and *E* and *C* can be determined by symmetry or by applying the equations of equilibrium to each truss. The main truss with the 1.5-kN loadings applied is shown in Fig. 3–31c. The force in each member can be found from the method of joints. For example, for joint *A* the free-body diagram is shown in Fig. 3–31d. We have

$$+\uparrow\Sigma F_y = 0; \qquad 4.62 - 1.5 \sin 45° - F_{AE} \sin 45° = 0$$
$$F_{AE} = 5.03 \text{ kN} \quad (C)$$
$$\xrightarrow{+}\Sigma F_x = 0; \qquad 1.5 \cos 45° - 5.03 \cos 45° + F_{AB} = 0$$
$$F_{AB} = 2.50 \text{ kN} \quad (T)$$

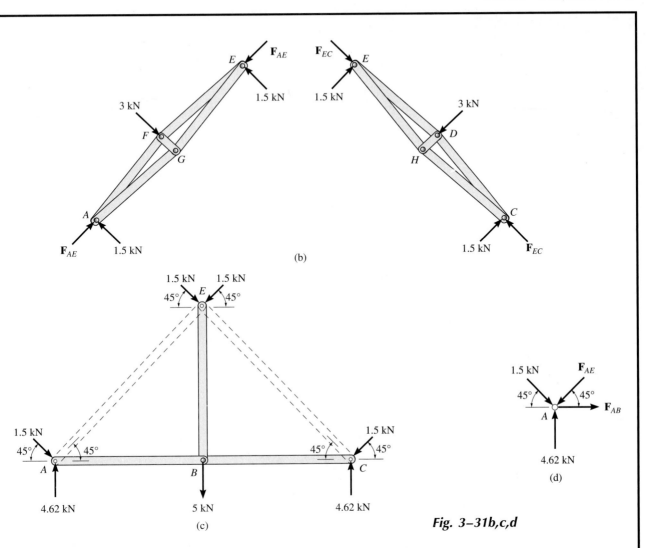

(b)

(c)

(d)

Fig. 3–31b,c,d

Having found F_{AE} = 5.03 kN (C), we can now apply this load numerically to the secondary truss $AGEF$ in Fig. 3–31b and then proceed to analyze the forces in the members of this truss by the method of joints.

Note that the truss in Fig. 3–31a can also be analyzed by using the method of sections. For example, pass a vertical section through FE, EG, and AB, then determine the force in each of these members. Proceed to determine the force in each of the other members using the method of joints.

3.7 Complex Trusses

The member forces in a complex truss can be determined using the method of joints; however, the solution will require writing the two equilibrium equations for each of the j joints of the truss and then solving the complete set of $2j$ equations *simultaneously*. (No doubt this can be readily accomplished using a computer.) This approach may be impractical for hand calculations, especially in the case of small trusses. Therefore, a more direct method for analyzing a complex truss, referred to as the *method of substitute members*, will be presented here.

Procedure for Analysis

With reference to the truss in Fig. 3–32a, the following steps are necessary to solve for the member forces using the substitute-member method.

Remove External Loading from Simple Truss Determine the reactions at the supports and begin by imagining how to analyze the truss by the method of joints, i.e., progressing from joint to joint and solving for each member force. If a joint is reached where there are *three un-*

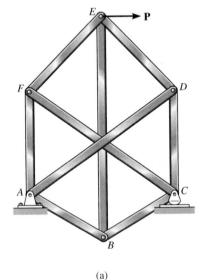

(a)

Fig. 3–32

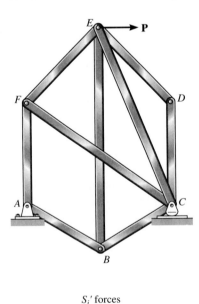

S_i' forces

(b)

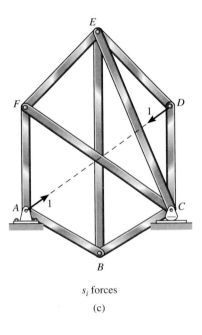

s_i forces

(c)

knowns, remove one of the members at the joint and replace it by an *imaginary* member elsewhere in the truss. By doing this, reconstruct the truss to be a stable simple truss.

In Fig. 3–32a it is observed that each joint will have three *unknown* member forces acting on it. Hence we will remove member AD and replace it with the imaginary member EC, Fig. 3–32b. This truss can now be analyzed by the method of joints for the two types of loading that follow.

Reduction to Stable Simple Truss   Load the simple truss with the actual loading P, then determine the force S_i' in each member i. For example, provided the reactions have been determined, one could start at joint A to determine the forces in AB and AF, then joint F to determine the forces in FE and FC, then joint D to determine the forces in DE and DC (both of which are zero), then joint E to determine EB and EC, and finally joint B to determine the force in BC.

External Loading on Simple Truss   Consider the simple truss without the external load P, Fig. 3–32c. Place equal but opposite collinear unit loads on the truss at the two joints from which the member was removed. If these forces develop a force s_i in the ith truss member, then by proportion an unknown force x in the removed member would exert a force xs_i in the ith member.

From Fig. 3–32c the equal but opposite unit loads will create *no reactions* at A and C when the equations of equilibrium are applied to the entire truss. The s_i forces can be determined by analyzing the joints in the same sequence as before, namely, joint A, then joints F, D, E, and finally B.

Superposition   If the effects of the above two loadings are combined, the force in the ith member of the truss will be

$$S_i = S_i' + xs_i \tag{1}$$

In particular, for the substituted member EC in Fig. 3–32b the force is $S_{EC} = S_{EC}' + xs_{EC} = 0$. Since member EC does not actually exist on the original truss, we will choose x to have a magnitude such that it yields *zero force* in EC. Hence,

$$S_{EC}' + xs_{EC} = 0 \tag{2}$$

or $x = -S_{EC}'/s_{EC}$. Once the value of x has been determined, the force in the other members i of the complex truss can be determined from Eq. (1).

The following example illustrates this procedure numerically.

123

Example 3–11

Determine the force in each member of the complex truss shown in Fig. 3–33a. Assume joints B, F, and D are on the same horizontal line. State whether the members are in tension or compression.

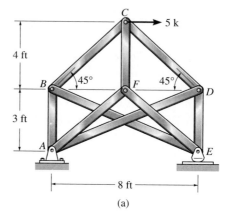

(a)

Solution

Reduction to Stable Simple Truss. By inspection, each joint has three unknown member forces. A joint analysis can be performed by hand if, for example, member CF is removed and member DB substituted, Fig. 3–33b. The resulting truss is stable and will not collapse.

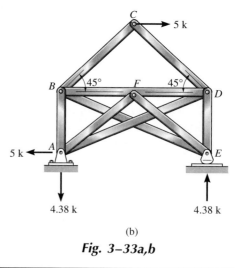

(b)

Fig. 3–33a,b

External Loading on Simple Truss.   As shown in Fig. 3–33b, the support reactions on the truss have been computed. Using the method of joints, we can first analyze joint C to determine the forces in members CB and CD; then joint F, where it is seen that FA and FE are zero-force members; then joint E to determine the forces in members EB and ED; then joint D to determine the forces in DA and DB; and finally joint B to determine the force in BA. Considering tension as positive and compression as negative, these S_i' forces are recorded in column 2 of Table 1.

Remove External Loading from Simple Truss.   The unit load acting on the truss is shown in Fig. 3–33c. These equal but opposite forces create no external reactions on the truss. The joint analysis follows the same sequence as discussed previously, namely, joints C, F, E, D, and B. The results of the s_i force analysis are recorded in column 3 of Table 1.

Superposition.   We require

$$S_{DB} = S_{DB}' + xs_{DB} = 0$$

Substituting the data for S_{DB}' and s_{DB}, where S_{DB}' is negative since the force is compressive, we have

$$(-2.50) + x(1.167) = 0 \qquad x = 2.142$$

The values of xs_i are recorded in column 4 of Table 1, and the actual member forces $S_i = S_i' + xs_i$ are listed in column 5.

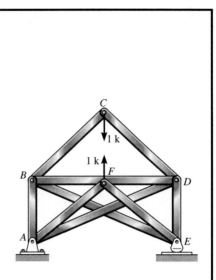

(c)

Fig. 3–33c

Table 1

Member	S_i'	s_i	xs_i	S_i
CB	3.54	−0.707	−1.51	2.02(T)
CD	−3.54	−0.707	−1.51	5.05(C)
FA	0	0.833	1.78	1.78(T)
FE	0	0.833	1.78	1.78(T)
EB	0	−0.712	−1.53	1.53(C)
ED	−4.38	−0.250	−0.536	4.91(C)
DA	5.34	−0.712	−1.52	3.81(T)
DB	−2.50	1.167	2.50	0
BA	2.50	−0.250	−0.535	1.96(T)

125

3.8   **Space Trusses**

A *space truss* consists of members joined together at their ends to form a stable three-dimensional structure. In Sec. 3.2 it was shown that the simplest form of a stable two-dimensional truss consists of the members arranged in the form of a triangle. We then built up the simple plane truss from this basic triangular element by adding two members at a time to form further elements. In a similar manner, the simplest element of a stable space truss is a *tetrahedron,* formed by connecting six members together with four joints as shown in Fig. 3–34. Any additional members added to this basic element would be redundant in supporting the force **P.** A simple space truss can be built from this basic tetrahedral element by adding three additional members and another joint forming multiconnected tetrahedrons.

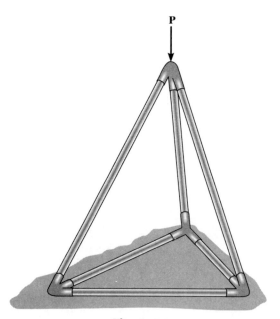

Fig. 3–34

Determinacy and Stability

Realizing that in three dimensions there are three equations of equilibrium available for each joint ($\Sigma F_x = 0$, $\Sigma F_y = 0$, $\Sigma F_z = 0$), then for a space truss with j number of joints, $3j$ equations are available. If the truss has b number of bars and r number of reactions, then like the case of a planar truss (Eq. 3–1) we can write

$$
\begin{array}{ll}
b + r < 3j & \text{unstable truss} \\
b + r = 3j & \text{statically determinate—check stability} \\
b + r > 3j & \text{statically indeterminate—check stability}
\end{array}
$$

$$(3\text{–}3)$$

   The *external stability* of the space truss requires that the lines of action of the reactions do not intersect a common axis, nor be parallel to one another. *Internal stability* can sometimes be checked by careful inspection of the member arrangement. Provided each joint is held fixed by its supports or connecting members, so that it cannot move with respect to the other joints, the truss can be classified as internally stable. Perhaps the easiest way to check truss stability, be it external or internal, is to perform a force analysis of the truss. If we obtain a *unique solution* for the member forces, the truss is *stable*, whereas inconsistent results indicate a truss configuration that is unstable or has a "critical form."

Assumptions for Design

The members of a space truss may be treated as axial-force members provided the external loading is applied at the joints and the joints consist of ball-and-socket connections. This assumption is justified provided the joined members at a connection intersect at a common point and the weight of the members can be neglected. In cases where the weight of a member is to be included in the analysis, it is generally satisfactory to apply it as a vertical force, half of its magnitude applied to each end of the member.
   For the force analysis the supports of a space truss are generally modeled as a short link, plane roller joint, slotted roller joint, or a ball-and-socket joint. Each of these supports and their reactive force components are shown in Table 3–1.

Table 3–1

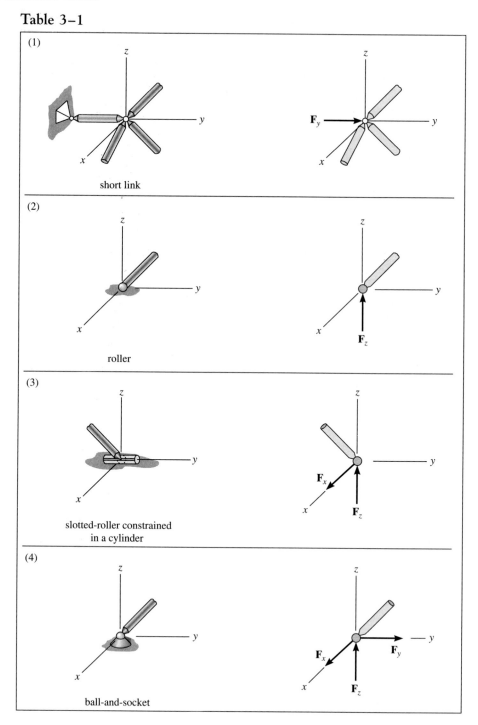

(1)

short link

(2)

roller

(3)

slotted-roller constrained
in a cylinder

(4)

ball-and-socket

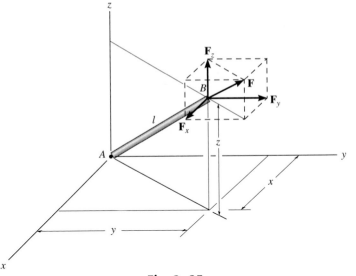

Fig. 3–35

x, y, z Force Components

Since the analysis of a space truss is three-dimensional, it will often be necessary to resolve the force F in a member into components acting along the x, y, z axes. This can be done using the geometry of the truss member as illustrated in Fig. 3–35. Here member AB has a length l and *known* projections x, y, z along the coordinate axes. These projections can be related to the member's length by the equation

$$l = \sqrt{x^2 + y^2 + z^2} \qquad (3\text{–}4)$$

If a force $\mathbf{F}$ acts along the axis of the member, Fig. 3–35, the components of $\mathbf{F}$ can be determined by *proportion* as follows:

$$F_x = F\left(\frac{x}{l}\right) \qquad F_y = F\left(\frac{y}{l}\right) \qquad F_z = F\left(\frac{z}{l}\right) \qquad (3\text{–}5)$$

Notice that this requires

$$F = \sqrt{F_x^2 + F_y^2 + F_z^2} \qquad (3\text{–}6)$$

Use of these equations will be illustrated in Example 3–12.

129

Procedure for Analysis

Either the method of sections or the method of joints can be used to determine the forces developed in the members of a space truss.

Method of Sections   If only a *few* member forces are to be determined, the method of sections may be used. When an imaginary section is passed through a truss, and the truss is separated into two parts, the force system acting on either one of the parts must satisfy the six scalar equilibrium equations: $\Sigma F_x = 0$, $\Sigma F_y = 0$, $\Sigma F_z = 0$, $\Sigma M_x = 0$, $\Sigma M_y = 0$, $\Sigma M_z = 0$. By proper choice of the section and axes for summing forces and moments, many of the unknown member forces in a space truss can be computed *directly*, using a single equilibrium equation. In this regard, recall that the *moment* of a force about an axis is *zero* provided *the force is parallel to the axis or its line of action passes through a point on the axis*.

Method of Joints   Generally, if the forces in *all* the members of the truss must be determined, the method of joints is most suitable for the analysis. When using the method of joints, it is necessary to solve the three scalar equilibrium equations $\Sigma F_x = 0$, $\Sigma F_y = 0$, $\Sigma F_z = 0$ at each joint. Since it is relatively easy to draw the free-body diagrams and apply the equations of equilibrium at each joint, the method of joints is very consistent in its application.

Zero-Force Members   In some cases the joint analysis of a truss can be simplified if one is able to spot the zero-force members by recognizing two common cases of joint geometry.

Case 1   If all but one of the members connected to a joint lie in the same plane, and provided no external load acts on the joint, then the member not lying in the plane of the other members must be subjected to zero force. The proof of this statement is shown in Fig. 3–36, where members A, B, C lie in the x–y plane. Since the z component of $\mathbf{F}_D$ must be zero to satisfy $\Sigma F_z = 0$, member D must be a zero-force member. By the same reasoning, member D will carry a load that can be determined from $\Sigma F_z = 0$ if an external force acts on the joint and has a component acting along the z axis.

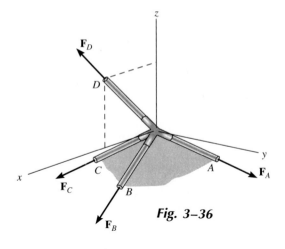

Fig. 3–36

Case 2   If it has been determined that all but two of several members connected at a joint support zero force, then the two remaining members must also support zero force, provided they do not lie along the same line. This situation is illustrated in Fig. 3–37, where it is known that A and C are zero-force members. Since F_D is collinear with the y axis, then application of $\Sigma F_x = 0$ or $\Sigma F_z = 0$ requires the x or z component of $\mathbf{F}_B$ to be zero. Consequently, $F_B = 0$. This being the case, $F_D = 0$ since $\Sigma F_y = 0$.

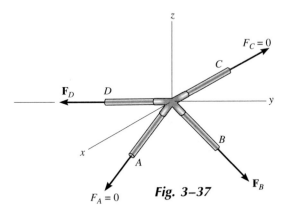

Fig. 3–37

Particular attention should be directed to the foregoing two cases of joint geometry and loading, since the analysis of a space truss can be considerably simplified by first spotting the zero-force members.

Example 3–12

Determine the force in each member of the space truss shown in Fig. 3–38a. The truss is supported by a ball-and-socket joint at A, a slotted roller joint at B, and a short link at C.

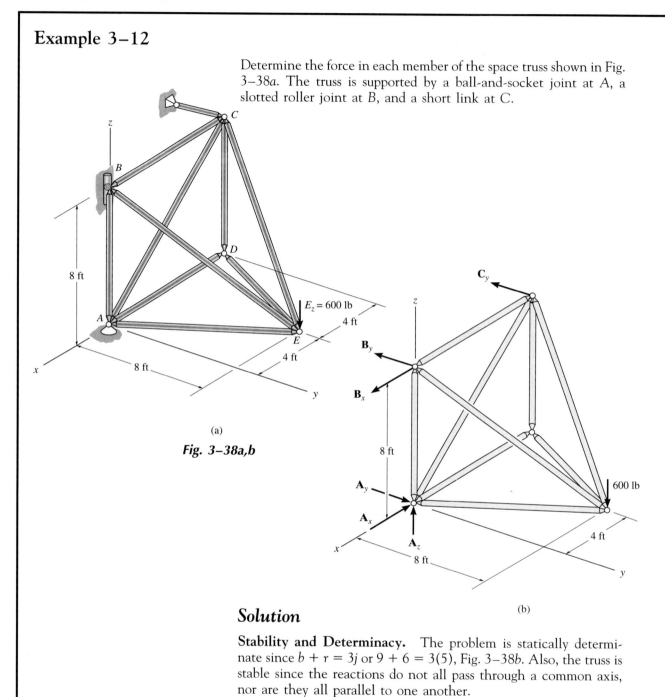

(a)

Fig. 3–38a,b

(b)

Solution

Stability and Determinacy. The problem is statically determinate since $b + r = 3j$ or $9 + 6 = 3(5)$, Fig. 3–38b. Also, the truss is stable since the reactions do not all pass through a common axis, nor are they all parallel to one another.

Support Reactions. We can obtain the support reactions from the free-body diagram of the entire truss, Fig. 3–38b, as follows:

$$\Sigma M_y = 0; \qquad -600(4) + B_x(8) = 0 \qquad B_x = 300 \text{ lb}$$
$$\Sigma M_z = 0; \qquad\qquad C_y = 0$$
$$\Sigma M_x = 0; \qquad B_y(8) - 600(8) = 0 \qquad B_y = 600 \text{ lb}$$
$$\Sigma F_x = 0; \qquad 300 - A_x = 0 \qquad A_x = 300 \text{ lb}$$
$$\Sigma F_y = 0; \qquad A_y - 600 = 0 \qquad A_y = 600 \text{ lb}$$
$$\Sigma F_z = 0; \qquad A_z - 600 = 0 \qquad A_z = 600 \text{ lb}$$

Joint B. We can begin the method of joints at B since there are three unknown member forces at this joint, Fig. 3–38c. The components of $\mathbf{F}_{BE}$ can be determined by proportion to the length of member BE, as indicated by Eqs. 3–5. We have

$$\Sigma F_y = 0; \quad -600 + F_{BE}(\tfrac{8}{12}) = 0 \qquad F_{BE} = 900 \text{ lb (T)} \quad \textbf{Ans.}$$
$$\Sigma F_x = 0; \quad 300 - F_{BC} - 900(\tfrac{4}{12}) = 0 \quad F_{BC} = 0 \qquad\qquad \textbf{Ans.}$$
$$\Sigma F_z = 0; \quad F_{BA} - 900(\tfrac{8}{12}) = 0 \qquad F_{BA} = 600 \text{ lb (C)} \quad \textbf{Ans.}$$

Joint A. Using the result for $F_{BA} = 600$ lb (C), the free-body diagram of joint A is shown in Fig. 3–38d. We have

$$\Sigma F_z = 0; \qquad 600 - 600 + F_{AC}\sin 45° = 0$$
$$F_{AC} = 0 \qquad\qquad \textbf{Ans.}$$

$$\Sigma F_y = 0; \qquad -F_{AE}\left(\frac{2}{\sqrt{5}}\right) + 600 = 0$$
$$F_{AE} = 670.8 \text{ lb (C)} \qquad \textbf{Ans.}$$

$$\Sigma F_x = 0; \qquad -300 + F_{AD} + 670.8\left(\frac{1}{\sqrt{5}}\right) = 0$$
$$F_{AD} = 0 \qquad\qquad \textbf{Ans.}$$

Joint D. By inspection the members at joint D, Fig. 3–38a, all support zero force, since the arrangement of the members is similar to either of the two cases discussed in reference to Figs. 3–36 and 3–37. Also, from Fig. 3–38e,

$$\Sigma F_x = 0; \qquad F_{DE} = 0 \qquad \textbf{Ans.}$$
$$\Sigma F_z = 0; \qquad F_{CD} = 0 \qquad \textbf{Ans.}$$

Joint C. By observation of the free-body diagram, Fig. 3–38f,

$$F_{CE} = 0 \qquad \textbf{Ans.}$$

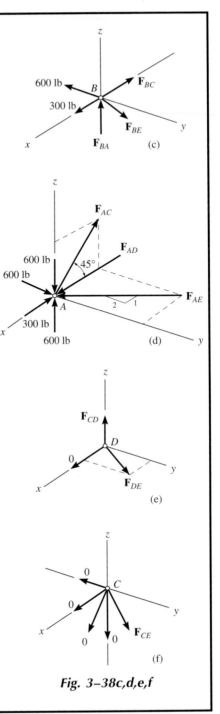

Fig. 3–38c,d,e,f

Example 3–13

Determine the zero-force members of the truss shown in Fig. 3–39a. The supports exert components of reaction on the truss as shown.

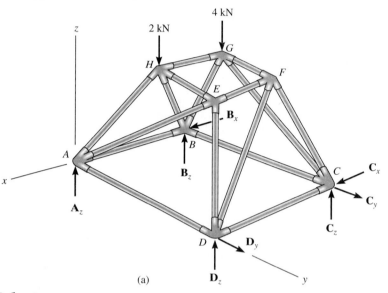

(a)

Fig. 3–39

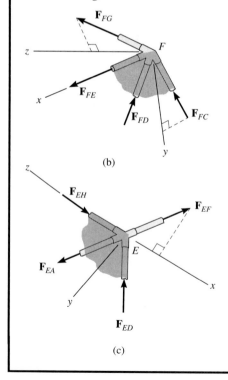

(b)

(c)

Solution

The free-body diagram, Fig. 3–39a, indicates there are eight unknown reactions for which only six equations of equilibrium are available for solution. In spite of this, the reactions can be determined, since $b + r = 3j$ or $16 + 8 = 3(8)$.

To spot the zero-force members, we must compare the conditions of joint geometry and loading to those of Figs. 3–36 and 3–37. In this regard, consider joint F, Fig. 3–39b. Since members FC, FD, FE lie in the x-y plane and FG is not in this plane, FG is a zero-force member. ($\Sigma F_z = 0$ must be satisfied.) In the same manner, from joint E, Fig. 3–39c, *EF is a zero-force member*, since it does not lie in the y-z plane. ($\Sigma F_x = 0$ must be satisfied.) Returning to joint F, Fig. 3–39b, it can be seen that $F_{FD} = F_{FC} = 0$, since $F_{FE} = F_{FG} = 0$, and there are no external forces acting on the joint.

The numerical force analysis of the joints can now proceed by analyzing joint G ($F_{GF} = 0$) to determine the forces in GH, GB, GC. Then analyze joint H to determine the forces in HE, HB, HA; joint E to determine the forces in EA, ED; joint A to determine the forces in AB, AD, and A_z; joint B to determine the force in BC and B_x, B_z; joint D to determine the force in DC and D_y, D_z; and finally, joint C to determine C_x, C_y, C_z.

REFERENCES

Henneberg, L., *Statik der Starren Systeme*. Darmstadt, 1886.

Timoshenko, S., and D. H. Young, *Theory of Structures*. McGraw-Hill Company, Inc., New York, 1965.

This system of space trusses is used to support the roof of a coliseum.
(Courtesy of Bethlehem Steel Corporation)

PROBLEMS

3–1. Classify each of the following trusses as statically determinate, statically indeterminate, or unstable.

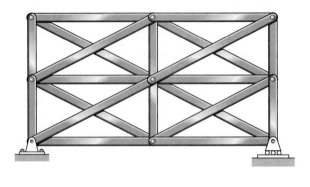

(c)

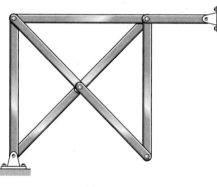

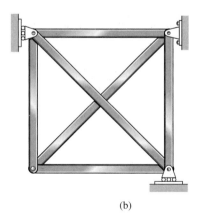

(a)

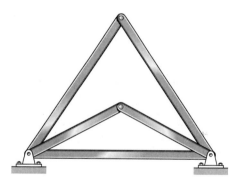

(d)

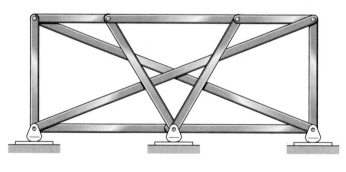

(b)

(e)

Prob. 3–1

136

3–2. Classify each of the following trusses as statically determinate, statically indeterminate, or unstable.

3–3. Classify each of the following trusses as statically determinate, statically indeterminate, or unstable.

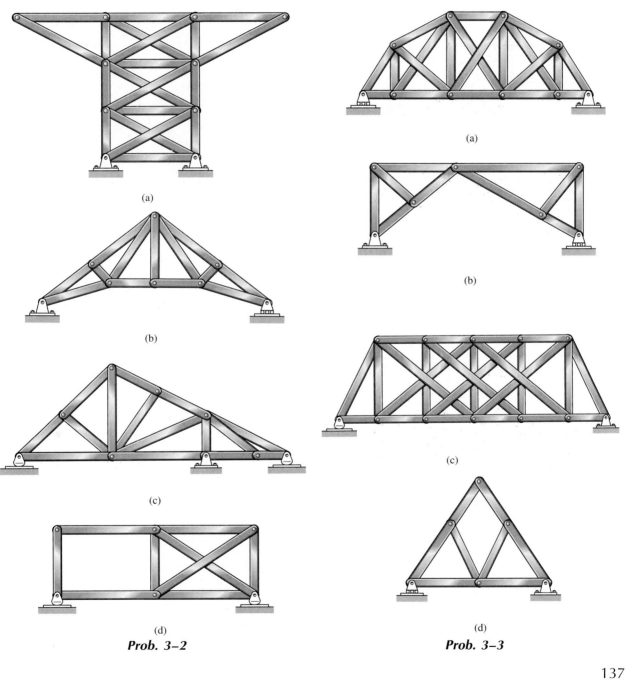

(a)

(b)

(c)

(d)

Prob. 3–2

(a)

(b)

(c)

(d)

Prob. 3–3

*3–4. Determine the force in each member of the truss. All interior angles are 60°. State if the members are in tension or compression. Assume all members are pin-connected.

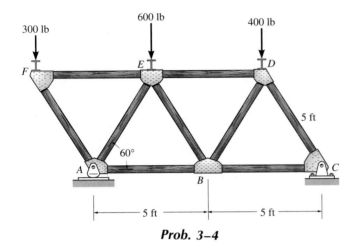

Prob. 3–4

3–5. Determine the force in each member of the roof truss. State if the members are in tension or compression.

3–6. Solve Prob. 3–5 assuming there is *no* external load on joints *J*, *H*, and *G* and only the vertical load of 3 k exists on joint *I*.

3–7. Determine the force in each member of the truss. Indicate if the members are in tension or compression. Assume all members are pin-connected.

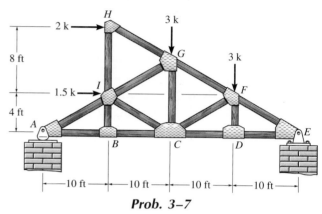

Prob. 3–7

*3–8. The truss shown is used to support the floor deck. The uniform load on the deck is 2.5 k/ft. This load is transferred from the deck to the floor beams, which rest on the top joints of the truss. Determine the force in each member of the truss, and state if the members are in tension or compression. Assume all members are pin-connected.

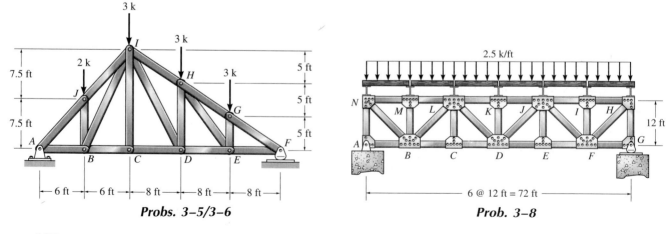

Probs. 3–5/3–6

Prob. 3–8

3–9. Determine the force in each member of the truss. State if the members are in tension or compression. Assume all members are pin-connected.

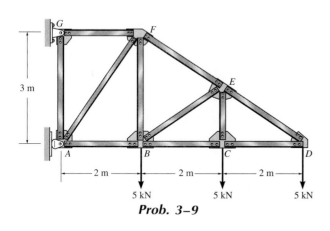

Prob. 3–9

5 kN   5 kN   5 kN

2 m   2 m   2 m

3 m

3–10. Determine the force in each member of the roof truss. State if the members are in tension or compression.

3–11. Determine the force in each member of the truss. State if the members are in tension or compression.

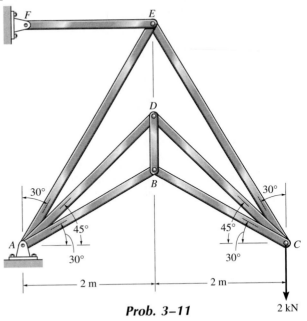

Prob. 3–11

2 kN

30°   45°   45°   30°
30°   30°

2 m   2 m

***3–12.** Determine the force in members KJ, JC, and CD. State if the members are in tension or compression. Assume the truss is pin-connected at F and roller-supported at C and the members are pin-connected.

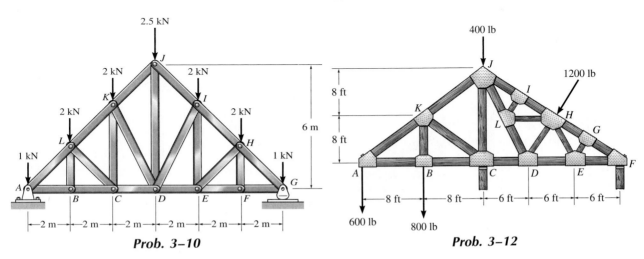

Prob. 3–10

2.5 kN
2 kN   2 kN
2 kN   2 kN
1 kN   1 kN

6 m

2 m 2 m 2 m 2 m 2 m 2 m

Prob. 3–12

400 lb
1200 lb

8 ft
8 ft

600 lb   800 lb

8 ft   8 ft   6 ft   6 ft   6 ft

139

3–13. Determine the force in members CB, BI, and IJ of the truss used to support an outdoor sign. The design wind pressure acting on the face of the sign creates the loading shown on the joints. Indicate whether the members are in tension or compression. Assume all members are pin-connected.

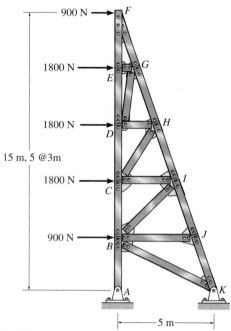

Prob. 3–13

3–14. Determine the force in members JI, DI, and DE of the *Pratt* truss. State if the members are in tension or compression. Assume all members are pin-connected.

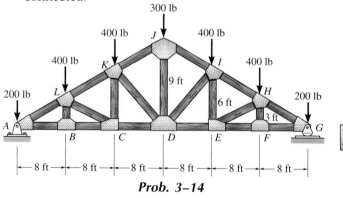

Prob. 3–14

3–15. Determine the force in members FC, BC, and FE. State if the members are in tension or compression. Assume all members are pin-connected.

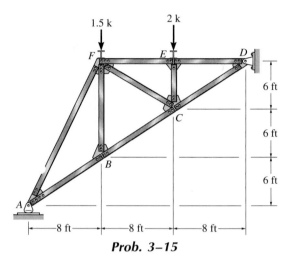

Prob. 3–15

*3–16. The *Howe* truss is subjected to the loading shown. Determine the forces in members GF, CD, and GC. State if the members are in tension or compression. Assume all members are pin-connected.

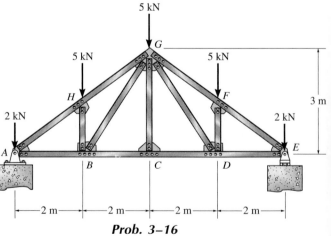

Prob. 3–16

3–17. The Warren truss is used to support the roof of an industrial building. The truss is simply supported on masonry walls at A and G so that only vertical reactions occur at these supports. Determine the force in members LD, CD, and KD. State if the members are in tension or compression. Assume all members are pin-connected.

3–19. Determine the force in members GF, GB, and BC of the truss. State if the members are in tension or compression. Assume the truss is pin-connected at A and roller-supported at D and the members are pin-connected.

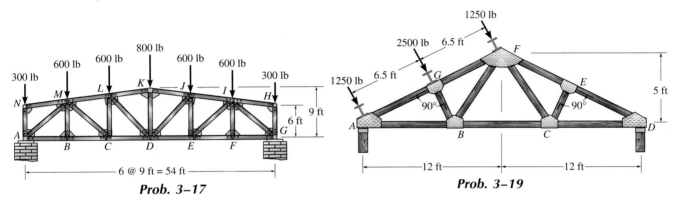

Prob. 3–17

Prob. 3–19

3–18. Determine the force in members GF, FC, and CD of the bridge truss. State if the members are in tension or compression. Assume all members are pin-connected.

*3–20. Determine the force in members IH, CD, and LH of the K-truss. State whether the members are in tension or compression. *Suggestion:* Section the truss through IH, IL, LC, and CD to determine IH and CD. Assume all members are pin-connected.

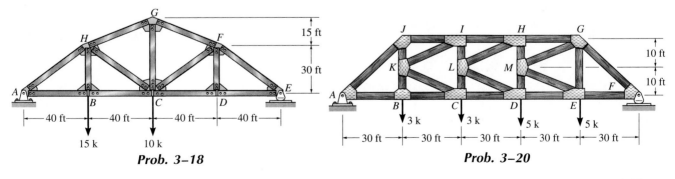

Prob. 3–18

Prob. 3–20

3–21.   Specify the type of compound truss and determine the force in members *JH*, *IH*, and *CD*. State if the members are in tension or compression. Assume all members are pin-connected.

3–23.   Specify the type of compound truss. Trusses *ACE* and *BDF* are connected by three bars *CF*, *ED*, and *CD*. Determine the force in each member and state if the members are in tension or compression.

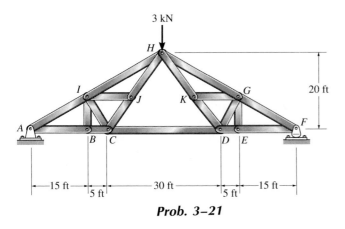

Prob. 3–21

3–22.   Specify the type of compound truss and determine the force in members *JH*, *BJ*, and *BI*. State if the members are in tension or compression. The internal angle between any two members is 60°. The truss is pin-supported at *A* and roller-supported at *F*. Assume all members are pin-connected.

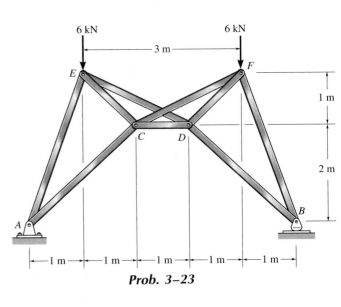

Prob. 3–23

*3–24.**   Determine the force in each member. State if the members are in tension or compression. Assume the supports are at *B* and *F* and the members are pin-connected.

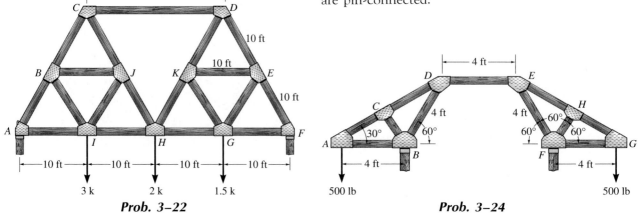

Prob. 3–22

Prob. 3–24

142

3–25. Specify the type of compound truss and determine the force in each member. State if the members are in tension or compression. Assume the members are pin-connected.

3–27. Specify the type of compound truss and determine the force in each member. State if the members are in tension or compression. Assume the members are pin-connected.

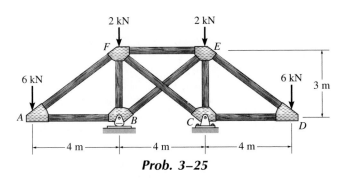

Prob. 3–25

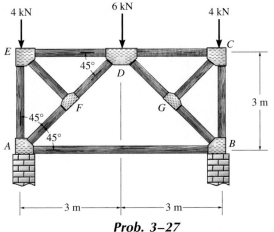

Prob. 3–27

*3–28. Show by analysis that the complex truss is unstable. *Suggestion:* Substitute member *EB* with one placed between *D* and *B*.

3–26. Determine the force in each member. State if the members are in tension or compression.

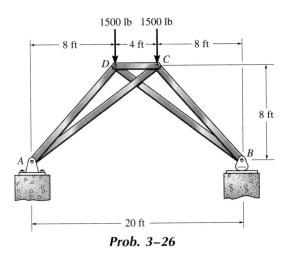

Prob. 3–26

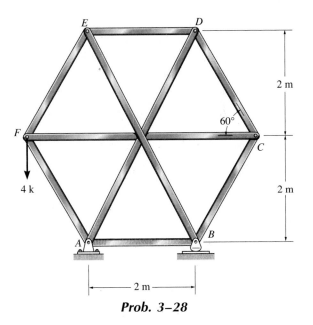

Prob. 3–28

3–29.   Determine the forces in all the members of the complex truss. State if the members are in tension or compression. *Suggestion:* Substitute member *AD* with one placed between *E* and *C*.

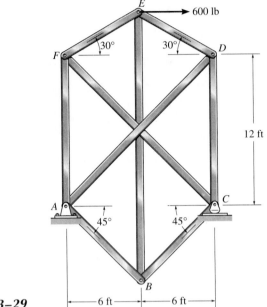

Prob. 3–29

3–30.   Determine the forces in all the members of the lattice (complex) truss. State if the members are in tension or compression. *Suggestion:* Substitute member *JE* by one placed between *K* and *F*.

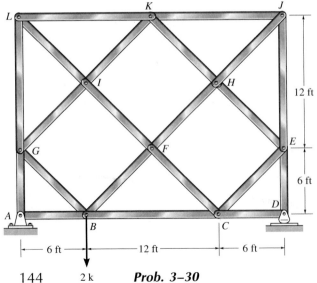

Prob. 3–30

3–31.   Determine the forces in all the members of the complex truss. State if the members are in tension or compression. *Suggestion:* Substitute member *AB* with one placed between *C* and *E*.

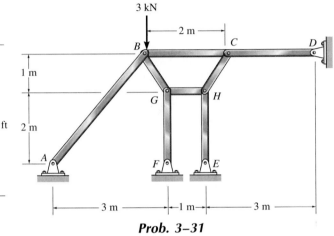

Prob. 3–31

*3–32.   Determine the reactions and the force in each member of the space truss. Indicate if the members are in tension or compression.

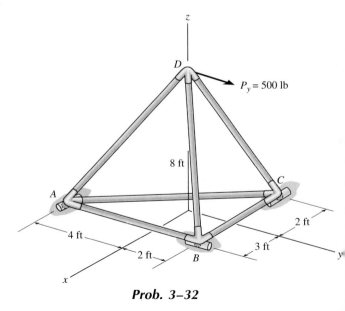

Prob. 3–32

3–33. Determine the force in each member of the space truss. Indicate if the members are in tension or compression.

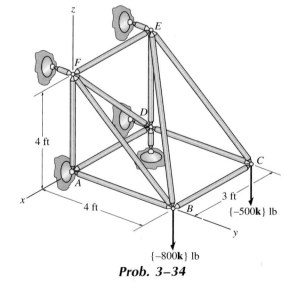

Prob. 3–33

3–34. Determine the force in each member of the space truss. Indicate if the members are in tension or compression.

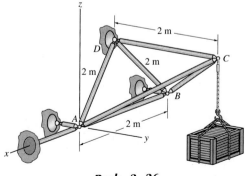

Prob. 3–34

3–35. Determine the force in members AB, AD, and AC of the space truss. Indicate if the members are in tension or compression.

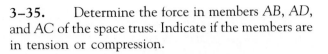

Prob. 3–35

***3–36.** Determine the reactions and the force in each member of the space truss. Indicate if the members are in tension or compression.

Prob. 3–36

145

Proper design of any structural element, such as this beam, requires a determination of the internal loadings acting throughout the beam. *(Photo courtesy of Portland Cement Association)*

4

Internal Loadings Developed in Structural Members

Before a structural member can be proportioned, it is necessary to determine the force and moment that act within the member. In this chapter we will develop the methods for finding these loadings at specified points along a member's axis and showing the variation graphically using the shear and moment diagrams. Applications are given for both beams and frames.

4.1 Internal Loadings at a Specified Point

As discussed in Sec. 2.3, the internal load at a specified point in a member can be determined by using the *method of sections*. In general, this loading for a coplanar structure will consist of a normal force **N**, shear force **V**, and bending moment **M**.* It should be realized, however, that these loadings actually represent the *resultants* of the *stress distribution* acting over the member's cross-sectional area at the cut section. Once the resultant internal loadings are known, the magnitude of the stress can be determined provided an assumed distribution of stress over the cross-sectional area is specified.

---

*Three-dimensional frameworks can also be subjected to a *torsional moment*, which tends to twist the member about its axis.

Sign Convention

Before presenting a method for finding the internal normal force, shear force, and bending moment, we will need to establish a sign convention to define their "positive" and "negative" values.* Although the choice is arbitrary, the sign convention to be adopted here has been widely accepted in structural engineering practice, and is illustrated in Fig. 4–1a. On the *left-hand face* of the cut member the normal force **N** acts to the right, the internal shear force **V** acts downward, and the moment **M** acts counterclockwise. In accordance with Newton's third law, an equal but opposite normal force, shear force, and bending moment must act on the right-hand face of the member at the section. Perhaps an easy way to remember this sign convention is to isolate a small segment of the member and note that *positive normal force tends to elongate the segment*, Fig. 4–1b; *positive shear tends to rotate the segment clockwise*, Fig. 4–1c; and *positive bending moment tends to bend the segment concave upward*, so as to "hold water," Fig. 4–1d.

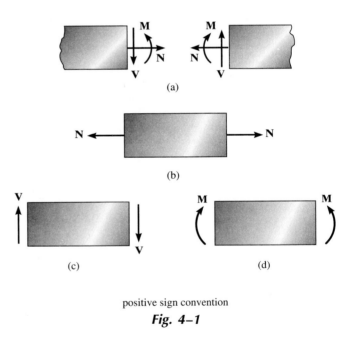

positive sign convention

Fig. 4–1

*This will be convenient later in Secs. 4.2 and 4.3 where we will express V and M as functions of x and then *plot* these functions. Having a sign convention is similar to assigning coordinate directions x positive to the right and y positive upward when plotting a function $y = f(x)$.

Procedure for Analysis

The following procedure provides a means for applying the method of sections to determine the internal normal force, shear force, and bending moment at a specific location in a structural member.

Support Reactions   Before the member is "cut" or sectioned, it may be necessary to determine the member's support reactions so that the equilibrium equations are used only to solve for the internal loadings when the member is sectioned. If the member is part of a pin-connected structure, the pin reactions can be determined using the methods of Sec. 2.5.

Free-Body Diagram   Keep all distributed loadings, couple moments, and forces acting on the member in their *exact location*, then pass an imaginary section through the member, perpendicular to its axis at the point where the internal loading is to be determined. Draw a free-body diagram of one of the "cut" segments on either side of the section and at the section indicate the unknown resultants **N, V,** and **M** acting in their *positive* directions (Fig. 4–1).

Equations of Equilibrium   Apply the three equations of equilibrium to obtain the unknowns **N, V,** and **M.** In most cases, moments should be summed at the section about axes that pass through the *centroid* of the member's cross-sectional area, in order to eliminate the unknowns **N** and **V** and thereby obtain a direct solution for **M.** If the solution of the equilibrium equations yields a quantity having a negative magnitude, the assumed directional sense of the quantity is opposite to that shown on the free-body diagram.

The following examples illustrate this procedure numerically.

Example 4–1

Determine the internal shear and moment acting in the cantilever beam shown in Fig. 4–2 at a section passing first through point C, then through point D.

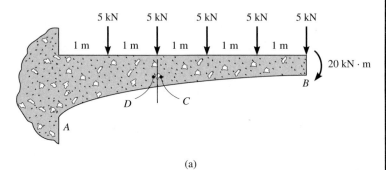

(a)

Fig. 4–2

Solution

Free-Body Diagram. If we consider a free-body diagram of a segment to the right of the section, the support reactions at A do not have to be calculated. These diagrams for segments CB and DB are shown in Fig. 4–2b and 4–2c. Note that the internal loadings act in their positive directions.

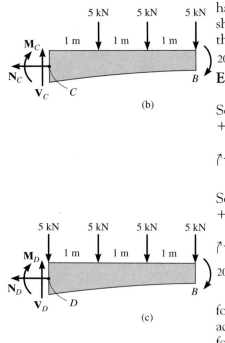

(b)

(c)

Equations of Equilibrium

Segment CB, Fig. 4–2b:
$$+\uparrow \Sigma F_y = 0; \qquad V_C - 5 - 5 - 5 = 0$$
$$V_C = 15 \text{ kN} \qquad \textbf{Ans.}$$
$$\curvearrowleft + \Sigma M_C = 0; \qquad M_C + 5(1) + 5(2) + 5(3) + 20 = 0$$
$$M_C = -50 \text{ kN} \cdot \text{m} \qquad \textbf{Ans.}$$

Segment DB, Fig 4–2c:
$$+\uparrow \Sigma F_y = 0; \qquad V_D - 5 - 5 - 5 - 5 = 0$$
$$V_D = 20 \text{ kN} \qquad \textbf{Ans.}$$
$$\curvearrowleft + \Sigma M_D = 0; \qquad M_D + 5(1) + 5(2) + 5(3) + 20 = 0$$
$$M_D = -50 \text{ kN} \cdot \text{m} \qquad \textbf{Ans.}$$

Here the shear force is different on either side of the concentrated force. Therefore, shear force must be defined at either one of these adjacent points, *never* at a point directly under the concentrated force.

Example 4-2

Determine the internal shear and moment acting at a section passing through point C in the beam shown in Fig. 4–3a.

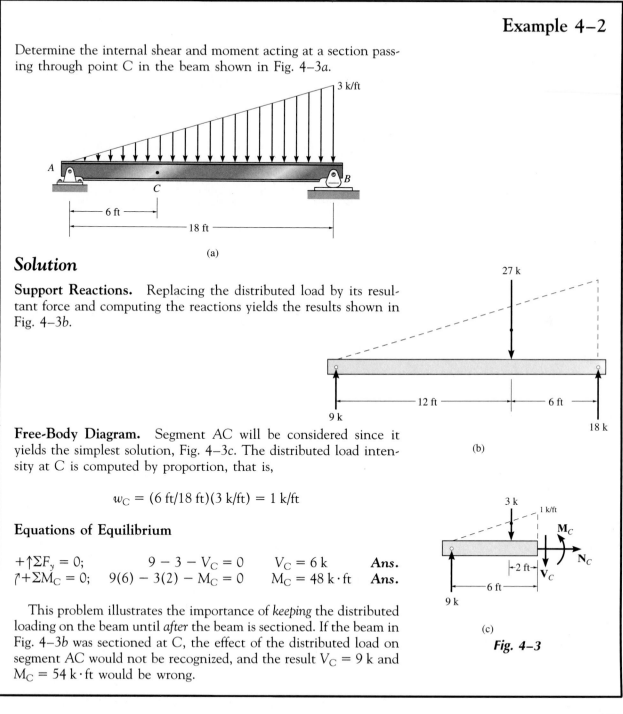

(a)

Solution

Support Reactions. Replacing the distributed load by its resultant force and computing the reactions yields the results shown in Fig. 4–3b.

Free-Body Diagram. Segment AC will be considered since it yields the simplest solution, Fig. 4–3c. The distributed load intensity at C is computed by proportion, that is,

$$w_C = (6 \text{ ft}/18 \text{ ft})(3 \text{ k/ft}) = 1 \text{ k/ft}$$

Equations of Equilibrium

$$+\uparrow\Sigma F_y = 0; \qquad 9 - 3 - V_C = 0 \qquad V_C = 6 \text{ k} \qquad \textbf{Ans.}$$
$$\curvearrowright+\Sigma M_C = 0; \quad 9(6) - 3(2) - M_C = 0 \qquad M_C = 48 \text{ k·ft} \qquad \textbf{Ans.}$$

(b)

(c)

Fig. 4–3

This problem illustrates the importance of *keeping* the distributed loading on the beam until *after* the beam is sectioned. If the beam in Fig. 4–3b was sectioned at C, the effect of the distributed load on segment AC would not be recognized, and the result $V_C = 9$ k and $M_C = 54$ k·ft would be wrong.

Example 4–3

The 9-k force in Fig. 4–4a is supported by the floor panel DE, which in turn is simply supported at its ends by floor beams. These beams transmit their loads to the simply supported girder AB. Determine the internal shear and moment acting at point C in the girder.

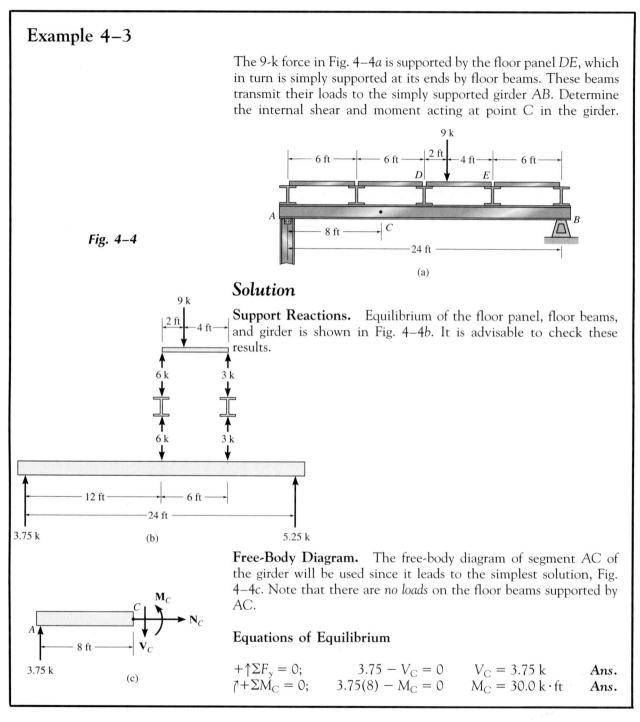

Fig. 4–4

(a)

Solution

Support Reactions. Equilibrium of the floor panel, floor beams, and girder is shown in Fig. 4–4b. It is advisable to check these results.

(b)

Free-Body Diagram. The free-body diagram of segment AC of the girder will be used since it leads to the simplest solution, Fig. 4–4c. Note that there are *no loads* on the floor beams supported by AC.

(c)

Equations of Equilibrium

$$+\uparrow \Sigma F_y = 0; \qquad 3.75 - V_C = 0 \qquad V_C = 3.75 \text{ k} \qquad \textbf{Ans.}$$
$$\curvearrowleft +\Sigma M_C = 0; \qquad 3.75(8) - M_C = 0 \qquad M_C = 30.0 \text{ k} \cdot \text{ft} \qquad \textbf{Ans.}$$

4.2   Shear and Moment Functions _____

Beams are structural members designed to support lateral loadings, i.e., those applied perpendicular to their axes. The design of such members requires a detailed knowledge of the *variations* of the internal shear force V and moment M acting at each point along the axis of the beam. The internal normal force is generally not considered for two reasons: (1) in most cases the loads applied to a beam act perpendicular to the beam's axis and hence produce only an internal shear force and bending moment; and (2), for design purposes the beam's resistance to shear, and particularly to bending, is more important than its ability to resist normal force. An important exception to this occurs, however, when beams are subjected to compressive axial forces, since the buckling or instability that may occur has to be investigated.

The variations of V and M as a function of the position x of an arbitrary point along the beam's axis can be obtained by using the method of sections discussed in Sec. 4.1. Here, however, it is necessary to locate the imaginary section or cut at an arbitrary distance x from one end of the beam rather than at a specific point.

In general, the internal shear and moment functions will be discontinuous, or their slope will be discontinuous, at points where the type or magnitude of the distributed load changes or where concentrated forces or couple moments are applied. Because of this, shear and moment functions must be determined for each region of the beam located between any two discontinuities of loading. For example, coordinates x_1, x_2, and x_3 will have to be used to describe the variation of V and M throughout the length of the beam in Fig. 4–5a. These coordinates will be valid only within regions from A to B for x_1, from B to C for x_2, and from C to D for x_3. Although each of these coordinates has the same origin, as noted here, this does not have to be the case.

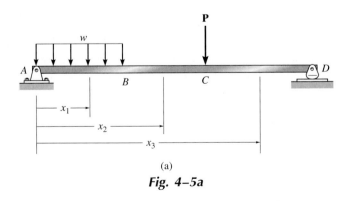

(a)

Fig. 4–5a

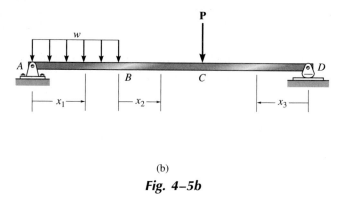

(b)

Fig. 4–5b

Indeed, it may be easier to develop the shear and moment functions using coordinates x_1, x_2, x_3 having origins at A, B, and D as shown in Fig. 4–5b. Here x_1 and x_2 are positive to the right and x_3 is positive to the left.

Procedure for Analysis

The following procedure provides a method for determining the variation of shear and moment in a beam as a function of position x.

Support Reactions Determine the support reactions on the beam and resolve all the external forces into components acting perpendicular and parallel to the beam's axis.

Shear and Moment Functions Specify separate coordinates x and associated origins, extending into regions of the beam between concentrated forces and/or couple moments, or where there is a discontinuity of distributed loading. Section the beam perpendicular to its axis at each distance x, and from the free-body diagram of one of the segments determine the unknowns V and M at the cut section as functions of x. On the free-body diagram, V and M should be shown acting in their *positive directions*, in accordance with the sign conventions given in Fig. 4–1. V is obtained from $\Sigma F_y = 0$ and M is obtained by summing moments about the point S located at the cut section, $\Sigma M_S = 0$. The results can be checked by noting that $dM/dx = V$ and $dV/dx = w$. These relationships are developed in Sec. 4.3.

The following examples illustrate this procedure numerically.

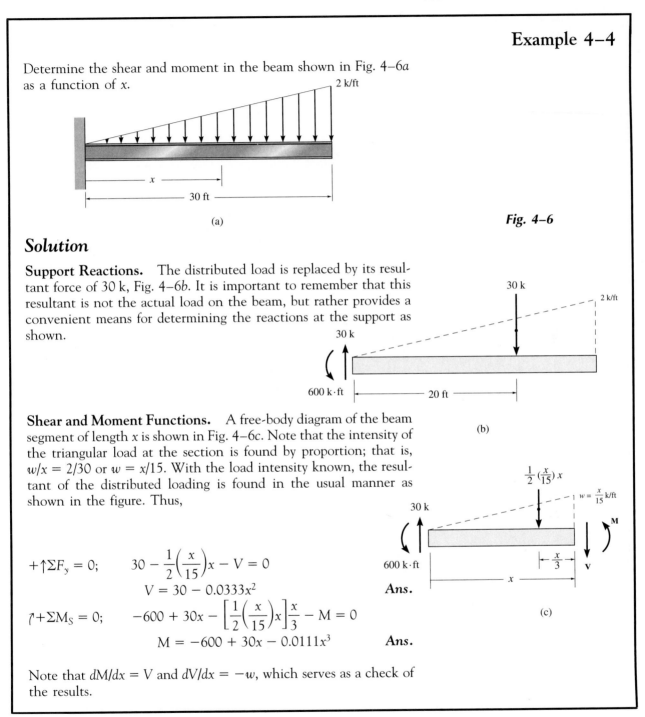

Example 4–4

Determine the shear and moment in the beam shown in Fig. 4–6a as a function of x.

Fig. 4–6

(a)

Solution

Support Reactions.   The distributed load is replaced by its resultant force of 30 k, Fig. 4–6b. It is important to remember that this resultant is not the actual load on the beam, but rather provides a convenient means for determining the reactions at the support as shown.

(b)

Shear and Moment Functions.   A free-body diagram of the beam segment of length x is shown in Fig. 4–6c. Note that the intensity of the triangular load at the section is found by proportion; that is, $w/x = 2/30$ or $w = x/15$. With the load intensity known, the resultant of the distributed loading is found in the usual manner as shown in the figure. Thus,

$$+\uparrow\Sigma F_y = 0; \qquad 30 - \frac{1}{2}\left(\frac{x}{15}\right)x - V = 0$$
$$V = 30 - 0.0333x^2 \qquad \textbf{Ans.}$$

$$\curvearrowright + \Sigma M_S = 0; \qquad -600 + 30x - \left[\frac{1}{2}\left(\frac{x}{15}\right)x\right]\frac{x}{3} - M = 0$$
$$M = -600 + 30x - 0.0111x^3 \qquad \textbf{Ans.}$$

(c)

Note that $dM/dx = V$ and $dV/dx = -w$, which serves as a check of the results.

155

Example 4–5

Determine the shear and moment in the beam shown in Fig. 4–7a as a function of x_1, x_2, x_3, and x_4.

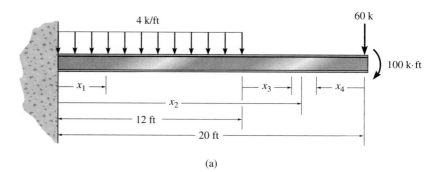

Fig. 4–7a,b,c

(a)

Solution

Support Reactions. The reactions at the fixed support have been computed and are shown on the free-body diagram of the beam, Fig. 4–7b.

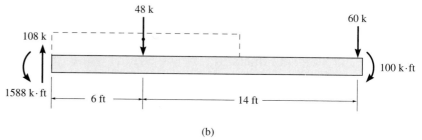

(b)

Shear and Moment Functions. Since there is a discontinuity of distributed load at $x = 12$ ft, two regions of x must be considered in order to describe the shear and moment functions for the entire beam. In this regard, x_1 is appropriate for the left 12 ft, and x_2, x_3, or x_4 can be used for the remaining segment.

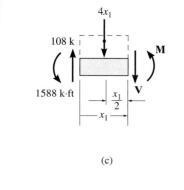

(c)

$0 \leq x_1 < 12$ ft, Fig. 4–7c:

$+\uparrow\Sigma F_y = 0;$ $108 - 4x_1 - V = 0$ $V = 108 - 4x_1$ **Ans.**

$\uparrow+\Sigma M_S = 0;$ $-1588 + 108x_1 - 4x_1\left(\dfrac{x_1}{2}\right) - M = 0$

$$M = -1588 + 108x_1 - 2x_1^2 \qquad \textbf{Ans.}$$

12 ft $< x_2 \leq$ 20 ft, Fig. 4–7d:

$+\uparrow\Sigma F_y = 0;$ $108 - 48 - V = 0$ $V = 60$ **Ans.**

$\gamma+\Sigma M_S = 0;$ $-1588 + 108x_2 - 48(x_2 - 6) - M = 0$

$$M = 60x_2 - 1300 \qquad \textbf{Ans.}$$

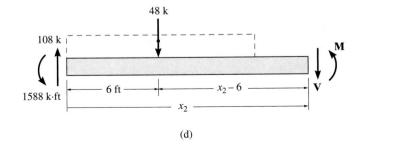

(d)

Fig. 4–7d,e,f

$0 < x_3 \leq$ 8 ft, Fig. 4–7e:

Show that

$$V = 60 \qquad \textbf{Ans.}$$
$$M = -1588 + 108(12 + x_3) - 48(6 + x_3)$$
$$M = 60x_3 - 580 \qquad \textbf{Ans.}$$

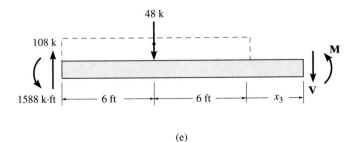

(e)

$0 \leq x_4 \leq$ 8 ft. Notice the correct positive directions for **V** and **M** in Fig. 4–7f. Show that

$$V = 60 \qquad \textbf{Ans.}$$
$$M = -60x_4 - 100 \qquad \textbf{Ans.}$$

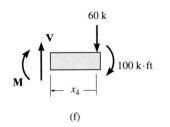

(f)

The above results can be partially checked by noting that when $x_2 = 15$ ft, $x_3 = 3$ ft, or $x_4 = 5$ ft, then $V = 60$ and $M = -400$. Also, note that $dM/dx = V$ and $dV/dx = -w$ in all cases.

Example 4–6

Determine the shear and moment in the beam shown in Fig. 4–8a as a function of x.

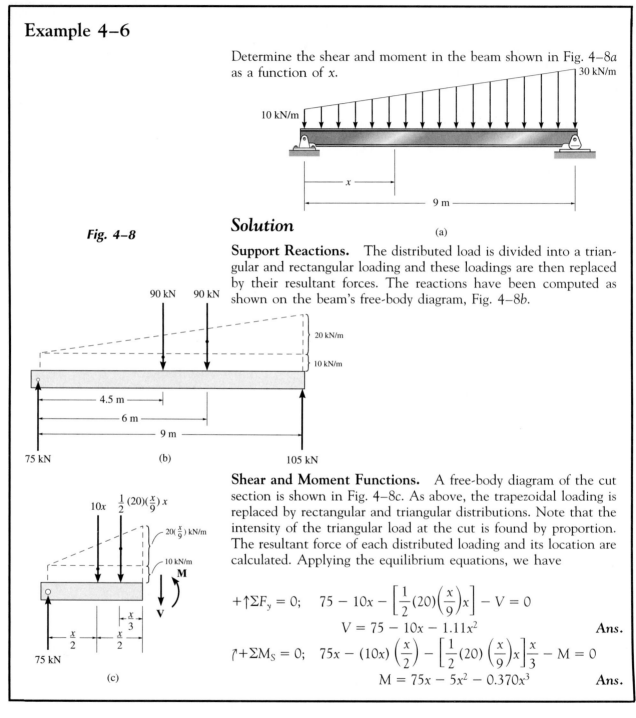

Fig. 4–8

Solution

(a)

Support Reactions. The distributed load is divided into a triangular and rectangular loading and these loadings are then replaced by their resultant forces. The reactions have been computed as shown on the beam's free-body diagram, Fig. 4–8b.

Shear and Moment Functions. A free-body diagram of the cut section is shown in Fig. 4–8c. As above, the trapezoidal loading is replaced by rectangular and triangular distributions. Note that the intensity of the triangular load at the cut is found by proportion. The resultant force of each distributed loading and its location are calculated. Applying the equilibrium equations, we have

$$+\uparrow\Sigma F_y = 0; \quad 75 - 10x - \left[\frac{1}{2}(20)\left(\frac{x}{9}\right)x\right] - V = 0$$

$$V = 75 - 10x - 1.11x^2 \qquad \textbf{Ans.}$$

$$\curvearrowright +\Sigma M_S = 0; \quad 75x - (10x)\left(\frac{x}{2}\right) - \left[\frac{1}{2}(20)\left(\frac{x}{9}\right)x\right]\frac{x}{3} - M = 0$$

$$M = 75x - 5x^2 - 0.370x^3 \qquad \textbf{Ans.}$$

4.3   Shear and Moment Diagrams for a Beam _____

If the variations of V and M as functions of x obtained in Sec. 4.2 are plotted, the graphs are termed the *shear diagram* and *moment diagram*, respectively. In cases where a beam is subjected to *several* concentrated forces, couples, and distributed loads, plotting V and M versus x can become quite tedious. In this section a simpler method for constructing these diagrams is discussed—a method based on differential relations that exist between the load, shear, and bending moment.

To derive these necessary relations, consider the beam AD shown in Fig. 4–9a, which is subjected to an arbitrary distributed loading $w = w(x)$ and a series of concentrated forces and couples. In the following discussion, *the distributed load will be considered positive when the loading acts downward* as shown. The free-body diagram for a small segment of the beam having a length Δx is shown in Fig. 4–9b. Since this segment has been chosen at a point x along the beam that is *not* subjected to a concentrated force or couple, any results obtained will not apply at points of concentrated loading. The internal shear force and bending moment shown on the free-body diagram are assumed to act in the *positive direction* according to the established sign convention, Fig. 4–1. Note that both the shear force and moment acting on the right-hand face must be increased by a small, finite amount in order to keep the segment in equilibrium. The distributed loading has been replaced by a concentrated force $w(x) \Delta x$ that acts at a fractional distance $\epsilon(\Delta x)$ from the right end, where $0 < \epsilon < 1$. [For example, if $w(x)$ is uniform or constant, then $w(x) \Delta x$ will act at $\frac{1}{2} \Delta x$, so $\epsilon = \frac{1}{2}$.] Applying the

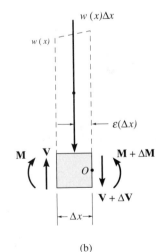

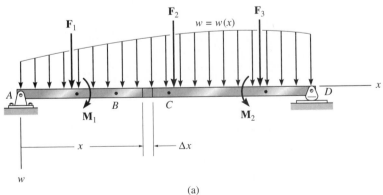

(a)

(b)

Fig. 4–9

159

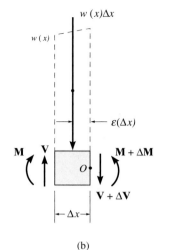

(b)

Fig. 4–9b

equations of equilibrium, we have

$$+\uparrow\Sigma F_y = 0; \qquad V - w(x)\,\Delta x - (V + \Delta V) = 0$$
$$\Delta V = -w(x)\,\Delta x$$
$$\zeta+\Sigma M_O = 0; \qquad -V\,\Delta x - M + w(x)\,\Delta x\epsilon(\Delta x) + (M + \Delta M) = 0$$
$$\Delta M = V\,\Delta x - w(x)\,\epsilon(\Delta x)^2$$

Dividing by Δx and taking the limit as $\Delta x \to 0$, the above equations become

$$\frac{dV}{dx} = -w(x)$$
$$\left.\begin{matrix}\text{Slope of}\\\text{Shear Diagram}\end{matrix}\right\} = \left\{\begin{matrix}-\text{Intensity of}\\\text{Distributed Load}\end{matrix}\right. \qquad (4\text{--}1)$$

$$\frac{dM}{dx} = V$$
$$\left.\begin{matrix}\text{Slope of}\\\text{Moment Diagram}\end{matrix}\right\} = \text{Shear} \qquad (4\text{--}2)$$

As noted, Eq. 4–1 states that the slope of the shear diagram at a point (dV/dx) is equal to the (negative) intensity of the distributed load $w(x)$ at the point. Likewise, Eq. 4–2 states that the slope of the moment diagram (dM/dx) is equal to the intensity of the shear at the point.

Equations 4–1 and 4–2 can be "integrated" from one point to another between concentrated forces or couples (such as from B to C in Fig. 4–9a), in which case

$$\Delta V = -\int w(x)\,dx$$
$$\left.\begin{matrix}\text{Change in}\\\text{Shear}\end{matrix}\right\} = \left\{\begin{matrix}-\text{Area under}\\\text{Distributed Loading}\\\text{Diagram}\end{matrix}\right. \qquad (4\text{--}3)$$

and

$$\Delta M = \int V(x)\,dx$$
$$\left.\begin{matrix}\text{Change in}\\\text{Moment}\end{matrix}\right\} = \left\{\begin{matrix}\text{Area under}\\\text{Shear Diagram}\end{matrix}\right. \qquad (4\text{--}4)$$

As noted, Eq. 4–3 states that the change in the shear between any two points on a beam equals the (negative) area under the distributed loading diagram between the points. Likewise, Eq. 4–4 states that the change in the moment between the two points equals the area under the shear diagram between the points. If the areas under the load and shear diagrams are easy to compute, Eqs. 4–3 and 4–4 provide a method for determining the numerical values of the shear and moment at various points along a beam.

From the above derivation it should be remembered that Eqs. 4–1 and 4–3 cannot be used at points where a concentrated force acts, since these equations do not account for the sudden change in shear at these points. Similarly, because of a discontinuity of moment, Eqs. 4–2 and 4–4 cannot be used at points where a couple is applied. In order to account for these two cases, we must consider the free-body diagrams of differential elements of the beam in Fig. 4–9a, which are located at concentrated force and couple moments. Examples of these elements are shown in Fig. 4–10a and 4–10b, respectively. From Fig. 4–10a it is seen that force equilibrium requires the change in shear to be

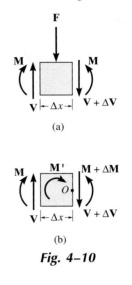

(a)

(b)

Fig. 4–10

$$+\uparrow\Sigma F_y = 0; \qquad\qquad \Delta V = -F \qquad\qquad (4\text{–}5)$$

Thus, when **F** acts *downward* on the beam, ΔV is negative so that the shear "jumps" *downward*. Likewise, if **F** acts *upward*, the jump (ΔV) is *upward*. From Fig. 4–10b, letting $\Delta x \to 0$, moment equilibrium requires the change in moment to be

$$\downarrow+\Sigma M_O = 0; \qquad\qquad \Delta M = M' \qquad\qquad (4\text{–}6)$$

In this case, if an external couple moment **M** is applied *clockwise*, ΔM is positive so that the moment diagram shows a jump *upward*, and when **M** acts *counterclockwise*, the jump (ΔM) must be *downward*.

Table 4–1 illustrates application of Eqs. 4–1, 4–2, 4–5, and 4–6 to some common loading cases assuming V and M retain positive values. None of these results should be memorized; rather, each should be studied carefully so that one becomes fully aware of how the shear and moment diagrams can be constructed on the basis of knowing the variation of the slope from the load and shear diagrams, respectively. It would be well worth the time and effort to self-test your understanding of these concepts by covering over the shear and moment diagram columns in the table and then trying to reconstruct these diagrams on the basis of knowing the loading.

Table 4–1

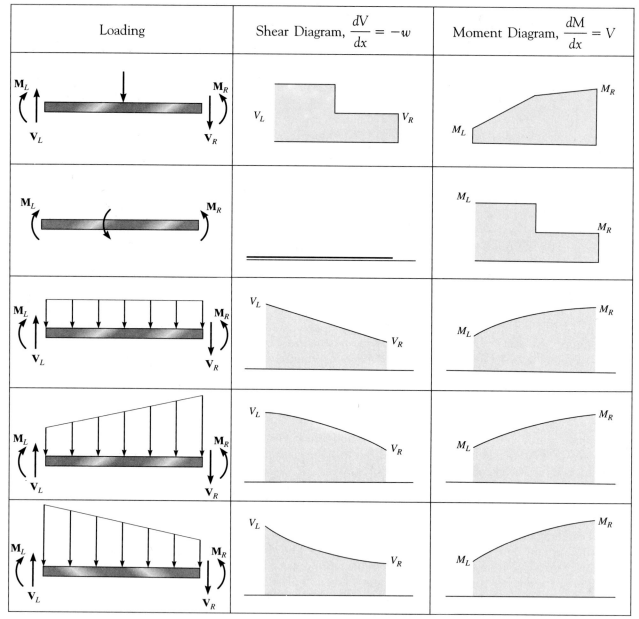

Procedure for Analysis

The following procedure provides a method for constructing the shear and moment diagrams for a beam using Eqs. 4–1 through 4–6.

Support Reactions   Determine the support reactions and resolve the forces acting on the beam into components which are perpendicular and parallel to the beam's axis.

Shear Diagram   Establish the V and x axes and plot the values of the shear at the two *ends* of the beam. Since $dV/dx = -w$, the *slope* of the *shear diagram* at any point is equal to the (negative) intensity of the *distributed loading* at the point. (Note that w is positive when it acts downward.) If a numerical value of the shear is to be determined at the point, one can find this value either by using the method of sections as discussed in Sec. 4.1 or by using Eq. 4–3, which states that the *change in the shear force* is equal to the (negative) *area under the distributed loading diagram*. Since $w(x)$ is *integrated* to obtain V, if $w(x)$ is a curve of degree n, then $V(x)$ will be a curve of degree $n + 1$. For example, if $w(x)$ is uniform, $V(x)$ will be linear.

Moment Diagram   Establish the M and x axes and plot the values of the moment at the ends of the beam. Since $dM/dx = V$, the *slope* of the *moment diagram* at any point is equal to the intensity of the *shear* at the point. In particular, note that at the point where the shear is zero, $dM/dx = 0$, and therefore this may be a point of maximum or minimum moment.

   If the numerical value of the moment is to be determined at a point, one can find this value either by using the method of sections as discussed in Sec. 4.1 or by using Eq. 4–4, which states that the *change in the moment* is equal to the *area under the shear diagram*. Since $V(x)$ is *integrated* to obtain M, if $V(x)$ is a curve of degree n, then $M(x)$ will be a curve of degree $n + 1$. For example, if $V(x)$ is linear, $M(x)$ will be parabolic.

The following examples illustrate this procedure.

Example 4–7

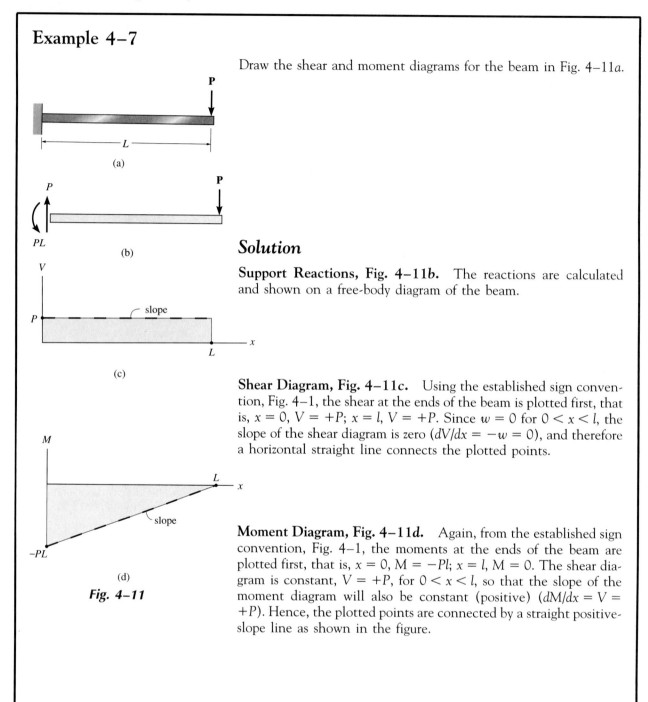

Draw the shear and moment diagrams for the beam in Fig. 4–11a.

Solution

Support Reactions, Fig. 4–11b. The reactions are calculated and shown on a free-body diagram of the beam.

Shear Diagram, Fig. 4–11c. Using the established sign convention, Fig. 4–1, the shear at the ends of the beam is plotted first, that is, $x = 0$, $V = +P$; $x = l$, $V = +P$. Since $w = 0$ for $0 < x < l$, the slope of the shear diagram is zero ($dV/dx = -w = 0$), and therefore a horizontal straight line connects the plotted points.

Moment Diagram, Fig. 4–11d. Again, from the established sign convention, Fig. 4–1, the moments at the ends of the beam are plotted first, that is, $x = 0$, $M = -Pl$; $x = l$, $M = 0$. The shear diagram is constant, $V = +P$, for $0 < x < l$, so that the slope of the moment diagram will also be constant (positive) ($dM/dx = V = +P$). Hence, the plotted points are connected by a straight positive-slope line as shown in the figure.

Fig. 4–11

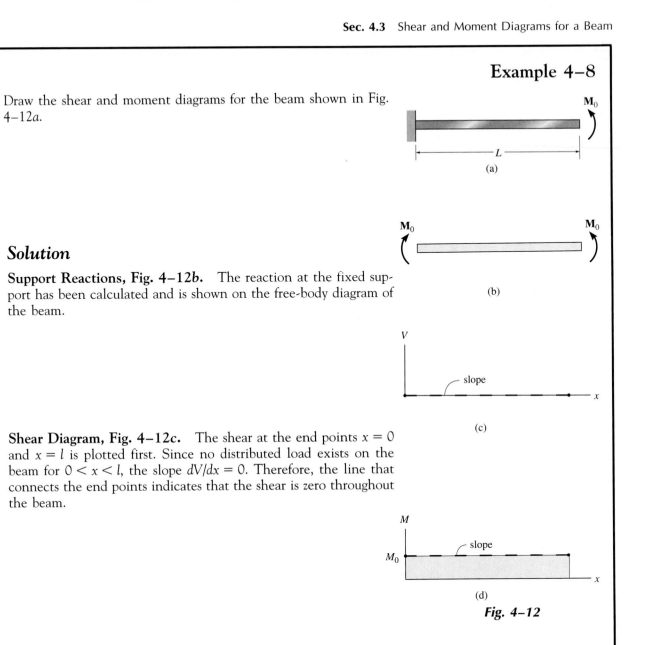

Example 4–8

Draw the shear and moment diagrams for the beam shown in Fig. 4–12a.

(a)

(b)

(c)

(d)

Fig. 4–12

Solution

Support Reactions, Fig. 4–12b. The reaction at the fixed support has been calculated and is shown on the free-body diagram of the beam.

Shear Diagram, Fig. 4–12c. The shear at the end points $x = 0$ and $x = l$ is plotted first. Since no distributed load exists on the beam for $0 < x < l$, the slope $dV/dx = 0$. Therefore, the line that connects the end points indicates that the shear is zero throughout the beam.

Moment Diagram, Fig. 4–12d. The moment M_O at the beam's end points $x = 0$ and $x = l$ is plotted first. The slope $dM/dx = 0$ for the moment diagram since the shear is zero for $0 < x < l$. Therefore, the moment diagram is rectangular.

165

Example 4–9

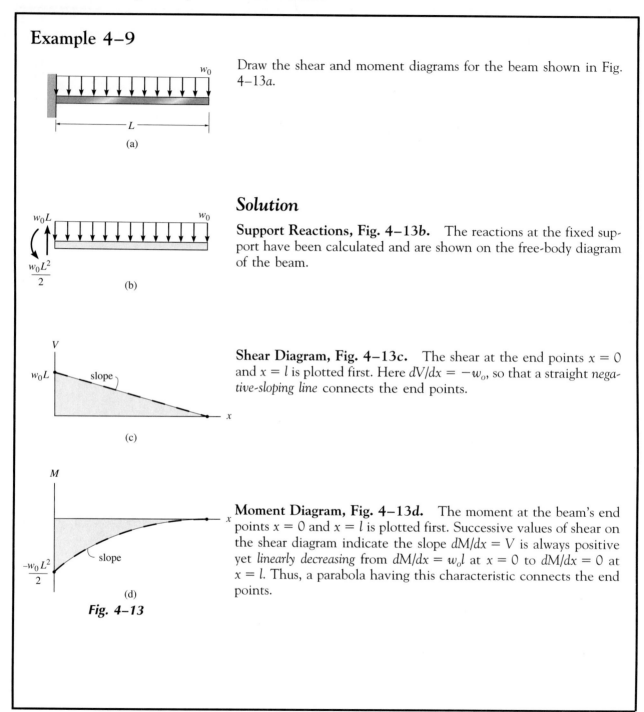

(a)

Draw the shear and moment diagrams for the beam shown in Fig. 4–13a.

Solution

Support Reactions, Fig. 4–13b. The reactions at the fixed support have been calculated and are shown on the free-body diagram of the beam.

(b)

Shear Diagram, Fig. 4–13c. The shear at the end points $x = 0$ and $x = l$ is plotted first. Here $dV/dx = -w_o$, so that a straight *negative-sloping line* connects the end points.

(c)

Moment Diagram, Fig. 4–13d. The moment at the beam's end points $x = 0$ and $x = l$ is plotted first. Successive values of shear on the shear diagram indicate the slope $dM/dx = V$ is always positive yet *linearly decreasing* from $dM/dx = w_o l$ at $x = 0$ to $dM/dx = 0$ at $x = l$. Thus, a parabola having this characteristic connects the end points.

(d)

Fig. 4–13

Example 4–10

Draw the shear and moment diagrams for the beam shown in Fig. 4–14a.

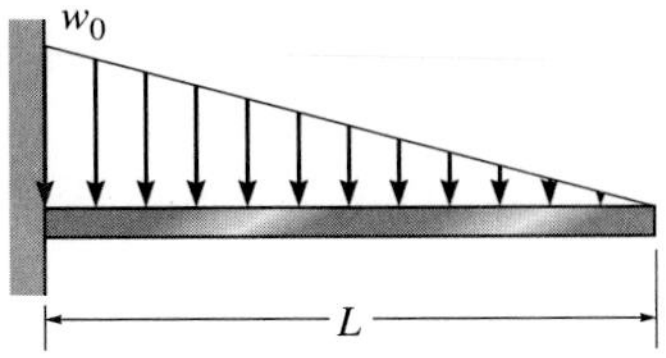

(a)

Solution

Support Reactions, Fig. 4–14b.   The reactions at the fixed support have been calculated and are shown on the free-body diagram of the beam.

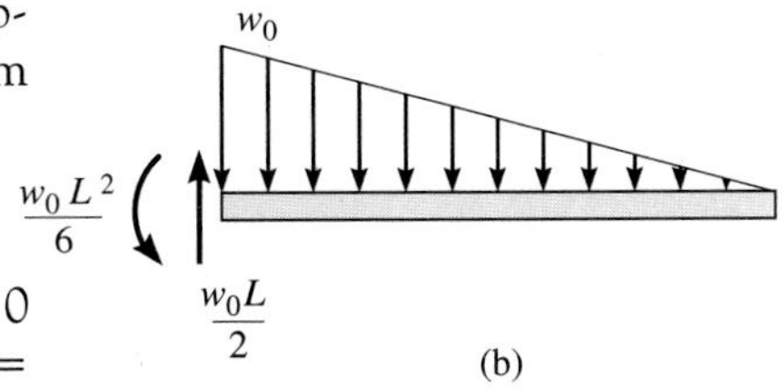

(b)

Shear Diagram, Fig. 4–14c.   The shear at the end points $x = 0$ and $x = l$ is plotted first. From the load diagram, the slope $dV/dx = -w$ is negatively decreasing from $dV/dx = -w_o$ at $x = 0$ to $dV/dx = 0$ at $x = l$. A parabola having this characteristic therefore connects the end points of the shear diagram.

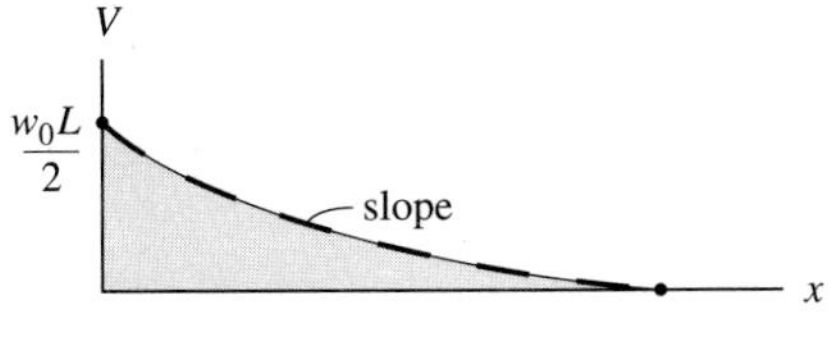

(c)

Moment Diagram, Fig. 4–14d.   The moment at the beam's end points $x = 0$ and $x = l$ is plotted first. From the shear diagram the (positive) slope is parabolically decreasing, from $dM/dx = +w_o l/2$ at $x = 0$ to $dM/dx = 0$ at $x = l$. The curve connecting the plotted end points that has this characteristic is a cubic function of x, as shown in the figure.

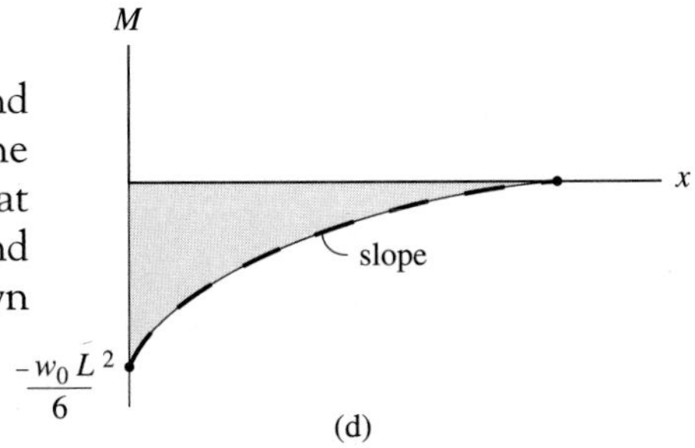

(d)

Fig. 4–14

Example 4–11

Draw the shear and moment diagrams for the beam in Fig. 4–15a.

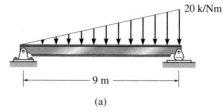

(a)

Solution

Support Reactions, Fig. 4–15b. The reactions have been calculated and are shown on the free-body diagram of the beam.

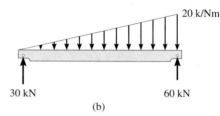

(b)

Shear Diagram, Fig. 4–15c. The end points $x = 0$, $V = +30$ and $x = 9$, $V = -60$ are plotted first. As shown from the beam's loading, the slope of the shear diagram varies from $dV/dx = 0$ at $x = 0$ to $dV/dx = -20$ at $x = 9$. For $0 \leq x \leq 9$ the slope is increasingly negative since the distributed load is increasing ($dV/dx = -w$).

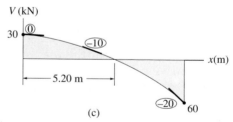

Fig. 4–15a,b,c (c)

The point of zero shear can be found by using the method of sections from a beam segment of length x, Fig. 4–15e. We require $V = 0$, so that

$$+\uparrow \Sigma F_y = 0; \quad 30 - \frac{1}{2}\left[20\left(\frac{x}{9}\right)\right]x = 0 \qquad x = 5.20 \text{ m}$$

Moment Diagram, Fig. 4–15d. The end points $x = 0$, $M = 0$ and $x = 9$, $M = 0$ are plotted first. From the shear diagram, the slope of the moment diagram is $dM/dx = +30$ at $x = 0$ and $dM/dx = 0$ at $x = 5.20$ m. For $0 \leq x < 5.20$ the slope is decreasingly positive since the shear is decreasingly positive. Likewise, for $5.20 < x \leq 9$ the slope is increasingly negative.

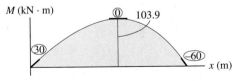

(d)

The maximum value of moment is at $x = 5.20$ m since $dM/dx = V = 0$ at this point. From the free-body diagram in Fig. 4–15e we have

$$\zeta + \Sigma M_S = 0; \quad 30(5.20) - \frac{1}{2}\left[20\left(\frac{5.20}{9}\right)\right](5.20)\left(\frac{5.20}{3}\right) - M = 0$$

$$M = 103.9 \text{ kN} \cdot \text{m}$$

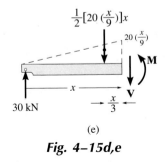

(e)

Fig. 4–15d,e

169

Example 4–12

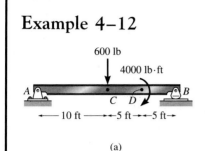

(a)

Draw the shear and moment diagrams for the beam shown in Fig. 4–16a.

Solution

Support Reactions. The reactions are calculated and indicated on the free-body diagram, Fig. 4–16b.

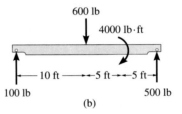

(b)

Shear Diagram. The values of the shear at the end points A and B are plotted first. From the sign convention, Fig. 4–1, $V_A = +100$ and $V_B = -500$, Fig. 4–16c. At an intermediate point between A and C the slope of the shear diagram is zero since $dV/dx = -w = 0$. Hence the shear retains its value of $+100$ within this region. At C the shear is *discontinuous* since there is a *concentrated force* of 600 there. The value of the shear just to the right of C (-500) can be found by sectioning the beam at this point. This yields the free-body diagram shown in equilibrium in Fig. 4–16e. This point ($V = -500$) is plotted on the shear diagram. As before, $w = 0$ from C to B, so the slope $dV/dx = 0$. The diagram closes to the value of -500 at B as shown. One might wonder why no jump or discontinuity in shear occurs at D, the point where the 4000-lb·ft couple moment is

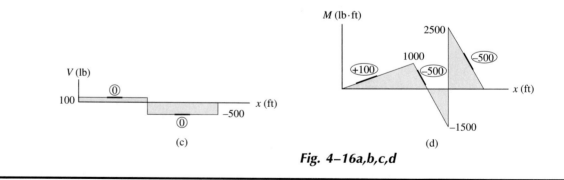

(c)

(d)

Fig. 4–16a,b,c,d

applied, Fig. 4–16a. Such is not the case as indicated by the equilibrium conditions on the free-body diagram in Fig. 4–16f.

It may also be noted that the shear diagram can be constructed quickly by "following the load" on the free-body diagram, Fig. 4–16b. In this regard, beginning at A the 100-lb force acts upward, so $V_A = +100$ lb. No load acts between A and C, so the shear remains constant. At C the 600-lb force is down, so the shear jumps down 600, from 100 to -500. Again the shear is constant (no load) and ends at -500, point B, which closes the diagram back to zero since the 500-lb force on the beam acts upward.

Moment Diagram.   The moment at each end of the beam is zero. Thus these two points are plotted first, Fig. 4–16d. The slope of the moment diagram from A to C is constant since $dM/dx = V = +100$. The value of the moment at C can be determined by the method of sections, Fig. 4–16e, or by computing the area under the shear diagram between A and C, that is, $\Delta M_{AC} = M_C - M_A = (100 \text{ lb})(10 \text{ ft}) = 1000$.   Since $M_A = 0$,   then $M_C 0 + 1000 \text{ lb·ft} = 1000$. From C to D the slope is $dM/dx = V = -500$, Fig. 4–16c. The area under the shear diagram between points C and D is $\Delta M_{CD} = M_D - M_C = (-500 \text{ lb})(5 \text{ ft}) = -2500$. Then $M_D = 1000 - 2500 = -1500$. A jump occurs at point D due to the concentrated couple moment of 4000. The method of sections, Fig. 4–16f, gives a value of $+2500$ just to the right of D. From this point, the slope of $dM/dx = -500$ is maintained until the diagram closes to zero at B, Fig. 4–16d.

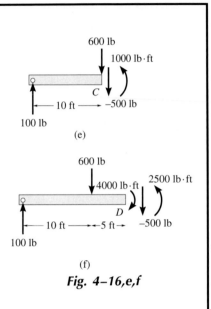

(e)

(f)

Fig. 4–16,e,f

Example 4–13

Draw the shear and moment diagrams for the beam shown in Fig. 4–17a.

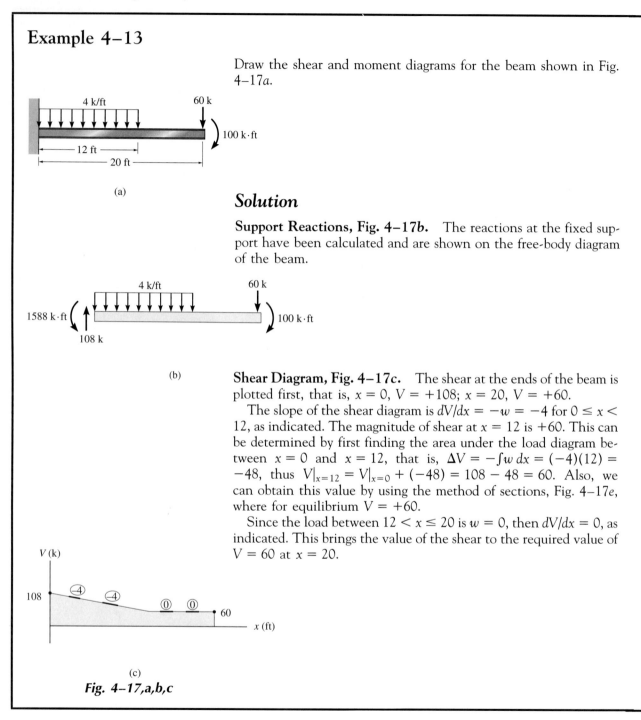

(a)

Solution

Support Reactions, Fig. 4–17b. The reactions at the fixed support have been calculated and are shown on the free-body diagram of the beam.

(b)

Shear Diagram, Fig. 4–17c. The shear at the ends of the beam is plotted first, that is, $x = 0$, $V = +108$; $x = 20$, $V = +60$.

The slope of the shear diagram is $dV/dx = -w = -4$ for $0 \leq x < 12$, as indicated. The magnitude of shear at $x = 12$ is $+60$. This can be determined by first finding the area under the load diagram between $x = 0$ and $x = 12$, that is, $\Delta V = -\int w\,dx = (-4)(12) = -48$, thus $V|_{x=12} = V|_{x=0} + (-48) = 108 - 48 = 60$. Also, we can obtain this value by using the method of sections, Fig. 4–17e, where for equilibrium $V = +60$.

Since the load between $12 < x \leq 20$ is $w = 0$, then $dV/dx = 0$, as indicated. This brings the value of the shear to the required value of $V = 60$ at $x = 20$.

(c)

Fig. 4–17,a,b,c

172

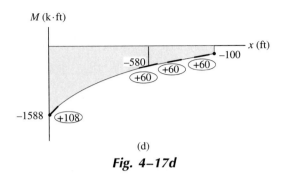

(d)

Fig. 4–17d

Moment Diagram, Fig. 4–17d. The moments at the ends of the beam are plotted first, that is, $x = 0$, $M = -1588$; $x = 20$, $M = -100$.

Numerical values of shear change from $+108$ to $+60$ for $0 \leq x < 12$. Each value of shear gives the slope of the moment diagram since $dM/dx = V$. As indicated, at $x = 0$, $dM/dx = +108$; and at $x = 12$, $dM/dx = +60$. For $0 \leq x < 12$ specific values of the shear diagram are positive but linearly decreasing. Hence, the moment diagram is parabolic with linearly decreasing slope.

The magnitude of moment at $x = 12$ is -580. This can be found from the area under the shear diagram, that is, $\Delta M = \int V dx = 60(12) + \frac{1}{2}(108/60)(12) = +1008$, so that $M|_{x=12} = M|_{x=0} + 1008 = -1588 + 1008 = -580$. The more "basic" method of sections can also be used, where equilibrium at $x = 12$ requires $M = -580$, Fig. 4–17e.

The moment diagram has a constant slope for $12 < x \leq 20$ since $dM/dx = V = +60$. This brings the value to $M = -100$ at $x = 20$, as required.

It should be noted that these diagrams can also be constructed by plotting the functions obtained in Example 4–6, although this would be a more tedious solution.

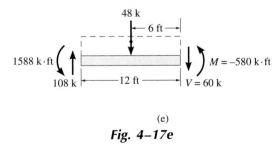

(e)

Fig. 4–17e

Example 4–14

Draw the shear and moment diagrams for the compound beam shown in Fig. 4–18a. Assume the supports at A and C are rollers and B and D are pin connections.

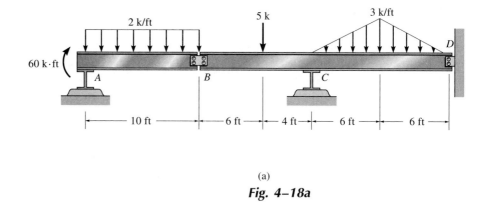

(a)

Fig. 4–18a

Solution

Support Reactions, Fig. 4–18b. Once the beam is disconnected from the pin at B, the support reactions can be calculated as shown in Fig. 4–18b.

Shear Diagram, Fig. 4–18c. As usual, we start by plotting the end shear at A and D. Using the equation $dV/dx = -w$, the curves are indicated. Try to establish the peak values using the appropriate areas under the load diagram (w curve) to find the change in shear. The zero value for shear at $x = 2$ ft can either be computed by proportional triangles, or by using statics as was done in Fig. 4–15e of Example 4–11.

Moment Diagram, Fig. 4–18d. The end moments $M_A = 60$ and $M_D = 0$ are plotted first. Study the diagram and note how the various curves are established using $dM/dx = V$. Verify the numerical values for the peaks using statics or by computing the appropriate areas under the shear diagram to find the change in moment.

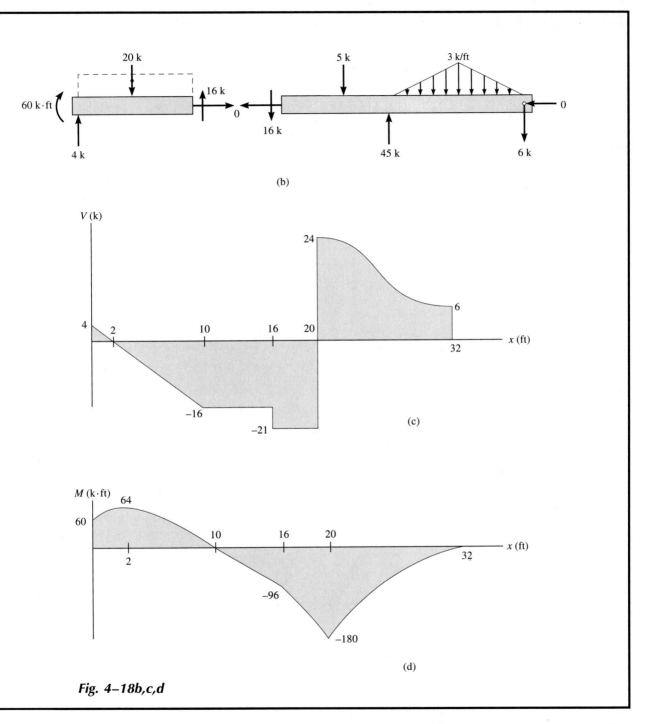

(b)

(c)

(d)

Fig. 4–18b,c,d

175

4.4   Shear and Moment Diagrams for a Frame

Recall that a *frame* is composed of several connected members that are either fixed- or pin-connected at their ends. The design of these structures often requires drawing the shear and moment diagrams for each of the members. To analyze any problem, we can use the procedure for analysis outlined in Sec. 4.3. This requires first determining the reactions at the frame supports. Then, using the method of sections, we find the axial force, shear force, and moment acting at the ends of each member. Provided all loadings are resolved into components acting parallel and perpendicular to the member's axis, the shear and moment diagrams for each member can then be drawn as described previously.

When drawing the moment diagram, one of two sign conventions is used in practice. In particular, if the frame is made of *reinforced concrete*, designers often draw the moment diagram on the tension side of the frame. In other words, if the moment produces tension on the outer surface of the frame, the moment diagram is drawn on this side. Since concrete easily fails in tension, it will be possible to tell at a glance on which side of the frame the reinforcement steel must be placed. In this text, however, we will use the opposite sign convention and *always draw the moment diagram on the compression side of the member*. This convention follows that used for beams discussed in Sec. 4.1.

The following examples illustrate this procedure numerically.

The simply supported girder of this building frame was designed by first drawing its shear and moment diagram.

Example 4–15

Draw the shear and moment diagrams for the frame shown in Fig. 4–19a. Assume A, C, and D are pinned and B is a fixed joint.

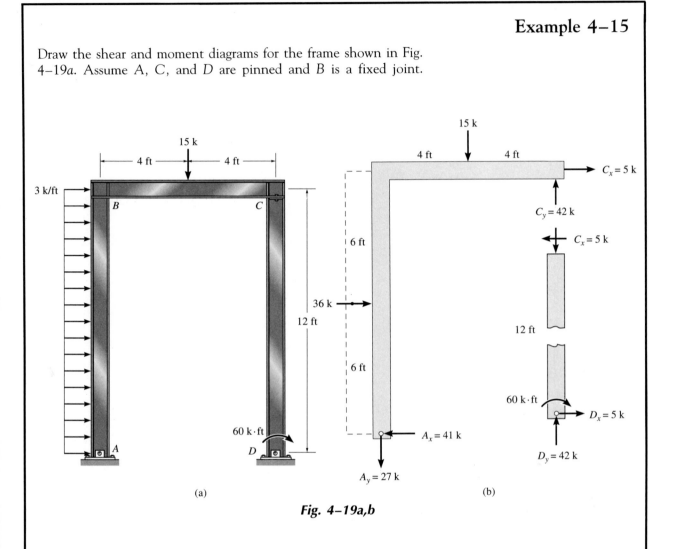

(a)   (b)

Fig. 4–19a,b

Solution

Support Reactions.   The frame is separated into two parts, ABC and DC, and the six reactions at the pins are calculated using the equations of equilibrium, Fig. 4–19b. Using these reactions, the frame is next sectioned at B and the internal shear, axial force, and bending moment are computed using the free-body diagram in either Fig. 4–19c or 4–19d.

(cont'd)

177

Example 4–15
(continued)

Shear and Moment Diagrams.   Shear and moment diagrams are plotted for each of the three members in Fig. 4–19c through 4–19e. Combining the results, the moment diagram is plotted on the line diagram of the frame, Fig. 4–19f.

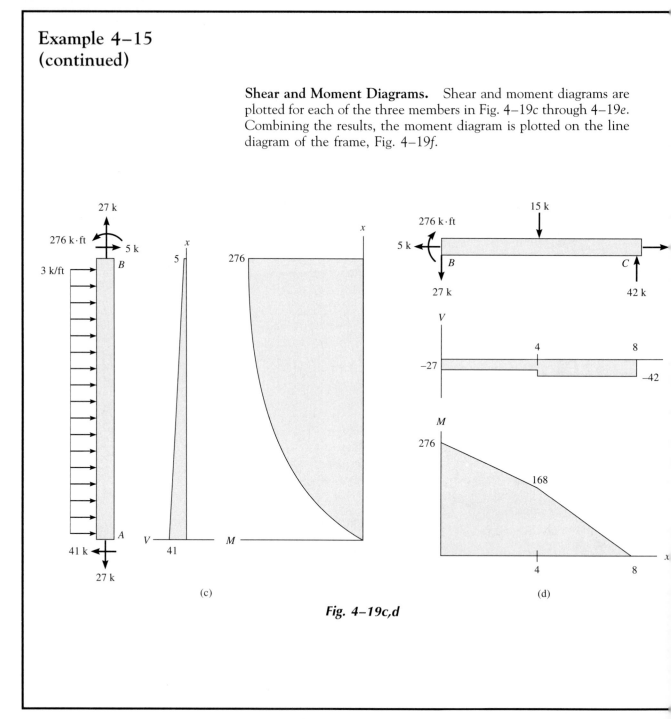

Fig. 4–19c,d

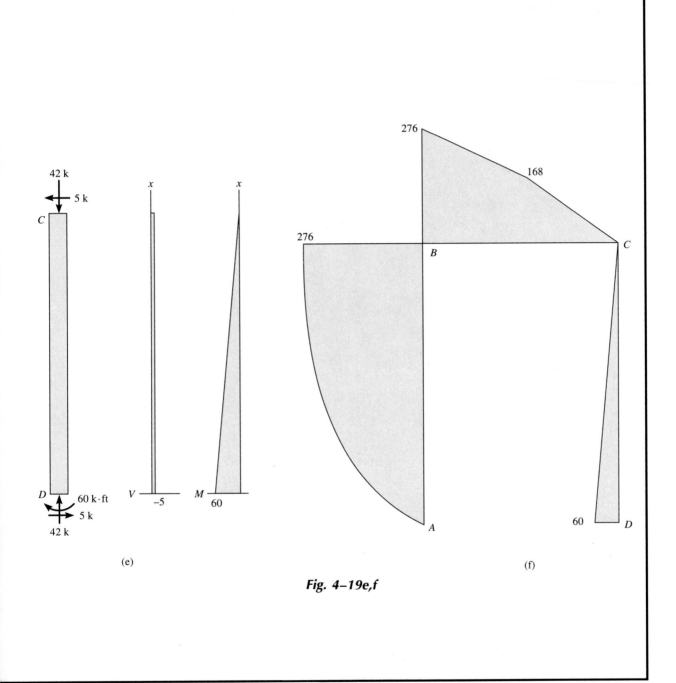

(e)

(f)

Fig. 4–19e,f

Example 4–16

Draw the moment diagram for the frame shown in Fig. 4–20a. Assume A is a pin, C is a roller, and B is a fixed joint.

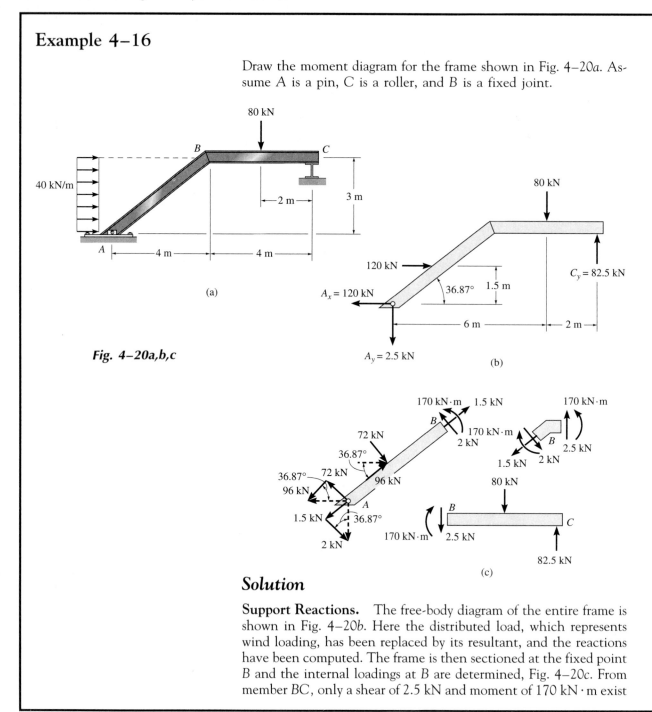

Fig. 4–20a,b,c

Solution

Support Reactions. The free-body diagram of the entire frame is shown in Fig. 4–20b. Here the distributed load, which represents wind loading, has been replaced by its resultant, and the reactions have been computed. The frame is then sectioned at the fixed point B and the internal loadings at B are determined, Fig. 4–20c. From member BC, only a shear of 2.5 kN and moment of 170 kN · m exist to the right of B. The loadings to the left of B are found from member AB. Here the reactions at A and the resultant of the distributed load have been resolved into components parallel and perpendicular to the member. The shear, 2 kN, normal force, 1.5 kN, and moment, 170 kN·m, have been computed using these components. As a check, equilibrium is satisfied at joint (or point) B, which is also shown in the figure.

Shear and Moment Diagrams.   The components of the distributed load, (72 kN)/(5 m) = 14.4 kN/m   and   (96 kN)/(5 m) = 19.2 kN/m, are shown on member AB, Fig. 4–20d. The associated shear and moment diagrams are drawn for each member as shown.

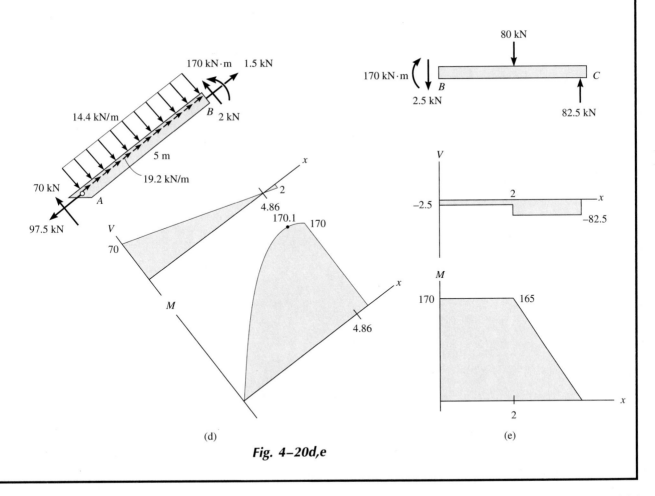

(d)   (e)

Fig. 4–20d,e

Example 4–17

Draw the moment diagram for the tapered frame shown in Fig. 4–21a. Assume the support at B is a pin.

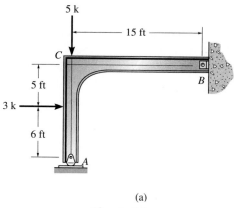

(a)

Fig. 4–21a

Solution

Support Reactions.   The support reactions are shown on the free-body diagram of the entire frame, represented as a centerline diagram in Fig. 4–21b. Using these results, the frame (centerline) is then sectioned into two members, and the internal reactions at the ends of the members are determined, Fig. 4–21c. In particular, note that the external 5-k load is shown only on the free-body diagram of the joint at C.

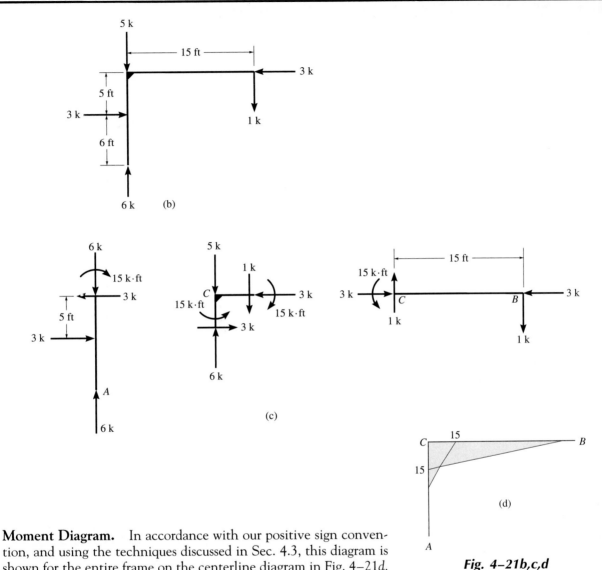

Fig. 4–21b,c,d

Moment Diagram.   In accordance with our positive sign convention, and using the techniques discussed in Sec. 4.3, this diagram is shown for the entire frame on the centerline diagram in Fig. 4–21d.

It should be noted that the calculations of the frame's support reactions are *independent* of the members' cross-sectional area since the frame is statically determinate. Had the frame been statically indeterminate, it would have been necessary to consider the variation of the members' cross-sectional area to obtain the reactions. The methods for doing this are discussed in later chapters of this book.

4.5   Moment Diagrams Constructed by the Method of Superposition

Since beams are used primarily to resist bending stress, it is important that the moment diagram accompany the solution for their design. In Sec. 4.3 the moment diagram was constructed by *first* drawing the shear diagram. If we use the principle of superposition, however, each of the loads on the beam can be treated separately and the moment diagram can then be constructed in a series of parts rather than a single and sometimes complicated shape. It will be shown later in the text that this can be particularly advantageous when applying geometric deflection methods to determine both the deflection of a beam and the reactions on statically indeterminate beams.

Most loadings on beams in structural analysis will be a combination of the loadings shown in Fig. 4–22. Construction of the associated

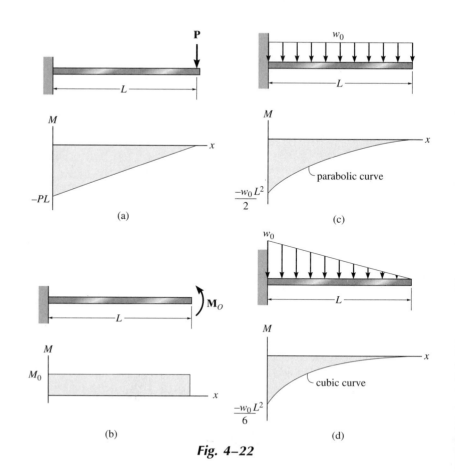

Fig. 4–22

moment diagrams has been discussed in Examples 4–7 through 4–10. To understand how to use the method of superposition to construct the moment diagram for a beam, consider the simply supported beam in Fig. 4–23a. Rather than consider *all the loads* on the beam *simultaneously* when constructing the moment diagram, we can instead consider a system of *statically equivalent loads*. For example, the three cantilevered beams shown in Fig. 4–23a are statically equivalent to the simply supported beam, since the load at each point in the simply supported beam is equal to the *superposition* or addition of the loadings on the cantilevered beams. Notice that the reaction at end A is indeed 15 k when the reactions on the cantilevered beams are added together. In the same manner, the internal moment at any point in the simply supported beam is equal to the sum of the moments at any point in the cantilevered beams. Thus, if the moment diagram for each cantilevered beam is drawn, Fig. 4–23b, the superposition of these diagrams yields the resultant moment diagram for the simply supported beam, shown at the top of Fig. 4–23b. For example, from each of the separate moment diagrams, the moment at end A is $M_A = -200 - 300 + 500 = 0$, as verified by the top moment diagram in Fig. 4–23b. In some cases it is often *easier* to construct and use a separate series of statically equiva-

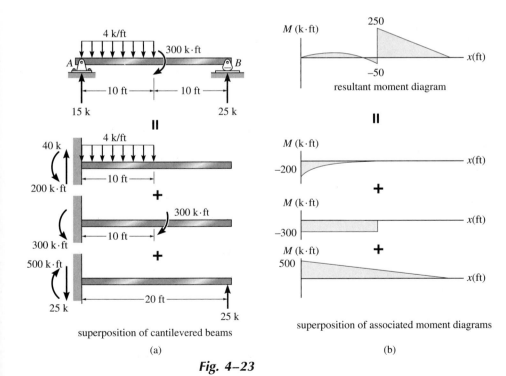

superposition of cantilevered beams

(a)

superposition of associated moment diagrams

(b)

Fig. 4–23

185

lent moment diagrams for a beam, *rather than* construct the beam's more complicated "resultant" moment diagram.

In a similar manner, we can also simplify construction of the "resultant" moment diagram for a beam by using a superposition of "simply supported" beams. For example, the loading on the beam shown at the top of Fig. 4–24a is equivalent to the beam loadings shown below it. Consequently, the separate moment diagrams for each of these three beams can be used *rather than* drawing the resultant moment diagram shown in Fig. 4–24b.

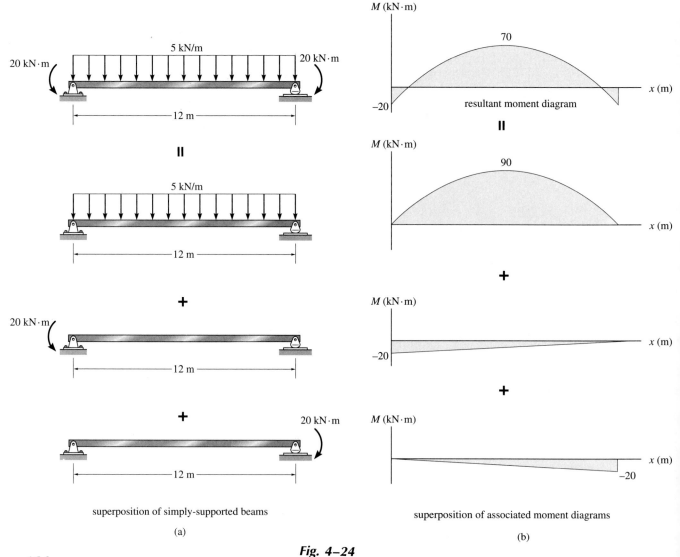

superposition of simply-supported beams

(a)

superposition of associated moment diagrams

(b)

Fig. 4–24

Example 4–18

Draw the moment diagrams for the beam shown in Fig. 4–25a using the method of superposition. Consider the beam to be cantilevered from the support at A. The beam reactions are given in Fig. 4–25a.

Solution

The superimposed cantilevered beams are shown in Fig. 4–25a together with their associated moment diagrams, Fig. 4–25b. (As an aid to their construction, refer to Fig. 4–22.) Although *not needed here*, the sum of these diagrams will yield the resultant moment diagram for the beam. For practice, try drawing this diagram and check the results.

Fig. 4–25

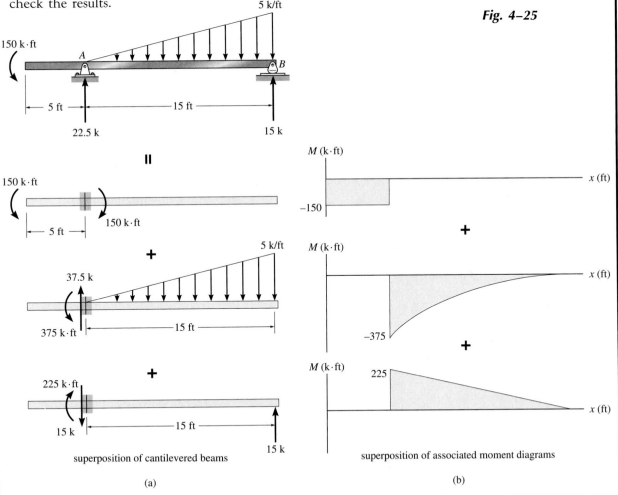

superposition of cantilevered beams

(a)

superposition of associated moment diagrams

(b)

187

PROBLEMS

4–1.   Determine the internal shear, axial load, and bending moment in the beam at points C and D. Assume the support at B is a roller. Point C is located just to the right of the 8-k load.

4–2.   Draw the shear and moment diagrams for the beam in Prob. 4–1.

4–5.   Determine the internal shear, axial load, and bending moment in the beam at points C and D. Assume the support at A is a roller and B is pinned.

4–6.   Draw the shear and moment diagrams for the beam in Prob. 4–5.

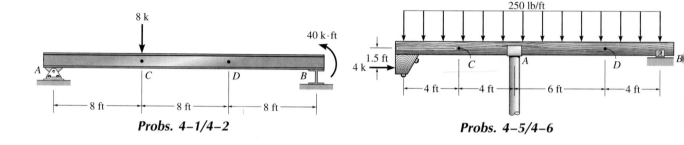

Probs. **4–1/4–2**

Probs. **4–5/4–6**

4–3.   Determine the internal shear, axial load, and bending moment in the beam at point B.

***4–4.**   Draw the shear and moment diagrams for the beam in Prob. 4–3.

4–7.   Determine the internal shear, axial load, and bending moment in the beam at points C and D. Assume A is a pin and B is a roller.

***4–8.**   Draw the shear and moment diagrams for the beam in Prob. 4–7.

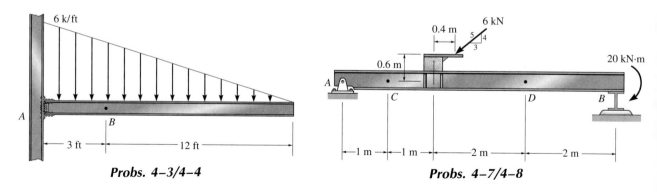

Probs. **4–3/4–4**

Probs. **4–7/4–8**

4–9. Determine the internal shear, axial load, and bending moment at point C.

4–10. Draw the shear and moment diagrams for the beam in Prob. 4–9.

4–13. Determine the internal shear, axial load, and bending moment at point C. Assume the support at A is a pin and B is a roller.

4–14. Draw the shear and moment diagrams of the beam in Prob. 4–13.

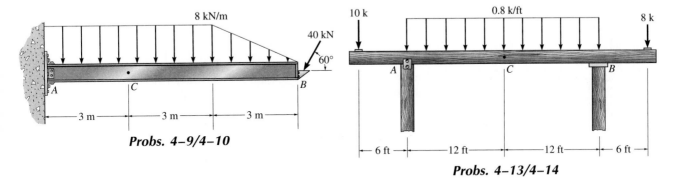

Probs. 4–9/4–10

Probs. 4–13/4–14

4–11 Determine the internal shear, axial load, and bending moment in the beam at points D and E. Point E is just to the right of the 4-k load. Assume A is a roller support, the splice at B is a pin, and C is a fixed support.

***4–12.** Draw the shear and moment diagrams for the beam in Prob. 4–11.

4–15. Determine the internal shear, axial load, and bending moment at (a) point C, which is just to the right of the roller at A, and (b) point D, which is just to the left of the 3000-lb concentrated force. Assume the support at B is a pin.

***4–16.** Draw the shear and moment diagrams for the beam in Prob. 4–15.

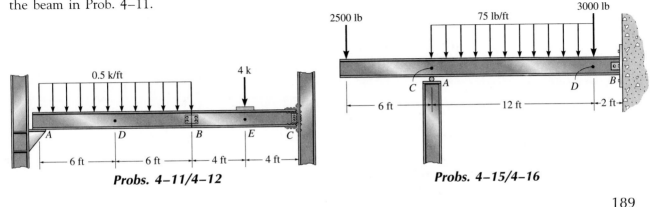

Probs. 4–11/4–12

Probs. 4–15/4–16

4–17. Determine the shear and moment in the tapered beam as a function of x.

4–18. Draw the shear and moment diagrams for the beam in Prob. 4–17.

4–21. Determine the shear and moment in the beam as a function of x. Assume the support at B is a roller.

4–22. Draw the shear and moment diagrams for the beam in Prob. 4–21.

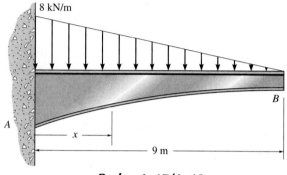

Probs. 4–17/4–18

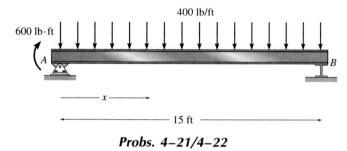

Probs. 4–21/4–22

4–19. Determine the shear and moment in the beam as a function of x.

*4–20. Draw the shear and moment diagrams for the beam in Prob. 4–19.

4–23. Determine the shear and moment in the beam as a function of x.

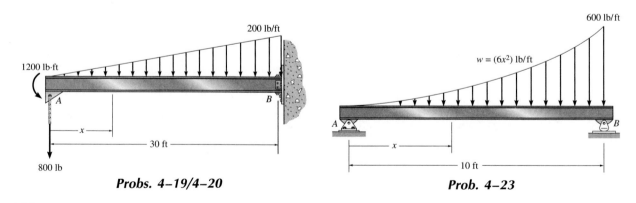

Probs. 4–19/4–20

Prob. 4–23

*4–24. Determine the shear and moment in the floor girder as a function of x, where 4 ft $< x <$ 8 ft. Assume the support at A is a roller and B is a pin. The floor boards are simply supported on the joists at C, D, E, F, and G.

4–25. Draw the shear and moment diagrams for the floor girder in Prob. 4–24.

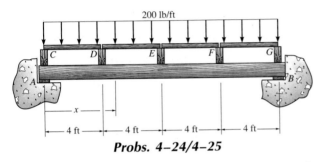

Probs. 4–24/4–25

4–26. The concrete girder supports the two column loads. If the soil pressure under the girder is assumed to be uniform, determine its intensity w and the placement d of the column at B. Draw the shear and moment diagrams for the girder.

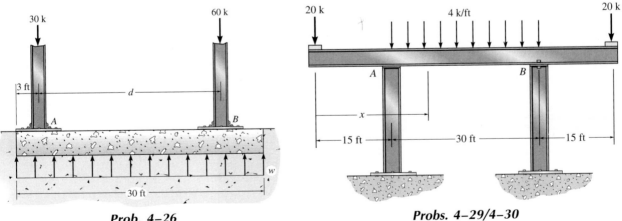

Prob. 4–26

4–27. Determine the shear and moment in the beam as a function of x, where 2 m $< x <$ 4 m.

*4–28. Draw the shear and moment diagrams for the beam in Prob. 4–27.

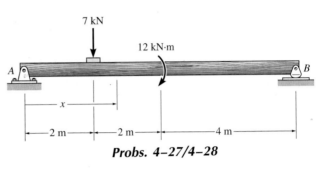

Probs. 4–27/4–28

4–29. Determine the shear and moment in the beam as a function of x, where 15 ft $< x <$ 45 ft. Assume the supports at A is a roller and B is a pin.

4–30. Draw the shear and moment diagrams for the beam in Prob. 4–29.

Probs. 4–29/4–30

4–31. Draw the shear and moment diagrams of the beam. Assume the support at A is a roller and B is a pin.

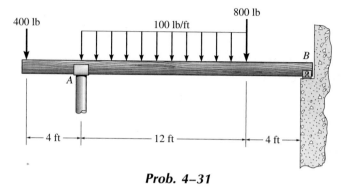

Prob. 4–31

*4–32. Draw the shear and moment diagrams for the beam. Assume the support at B is a pin.

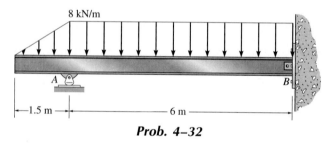

Prob. 4–32

4–33. Draw the shear and moment diagrams for the beam. There is a short link at C. Assume the support at A is fixed and B is a roller.

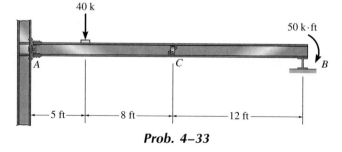

Prob. 4–33

4–34. Draw the shear and moment diagrams for the beam.

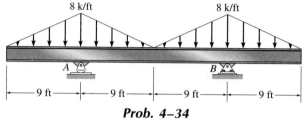

Prob. 4–34

4–35. Draw the shear and moment diagrams for the beam.

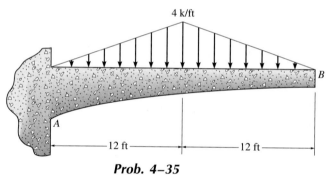

Prob. 4–35

*4–36. Draw the shear and moment diagrams for the beam.

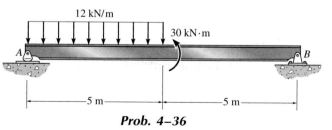

Prob. 4–36

4–37. Draw the shear and moment diagrams for the beam.

4–39. Draw the shear and moment diagrams for *each* of the three members of the frame. Assume the support at A is a pin.

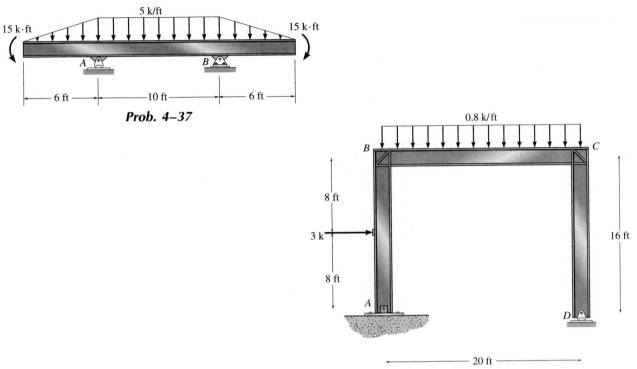

Prob. 4–37

Prob. 4–39

4–38. The boards *ABC* and *BCD* are loosely bolted together as shown. If the bolts exert only vertical reactions on the boards, determine the reactions at the supports and draw the shear and moment diagrams for each board.

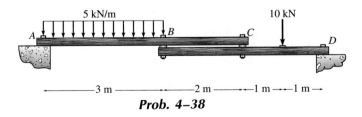

Prob. 4–38

*4–40. Draw the shear and moment diagrams for each of the three members of the frame. Assume the frame is pin-connected at A, C, and D and there is a fixed joint at B.

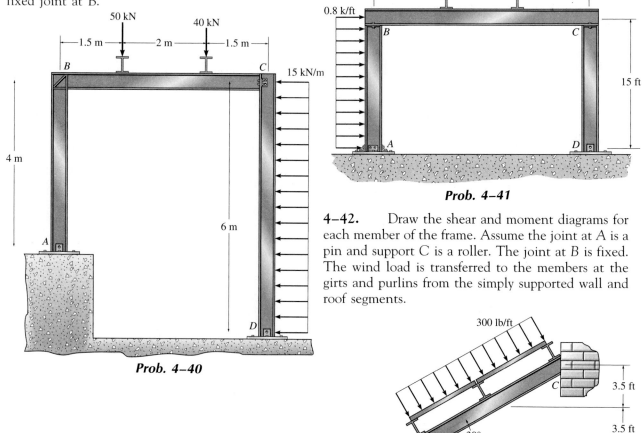

Prob. 4–40

Prob. 4–41

4–42. Draw the shear and moment diagrams for each member of the frame. Assume the joint at A is a pin and support C is a roller. The joint at B is fixed. The wind load is transferred to the members at the girts and purlins from the simply supported wall and roof segments.

Prob. 4–42

4–41. Draw the shear and moment diagrams for each of the three members of the frame. Assume the frame is pin-connected at B, C, and D and A is fixed.

4–43. Draw the shear and moment diagrams for each member of the frame. Assume the frame is pin-connected at A, B, and C.

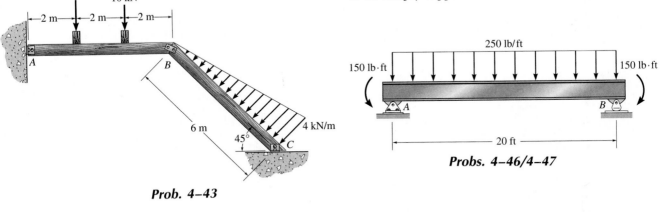

Prob. 4–43

4–46. Draw the moment diagrams for the beam using the method of superposition. Consider the beam to be cantilevered from end A.

4–47. Draw the moment diagrams for the beam using the method of superposition. Consider the beam to be simply supported at A and B as shown.

Probs. 4–46/4–47

*4–44. Draw the moment diagrams for the beam using the method of superposition. Consider the beam to be simply supported. Assume A is a pin and B is a roller.

4–45. Solve Prob. 4–44 by considering the beam to be cantilevered from the support at A.

*4–48. Draw the moment diagrams for the beam using the method of superposition. Consider the beam to be cantilevered from the pin support at A.

4–49. Draw the moment diagrams for the beam using the method of superposition. Consider the beam to be cantilevered from the rocker at B.

4–50. Draw the moment diagrams for beam using the method of superposition. Consider the beam to be cantilevered from end C.

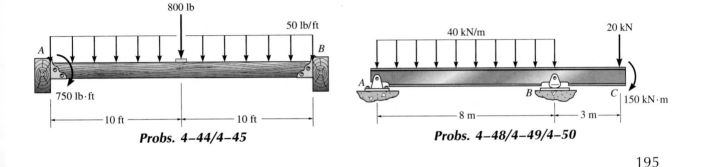

Probs. 4–44/4–45

Probs. 4–48/4–49/4–50

An example of a two-ribbed, open-spandrel fixed-arched bridge having a main span of 1,000 ft.
(Courtesy of Bethlehem Steel Corporation)

5

Cables and Arches

Cables and arches often form the main load-carrying element in many types of structures, and in this chapter we will discuss some of the important aspects related to their structural analysis. The chapter begins with a general discussion of cables, followed by an analysis of cables subjected to a concentrated load and to a uniform distributed load. Since most arches are statically indeterminate, only the special case of a three-hinged arch will be considered. The analysis of this structure will provide some insight regarding the fundamental behavior of all arched structures.

5.1 Cables

Cables are often used in engineering structures for support and to transmit loads from one member to another. When used to support suspension roofs, bridges, and trolley wheels, cables form the main load-carrying element in the structure. In the force analysis of such systems, the weight of the cable itself may be neglected; however, when cables are used as guys for radio antennas, electrical transmission lines, and derricks, the cable weight may become important and must be included in the structural analysis. Two cases will be considered in the sections that follow: a cable subjected to concentrated loads, and a cable subjected to a distributed load. Regardless of which loading conditions are present, provided the loading is coplanar with the cable, the requirements for equilibrium are formulated in an identical manner.

When deriving the necessary relations between the force in the cable and its slope, we will make the assumption that the cable is *perfectly flexible* and *inextensible*. Due to its flexibility, the cable offers no resistance to shear or bending and, therefore, the force acting in the cable is always tangent to the cable at points along its length. Being inextensible, the cable has a constant length both before and after the load is applied. As a result, once the load is applied, the geometry of the cable remains fixed, and the cable or a segment of it can be treated as a rigid body.

5.2   Cable Subjected to Concentrated Loads

When a cable of negligible weight supports several concentrated loads, the cable takes the form of several straight-line segments, each of which is subjected to a constant tensile force. Consider, for example, the cable shown in Fig. 5–1. Here θ specifies the angle of the cable's *cord*, and $\mathscr{L}$ is the cable's span. If the distances L_1, L_2, and L_3, and the loads $\mathbf{P}_1$ and $\mathbf{P}_2$ are known, then the problem is to determine the *nine unknowns consisting of the tension in each of the three* segments, the *four* components of reaction at A and B, and the sags y_C and y_D at the *two* points C and D. For the solution we can write *two* equations of force equilibrium at each of points A, B, C, and D. This results in a total of *eight equations*. To complete the solution, it will be necessary to know something about the geometry of the cable in order to obtain the necessary ninth equation. For example, if the cable's total *length* $\mathscr{L}$ is specified, then the Pythagorean theorem can be used to relate each of the three segmental lengths, written in terms of θ, y_C, y_D, L_1, L_2, and L_3, to the total length $\mathscr{L}$. Unfortunately, this type of problem cannot

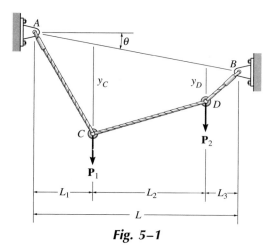

Fig. 5–1

be solved easily by hand. Another possibility, however, is to specify one of the sags, either y_C or y_D, instead of the cable length. By doing this, the equilibrium equations are then sufficient for obtaining the unknown forces and the remaining sag. Once the sag at each point of loading is obtained, the length of the cable can then be determined by trigonometry.

When performing an equilibrium analysis for a problem of this type, the forces in the cable can also be obtained by writing the equations of equilibrium for the entire cable or any portion thereof. The following example numerically illustrates these concepts.

Example 5–1

Determine the tension in each segment of the cable shown in Fig. 5–2a. Also, what is the dimension h?

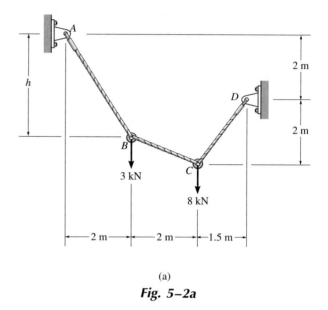

(a)

Fig. 5–2a

Solution

By inspection, there are four unknown external reactions (A_x, A_y, D_x, and D_y) and three unknown cable tensions, one in each cable segment. These seven unknowns along with the distance h can be determined from the eight available equilibrium equations ($\Sigma F_x = 0$, $\Sigma F_y = 0$) applied to points A through D.

A more direct approach to the solution is to recognize that the slope of cable CD is specified, and so a free-body diagram of the entire cable is shown in Fig. 5–2b. We can obtain the tension in segment CD as follows:

$$T_{CD}(\cos 45°)(2\ m) + T_{CD}(\sin 45°)(6\ m) - 3\ kN(2\ m) - 8\ kN(4\ m) = 0$$
$$T_{CD} = 6.72\ kN \qquad \textbf{Ans.}$$

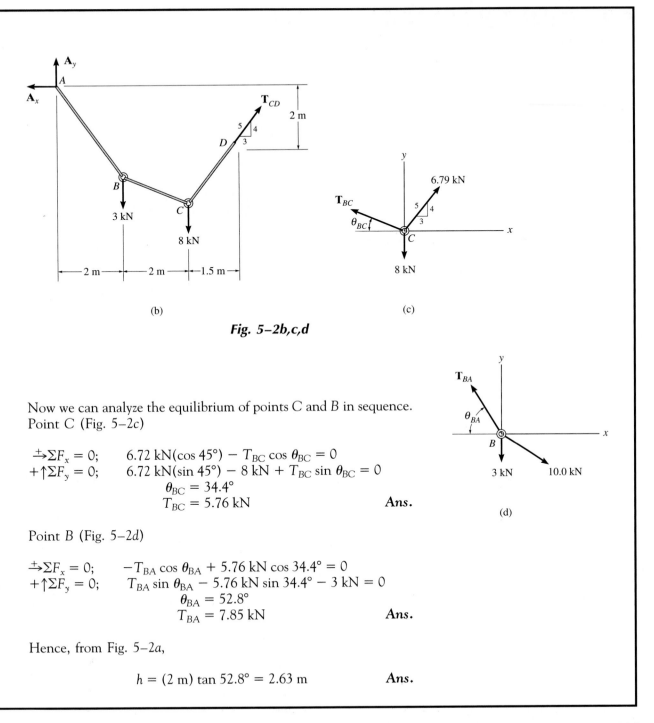

(b)   (c)

Fig. 5–2b,c,d

Now we can analyze the equilibrium of points C and B in sequence.
Point C (Fig. 5–2c)

$\xrightarrow{+}\Sigma F_x = 0;$ $6.72 \text{ kN}(\cos 45°) - T_{BC} \cos \theta_{BC} = 0$
$+\uparrow\Sigma F_y = 0;$ $6.72 \text{ kN}(\sin 45°) - 8 \text{ kN} + T_{BC} \sin \theta_{BC} = 0$
$\theta_{BC} = 34.4°$
$T_{BC} = 5.76 \text{ kN}$ **Ans.**

Point B (Fig. 5–2d)

$\xrightarrow{+}\Sigma F_x = 0;$ $-T_{BA} \cos \theta_{BA} + 5.76 \text{ kN} \cos 34.4° = 0$
$+\uparrow\Sigma F_y = 0;$ $T_{BA} \sin \theta_{BA} - 5.76 \text{ kN} \sin 34.4° - 3 \text{ kN} = 0$
$\theta_{BA} = 52.8°$
$T_{BA} = 7.85 \text{ kN}$ **Ans.**

Hence, from Fig. 5–2a,

$$h = (2 \text{ m}) \tan 52.8° = 2.63 \text{ m}$$ **Ans.**

5.3 Cable Subjected to a Uniform Distributed Load

Cables provide a very effective means of supporting the dead weight of girders or bridge decks having very long spans. A suspension bridge is a typical example, in which the deck is suspended from the cable using a series of close and equally spaced hangers.

In order to analyze this problem, we will first determine the shape of a cable subjected to a uniform *horizontally* distributed vertical load w_0, Fig. 5–3a. Here the x,y axes have their origin located at the lowest point on the cable, such that the slope is zero at this point. The free-body diagram of a small segment of the cable having a length ds is shown in Fig. 5–3b. Since the tensile force in the cable changes continuously in both magnitude and direction along the cable's length, the change is denoted on the free-body diagram by ΔT. The distributed load is represented by its resultant force $w_0 \Delta x$, which acts at $\Delta x/2$ from point O. Applying the equations of equilibrium yields

$$\xrightarrow{+} \Sigma F_x = 0; \qquad -T \cos \theta + (T + \Delta T) \cos(\theta + \Delta \theta) = 0$$
$$+\uparrow \Sigma F_y = 0; \qquad -T \sin \theta - w_0(\Delta x) + (T + \Delta T) \sin(\theta + \Delta \theta) = 0$$
$$\zeta + \Sigma M_O = 0; \qquad w_0(\Delta x)(\Delta x/2) - T \cos \theta \, \Delta y + T \sin \theta \, \Delta x = 0$$

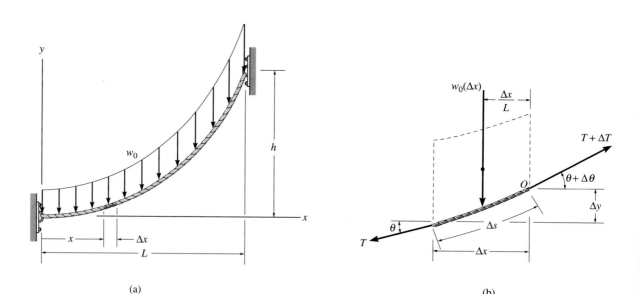

(a)

(b)

Fig. 5–3

Dividing each of these equations by Δx and taking the limit as $\Delta x \rightarrow 0$, and hence $\Delta y \rightarrow 0$, $\Delta \theta \rightarrow 0$, and $\Delta T \rightarrow 0$, we obtain

$$\frac{d(T \cos \theta)}{dx} = 0 \tag{5–1}$$

$$\frac{d(T \sin \theta)}{dx} = w_0 \tag{5–2}$$

$$\frac{dy}{dx} = \tan \theta \tag{5–3}$$

Integrating Eq. 5–1, where $T = F_H$ at $x = 0$, we have:

$$T \cos \theta = F_H \tag{5–4}$$

Hence, the horizontal component of force at *any point* along the cable remains *constant*.

Integrating Eq. 5–2, realizing that $T \sin \theta = 0$ at $x = 0$, gives

$$T \sin \theta = w_0 x \tag{5–5}$$

Dividing Eq. 5–5 by Eq. 5–4 eliminates T. Then using Eq. 5–3, we can obtain the slope at any point,

$$\tan \theta = \frac{dy}{dx} = \frac{w_0 x}{F_H} \tag{5–6}$$

Performing a second integration with $y = 0$ at $x = 0$ yields

$$y = \frac{w_0}{2F_H} x^2 \tag{5–7}$$

This is the equation of a *parabola*. The constant F_H may be obtained by using the boundary condition $y = h$ at $x = L$. Thus,

$$F_H = \frac{w_0 L^2}{2h} \tag{5–8}$$

Finally, substituting into Eq. 5–7 yields

$$y = \frac{h}{L^2} x^2 \tag{5–9}$$

From Eq. 5–4, the maximum tension in the cable occurs when θ is maximum; i.e., at $x = L$. Hence, from Eqs. 5–4 and 5–5,

$$T_{max} = \sqrt{F_H^2 + (w_0L)^2} \qquad (5\text{–}10)$$

Or, using Eq. 5–8, we can express T_{max} in terms of w_0, i.e.,

$$T_{max} = w_0L\sqrt{1 + (L/2h)^2} \qquad (5\text{–}11)$$

Note that we have neglected the weight of the cable, which is *uniform* along the *length* of the cable, and not along its horizontal projection. Actually, a cable subjected to its own weight and free of any other loads will take the form of a *catenary* curve. However, if the sag-to-span ratio is small, which is the case for most structural applications, this curve closely approximates a parabolic shape, as determined here.

From the results of this analysis, it follows that if a cable *maintains* a parabolic shape, the dead load of the deck for a suspension bridge or a suspended girder will be *uniformly distributed* over the horizontal projected length of the cable. Hence, if the girder in Fig. 5–4a is supported by a series of hangers, which are close and uniformly spaced, the load in each hanger must be the *same* so as to ensure that the cable has a parabolic shape. Furthermore, if we assume the girder is *rigid*, and the parabolic slope of the cable is maintained, then any moving load **P** must be equally shared by each hanger.

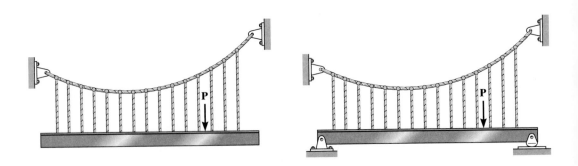

(a) (b)

Fig. 5–4a,b

Using this assumption, we can perform the structural analysis of the girder or any other framework which is freely suspended from the cable. In particular, if the girder is simply supported as well as supported by the cable, the analysis will be statically indeterminate to the first degree, Fig. 5–4b. However, if the girder has an internal pin at some intermediate point along its length, Fig. 5–4c, then this would provide a condition of zero moment, and so a determinate structural analysis of the girder can be performed.

The following examples numerically illustrate these concepts.

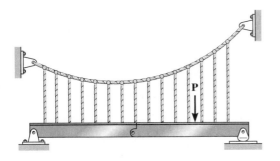

(c)

Fig. 5–4c

Example 5–2

The cable in Fig. 5–5a supports a girder which weighs 850 lb/ft. Determine the tension in the cable at points A, B, and C. Also, what is the force in each of the hangers?

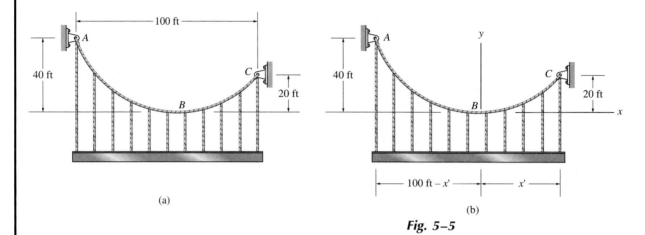

(a)

(b)

Fig. 5–5

Solution [b]

The origin of the coordinate axes is established at point B, the lowest point on the cable, where the slope is zero; Fig. 5–5b. From Eq. 5–7, the parabolic equation for the cable is:

$$y = \frac{w_0}{2F_H}x^2 = \frac{850 \text{ lb/ft}}{2F_H}x^2 = \frac{425}{F_H}x^2 \qquad (1)$$

Assuming point C is located x' from B, we have

$$20 = \frac{425}{F_H}x'^2$$
$$F_H = 21.25x'^2 \qquad (2)$$

Also, for point A,

$$40 = \frac{425}{F_H}[-(100 - x')]^2$$
$$40 = \frac{425}{21.25x'^2}[-(100 - x')]^2$$
$$x'^2 + 200x' - 10,000 = 0$$
$$x' = 41.42 \text{ ft}$$

Thus, from Eqs. 2 and 1 (or Eq. 5–6) we have

$$F_H = 21.25(41.42)^2 = 36{,}459.2 \text{ lb}$$

$$\frac{dy}{dx} = \frac{850}{36{,}459.2} x = 0.02331x \qquad\qquad (3)$$

At point A,

$$x = -(100 - 41.42) = -58.58 \text{ ft}$$

$$\tan \theta_A = \left.\frac{dy}{dx}\right|_{x=-58.58} = 0.02331(-58.58) = -1.366$$

$$\theta_A = -53.79°$$

Using Eq. 5–4,

$$T_A = \frac{F_H}{\cos \theta} = \frac{36{,}459.2}{\cos(-53.79°)} = 61.7 \text{ k} \qquad\qquad \textbf{Ans.}$$

At point B, $x = 0$,

$$\tan \theta_B = \left.\frac{dy}{dx}\right|_{x=0} = 0, \qquad \theta_B = 0$$

$$T_B = \frac{F_H}{\cos \theta_B} = \frac{36{,}459.2}{\cos 0} = 36.5 \text{ k} \qquad\qquad \textbf{Ans.}$$

At point C,

$$x = 41.42 \text{ ft}$$

$$\tan \theta_C = \left.\frac{dy}{dx}\right|_{x=41.42} = 0.02331(41.42) = 0.9657$$

$$\theta_C = 44.0°$$

$$T_C = \frac{F_H}{\cos \theta_C} = \frac{36{,}459.2}{\cos 44.0°} = 50.7 \text{ k} \qquad\qquad \textbf{Ans.}$$

Since the girder is supported by 11 hangers, the force in each hanger is

$$F_h = \frac{850 \text{ lb/ft } (100 \text{ ft})}{11} = 7.73 \text{ k} \qquad\qquad \textbf{Ans.}$$

Example 5–3

The suspension bridge in Fig. 5–6a is constructed using the two stiffening trusses that are pin-connected at their ends C and supported by a pin at A and the rocker at B. Determine the maximum tension in the cable IH and the force in each hanger. The cable has a parabolic shape and the bridge is subjected to the single load of 50 kN.

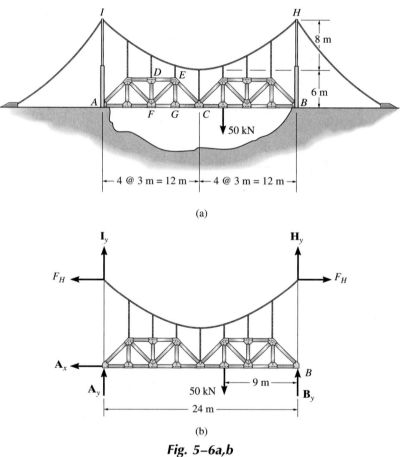

(a)

(b)

Fig. 5–6a,b

Solution

The free-body diagram of the cable-truss system is shown in Fig. 5–6b. According to Eq. 5–4, the horizontal component of cable tension at I and H must be constant, F_H. Taking moments about B, we have

$\zeta + \Sigma M_B = 0; \qquad -I_y(24\text{ m}) - A_y(24\text{ m}) + 50\text{ kN}(9\text{ m}) = 0$
$$I_y + A_y = 18.75$$

If only half the suspended structure is considered, Fig. 5–6c, then summing moments about the pin at C, we have

$\zeta + \Sigma M_C = 0; \qquad F_H(14\text{ m}) - F_H(6\text{ m}) - I_y(12\text{ m}) - A_y(12\text{ m}) = 0$
$$I_y + A_y = 0.667F_H$$

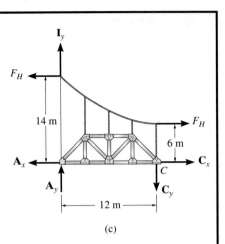

(c)

$$18.75 = 0.667F_H$$
$$F_H = 28.125\text{ kN}$$

To obtain the maximum tension in the cable, it is first necessary to determine w_0 from Eq. 5–8:

$$w_0 = \frac{2F_H h}{L^2} = \frac{2(28.125\text{ kN})(6\text{ m})}{(12\text{ m})^2} = 2.344\text{ kN/m}$$

Thus, using Eq. 5–11, we have

$$\begin{aligned} T_{max} &= w_0 L\sqrt{1 + (L/2h)^2} \\ &= 2.344(12\text{ m})\sqrt{1 + (12\text{ m}/2(6\text{ m}))^2} \\ &= 39.8\text{ kN} \end{aligned}$$

Ans.

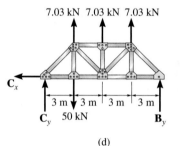

(d)

Fig. 5–6c,d

Each hanger supports the load w_0 over an effective length of 3 m. Hence the load per hanger is

$$F_h = (2.344\text{ kN/m})(3\text{ m}) = 7.03\text{ kN} \qquad \textbf{Ans.}$$

Note: To determine the member forces in the trusses, it is necessary first to determine the support reactions at A, B, and C. For example, the free-body diagram of truss CB is shown in Fig. 5–6d. Thus,

$$7.03\text{ kN}(3\text{ m} + 6\text{ m} + 9\text{ m}) - 50\text{ kN}(3\text{ m}) + B_y(12\text{ m}) = 0$$
$$B_y = 1.95\text{ kN}$$
$\xrightarrow{+} \Sigma F_x = 0; \qquad C_x = 0$
$+\uparrow \Sigma F_y = 0; \qquad 3(7.03\text{ kN}) - 50\text{ kN} + 1.95\text{ kN} + C_y = 0$
$$C_y = 27.0\text{ kN}$$

Finally, we can now use the method of joints or the method of sections to determine the member forces in the usual manner.

5.4 Arches

Like cables, arches can be used to reduce the bending moments in long-span structures. Essentially, an arch acts as an inverted cable, so it receives its load mainly in compression although, because of its rigidity, it must also resist some bending and shear depending upon how it is loaded and shaped. In particular, if the arch has a *parabolic shape* and it is subjected to a *uniform* horizontally distributed vertical load, then from the analysis of cables it follows that only compressive forces will be resisted by the arch.

A typical arch is shown in Fig. 5–7, which specifies some of the nomenclature used to define its geometry. Depending upon the application, several types of arches can be selected to support a loading. A *fixed arch*, Fig. 5–8a, is often made from reinforced concrete. Although it may require less material to construct than other types of arches, it must have solid foundation abutments since it is indeterminate to the third degree and, consequently, additional stresses can be introduced into the arch due to relative settlement of its supports. A *two-hinged arch*, Fig. 5–8b, is commonly made from metal or timber. It is indeterminate to the first degree, and although it is not as rigid as a fixed arch, it is somewhat insensitive to settlement. We could make this structure statically determinate by replacing one of the hinges with a roller. Doing so, however, would remove the capacity of the structure to resist bending along its span, and as a result it would serve as a curved beam, and *not* as an arch. A *three-hinged arch*, Fig. 5–8c, which is also made from metal or timber, is statically determinate. Unlike statically indeterminate arches, it is not affected by settlement or temperature changes. Finally, if two- and three-hinged arches are to be constructed without the need for larger foundation abutments and if clearance is not a problem, then the supports can be connected with a tie rod, Fig. 5–8d. A *tied arch* allows the structure to behave as a rigid unit, since the tie rod carries the horizontal component of thrust at the supports. It is also unaffected by relative settlement of the supports.

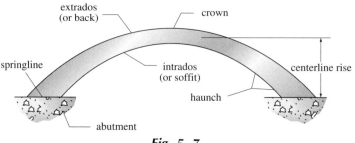

Fig. 5–7

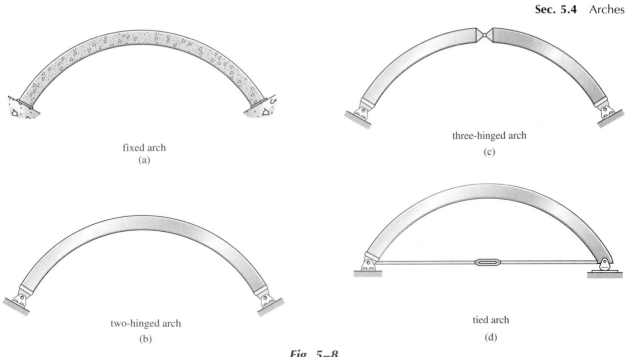

fixed arch
(a)

two-hinged arch
(b)

three-hinged arch
(c)

tied arch
(d)

Fig. 5–8

A trussed through-arch bridge having a main span of 1,344 ft. *(Courtesy of Bethlehem Steel Corporation)*

5.5 **Three-Hinged Arch**

To provide some insight as to how arches transmit loads, we will now consider the analysis of a three-hinged arch such as the one shown in Fig. 5–9a. In this case, the third hinge is located at the crown and the supports are located at different elevations. In order to determine the reactions at the supports, the arch is disassembled and the free-body diagram of each member is shown in Fig. 5–9b. Here there are six unknowns for which six equations of equilibrium are available. One method of solving this problem is to apply moment equations about points A and B. Simultaneous solution will yield the reactions C_x and C_y. The support reactions are then determined from the force equations of equilibrium. Once obtained, the internal axial, shear, and moment loadings at any point along the arch can be found using the method of sections. Here, of course, the section should be taken perpendicular to the axis of the arch at the point considered. For example, the free-body diagram for segment AD is shown in Fig. 5–9c.

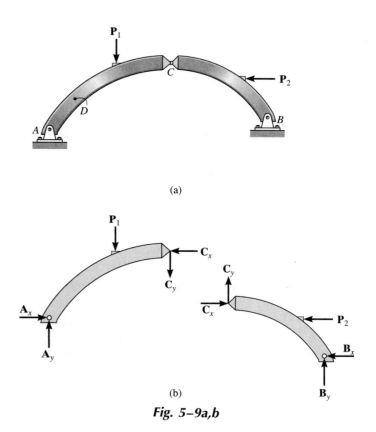

(a)

(b)

Fig. 5–9a,b

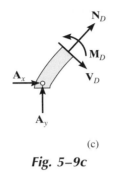

(c)

Fig. 5–9c

Three-hinged arches can also take the form of two pin-connected trusses, each of which would replace the arch ribs *AC* and *CB* in Fig. 5–9a. The analysis of this form follows the same procedure outlined above. The following examples numerically illustrate these concepts.

Typical three-hinged open spandrel arch bridge

213

Example 5–4

The three-hinged open spandrel arch bridge shown in Fig. 5–10a has a parabolic shape and supports the uniform load. Show that the parabolic arch is subjected only to axial compression at an intermediate point D along its axis. Assume the load is uniformly transmitted to the arch ribs.

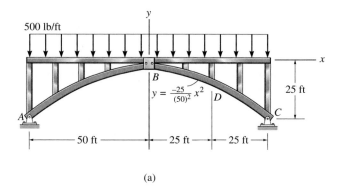

(a)

Solution

Since the supports are at the same elevation, the free-body diagrams of the entire arch and each part are shown in Fig. 5–10b. Applying the equations of equilibrium, we have:

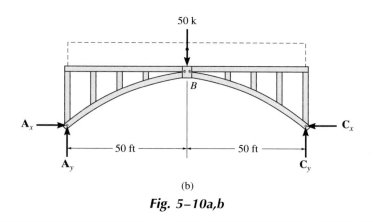

(b)

Fig. 5–10a,b

Entire arch

$$\zeta+\Sigma M_A = 0; \qquad C_y(100 \text{ ft}) - 50 \text{ k}(50 \text{ ft}) = 0$$
$$C_y = 25 \text{ k}$$

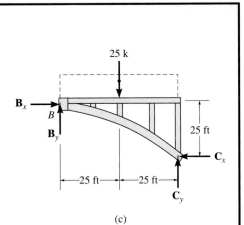

(c)

Arch segment

$$\zeta+\Sigma M_B = 0; \qquad -25 \text{ k}(25 \text{ ft}) + 25 \text{ k}(50 \text{ ft}) - C_x(25 \text{ ft}) = 0$$
$$C_x = 25 \text{ k}$$
$$\xrightarrow{+}\Sigma F_x = 0; \qquad B_x = 25 \text{ k}$$
$$+\uparrow\Sigma F_y = 0; \qquad B_y - 25 \text{ k} + 25 \text{ k} = 0$$
$$B_y = 0$$

A section of the arch taken at point D, $x = 25$ ft, $y = -25(25)^2/(50)^2 = -6.25$ ft, is shown in Fig. 5–10c. The slope of the segment at D is

$$\tan \theta = \frac{dy}{dx} = \frac{-50}{(50)^2}x \bigg|_{x=25 \text{ ft}} = -0.5$$
$$\theta = -26.6°$$

Applying the equations of equilibrium, Fig. 5–10d, we have

$$\xrightarrow{+}\Sigma F_x = 0; \qquad 25 \text{ k} - N_D \cos 26.6° + V_D \sin 26.6° = 0$$
$$+\uparrow\Sigma F_y = 0; \qquad -12.5 \text{ k} + N_D \sin 26.6° - V_D \cos 26.6° = 0$$
$$\zeta+\Sigma M_D = 0; \qquad M_D + 12.5 \text{ k}(12.5 \text{ ft}) - 25 \text{ k}(6.25 \text{ ft}) = 0$$

$$N_D = 28.0 \text{ k} \qquad \textit{Ans.}$$
$$V_D = 0 \qquad \textit{Ans.}$$
$$M_D = 0 \qquad \textit{Ans.}$$

Note: If the arch had a different shape or if the load was nonuniform, then the internal shear and moment would be nonzero. Also, if a simply supported beam was used to support the load, it would have to resist a bending moment of $M = 469$ k·ft. By comparison, it is more efficient to structurally resist a load in direct compression (although one must consider the possibility of buckling) than to resist the load by a bending moment.

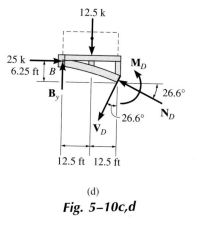

(d)

Fig. 5–10c,d

Example 5–5

The three-hinged tied arch is subjected to the loading shown in Fig. 5–11a. Determine the force in members CH and CB. The dashed member GF of the truss is intended to carry no force.

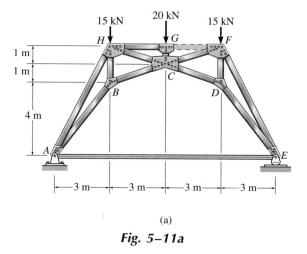

(a)

Fig. 5–11a

Solution

The support reactions can be obtained from a free-body diagram of the entire arch, Fig. 5–11b:

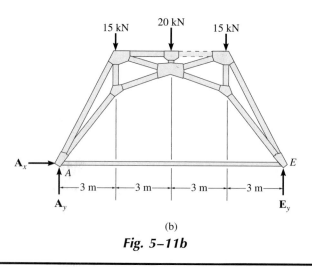

(b)

Fig. 5–11b

$\zeta+\Sigma M_A = 0;$
$$E_y(12 \text{ m}) - 15 \text{ kN}(3 \text{ m}) - 20 \text{ kN}(6 \text{ m}) - 15 \text{ kN}(9 \text{ m}) = 0$$
$$E_y = 25 \text{ kN}$$

$\xrightarrow{+}\Sigma F_x = 0; \qquad A_x = 0$
$+\uparrow\Sigma F_y = 0; \qquad A_y - 15 \text{ kN} - 20 \text{ kN} - 15 \text{ kN} + 25 \text{ kN} = 0$
$$A_y = 25 \text{ kN}$$

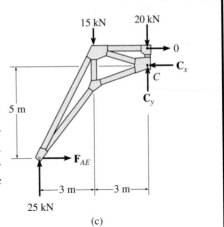

(c)

The force components acting at joint C can be determined by considering the free-body diagram of the left part of the arch, Fig. 5–11c. (These components represent the resultant effect of members CF and CD acting at C.) First, we determine the force in the tie rod:

$\zeta+\Sigma M_C = 0; \qquad F_{AE}(5 \text{ m}) - 25 \text{ kN}(6 \text{ m}) + 15 \text{ kN}(3 \text{ m}) = 0$
$$F_{AE} = 21.0 \text{ kN}$$

Then,

$\xrightarrow{+}\Sigma F_x = 0; \qquad -C_x + 21.0 \text{ kN} = 0; \qquad C_x = 21.0 \text{ kN}$
$+\uparrow\Sigma F_y = 0; \qquad 25 \text{ kN} - 15 \text{ kN} - 20 \text{ kN} + C_y = 0$
$$C_y = 10 \text{ kN}$$

To obtain the forces in CH and CB, we can use the method of joints as follows:

Joint G; Fig. 5–11d,

$+\uparrow\Sigma F_y = 0; \qquad F_{GC} - 20 \text{ kN} = 0$
$$F_{GC} = 20 \text{ kN(C)}$$

(d)

Joint C; Fig. 5–11e,

$\xrightarrow{+}\Sigma F_x = 0; \qquad F_{CB}\left(\dfrac{3}{\sqrt{10}}\right) - 21.0 \text{ kN} - F_{CH}\left(\dfrac{3}{\sqrt{10}}\right) = 0$

$+\uparrow\Sigma F_y = 0; \qquad F_{CB}\left(\dfrac{1}{\sqrt{10}}\right) + F_{CH}\left(\dfrac{1}{\sqrt{10}}\right) - 20 \text{ kN} + 10 \text{ kN} = 0$

(e)

Fig. 5–11c,d,e

Thus,

$$F_{CB} = 26.9 \text{ kN(C)} \qquad\qquad \textbf{Ans.}$$
$$F_{CH} = 4.74 \text{ kN(T)} \qquad\qquad \textbf{Ans.}$$

217

PROBLEMS

5–1. Determine the horizontal and vertical components of reaction at A, B, and C of the three-hinged arch. Assume A, B, and C are pin-connected.

5–3. The bridge is constructed as a three-hinged trussed arch. Determine the horizontal and vertical components of reaction at the hinges (pins) at A, B, and C. The dashed member DE is intended to carry *no force.*

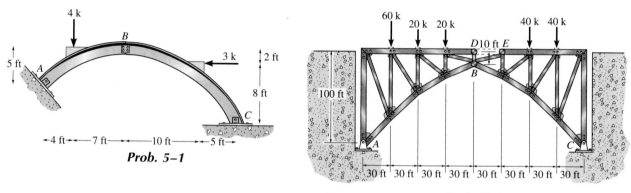

Prob. 5–1

Prob. 5–3

5–2. Determine the resultant forces at the pins A, B, and C of the three-hinged arched roof truss.

*5–4. The three-hinged spandrel arch is subjected to the uniform load of 20kN/m. Determine the internal moment in the arch at point D.

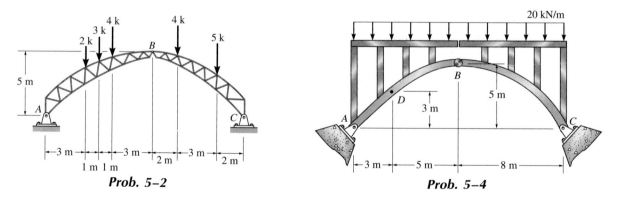

Prob. 5–2

Prob. 5–4

5–5. The tied three-hinged truss arch is subjected to the loading shown. Determine the components of reaction at the supports A and B, and the tension in the tie rod.

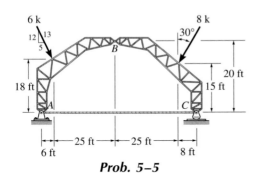

Prob. 5–5

5–6. The laminated-wood three-hinged arch is subjected to the loading shown. Determine the horizontal and vertical components of reactions at the pins A, B, and C, and draw the moment diagram for member AB.

5–7. Determine the tension in each segment of the cable and the cable's total length.

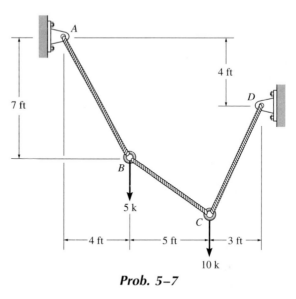

Prob. 5–7

*5–8. Cable ABCD supports the loading shown. Determine the maximum tension in the cable and the sag of point B.

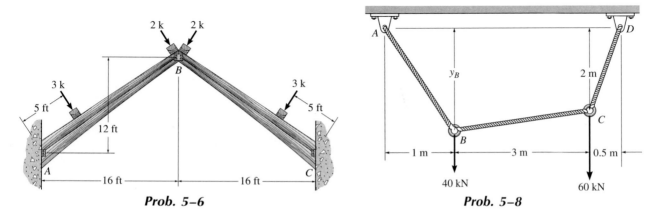

Prob. 5–6

Prob. 5–8

5–9. Determine the force **P** needed to hold the cable in the position shown, i.e., so segment BC remains horizontal. Also, compute the sag y_B and the maximum tension in the cable.

5–11. The cable will break when the maximum tension reaches $T_{max} = 12$ kN. Determine the uniform distributed load w required to develop this maximum tension.

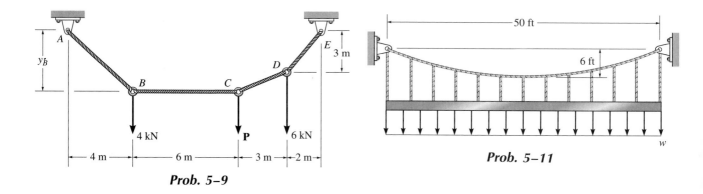

Prob. 5–9

Prob. 5–11

*5–12. The beams AB and BC are supported by the cable that has a parabolic shape. Determine the tension in the cable at points D, F, and E, and the force in each of the equally spaced hangers.

5–13. Draw the shear and moment diagrams for beams AB and BC in Prob. 5–12.

5–10. Determine the maximum uniform loading w lb/ft that the cable can support if it is capable of sustaining a maximum tension of 3000 lb before it will break.

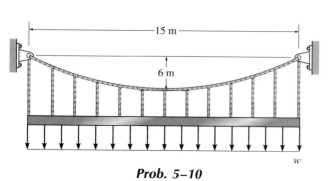

Prob. 5–10

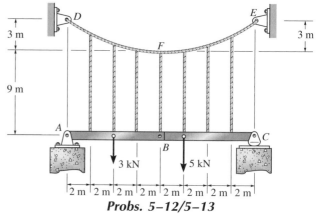

Probs. 5–12/5–13

5–14. Determine the maximum and minimum tension in the parabolic cable and the force in each of the hangers. The girder is subjected to the uniform load and is pin-connected at *B*.

5–15. Draw the shear and moment diagrams for girders *AB* and *BC* in Prob. 5–14.

*5–16. The trusses are pin-connected and suspended from the parabolic cable. Determine the force in members *KJ* and *KG* when the structure is subjected to the loading shown.

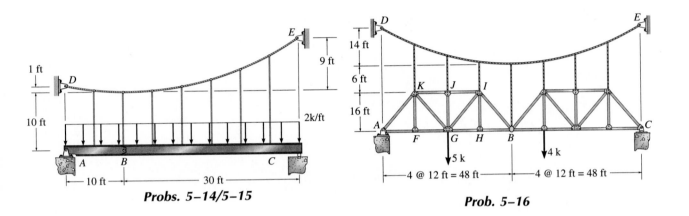

Probs. 5–14/5–15

Prob. 5–16

Moving loads caused by traffic must be considered when designing girders for the bridges shown here. The influence line for a girder thereby becomes an important part of the structural analysis. *(Photo courtesy Bethlehem Steel Corp.)*

6

Influence Lines for Statically Determinate Structures

Influence lines have important application for the design of structures that resist large live loads. In this chapter we will discuss how to draw the influence line for a statically determinate structure. The theory is applied to structures subjected to a distributed load or a series of concentrated forces, and specific applications to floor girders and bridge trusses are given. The determination of the absolute maximum live shear and moment in a member is discussed at the end of the chapter.

6.1 Influence Lines

In the previous chapters we developed techniques for analyzing the forces in structural members due to *dead* or *fixed loads*. It was shown that the *shear* and *moment diagrams* represent the most descriptive methods for displaying the variation of these loads in a member. If a structure is subjected to a *live* or *moving load*, however, the variation of the shear and bending moment in the member is best described using the *influence line*. An influence line represents the variation of either the reaction, shear, moment, or deflection at a *specific point* in a member as a concentrated force moves over the member. Once this line is constructed, one can tell at a glance where a live load should be placed on the structure so that it creates the greatest influence at the specified point. Furthermore, the magnitude of the associated reaction, shear, moment, or deflection at the point can then be calculated from the ordinates of the influence-line diagram. For these reasons, influence lines play an important part in the design of bridges, industrial crane rails, conveyors, and other structures where loads move across their span.

Procedure for Analysis

Either of the following procedures can be used to construct the influence line at a specific point P in a member for any function (reaction, shear, or moment). For both of these procedures we will choose the moving force to have a *dimensionless magnitude of unity*.*

Tabulate Values. Place a unit load at various locations, x, along the member, and at *each* location use statics to determine the value of the function (reaction, shear, or moment) at the specified point P. For example, if the influence line for a vertical force *reaction* at point P on a beam is to be constructed, consider the reaction to be *positive* at P when it acts *upward* on the beam. If a shear or moment influence line is to be drawn for point P, take positive shear and moment at P as acting in the conventional sense as that used for drawing shear and moment diagrams. See Fig. 4–1. All statically determinate beams will have influence lines that consist of straight line segments. After some practice one should be able to minimize computations and locate the unit load *only* at points representing the *end points* of each line segment. In order to avoid errors, it is recommended that one first construct a table, listing "unit load at x" versus the corresponding value of the function calculated at the specific point P, that is, "reaction R," "shear V" or "moment M." Once the load has been placed at various points along the span of the member, the tabulated values can be plotted and the influence-line segments constructed.

Influence-Line Equations. The influence line can also be constructed by placing the unit load at a *variable* position x on the member and then computing the value of R, V, or M at point P as a function of x. In this manner, the equations of the various line segments composing the influence line can be determined and plotted.

*The reason for this choice will be explained in Sec. 6.2.

Although the procedure for constructing an influence line is rather basic, one should clearly be aware of the *difference* between constructing an influence line and constructing a shear or moment diagram. Influence lines represent the effect of a *moving load* only at a *specified point* on a member, whereas shear and moment diagrams represent the effect of fixed loads at *all points* along the axis of the member.

The following examples illustrate the methods for constructing influence lines numerically.

Influence lines must be drawn for these steel and concrete bridge girders in order to design them properly.

Example 6–1

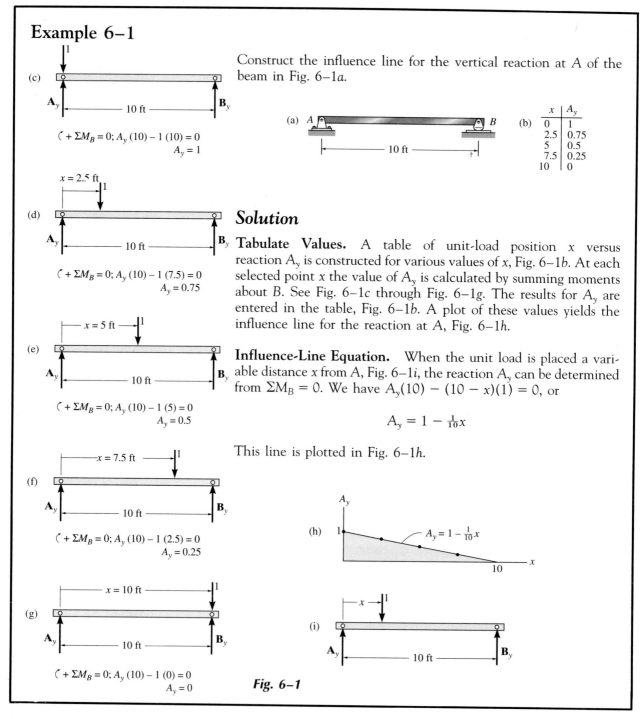

Construct the influence line for the vertical reaction at A of the beam in Fig. 6–1a.

(c)

$(+ \Sigma M_B = 0; A_y (10) - 1 (10) = 0$
$A_y = 1$

(a)

(b)

x	A_y
0	1
2.5	0.75
5	0.5
7.5	0.25
10	0

$x = 2.5$ ft

(d)

$(+ \Sigma M_B = 0; A_y (10) - 1 (7.5) = 0$
$A_y = 0.75$

Solution

Tabulate Values. A table of unit-load position x versus reaction A_y is constructed for various values of x, Fig. 6–1b. At each selected point x the value of A_y is calculated by summing moments about B. See Fig. 6–1c through Fig. 6–1g. The results for A_y are entered in the table, Fig. 6–1b. A plot of these values yields the influence line for the reaction at A, Fig. 6–1h.

$x = 5$ ft

(e)

$(+ \Sigma M_B = 0; A_y (10) - 1 (5) = 0$
$A_y = 0.5$

Influence-Line Equation. When the unit load is placed a variable distance x from A, Fig. 6–1i, the reaction A_y can be determined from $\Sigma M_B = 0$. We have $A_y(10) - (10 - x)(1) = 0$, or

$$A_y = 1 - \tfrac{1}{10}x$$

This line is plotted in Fig. 6–1h.

$x = 7.5$ ft

(f)

$(+ \Sigma M_B = 0; A_y (10) - 1 (2.5) = 0$
$A_y = 0.25$

(h)

$A_y = 1 - \tfrac{1}{10}x$

$x = 10$ ft

(g)

$(+ \Sigma M_B = 0; A_y (10) - 1 (0) = 0$
$A_y = 0$

(i)

Fig. 6–1

Example 6–2

Construct the influence line for the vertical reaction at B of the beam in Fig. 6–2a.

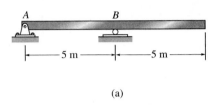

(a)

Solution

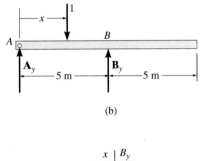

(b)

Tabulate Values.   Using statics, verify that the values for the reaction B_y listed in the table, Fig. 6–2c, are correctly computed for each position x of the unit load, Fig. 6–2b. A plot of the values in Fig. 6–2c yields the influence line in Fig. 6–2d.

Influence-Line Equation.   Using the equation $\Sigma M_A = 0$ in Fig. 6–2b, we can write

$$B_y = \tfrac{1}{5}x$$

which is plotted in Fig. 6–2d.

x	B_y
0	0
2.5	0.5
5	1
7.5	1.5
10	2

(c)

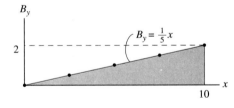

(d)

Fig. 6–2

227

Example 6–3

Construct the influence line for the shear at point C of the beam in Fig. 6–3*a*.

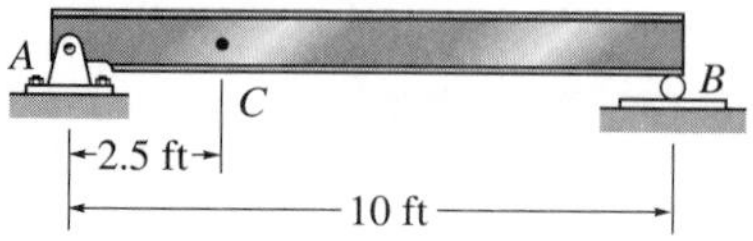

(a)

(b)

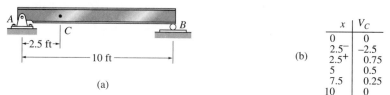

x	V_C
0	0
2.5^-	−2.5
2.5^+	0.75
5	0.5
7.5	0.25
10	0

Fig. 6–3a,b

Solution

Tabulate Values. A table of unit-load position x versus shear V_C at point C is constructed, Fig. 6–3*b*, and at each selected position x of the unit load the method of sections is used to calculate the value of V_C as shown on the free-body diagrams in Fig. 6–3*c* through 6–3*h*. Note in particular that the unit load must be placed just to the left ($x = 2.5^-$) and just to the right ($x = 2.5^+$) of point C since the shear is discontinuous at C. A plot of the values in Fig. 6–3*b* yields the influence line for the shear at C, Fig. 6–3*i*.

Influence-Line Equations. Here two equations have to be determined since there are two segments for the influence line. By the method of sections, Fig. 6–3*j*, we have

$$V_C = -\tfrac{1}{10}x \qquad 0 \le x < 2.5\text{ ft}$$
$$V_C = 1 - \tfrac{1}{10}x \qquad 2.5\text{ ft} < x \le 10\text{ ft}$$

These equations are plotted in Fig. 6–3*i*.

(j)

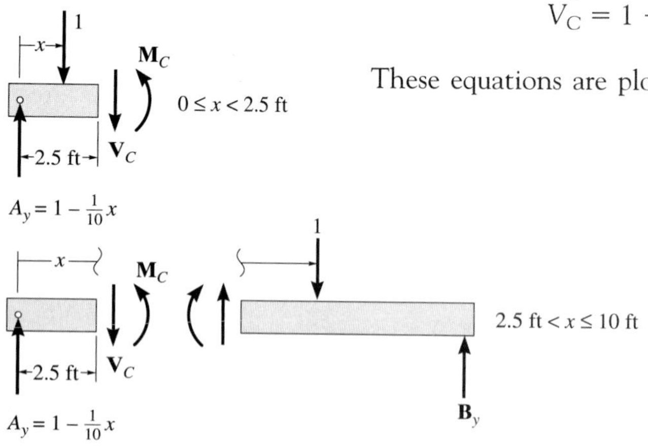

(i)

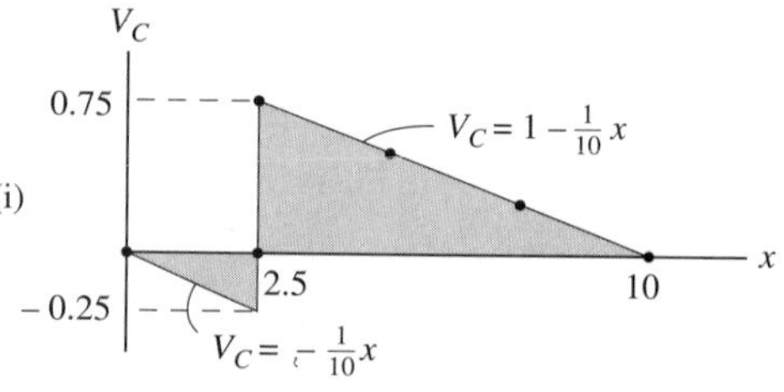

Fig. 6–3i,j

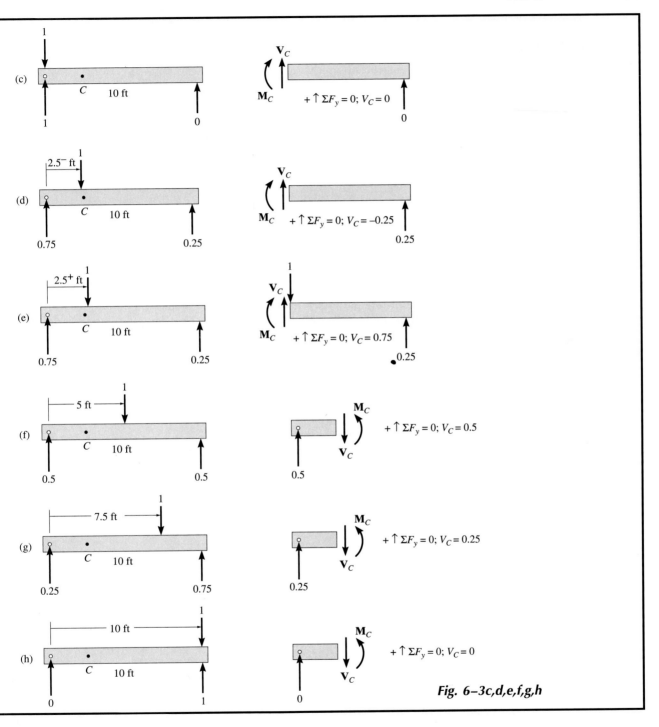

Fig. 6–3c,d,e,f,g,h

Example 6–4

Construct the influence line for the shear at point C of the beam in Fig. 6–4a.

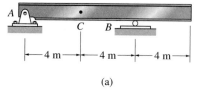

(a)

Solution

Tabulate Values.   Using statics and the method of sections, verify that the values of the shear V_C at point C in Fig. 6–4c correspond to each position x of the unit load on the beam, Fig. 6–4b. A plot of the values in Fig. 6–4c yields the influence line in Fig. 6–4d.

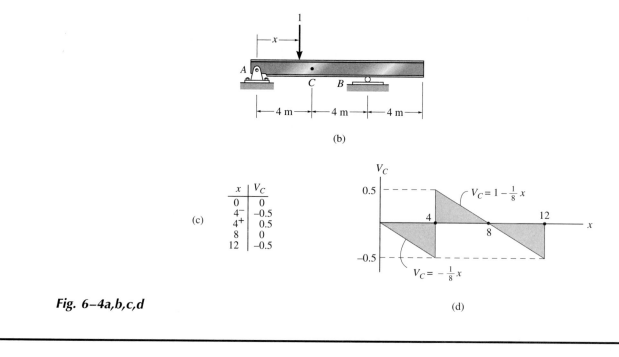

(b)

(c)

x	V_C
0	0
4⁻	−0.5
4⁺	0.5
8	0
12	−0.5

Fig. 6–4a,b,c,d

(d)

230

Influence-Line Equations. From Fig. 6–4e, verify that

$$V_C = -\tfrac{1}{8}x \qquad\qquad 0 \le x < 4 \text{ m}$$
$$V_C = 1 - \tfrac{1}{8}x \qquad 4 \text{ m} < x \le 12 \text{ m}$$

These equations are plotted in Fig. 6–4d.

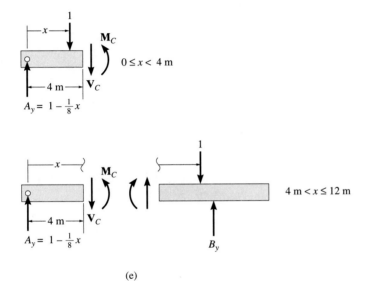

(e)

Fig. 6–4e

Example 6–5

Construct the influence line for the moment at point C of the beam in Fig. 6–5a.

(a)

Fig. 6–5a,b

(b)

x	M_C
0	0
2.5	1.25
5	2.5
7.5	1.25
10	0

Solution

Tabulate Values. A table of unit-load position x versus moment M_C at point C is constructed in Fig. 6–5b. At each selected position the value of M_C is calculated using the method of sections as shown on the free-body diagrams in Fig. 6–5c through 6–5g. A plot of the values in Fig. 6–5b yields the influence line for the moment at C, Fig. 6–5h.

Influence-Line Equations. The two line segments for the influence line can be determined using $\Sigma M_C = 0$ along with the method of sections shown in Fig. 6–5i. We have

$$M_C = \tfrac{1}{2}x \qquad\qquad 0 \le x < 5 \text{ ft}$$
$$M_C = 5 - \tfrac{1}{2}x \qquad 5 \text{ ft} < x \le 10 \text{ ft}$$

These equations when plotted yield the influence line shown in Fig. 6–5h.

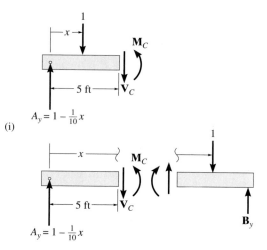

(i)

Fig. 6–5i

$A_y = 1 - \tfrac{1}{10}x$

232

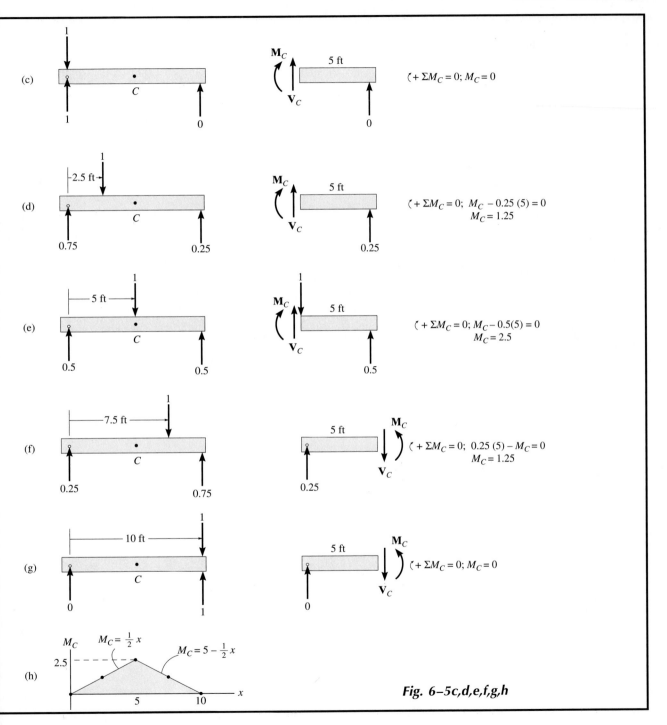

(c) $(+ \Sigma M_C = 0;\ M_C = 0$

(d) $(+ \Sigma M_C = 0;\ M_C - 0.25\ (5) = 0$
$M_C = 1.25$

(e) $(+ \Sigma M_C = 0;\ M_C - 0.5(5) = 0$
$M_C = 2.5$

(f) $(+ \Sigma M_C = 0;\ 0.25\ (5) - M_C = 0$
$M_C = 1.25$

(g) $(+ \Sigma M_C = 0;\ M_C = 0$

(h) $M_C = \frac{1}{2} x$ $M_C = 5 - \frac{1}{2} x$

Fig. 6–5c,d,e,f,g,h

Example 6–6

Construct the influence line for the moment at point C of the beam in Fig. 6–6a.

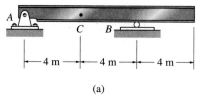

(a)

Solution

Tabulate Values. Using statics and the method of sections, verify that the values of the moment M_C at point C in Fig. 6–6c correspond to each position x of the unit load, Fig. 6–6b. A plot of the values in Fig. 6–6c yields the influence line in Fig. 6–6d.

Influence-Line Equations. From Fig. 6–6e verify that

$$M_C = \tfrac{1}{2}x \qquad\qquad 0 \le x < 4\text{ m}$$
$$M_C = 4 - \tfrac{1}{2}x \qquad 4\text{ m} < x \le 12\text{ m}$$

These equations are plotted in Fig. 6–6d.

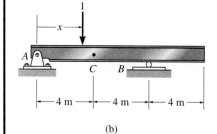

(b)

x	M_C
0	0
4	2
8	0
12	−2

(c)

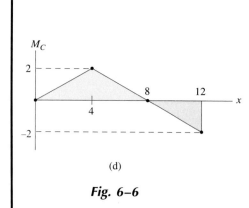

(d)

Fig. 6–6

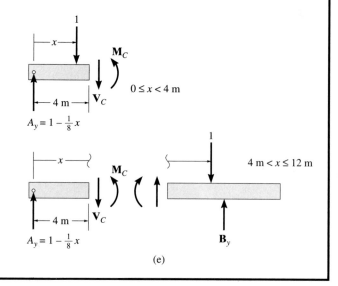

(e)

6.2 Influence Lines for Beams

Since beams (or girders) often form the main load-carrying elements of a floor system or bridge deck, it is important to be able to construct the influence lines for the reactions, shear, or moment at any specified point in a beam.

Loadings

Once the influence line for a function (reaction, shear, or moment) has been constructed, it will then be possible to *locate* the live loads on the beam which will produce the maximum value of the function. In this regard, two types of loadings will now be considered.

Concentrated Force. Since the numerical values of a function for an influence line are determined using a dimensionless unit load, then for any concentrated force **F** acting on the beam at any position x, *the value of the function can be found by multiplying the ordinate of the influence line at the position x by the magnitude of* **F.** For example, the influence line for the reaction at A on the beam AB, Fig. 6–7a, is given in Fig. 6–7b. As shown, when the unit load is at $x = \frac{1}{2}l$, the reaction at A is $A_y = \frac{1}{2}$. Hence, if the force F lb is at this same point, Fig. 6–7a, the reaction is $A_y = (\frac{1}{2})(F)$ lb. Of course, this same value can also be determined by statics. Obviously, the *maximum influence* caused by **F** occurs when it is placed on the beam at the same location as the *peak* of the influence line—in this case at $x = 0$, where the reaction would be $(1)(F)$ lb.

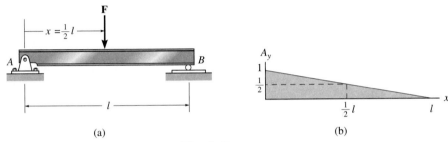

(a)   (b)

Fig. 6–7

Uniform Load.   Consider a portion of a beam subjected to a uniform load w, Fig. 6–8a. As shown, each dx segment of the load w creates a concentrated force of $dF = w\, dx$ on the beam. If $d\mathbf{F}$ is located at x, where the beam's influence-line ordinate for some function (reaction, shear, moment) is y, Fig. 6–8b, then the value of the function is $(dF)(y) = (w\, dx)y$. The effect of all the concentrated forces $d\mathbf{F}$ is determined by integrating over the entire length of the beam, that is, $\int wy\, dx = w\int y\, dx$, since w is constant. Also, since $\int y\, dx$ is equivalent to the *area* under the influence line, Fig. 6–8b, then, in general, *the value of a function caused by a uniform distributed load is simply the area under the influence line for the function multiplied by the intensity of the uniform load.* For example, in the case of a uniformly loaded beam shown in Fig. 6–9a, the reaction $\mathbf{A}_y$ can be determined from the influence line, Fig. 6–9b, as $A_y = (\text{area})(w) = [\frac{1}{2}(1)(l)]w = \frac{1}{2}wl$. This value can of course also be determined from statics, Fig. 6–9c.

The following examples illustrate these concepts numerically.

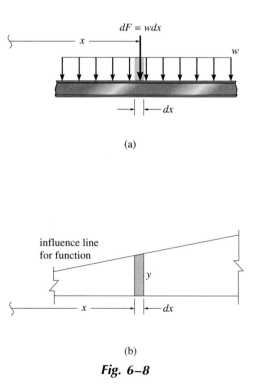

(a)

influence line
for function

(b)

Fig. 6–8

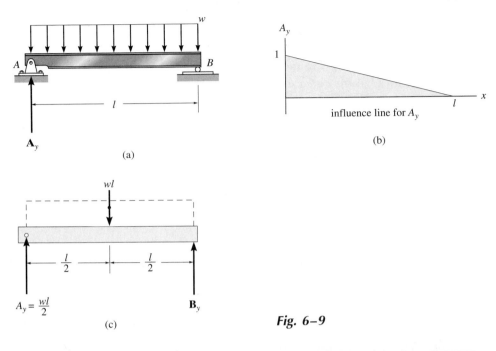

(a)

influence line for A_y

(b)

(c)

Fig. 6–9

The girders of this highway bridge must be designed to support unusually heavy loads such as this truck. Influence lines must be used for this purpose. *(Joseph Schuyler/Stock, Boston)*

Example 6–7

Determine the maximum *positive* live shear that can be developed at point C in the beam shown in Fig. 6–10a due to a concentrated live load of 4000 lb and a uniform live load of 2000 lb/ft.

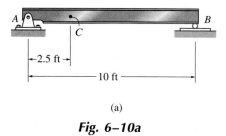

(a)

Fig. 6–10a

Solution

The influence line for the shear at C has been established in Example 6–3, and is shown in Fig. 6–10b.

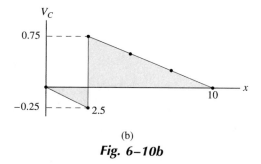

(b)

Fig. 6–10b

Concentrated Force. The maximum positive shear at C will occur when the 4-k live force is placed at $x = 2.5^+$ ft, since this is the location of the positive peak of the influence line. The ordinate of this peak is $+0.75$; thus

$$V_C = 0.75(4000 \text{ lb}) = 3000 \text{ lb}$$

Uniform Load. The uniform live load creates the maximum positive influence for V_C when the load acts on the beam between $x = 2.5^+$ ft and $x = 10$ ft, since within this region the influence line has a positive area. The magnitude of V_C due to this loading is

$$V_C = [\tfrac{1}{2}(10 \text{ ft} - 2.5 \text{ ft})(0.75)]2000 \text{ lb/ft} = 5625 \text{ lb}$$

Total Maximum Shear at C

$$(V_C)_{max} = 3000 \text{ lb} + 5625 \text{ lb} = 8625 \text{ lb} \qquad \textbf{Ans.}$$

Notice that once the *positions* of the loads have been established using the influence line, Fig. 6–10c, the value of $(V_C)_{max}$ can *also* be determined using statics and the method of sections. Show that this is the case.

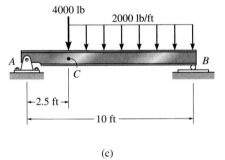

(c)

Fig. 6–10c

239

Example 6–8

Determine the maximum *positive* moment that can be developed at point C on the beam shown in Fig. 6–11a due to a single concentrated live load of 8000 N, a uniform live load of 3000 N/m, and a beam weight (dead load) of 1000 N/m.

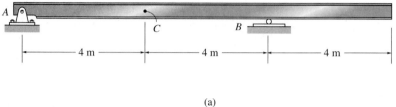

(a)

Fig. 6–11a

Solution

The influence line for the moment at C has been established in Example 6–6 and is shown in Fig. 6–11b.

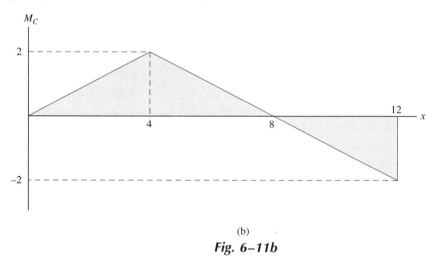

(b)

Fig. 6–11b

Concentrated Force.   The 8000-N live force must be placed at $x = 4$ m for maximum positive moment M_C. Why? Hence,

$$M_C = 2(8000) = 16 \text{ kN} \cdot \text{m}$$

Uniform Live Load. The uniform 3000-N/m live load must be placed on the beam from $x = 0$ to $x = 8$ m for maximum positive moment M_C. Why? Hence,

$$M_C = [\tfrac{1}{2}(8)(2)](3000) = 24 \text{ kN} \cdot \text{m}$$

Uniform Dead Load. It is *required* that the 1000-N/m dead load act over the *entire length* of the beam. We can find the moment at C due to this load using the method of sections and statics, or for convenience we can use the influence line. In this case the *total area* of the influence line must be used. Since the area from $x = 8$ to $x = 12$ is negative, we have

$$M_C = [\tfrac{1}{2}(8)(2) - \tfrac{1}{2}(12 - 8)(2)]1000 = 4 \text{ kN} \cdot \text{m}$$

Total Maximum Moment at C

$$(M_C)_{\text{max}} = 16 + 24 + 4 = 44 \text{ kN} \cdot \text{m} \qquad \textbf{Ans.}$$

   Knowing the *position* of the live loads using the influence line, Fig. 6–11c, show that $(M_C)_{\text{max}} = 44$ kN $\cdot$ m can also be found using statics and the method of sections.

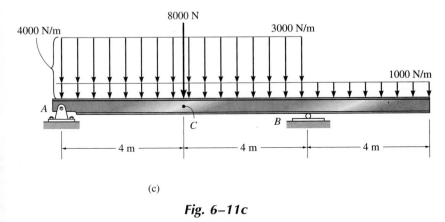

(c)

Fig. 6–11c

6.3   Qualitative Influence Lines

In 1886, Heinrich Müller-Breslau developed a technique for rapidly constructing the shape of an influence line. Referred to as the *Müller-Breslau principle, it states that the influence line for a function (reaction, shear, or moment) is to the same scale as the deflected shape of the beam when the beam is acted upon by the function.* In order to draw the deflected shape properly, the capacity of the beam to resist the applied function must be *removed* so the beam can deflect when the function is applied. For example, consider the beam in Fig. 6–12a. If the shape of the influence line for the *vertical reaction* at A is to be determined, the pin is first replaced by a *roller guide* as shown in Fig. 6–12b. A roller guide is necessary since the beam must still resist a horizontal force at A but *no vertical force.* When the positive (upward) force A_y is then applied at A, the beam deflects to the dashed position,* which represents the general shape of the influence line for A_y, Fig. 6–12c. (Numerical values for this specific case have been calculated in Example 6–1.) If the shape of the influence line for the *shear* at C is to be determined, Fig. 6–13a, the connection at C may be symbolized by a *roller guide* as shown in Fig. 6–13b. This device will resist a moment

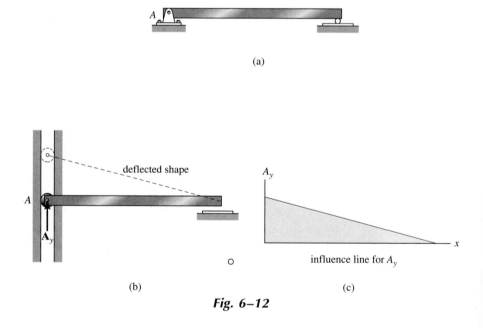

(a)

(b)

(c)

Fig. 6–12

---

*Throughout the discussion all deflected positions are drawn to an exaggerated scale to illustrate the concept.

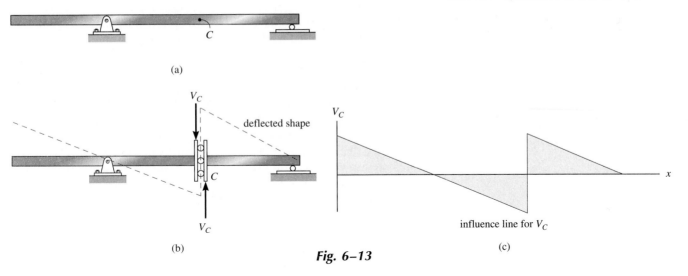

(a)

(b)

(c)

Fig. 6–13

and axial force but no *shear*.† Applying a positive shear force $\mathbf{V}_C$ to the beam at C and allowing the beam to deflect to the dashed position, we find the influence-line shape as shown in Fig. 6–13c. Finally, if the shape of the influence line for the *moment* at C, Fig. 6–14a, is to be determined, an internal *hinge* or *pin* is placed at C, since this connection resists axial and shear forces but *cannot resist a moment*, Fig. 6–14b. Applying positive moments $\mathbf{M}_C$ to the beam, the beam then deflects to the dashed position, which is the shape of the influence line, Fig. 6–14c.

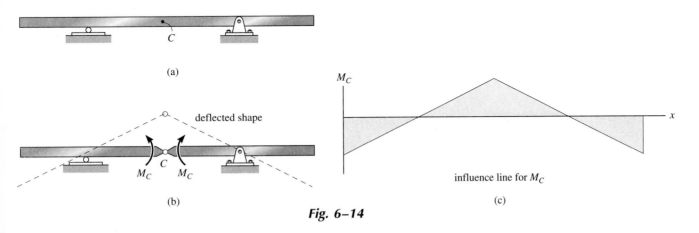

(a)

(b)

(c)

Fig. 6–14

† Here the rollers *symbolize* supports that carry loads both in tension or compression. See Table 2–1, support (2).

243

The proof of the Müller-Breslau principle can be established using the principle of virtual work. Recall that *work is the product of either a linear displacement and force in the direction of linear displacement or a rotational displacement and moment in the direction of rotational displacement*. If a rigid body (beam) is in equilibrium, the resultant force and moment on it are equal to zero. Consequently, if the body is given an *imaginary* or *virtual displacement*, the work done by *all* the forces and couple moments acting on it must also be equal to zero. Consider, for example, the simply-supported beam shown in Fig. 6–15a, which is subjected to the unit load placed at an arbitrary point along its length. If the beam is given a virtual (or imaginary) displacement δy at the support A, Fig. 6–15b, then only the support reaction $\mathbf{A}_y$ and the unit load do virtual work. Specifically, A_y does positive work $A_y\ \delta y$ and the unit load does negative work, $-1\ \delta y'$. (The support at B does not move and therefore the force at B does no work.) Since the beam is in equilibrium and therefore does not actually move, the virtual work sums to zero, i.e.,

$$A_y\ \delta y - 1\ \delta y' = 0$$

Notice that if δy is set equal to 1, then

$$A_y = \delta y'$$

In other words, the numerical value of the reaction at A is equivalent to the displacement at the position of the unit load so that the *shape* of the influence line for the reaction at A has been established (see Example 6–1). This proves the Müller-Breslau principle for reactions.

In the same manner, if the beam is sectioned at C, and the beam undergoes a virtual displacement δy at this point, Fig. 6–15c, such that

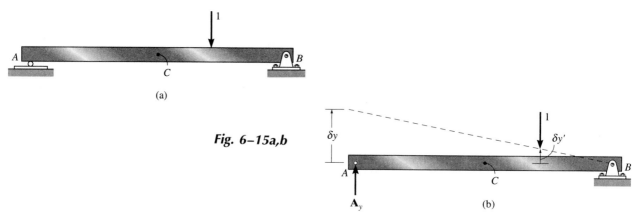

(a)

Fig. 6–15a,b

(b)

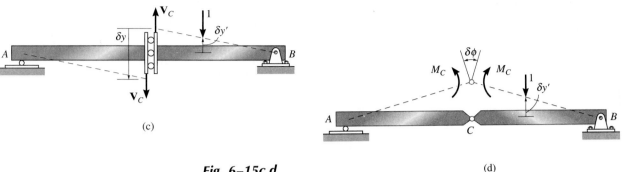

Fig. 6–15c,d

(c)

(d)

segments AC and BC remain *parallel*, then only the internal shear at C and the unit load do work. Thus, the virtual work equation is

$$V_C \, \delta y - 1 \, \delta y' = 0$$

Again, if δy is set equal to 1, then

$$V_C = \delta y'$$

and the *shape* of the influence line for the shear at C has been established (see Example 6–3).

Lastly, assume a hinge or pin is introduced into the beam at point C, Fig. 6–15d. If a virtual rotation $\delta\phi$ is introduced at the pin, virtual work will be done only by the internal moment $\mathbf{M}_C$ and the unit load. So

$$M_C \, \delta\phi - 1 \, \delta y' = 0$$

Setting $\delta\phi$ equal to 1, it is seen that

$$M_C = \delta y'$$

which indicates that the deflected beam has the same *shape* as the influence line for the internal moment at point C (see Example 6–5).

Obviously, the Müller-Breslau principle provides a quick method for establishing the *shape* of the influence line. Once this is known, the ordinates at the peaks can be determined by using the basic method discussed in Sec. 6.1. Also, by simply knowing the general shape of the influence line, it is possible to *locate* the live load on the beam and then determine the maximum value of the function by *using statics*. Example 6–12 illustrates this technique.

245

Example 6–9

For each beam in Fig. 6–16a through 6–16c, sketch the influence line for the vertical reaction at A.

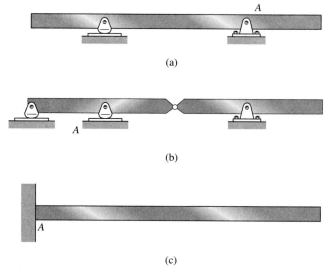

Solution

Fig. 6–16a. The support is replaced by a roller guide at A and the force A_y is applied as shown in Fig. 6–16d.

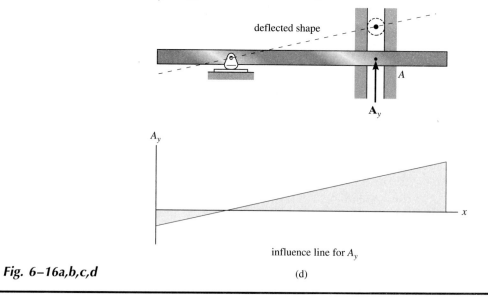

Fig. 6–16a,b,c,d

Fig. 6–16b.   Again, a roller guide is placed at A and the force **A**$_y$ is applied as shown in Fig. 6–16e.

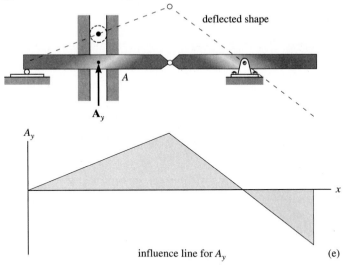

influence line for A_y (e)

Fig. 6–16c.   A *double-roller guide* must be used at A in this case, since this type of support will then transmit both a moment **M**$_A$ at the fixed support and axial load **A**$_x$, but will not transmit **A**$_y$, Fig. 6–16f.

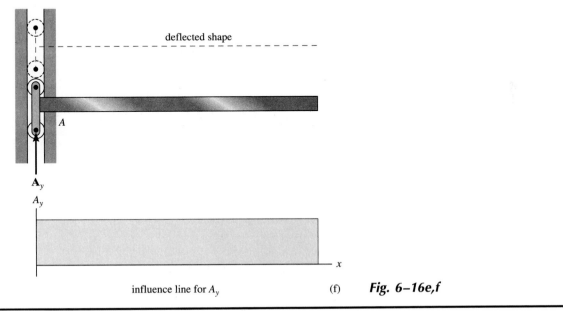

influence line for A_y (f)   ***Fig. 6–16e,f***

247

Example 6–10

For each beam in Fig. 6–17a through 6–17c, sketch the influence line for the shear at B.

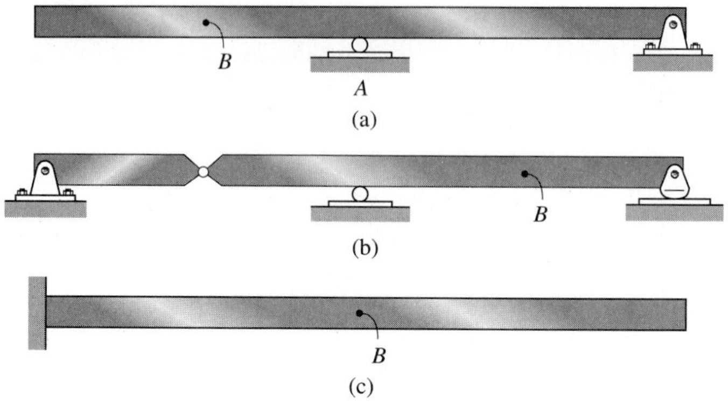

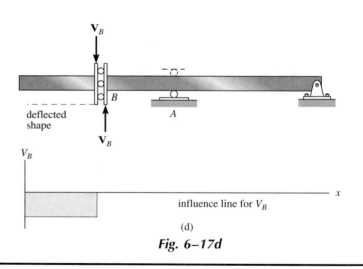

Fig. 6–17a,b,c

Solution

Fig. 6–17a. The roller guide is introduced at B and the positive shear V_B is applied as shown in Fig. 6–17d. Notice that the right segment of the beam will *not deflect* since the roller at A in Fig. 6–17a actually constrains the beam from moving vertically, either up or down. [See support (2) in Table 2–1.]

influence line for V_B

(d)

Fig. 6–17d

Fig. 6–17b.   Placing the roller guide at B and applying the positive shear at B yields the deflected shape and corresponding influence line as shown in Fig. 6–17e.

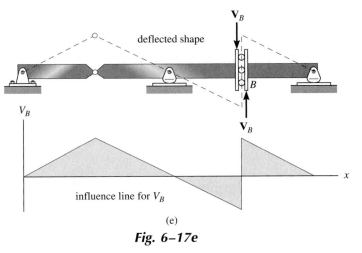

(e)

Fig. 6–17e

Fig. 6–17c.   Again, the roller guide is placed at B, the positive shear is applied, and the deflected shape and corresponding influence line are shown in Fig. 6–17f. Note that the left segment of the beam does not deflect, due to the fixed support.

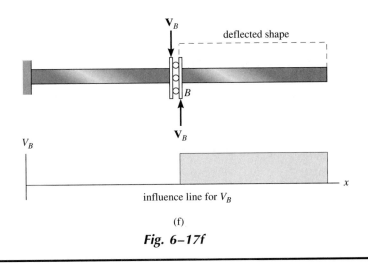

(f)

Fig. 6–17f

Example 6–11

For each beam in Fig. 6–18a through 6–18c, sketch the influence line for the moment at B.

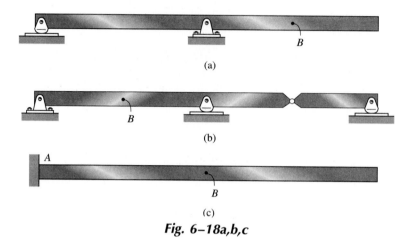

(a)

(b)

(c)

Fig. 6–18a,b,c

Solution

Fig. 6–18a. A hinge is introduced at B and positive moments M_B are applied to the beam. The deflected shape and corresponding influence line are shown in Fig. 6–18d.

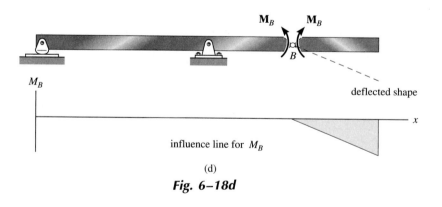

influence line for M_B

(d)

Fig. 6–18d

Fig. 6–18b. Placing a hinge at B and applying positive moments M_B to the beam yields the deflected shape and influence line shown in Fig. 6–18e.

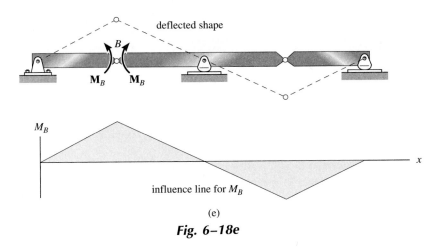

(e)

Fig. 6–18e

Fig. 6–18c. With the hinge and positive moment at B the deflected shape and influence line are as shown in Fig. 6–18f. The left segment of the beam is constrained from moving due to the fixed wall at A.

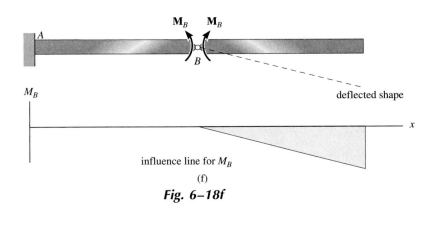

(f)

Fig. 6–18f

Example 6–12

Determine the maximum positive moment that can be developed at point D in the beam shown in Fig. 6–19a due to a concentrated live load of 4000 lb, a uniform live load of 300 lb/ft, and a dead load of 200 lb/ft.

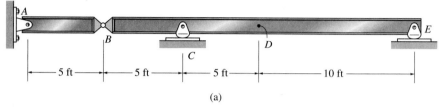

Fig. 6–19a

(a)

Solution

A hinge is placed at D and positive moments $\mathbf{M}_D$ are applied to the beam. The deflected shape and corresponding influence line are shown in Fig. 6–19b. Immediately one recognizes that the live load of 4000 lb creates a maximum *positive* moment at D when it is placed at D, i.e., the peak of the influence line. Also, the uniform live load of 300 lb/ft must extend from C to E in order to cover the region where the area of the influence line is positive. Finally, the uniform dead load of 200 lb/ft acts over the entire length of the beam. The entire loading is shown on the beam in Fig. 6–19c. Knowing the position of the loads, we can now determine the maximum moment at D using statics. In Fig. 6–19d the reactions on BE have been computed. Sectioning the beam at D and using segment DE, Fig. 6–19e, we have

$$\zeta + \Sigma M_D = 0; \qquad M_D + 5000(5) - 4750(10) = 0$$
$$M_D = 22{,}500 \text{ lb} \cdot \text{ft} = 22.5 \text{ k} \cdot \text{ft} \qquad \textbf{Ans.}$$

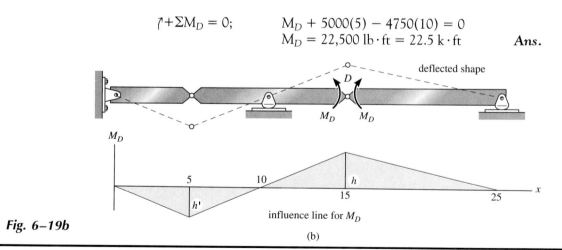

Fig. 6–19b

(b)

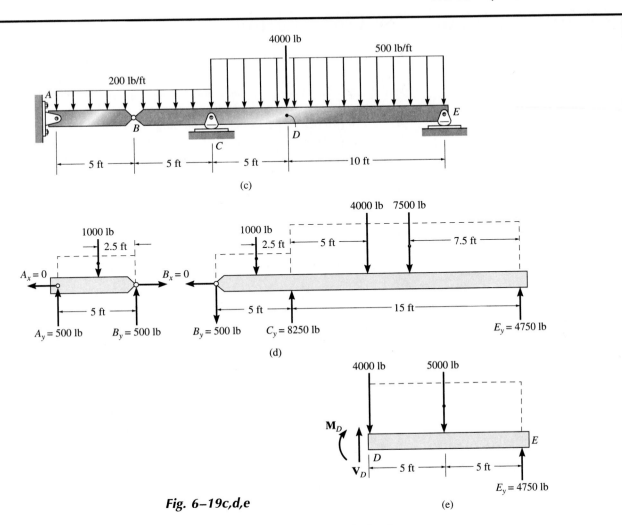

Fig. 6–19c,d,e

This problem can also be worked by computing *numerical values* for the influence line as in Sec. 6.2. Actually, by inspection of Fig. 6–19b, only the peak value h at D must be computed. This requires placing a unit load on the beam at D in Fig. 6–19a and then solving for the internal moment in the beam at D. Show that the ordinate value obtained is $h = M_D = 3.33$. By proportional triangles, $h'/(10 - 5) = 3.33/(15 - 10)$ or $h' = 3.33$. Hence, the maximum moment is

$$M_D = 500[\tfrac{1}{2}(25 - 10)(3.33)] + 4000(3.33) - 200[\tfrac{1}{2}(10)(3.33)]$$
$$= 22{,}500 \text{ lb} \cdot \text{ft} = 22.5 \text{ k} \cdot \text{ft} \qquad \textbf{\textit{Ans.}}$$

253

6.4 Influence Lines for Floor Girders

Occasionally, floor systems are constructed as shown in Fig. 6–20a, where it can be seen that floor loads are first transmitted from *slabs* to *floor beams*, then to *side girders*, and finally supporting *columns*. An idealized model of this system is shown in plane view, Fig. 6–20b. Here the slab is assumed to be a one-way slab and is segmented into simply-supported spans resting on the floor beams. Furthermore, the girder is simply-supported on the columns. Since the girders are main load-carrying members in this system, it is sometimes necessary to construct their shear and moment influence lines. This is especially true for industrial buildings subjected to heavy concentrated loads. In this regard, notice that a unit load on the floor slab is transferred to the girder only at points where it is in contact with the floor beams, i.e., points A, B, C, and D. These points are called *panel points*, and the region between these points is called a *panel*, such as BC in Fig. 6–20b.

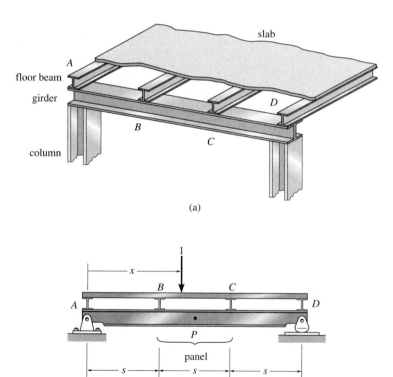

(a)

(b)

Fig. 6–20a,b

The influence line for specified points in the girder can be determined using the same statics procedure as in Sec. 6.2; i.e., place the unit load at various points x on the floor slab and always compute the function (shear or moment) at the specified point P in the girder, Fig. 6–20b. Plotting these values versus x yields the influence line for the function at P. In particular, the value for the internal moment in a girder panel will depend upon where point P is chosen for the influence line, since the magnitude of $\mathbf{M}_P$ depends upon the point's location d from the end of the girder. For example, if the unit load acts on the floor slab as shown in Fig. 6–20c, one first finds the reactions $\mathbf{F}_B$ and $\mathbf{F}_C$ on the slab, then calculates the support reactions $\mathbf{F}_1$ and $\mathbf{F}_2$ on the girder. The internal moment at P is then determined by the method of sections, Fig. 6–20d. This gives $M_P = F_1 d - F_B(d - s)$. Using a similar analysis, the internal shear $\mathbf{V}_P$ can be determined. In this case, however, $\mathbf{V}_P$ will be *constant* throughout the panel $BC(V_P = F_1 - F_B)$ and as seen does not depend upon the exact location of P within the panel. For this reason, influence lines for shear in floor girders are specified for *panels* in the girder and not specific points along the girder. The shear is then referred to as *panel shear*. It should also be noted that since the girder is affected only by the loadings transmitted by the floor beams, the unit load has only to be placed at each floor-beam location to establish the necessary data used to draw the influence line.

The following numerical examples should clarify the force analysis.

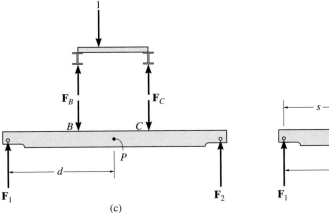

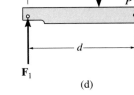

Fig. 6–20c,d

Example 6–13

Draw the influence line for the shear in panel CD of the floor girder in Fig. 6–21a.

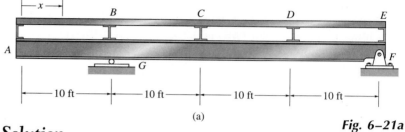

(a)

Fig. 6–21a

Solution

x	V_{CD}
0	0.333
10	0
20	−0.333
30	0.333
40	0

(b)

Fig. 6–21b

Tabulate Values. The unit load is placed at each floor beam location and the shear in panel CD is calculated. A table of the results is shown in Fig. 6–21b. The details for the calculations when $x = 0$ and $x = 20$ ft are given in Fig. 6–21c and 6–21d, respectively. Notice how in each case the reactions of the floor beams on the girder are calculated first, followed by a determination of the girder support reaction at F ($\mathbf{G}_y$ is not needed), and finally, a segment of the girder is considered and the internal panel shear V_{CD} is calculated. As an exercise, verify the values for V_{CD} when $x = 10$ ft, 30 ft, and 40 ft.

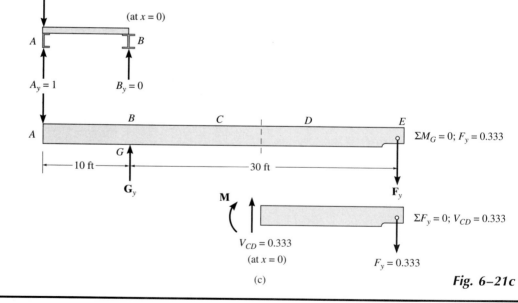

(c)

Fig. 6–21c

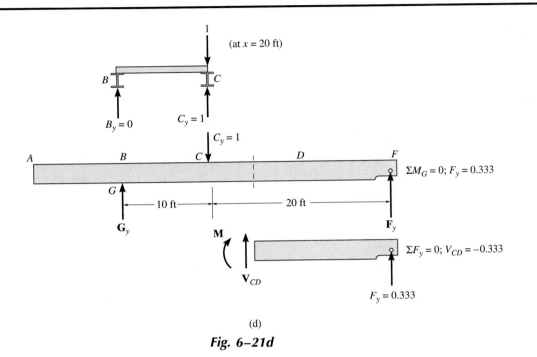

(d)

Fig. 6–21d

Influence Line.   When the values in Fig. 6–21b are plotted and the points connected with straight-line segments, the resulting influence line for V_{CD} is as shown in Fig. 6–21e.

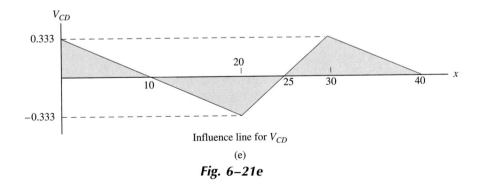

Influence line for V_{CD}

(e)

Fig. 6–21e

Example 6–14

Draw the influence line for the moment at point F for the floor girder in Fig. 6–22a.

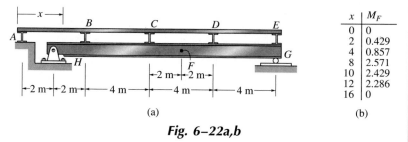

x	M_F
0	0
2	0.429
4	0.857
8	2.571
10	2.429
12	2.286
16	0

(a)   (b)

Fig. 6–22a,b

Solution

Tabulate Values.   The unit load is placed at $x = 0$ and each panel point thereafter. The corresponding values for M_F are calculated and shown in the table, Fig. 6–22b. Details of the calculations for $x = 2$ m are shown in Fig. 6–22c. As in the previous example, it is first necessary to determine the reactions of the floor beams on the girder, followed by a determination of the girder support reaction $\mathbf{G}_y$ ($\mathbf{H}_y$ is not needed), and finally segment GF of the girder is considered and the internal moment $\mathbf{M}_F$ is calculated. As an exercise, determine the other values of M_F listed in Fig. 6–22b.

Influence Line.   A plot of the values in Fig. 6–22b yields the influence line for M_F, Fig. 6–22d.

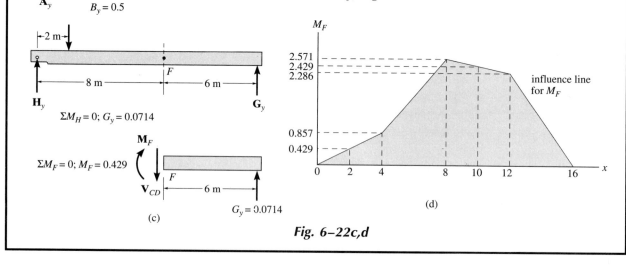

(c)   (d)

Fig. 6–22c,d

6.5 Influence Lines for Trusses

Trusses are often used as primary load-carrying elements for bridges. Hence, for design it is important to be able to construct the influence lines for each of the members. As shown in Fig. 6–23, the loading on the bridge deck is first transmitted to stringers, which in turn transmit the loading to floor beams and then to the *joints* along the bottom cord of the truss. Since the truss members are affected only by the joint loading, we can therefore obtain the ordinate values of the influence line for a member by loading each joint along the deck with a unit load and then use the method of joints or the method of sections to calculate the force in the member. The data can be arranged in tabular form, listing "unit load at joint" versus "force in member." As a convention, if the member force is *tensile* it is considered a *positive* value; if it is *compressive* it is *negative*. The influence line for the member is constructed by plotting the data and drawing straight lines between the points.

The following examples illustrate the method of construction.

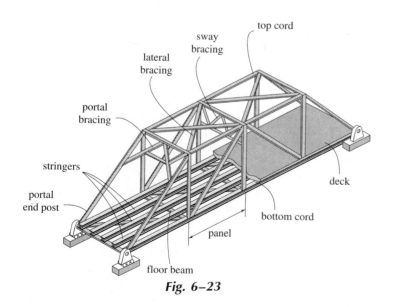

Fig. 6–23

259

Example 6–15

Draw the influence line for the force in member GB of the bridge truss shown in Fig. 6–24a.

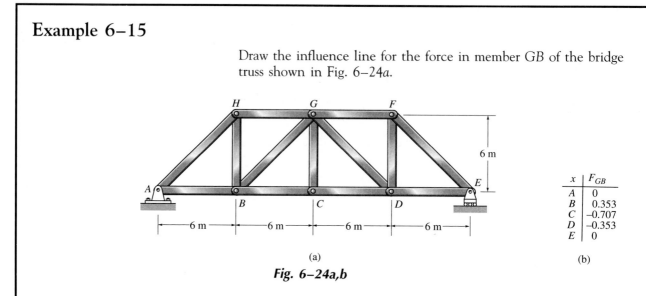

(a)

x	F_{GB}
A	0
B	0.353
C	–0.707
D	–0.353
E	0

(b)

Fig. 6–24a,b

Solution

Tabulate Values.   A table of unit-load position x at the joints along the bottom cord versus the force in member GB is shown in Fig. 6–24b. Here each successive joint at the bottom cord is loaded and the force in member GB is calculated using the method of sections. For example, placing the unit load at $x = 6$ m (joint B), the support reaction at E is calculated first, Fig. 6–24c, then passing a section through HG, BG, BC and isolating the right-hand segment, the force in GB is determined, Fig. 6–24d. In the same manner, determine the other values listed in Fig. 6–24b.

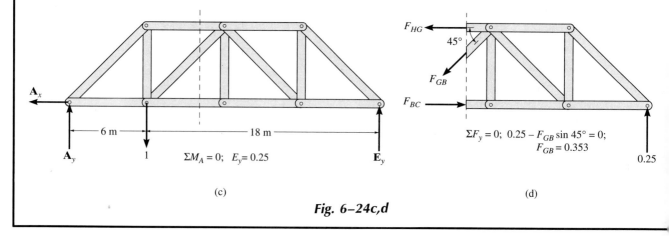

(c)

$\Sigma M_A = 0;$ $E_y = 0.25$

$\Sigma F_y = 0;$ $0.25 - F_{GB} \sin 45° = 0;$
$F_{GB} = 0.353$

(d)

Fig. 6–24c,d

Influence Line. Plotting the data in Fig. 6–24b and connecting the points yields the influence line for member GB as shown in Fig. 6–24e. Since the influence line extends over the entire span of the truss, member GB is referred to as a *primary member*. This means GB is subjected to a force regardless of where the bridge deck (roadway) is loaded. The point of zero force, $x = 8$ m, is determined by similar triangles between $x = 6$ m and $x = 12$ m, that is, $(0.353 + 0.707)/(12 - 6) = 0.353/x'$, $x' = 2$ m, so $x = 6 + 2 = 8$ m.

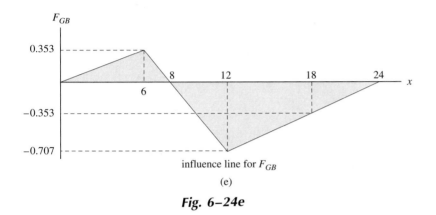

influence line for F_{GB}

(e)

Fig. 6–24e

Example 6–16

Draw the influence line for the force in member GC of the bridge truss shown in Fig. 6–25a.

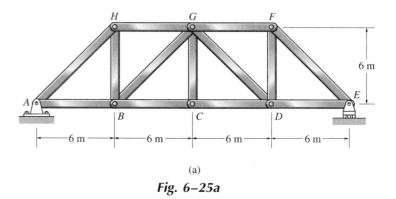

(a)

Fig. 6–25a

Solution

Tabulate Values.   A table of unit-load position at the joints of the bottom cord versus the force in member GC is shown in Fig. 6–25b. These values are easily obtained by isolating joint C, Fig. 6–25c. Here it is seen that GC is a zero-force member unless the unit load is applied at joint C, in which case $F_{GC} = 1$ (T).

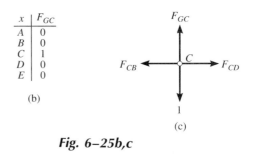

x	F_{GC}
A	0
B	0
C	1
D	0
E	0

(b)

(c)

Fig. 6–25b,c

Influence Line.   Plotting the data in Fig. 6–25*b* and connecting the points yields the influence line for member GC as shown in Fig. 6–25*d*. In particular, notice that when the unit load is at $x = 9$ m, the force in member GC is $F_{GC} = 0.5$. This situation requires the unit load to be placed on the bridge deck *between* the joints. The transference of this load from the deck to the truss is shown in Fig. 6–25*e*. From this one can see that indeed $F_{GC} = 0.5$ by analyzing the equilibrium of joint C, Fig. 6–25*f*. Since the influence line for GC does *not* extend over the entire span of the truss, Fig. 6–25*d*, member GC is referred to as a *secondary member*.

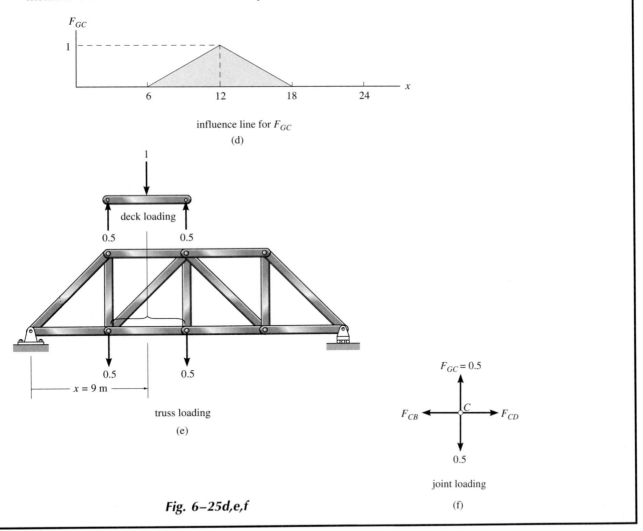

influence line for F_{GC}

(d)

truss loading

(e)

joint loading

(f)

Fig. 6–25d,e,f

Example 6–17

Determine the largest force that can be developed in member BC of the bridge truss shown in Fig. 6–26a due to a concentrated live load of 20 k and a uniform live load of 0.6 k/ft. The loading is applied at the top cord.

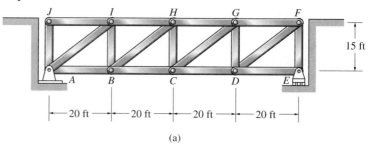

(a)

Solution

Tabulate Values.   A table of unit-load position x at the joints along the top cord versus the force in member BC is shown in Fig. 6–26b. The method of sections can be used for the calculations. For example, when the unit load is at joint I ($x = 20$ ft), Fig. 6–26c, the reaction $\mathbf{E}_y$ is determined first. Then the truss is sectioned through BC and the right segment is isolated. One obtains $\mathbf{F}_{BC}$ by summing moments about point H. In a similar manner determine the other values in Fig. 6–26b.

x	F_{BC}
J	0
I	0.667
H	1.33
G	0.667
F	0

(b)

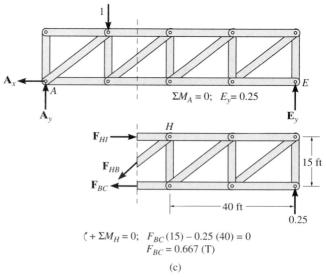

$$(\,+\ \Sigma M_H = 0; \quad F_{BC}(15) - 0.25(40) = 0$$
$$F_{BC} = 0.667\ (\text{T})$$

(c)

Fig. 6–26a,b,c

Influence Line.   A plot of the values in Fig. 6–26*b* yields the influence line, Fig. 6–26*d*. By inspection, *BC* is a primary member. Why?

Concentrated Force.   The largest force in member *BC* is created when the concentrated live force of 20 k is placed at $x = 40$ ft. Thus,

$$F_{BC} = (1.33)(20) = 26.7 \text{ k}$$

Uniform Live Load.   The uniform live load must be placed over the entire deck of the truss to create the largest tensile force in *BC*.* Thus,

$$F_{BC} = [\tfrac{1}{2}(80)(1.33)]0.6 = 32.0 \text{ k}$$

Total Maximum Force

$$(F_{BC})_{\max} = 26.7 \text{ k} + 32.0 \text{ k} = 58.7 \text{ k} \qquad \textbf{Ans.}$$

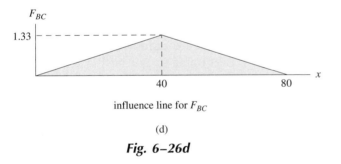

influence line for F_{BC}

(d)

Fig. 6–26d

*The largest *tensile* force in member *GB* of Example 6–15 is created when the distributed load acts on the deck of the truss from $x = 0$ to $x = 8$ m, Fig. 6–24*e*.

6.6   Live Loads for Bridges

Highway Bridges

In Sec. 1.3 we mentioned that the primary live loads on bridge spans are those due to traffic. The heaviest vehicle loading encountered is that caused by a series of trucks. Specifications for truck loadings on highway bridges are reported in the code of the American Association of State and Highway Transportation Officials (AASHTO). For two-axle trucks, these loads are designated with an H, followed by the weight of the truck in tons and another number, which gives the year of the specifications in which the load was reported. For example, an H15-44 is a 15-ton truck as reported in the 1944 specifications. H-series truck weights vary from 10 to 20 tons. However, bridges located on major highways, which carry a great deal of traffic, are often designed for two-axle trucks plus a one-axle semitrailer. These are designated as HS loadings, for example, HS 20-44. In general, a truck loading selected for design depends upon the type of bridge, its location, and the type of traffic anticipated.

The size of the "standard truck" and the distribution of its weight is also reported in the AASHTO specifications. For example, the HS 20-44 is shown in Fig. 6–27. Although trucks are assumed to occupy 10-ft lanes, all lanes on the bridge need not be fully loaded with a row of trucks to obtain the critical load, since such a loading would be highly improbable. Furthermore, rather than determine the critical load using a series of concentrated truck-wheel loads, in some cases the specifications allow a simplification by representing a lane loading as a uniform load plus a single concentrated force. This is meant to represent a distribution of medium-weight traffic with a heavy truck placed at the peak of the influence line.

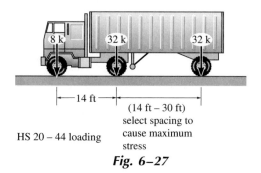

HS 20 – 44 loading

14 ft

(14 ft – 30 ft) select spacing to cause maximum stress

Fig. 6–27

Railroad Bridges

The loadings on railroad bridges are specified by the code of the American Railroad Engineers Association (AREA). Normally, E loads, as originally devised by Theodore Cooper in 1894, are used for design. For example, a modern train having a 72-k loading on the driving axle of the engine is designated as an E-72 loading. The entire E-72 loading for design is distributed as shown in Fig. 6–28. D. B. Steinmann has since updated Cooper's load distribution and has devised a series of M loadings, which are also acceptable for design. Since train loadings involve a complicated series of concentrated forces, to simplify hand calculations tables and graphs are sometimes used in conjunction with influence lines to obtain the critical load.*

Impact Loads

Moving vehicles may bounce or sidesway as they move over a bridge, and therefore they impart an *impact* to the deck. The percentage increase of the live loads due to impact is called the *impact factor*, I. This factor is generally obtained from formulas developed from experimental evidence. For example, for highway bridges the AASHTO specifications require that

$$I = \frac{50}{L + 125} \qquad \text{but not larger than 0.3}$$

where L is the length of the span in feet that is subjected to the live load. For example, member BC in Example 6–17 has an impact factor computed for $L = 80$ ft, since the influence line (and load) extends over the entire length of the truss, Fig. 6–26d.† Hence, $I = 50/(80 + 125) = 0.244 \le 0.3$. The additional load in member BC due to impact is thus $I(F_{BC})_{\max} = 0.244(58.7 \text{ k}) = 14.3 \text{ k}$. When added to the "static" placement of the live load, the "total" force in BC is therefore $58.7 \text{ k} + 14.3 \text{ k} = 73.0 \text{ k}$. Similar types of formulas and their application can be found in the AREA code.

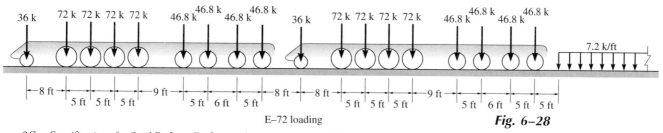

E–72 loading

Fig. 6–28

*See *Specifications for Steel Railway Bridges*, referenced at the end of the chapter.

†For member GB in Example 6–15, $L = 8$ m for tensile live loadings and $L = 24$ m $- 8$ m $= 16$ m for compressive live loads.

6.7   Maximum Influence at a Point Due to a Series of Concentrated Loads

Once the influence line of a function has been established for a point in a structure, the maximum effect caused by a live concentrated force is determined by multiplying the peak ordinate of the influence line by the magnitude of the force. In some cases, however, *several* concentrated forces must be placed on the structure. An example would be the wheel loadings of a truck or train as shown in Figs. 6–27 and 6–28. In order to determine the maximum effect in this case, either a trial-and-error procedure can be used or a method that is based on the change in the function that takes place as the load is moved. Each of these methods will now be explained specifically as it applies to shear and moment.

Shear

Consider the simply supported beam with the associated influence line for the shear at point C in Fig. 6–29a. The maximum positive shear at point C is to be determined due to the series of concentrated (wheel) loads shown in Fig. 6–29b, which move from right to left over the beam. The critical loading will occur when one of the loads is placed *just to the right* of point C, which is coincident with the positive peak of the influence line. By trial and error each of three possible cases can therefore be investigated, Fig. 6–29c. We have

Case 1: $(V_C)_1 = 1(0.75) + 4(0.625) + 4(0.5) = 5.25 \text{ k}$
Case 2: $(V_C)_2 = 1(-0.125) + 4(0.75) + 4(0.625) = 5.375 \text{ k}$
Case 3: $(V_C)_3 = 1(0) + 4(-0.125) + 4(0.75) = 2.5 \text{ k}$

Case 2, with the 1-k force located 5 ft from the left support, yields the largest value for V_C and therefore represents the critical loading. Notice that investigation of Case 3 is not actually needed here, since by inspection such an arrangement of loads would not yield a value of $(V_C)_3$ that would be larger than $(V_C)_2$.

When many concentrated loads act on the span, as in the case of the E-72 load of Fig. 6–28, the trial-and-error computations used above can be tedious. Instead, the critical position of the loads can be determined in a more direct manner by computing the change in shear, ΔV, which occurs when the loads are moved from Case 1 to Case 2, then from Case 2 to Case 3, and so on. As long as each computed ΔV is *positive*, the new position will yield a larger shear in the beam at C than the previous position. Each movement is investigated until a negative

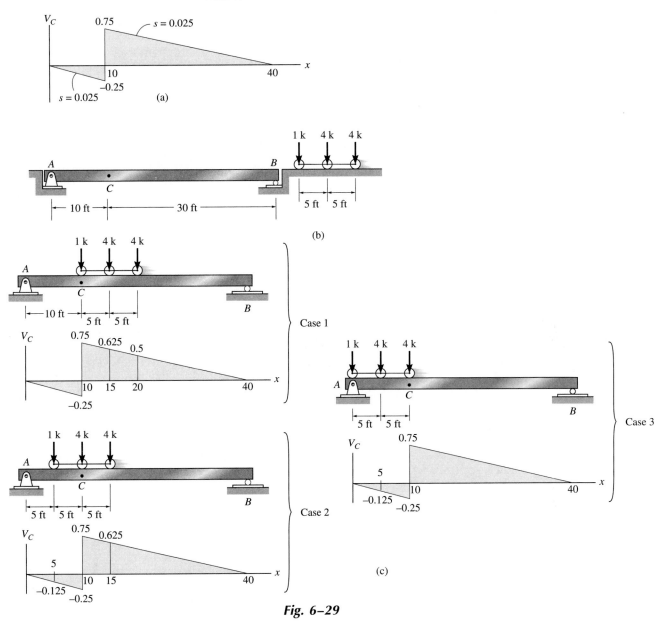

Fig. 6–29

change in shear is computed. When this occurs, the previous position of the loads will give the critical value. The change in shear ΔV for a load P that moves from position x_1 to x_2 over a beam can be determined by multiplying P by the change in the ordinate of the influence line, that is, $(y_2 - y_1)$. If the slope of the influence line is s, then $(y_2 - y_1) = s(x_2 - x_1)$, and therefore

$$\Delta V = Ps(x_2 - x_1)$$
$$\text{Sloping Line}$$

(6–1)

If the load moves past a point where there is a discontinuity or "jump" in the influence line, as point C in Fig. 6–29a, then the change in shear is simply

$$\Delta V = P(y_2 - y_1)$$
$$\text{Jump}$$

(6–2)

Use of the above equations will be illustrated with reference to the beam, loading, and influence line for V_C, shown in Fig. 6–29a and b. Notice that the slope of the influence line is $s = 0.75/(40 - 10) = 0.25/10 = 0.025$, and the jump at C has a magnitude of $0.75 + 0.25 = 1$. Consider the loads of Case 1 moving 5 ft to Case 2, Fig. 6–29c. When this occurs, the 1-k load jumps *down* (-1) and *all* the loads move *up* the slope of the influence line. This causes a change of shear,

$$\Delta V_{1-2} = 1(-1) + [1 + 4 + 4](0.025)(5) = +0.125 \text{ k}$$

Since ΔV_{1-2} is positive, Case 2 will yield a larger value for V_C than Case 1. [Compare the answers for $(V_C)_1$ and $(V_C)_2$ previously computed, where indeed $(V_C)_2 = (V_C)_1 + 0.125$.] Investigating ΔV_{2-3}, which occurs when Case 2 moves to Case 3, Fig. 6–29c, we must account for the downward (negative) jump of the 4-k load and the 5-ft horizontal movement of all the loads *up* the slope of the influence line. We have

$$\Delta V_{2-3} = 4(-1) + (1 + 4 + 4)(0.025)(5) = -2.875 \text{ k}$$

Since ΔV_{2-3} is negative, Case 2 is the position of the critical loading, as determined previously.

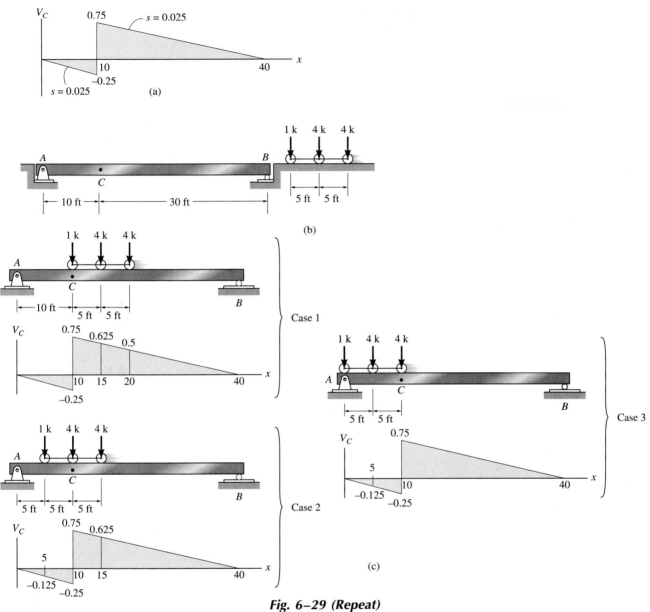

Fig. 6–29 (Repeat)

271

Moment

We can also use the foregoing methods to determine the critical position of a series of concentrated forces so that they create the largest internal moment at a specific point in a structure. Of course, it is first necessary to draw the influence line for the moment at the point and determine the slopes s of its line segments. For a horizontal movement $(x_2 - x_1)$ of a concentrated force P, the change in moment, ΔM, is equivalent to the magnitude of the force times the change in the influence-line ordinate under the load, that is,

$$\Delta M = Ps(x_2 - x_1)$$
$$\text{Sloping Line}$$

$$(6\text{--}3)$$

As an example, consider the beam, loading, and influence line for the moment at point C in Fig. 6–30a. If each of the three concentrated forces is placed on the beam, coincident with the peak of the influence line, we will obtain the greatest influence from each force. The three cases of loading are shown in Fig. 6–30b. When the loads of Case 1 are moved 4 ft to the left to Case 2, it is observed that the 2-k load *decreases* ΔM_{1-2}, since the *slope* (7.5/10) is *downward*, Fig. 6–30a. Likewise, the 4-k and 3-k forces cause an *increase* in ΔM_{1-2}, since the *slope* $[7.5/(40 - 10)]$ is *upward*. We have

$$\Delta M_{1-2} = -2\left(\frac{7.5}{10}\right)(4) + (4 + 3)\left(\frac{7.5}{40 - 10}\right)(4) = 1.0 \, \text{k} \cdot \text{ft}$$

Since ΔM_{1-2} is positive, we must further investigate moving the loads 6 ft from Case 2 to Case 3.

$$\Delta M_{2-3} = -(2 + 4)\left(\frac{7.5}{10}\right)(6) + 3\left(\frac{7.5}{40 - 10}\right)(6) = -22.5 \, \text{k} \cdot \text{ft}$$

Here the change is negative, so the greatest moment at C will occur when the beam is loaded as shown in Case 2, Fig. 6–30c. The maximum moment at C is therefore

$$(M_C)_{\text{max}} = 2(4.5) + 4(7.5) + 3(6.0) = 57.0 \, \text{k} \cdot \text{ft}$$

The following examples further illustrate this method.

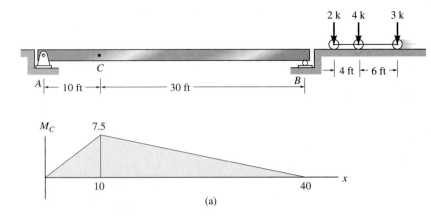

(a)

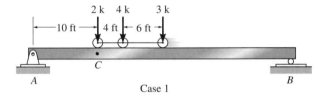

Case 1

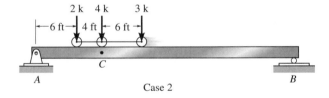

Case 2

Case 3

(b)

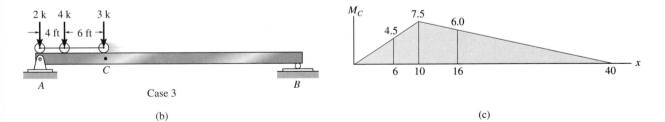

(c)

Fig. 6–30

273

Example 6–18

Determine the maximum positive shear created at point B in the beam shown in Fig. 6–31a due to the wheel loads of the moving truck. The truck travels toward the left.

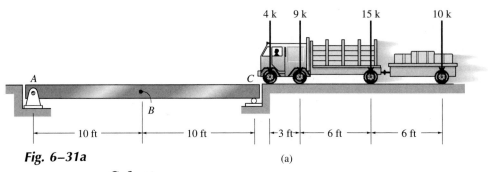

Fig. 6–31a

(a)

Solution

The influence line for the shear at B is shown in Fig. 6–31b.

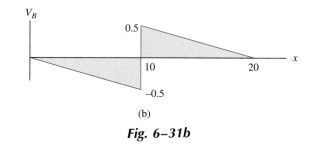

(b)

Fig. 6–31b

3-ft Movement of 4-k Load. Imagine that the 4-k load acts just to the right of point B so that we obtain its maximum positive influence. Since the beam segment BC is 10 ft long, the 10-k load is not as yet on the beam. When the truck moves 3 ft to the left, the 4-k load jumps *downward* on the influence line 1 unit and the 4-k, 9-k, and 15-k loads create a positive increase in ΔV_B, since the slope is upward to the left. Although the 10-k load also moves forward 3 ft, it is still not on the beam. Thus,

$$\Delta V = 4(-1) + (4 + 9 + 15)\left(\frac{0.5}{10}\right)3 = +0.2 \text{ k}$$

6-ft Movement of 9-k Load.   When the 9-k load acts just to the right of B, and then the truck moves 6 ft to the left, we have

$$\Delta V = 9(-1) + (4 + 9 + 15)\left(\frac{0.5}{10}\right)(6) + 10\left(\frac{0.5}{10}\right)(4) = +1.4\,\text{k}$$

Note in the calculation that the 10-k load only moves 4 ft on the beam.

6-ft Movement of 15-k Load.   If the 15-k load is positioned just to the right of B and then the truck moves 6 ft to the left, the 4-k load moves only 1 ft until it is off the beam, and likewise the 9-k load moves only 4 ft until it is off the beam. Hence,

$$\Delta V = 15(-1) + 4\left(\frac{0.5}{10}\right)(1) + 9\left(\frac{0.5}{10}\right)(4) + (15 + 10)\left(\frac{0.5}{10}\right)(6)$$
$$= -5.5\,\text{k}$$

Since ΔV is now negative, the correct position of the loads occurs when the 15-k load is just to the right of point B, Fig. 6–30c. Consequently,

$$(V_B)_{\text{max}} = 4(-0.05) + 9(-0.2) + 15(0.5) + 10(0.2)$$
$$= 7.5\,\text{k} \qquad\qquad\qquad\qquad \textbf{\textit{Ans.}}$$

In practice one also has to consider motion of the truck from left to right and then choose the maximum value between these two situations.

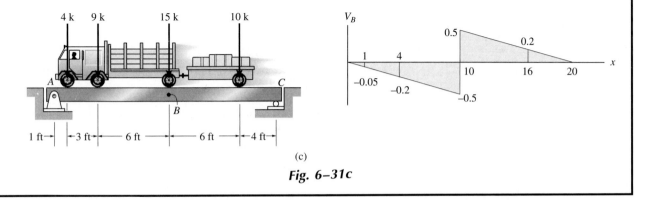

(c)

Fig. 6–31c

275

Example 6–19

Determine the maximum positive moment created at point B in the beam shown in Fig. 6–32a due to the wheel loads of the crane.

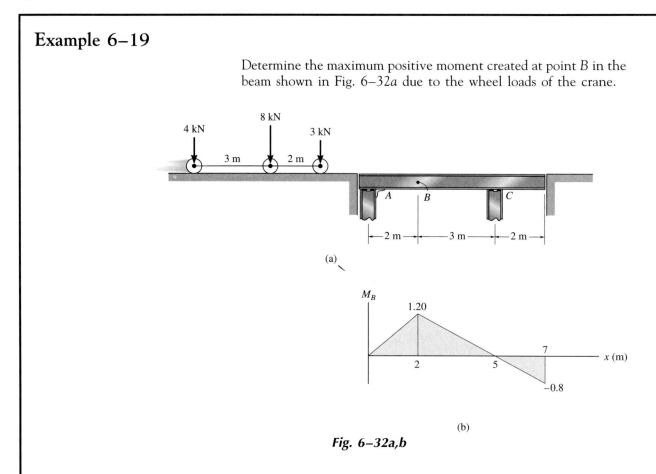

(a)

(b)

Fig. 6–32a,b

Solution

The influence line for the moment at B is shown in Fig. 6–32b.

2-m Movement of 3-kN Load. If the 3-kN load is assumed to act at B and then moves 2 m to the right, the change in moment is

$$\Delta M = -3\left(\frac{1.20}{3}\right)(2) + 8\left(\frac{1.20}{2}\right)(2) = 7.20 \text{ kN} \cdot \text{m}$$

Why is the 4-kN load not included in the calculations?

3-m Movement of 8-kN Load.   If the 8-kN load is assumed to act at B and then moves 3 m to the right, the change in moment is

$$\Delta M = -3\left(\frac{1.20}{3}\right)(3) - 8\left(\frac{1.20}{3}\right)(3) + 4\left(\frac{1.20}{2}\right)(2)$$
$$= -8.40 \text{ kN} \cdot \text{m}$$

Notice here that the 4-kN load was initially 1 m off the beam, and therefore moves only 2 m on the beam.

Since there is a sign change in ΔM, the correct position of the loads for maximum positive moment at B occurs when the 8-kN force is at B, Fig. 6–32c. Therefore,

$$(M_B)_{\text{max}} = 8(1.20) + 3(0.4) = 10.8 \text{ kN} \cdot \text{m} \qquad \textbf{\textit{Ans.}}$$

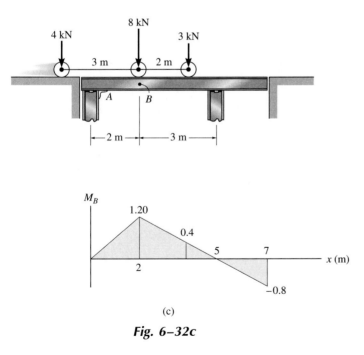

(c)

Fig. 6–32c

Example 6–20

Determine the maximum compressive force developed in member BG of the truss in Fig. 6–33a due to the wheel loads of the car and trailer. Assume the loads are applied directly to the truss and move only to the right.

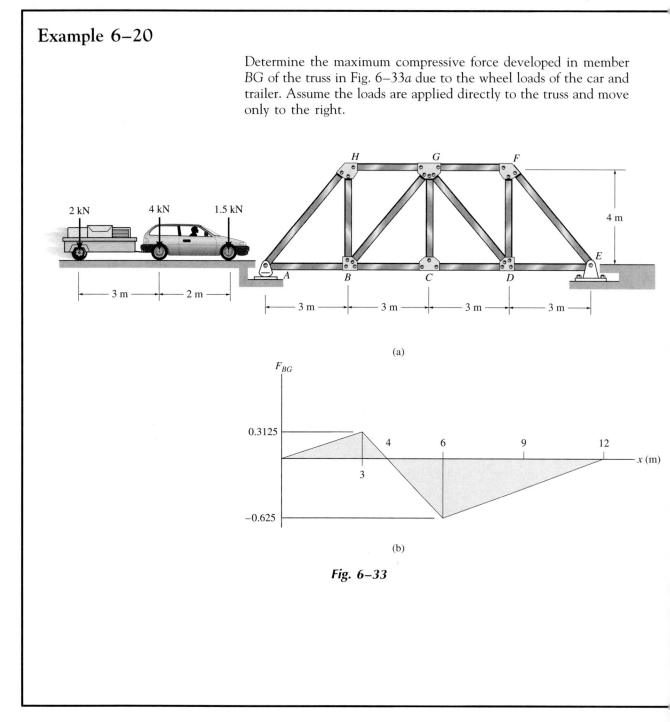

(a)

(b)

Fig. 6–33

Solution

The influence line for the force in member BG is shown in Fig. 6–33b. Here a trial-and-error approach for the solution will be used. Since we want the greatest negative (compressive) force in BG, we begin as follows:

1.5-kN Load at Point C. In this case

$$F_{BC} = 1.5 \text{ kN}(-0.625) + 4(0) + 2 \text{ kN}\left(\frac{0.3125}{3 \text{ m}}\right)(1 \text{ m})$$
$$= -0.729 \text{ kN}$$

4-kN Load at Point C. By inspection this would seem a more reasonable case than the previous one.

$$F_{BC} = 4 \text{ kN}(-0.625) + 1.5 \text{ kN}\left(\frac{-0.625}{6 \text{ m}}\right)(4 \text{ m}) + 2 \text{ kN}(0.3125)$$
$$= -2.50 \text{ kN}$$

2-kN Load at Point C. In this case all loads will create a compressive force in BC.

$$F_{BC} = 2 \text{ kN}(-0.625) + 4 \text{ kN}\left(\frac{-0.625}{6 \text{ m}}\right)(3 \text{ m})$$
$$+ 1.5 \text{ kN}\left(\frac{-0.625}{6 \text{ m}}\right)(1 \text{ m})$$
$$= -2.66 \text{ kN} \qquad \qquad \textbf{\textit{Ans.}}$$

Since this final case results in the largest answer, the critical loading occurs when the 2-kN load is at C.

6.8   Absolute Maximum Shear and Moment _____

In Sec. 6.7 we developed the methods for computing the maximum shear and moment at a *specified point* in a beam due to a series of concentrated moving loads. A more general problem involves the determination of both the *location of the point* in the beam *and the position of the loading* on the beam so that one can obtain the *absolute maximum* shear and moment caused by the loads. If the beam is cantilevered or simply supported, this problem can be readily solved.

Shear

For a *cantilevered beam* the absolute maximum shear will occur at a point located just next to the fixed support. The maximum shear is found by the method of sections, with the loads positioned close to the support, the first load being just next to the section as in Fig. 6–34.

For *simply supported beams* the absolute maximum shear will occur just next to one of the supports. In this case the loads are positioned so that the first one in sequence is placed close to the support, as in Fig. 6–35.

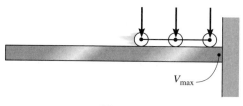

V_{max}

Fig. 6–34

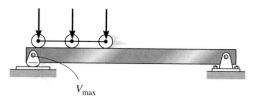

V_{max}

Fig. 6–35

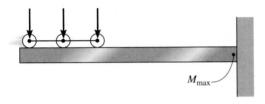

Fig. 6–36

Moment

The absolute maximum moment for a *cantilevered beam* occurs at the same point where absolute maximum shear occurs, although in this case the concentrated loads should be positioned at the *far end* of the beam, as in Fig. 6–36.

For a *simply supported beam* the critical positions of the loads and the associated absolute maximum moment cannot, in general, be determined by inspection. We can, however, determine the position analytically. For purposes of discussion, consider a beam subjected to the forces $\mathbf{F}_1$, $\mathbf{F}_2$, $\mathbf{F}_3$ shown in Fig. 6–37a. Since the moment diagram for a series of concentrated forces consists of straight line segments having peaks at each force, the absolute maximum moment will occur under one of the forces. Assume this maximum moment occurs under $\mathbf{F}_2$. The position of the loads $\mathbf{F}_1$, $\mathbf{F}_2$, $\mathbf{F}_3$ on the beam will be specified by the distance x, measured from $\mathbf{F}_2$ to the beam's center line as shown. To determine a specific value of x, we first obtain the resultant force of the system, $\mathbf{F}_R$, and its distance $\bar{x}'$ measured from $\mathbf{F}_2$. Once this is done, moments are summed about B, which yields the beam's left reaction, $\mathbf{A}_y$, that is,

$$\circlearrowleft + \Sigma M_B = 0; \qquad A_y = \frac{1}{L}(F_R)\left[\frac{L}{2} - (\bar{x}' - x)\right]$$

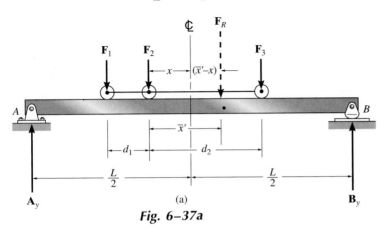

(a)

Fig. 6–37a

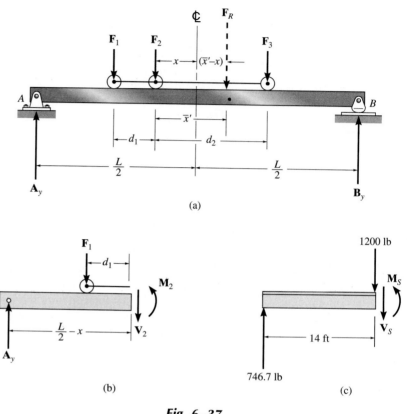

Fig. 6–37

If the beam is sectioned between the support at A and $\mathbf{F}_2$, the resulting free-body diagram is shown in Fig. 6–37b. The moment $\mathbf{M}_2$ under $\mathbf{F}_2$ is therefore

$$
\begin{aligned}
M_2 &= A_y\left(\frac{L}{2} - x\right) - F_1 d_1 \\
&= \frac{1}{L}(F_R)\left[\frac{L}{2} - (\bar{x}' - x)\right]\left(\frac{L}{2} - x\right) - F_1 d_1 \\
&= \frac{F_R L}{4} - \frac{F_R \bar{x}'}{2} - \frac{F_R x^2}{L} + \frac{F_R x \bar{x}'}{L} - F_1 d_1
\end{aligned}
$$

For maximum M_2 we require

$$\frac{dM_2}{dx} = \frac{-2F_R x}{L} + \frac{F_R \overline{x}'}{L} = 0$$

or

$$x = \frac{\overline{x}'}{2}$$

Hence, we may conclude that the *absolute maximum moment in a simply supported beam occurs under one of the concentrated forces, such that this force is positioned on the beam so that it and the resultant force of the system are equidistant from the beam's center line.* Since there are a series of loads on the span (for example, F_1, F_2, F_3 in Fig. 6–37a), this principle will have to be applied to each load in the series and the corresponding maximum moment computed. By comparison, the largest moment is the absolute maximum. As a general rule, though, the absolute maximum moment often occurs under the largest force lying nearest the resultant force of the system.

Envelope of Maximum Influence-Line Values

Rules or formulations for determining the absolute maximum shear or moment are difficult to establish for beams supported in ways other than the cantilever or simple support discussed here. An elementary way to proceed to solve this problem, however, requires constructing influence lines for the shear or moment at selected points along the entire length of the beam and then computing the maximum shear or moment in the beam for each point using the methods of Sec. 6.7. These values when plotted yield an "envelope of maximums," from which both the absolute maximum value of shear or moment and its location can be found. Obviously, a computer solution for this problem is desirable, since the work can be rather tedious if carried out by hand calculations.

Example 6–21

Determine the absolute maximum moment on the simply supported beam caused by the wheel loads of the automobile shown in Fig. 6–38a.

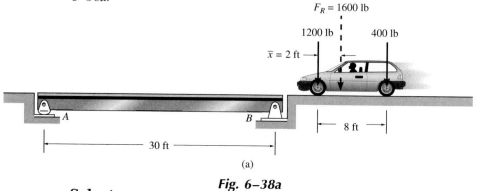

(a)

Fig. 6–38a

Solution

The magnitude and position of the resultant force of the two forces are determined first, Fig. 6–38a. We have

$$+\downarrow F_R = \Sigma F; \qquad F_R = 1200 + 400 = 1600 \text{ lb}$$
$$\curvearrowleft + M_{R_C} = \Sigma M_C; \quad 1600\bar{x} = 1200(0) + 400(8) \qquad \bar{x} = 2 \text{ ft}$$

We will first assume the absolute maximum moment occurs under the 1200-lb force. Positioning this force and the resultant force equidistant from the beam's center line, Fig. 6–38b, we can then

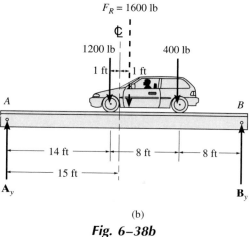

(b)

Fig. 6–38b

284

calculate the internal moment under the 1200-lb force using the method of sections. This requires computing the reaction at A first. The simplest way to do this is to use *only* the resultant force for the calculation, Fig. 6–38b, that is,

$$\zeta + \Sigma M_B = 0; \qquad A_y(30) - 1600(14) = 0 \qquad A_y = 746.7 \text{ lb}$$

Now using the left section of the beam, Fig. 6–38c, we have

$$\zeta + \Sigma M_S = 0; \qquad 746.7(14) - M_S = 0 \qquad M_S = 10.45 \text{ k·ft}$$

There is the possibility that the absolute maximum moment may occur under the 400-lb force. In this case the 400-lb force and F_R are positioned equidistant from the beam's center line, Fig. 6–38d. Show that $B_y = 640$ lb as indicated in Fig. 6–38e and that

$$M_S = 7.68 \text{ k·ft}$$

By comparison, the absolute maximum moment is

$$M_S = 10.45 \text{ k·ft} \qquad\qquad \textbf{Ans.}$$

which occurs under the 1200-lb force, when the wheels of the car are positioned on the beam as shown in Fig. 6–38b.

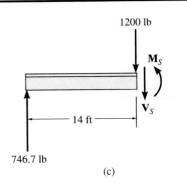

(c)

Fig. 6–38c

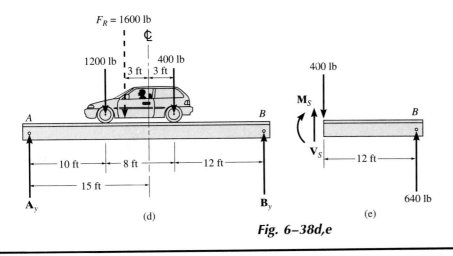

(d)

(e)

Fig. 6–38d,e

Example 6–22

Determine the absolute maximum moment in the simply supported beam shown in Fig. 6–39a.

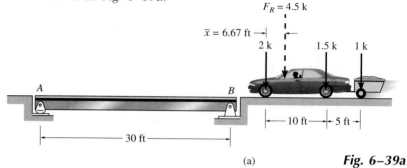

(a)

Fig. 6–39a

Solution

The magnitude and position of the resultant force of the system are determined first, Fig. 6–39a. We have

$$+\downarrow F_R = \Sigma F; \qquad F_R = 2000 + 1500 + 1000 = 4500 \text{ lb}$$
$$\curvearrowright + M_{R_C} = \Sigma M_C; \quad 4500\bar{x} = 1500(10) + 1000(15)$$
$$\bar{x} = 6.67 \text{ ft}$$

Let us first assume the absolute maximum moment occurs under the 1.5-k load. The load and the resultant force are positioned equidistant from the beam's center line, Fig. 6–39b. Calculating A_y first, Fig. 6–39b, we have

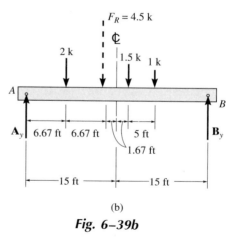

(b)

Fig. 6–39b

$\uparrow + \Sigma M_B = 0; \qquad A_y(30) - 4.5(16.67) = 0 \qquad A_y = 2.50 \text{ k}$

Now using the left section of the beam, Fig. 6–39c, yields

$\uparrow + \Sigma M_S = 0; \qquad 2.50(16.67) - 2(10) - M_S = 0$
$$M_S = 21.67 \text{ k} \cdot \text{ft}$$

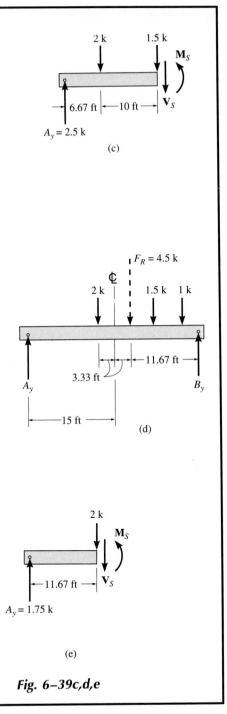

There is a possibility that the absolute maximum moment may occur under the 2-k load, since 2 k > 1.5 k and $\mathbf{F}_R$ is between both 2 k and 1.5 k. To investigate this case, the 2-k load and $\mathbf{F}_R$ are positioned equidistant from the beam's center line, Fig. 6–39d. Show that $A_y = 1.75$ k as indicated in Fig. 6–39e and that

$$M_S = 20.42 \text{ k} \cdot \text{ft}$$

By comparison, the absolute maximum moment is

$$M_S = 21.67 \text{ k} \cdot \text{ft} \qquad \qquad \textbf{\textit{Ans.}}$$

which occurs under the 1.5-k load, when the loads are positioned on the beam as shown in Fig. 6–39b.

Fig. 6–39c,d,e

287

REFERENCES

American Society of Civil Engineers, "Locomotive Loadings for Railroad Bridges," *Transactions*, ASCE, Vol. 86, 1923.

Müller-Breslau, H. F. B., *Die Neueren Methoden der Festigkeitslehre und der Statik der Baukonstruktionen.* Berlin, 1886.

Specifications for Steel Railway Bridges, American Railway Engineering Association, Chicago, 1965.

Standard Specifications for Highway Bridges, American Association of State Highway and Transportation Officials, 12th ed., Washington, D.C., 1973.

The members of this trussed bridge were designed using influence lines. *(Dean Abramson/Stock, Boston)*

PROBLEMS

6–1. Draw the influence lines for (a) the vertical reaction at A, (b) the shear at C, and (c) the moment at D. Assume the support at A is a roller and B is a pin. Solve this problem using the basic method of Sec. 6–1.

6–2. Solve Prob. 6–1. using Müller-Breslau's principle.

6–5. Draw the influence lines or (a) the shear at the fixed support A, and (b) the moment at B.

6–6. Solve Prob. 6–5 using Müller-Breslau's principle.

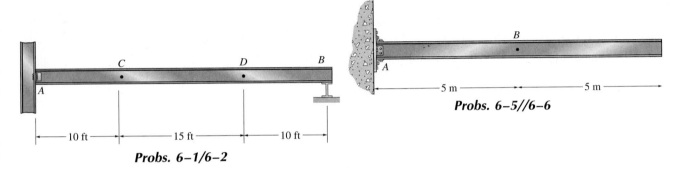

Probs. 6–5//6–6

Probs. 6–1/6–2

6–3. Draw the influence lines for (a) the vertical reaction at B, (b) the shear at C, and (c) the moment at C. Solve this problem using the basic method of Sec. 6–1.

*6–4. Solve Prob. 6–3 using Müller-Breslau's principle.

6–7. Draw the influence lines for (a) the vertical reaction at A, (b) the shear at C, and (c) the moment at C. Solve this problem using the basic method of Sec. 6–1.

*6–8. Solve Prob. 6–7 using Müller-Breslau's principle.

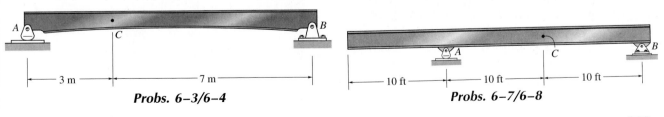

Probs. 6–3/6–4

Probs. 6–7/6–8

289

6–9. Draw the influence line for the moment at C. Assume the support at A is a pin and B is a roller.

6–10. Solve Prob. 6–9 using Müller-Breslau's principle.

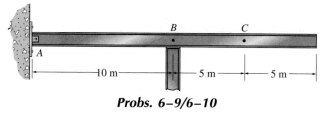

Probs. 6–9/6–10

6–11. Draw the influence lines for (a) the vertical reaction at A, (b) the moment at A, and (c) the shear at B. Assume the support at A is fixed. Solve this problem using the basic method of Sec. 6–1.

***6–12.** Solve Prob. 6–11 using Müller-Breslau's principle.

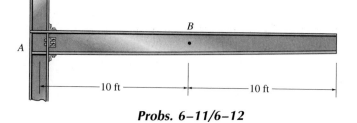

Probs. 6–11/6–12

6–13. Draw the influence lines for (a) the shear at C, (b) the moment at C, and (c) the vertical reaction at D. Indicate numerical values for the peaks. There is a short vertical link at B, and A is a pin support. Solve this problem using the basic method of Sec. 6–1.

6–14. Solve Prob. 6–13 using Müller-Breslau's principle.

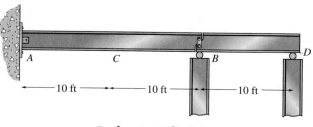

Probs. 6–13/6–14

6–15. Draw the influence line for the shear at C. Assume A is a roller support and B is a pin. Solve this problem using the basic method of Sec. 6–1.

***6–16.** Solve Prob. 6–15 using Müller-Breslau's principle.

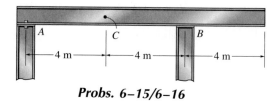

Probs. 6–15/6–16

6–17. Draw the influence line for the moment at C. Assume A is a roller support and B is a pin. Solve this problem using the basic method of Sec. 6–1.

6–18. Solve Prob. 6–17 using Müller-Breslau's principle.

6–21. A uniform live load of 0.2 k/ft and a single live concentrated force of 10 k can be placed on the haunched girder. Determine (a) the maximum live negative moment at A, (b) the maximum live positive shear at B, and (c) the maximum live negative moment at B.

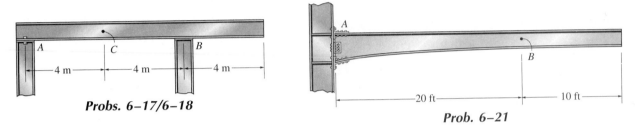

Probs. 6–17/6–18

Prob. 6–21

6–19. Draw the influence lines for (a) the vertical reaction at B, and (b) the moment at E. Assume the supports at B and D are rollers. There is a short link at C. Solve this problem using the basic method of Sec. 6–1.

*6–20. Solve Prob. 6–19 using Müller-Breslau's principle.

6–22. The beam supports a uniform dead load of 500 N/m and a single live concentrated force of 3000 N. Determine (a) the maximum live moment that can be developed at C, and (b) the maximum live positive shear that can be developed at C. Assume the support at A is a roller and B is a pin.

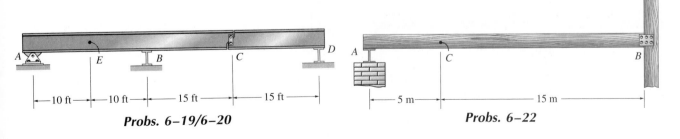

Probs. 6–19/6–20

Probs. 6–22

6–23. A uniform live load of 0.5 k/ft and a single concentrated force of 20 k are to be placed on the beam. Determine (a) the maximum positive live vertical reaction at support B, (b) the maximum positive live shear at point C, and (c) the maximum positive live moment at point C. Assume the support at A is a pin and B is a roller.

6–25. The compound beam is subjected to a uniform dead load of 200 lb/ft and a uniform live load of 150 lb/ft. Determine (a) the maximum negative live moment these loads develop at C, and (b) the maximum positive live shear at B. Assume C is a fixed support, D is a pin, and A is a roller.

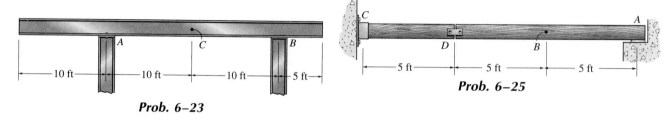

Prob. 6–25

6–26. Where should a single 500-lb live load be placed on the beam so it causes the largest live moment at D? What is this moment? Assume the support at A is fixed, B is pinned, and C is a roller.

*6–24. The beam supports a uniform live load of 60 lb/ft and a live concentrated force of 200 lb. Determine (a) the maximum positive live moment that can be developed at B, and (b) the maximum positive live vertical reaction at D. Assume C is a pin support and D is a roller.

6–27. Where should the beam ABC be loaded with a 300-lb/ft uniform distributed live load so it causes (a) the largest live moment at point A and (b) the largest live shear at D? Calculate the values of the moment and shear. Assume the support at A is fixed, B is pinned and C is a roller.

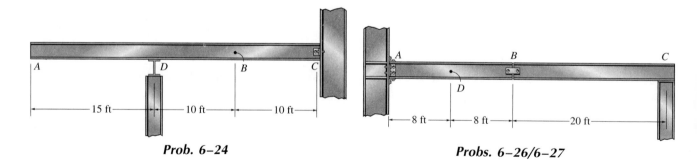

Prob. 6–24

Probs. 6–26/6–27

*6–28.   The compound beam is subjected to a uniform dead load of 0.8 kN/m and a single concentrated live load of 4 kN. Determine (a) the maximum moment created by these loads at C, and (b) the maximum pin reaction at B. Assume A is a roller, B is a pin, and C is fixed.

6–30.   Draw the influence line for the moment at point F in the girder. Assume the support at A is a pin and B is a roller. Compute the maximum live moment at F in the girder if the floor slabs are subjected to a uniform distributed live load of 1.5 k/ft.

6–31.   Draw the influence line for the shear in the end panel CA of the girder. Assume the support at A is a pin and B is a roller. Compute the maximum live shear in the girder panel CA if the floor slabs are subjected to a uniform distributed load of 1.5 k/ft.

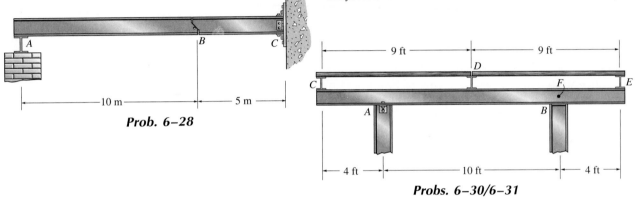

Prob. 6–28

Probs. 6–30/6–31

6–29.   Draw the influence lines for (a) the shear in panel DE of the girder, and (b) the moment at C.

*6–32.   A uniform live load of 0.4 k/ft and a concentrated live force of 2 k are to be placed on the floor slabs. Determine (a) the maximum live shear in panel CD, and (b) the maximum live moment at B.

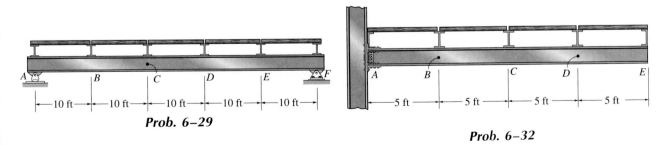

Prob. 6–29

Prob. 6–32

293

6–33. Draw the influence line for (a) the shear in panel *BC* of the girder, and (b) the moment at *C*. Assume the support at *B* is a pin and *D* is a roller.

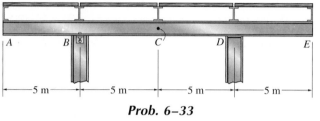

Prob. 6–33

6–34. A uniform live load of 250 lb/ft and a single concentrated live force of 1.5 k are to be placed on the floor beams. Determine (a) the maximum positive live shear in panel *AB*, and (b) the maximum live moment at *D*. Assume only vertical reaction occur at the supports.

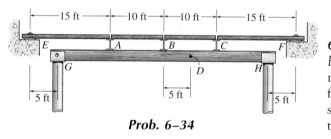

Prob. 6–34

6–35. Draw the influence lines for (a) the moment at *D* in the girder, and (b) the shear in panel *BC*.

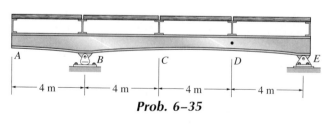

Prob. 6–35

*6–36. A uniform live load of 0.6 k/ft and a single concentrated live force of 5 k are to be placed on the top beams. Determine (a) the maximum positive live shear in panel *BC* of the girder, and (b) the maximum positive live moment at *C*. Assume the support at *B* is a roller and at *D* a pin.

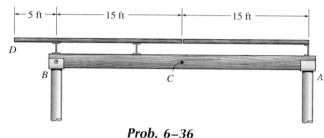

Prob. 6–36

6–37. Draw the influence line for the moment at *F* in the girder. Determine the maximum positive live moment in the girder at *F* if a single concentrated live force of 8 kN moves across the top floor beams. Assume the supports for all members can only exert either upward or downward forces on the members.

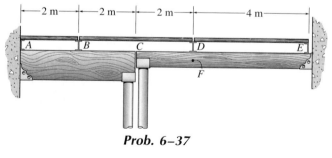

Prob. 6–37

6–38. A uniform live load of 0.2 k/ft and a single concentrated live force of 4 k are to be placed on the floor slabs. Determine (a) the maximum live vertical reaction at the support D, (b) the maximum live shear in panel DE of the girder, and (c) the maximum positive live moment at E. Assume D is a roller and G is a pin.

6–41. Draw the influence line for the force in member BC of the Warren truss. Indicate numerical values for the peaks. All members have the same length. The load moves along the bottom cord.

6–42. Draw the influence line for the force in member BF of the Warren truss. Indicate numerical values for the peaks. All members have the same length. The load moves along the bottom cord.

6–43. Draw the influence line for the force in member FE of the Warren truss. Indicate numerical values for the peaks. All members have the same length. The load moves along the bottom cord.

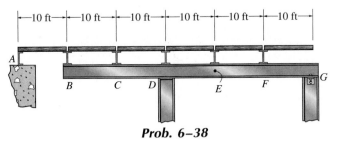

Prob. 6–38

6–39. Draw the influence line for the shear in panel CD of the girder. Determine the maximum negative live shear in panel CD due to a uniform live load of 500 lb/ft acting on the top beams.

***6–40.** Draw the influence line for the moment at E in the girder. Determine the maximum positive live moment in the girder at E if a concentrated live force of 8 kip moves across the top beams.

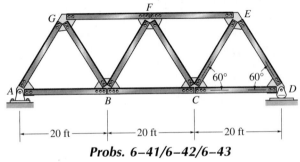

Probs. 6–41/6–42/6–43

***6–44.** Draw the influence line for the force in (a) member CJ, and (b) member DJ.

6–45. Draw the influence line for the force in member CD.

6–46. Draw the influence line for the force in member KJ.

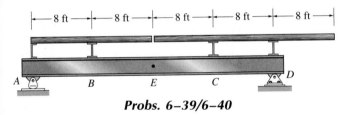

Probs. 6–39/6–40

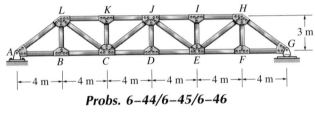

Probs. 6–44/6–45/6–46

295

6–47. Draw the influence line for the force in member CF, and then compute the maximum live force (tension or compression) that can be developed in this member due to a uniform live load of 800 lb/ft which is transmitted to the truss along the bottom cord.

*6–48. Draw the influence line for the force in member CD, and then compute the maximum live force (tension or compression) that can be developed in this member due to a uniform live load of 800 lb/ft. which is transmitted to the truss along the bottom cord.

6–49. Draw the influence line for the force in member GF, and then compute the maximum live force (tension or compression) that can be developed in this member due to a uniform live load of 800 lb/ft which is transmitted to the truss along the bottom cord.

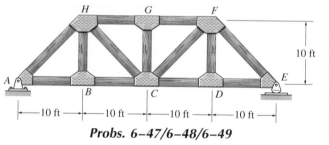

Probs. 6–47/6–48/6–49

6–50. Draw the influence line for the force in member KJ, then compute the maximum live force (tension or compression) that can be developed in this member due to a uniform live load of 3 kN/m which is transmitted to the truss along the bottom cord.

6–51. Draw the influence line for the force in member CJ, then compute the maximum live force (tension or compression) that can be developed in this member due to a uniform live load of 3 kN/m which is transmitted to the truss along the bottom cord.

*6–52. Draw the influence line for the force in member CD, then compute the maximum live force (tension or compression) that can be developed in this member due to a uniform live load of 3 kN/m which is transmitted to the truss along the bottom cord.

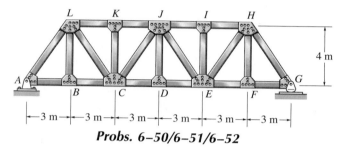

Probs. 6–50/6–51/6–52

6–53. The roof truss serves to support a crane rail which is attached to the bottom cord of the truss as shown. Determine the maximum live force (tension or compression) that can be developed in member GF, due to the crane load of 15 k. Specify the position x of the load. Assume the truss is supported at A by a pin and at E by a roller. Also, assume all members are sectioned and pin-connected at the gusset plates.

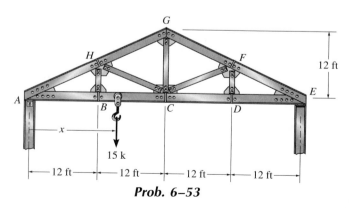

Prob. 6–53

6–54. Determine the maximum live force (tension or compression) that can be developed in member CF of the truss due to the crane load of 15 k. Specify the position x of the load. Assume the truss is supported at A by a pin and at E by a roller. Also, assume all members are sectioned and pin-connected at the gusset plates.

*6–56. Determine the maximum live moment at point C on the single girder caused by the moving dolly that has a mass of 2 Mg and a mass center at G. Assume A is a roller.

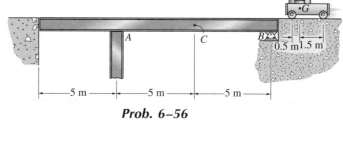

Prob. 6–56

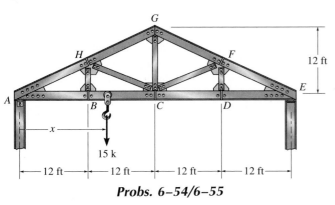

Probs. 6–54/6–55

6–55. Determine the maximum live force (tension or compression) that can be developed in member GC of the truss due to the crane load of 15 k. Specify the position x of the load. Assume the truss is supported at A by a pin and at E by a roller. Also, assume all members are sectioned and pin-connected at the gusset plates.

6–57. The 12,000-lb truck exerts the wheel reactions shown on the deck of a girder bridge. Determine (a) the largest live shear it creates in the splice joint at C, and (b) the largest moment it exerts at the splice. Assume the truck travels in *either direction* along the *center* of the deck, and therefore transfers *half* of its load to each of the two side girders. Assume the splice is a fixed connection and like the girder it can support both shear and moment.

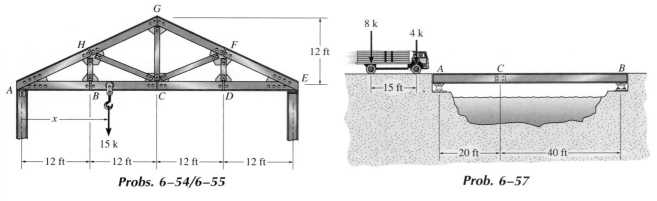

Probs. 6–54/6–55

Prob. 6–57

6–58.   Determine the maximum live moment in the suspended rail at point B if the rail supports the load of 2.5 k on the trolley.

6–59.   Determine the maximum positive live shear at point B if the rail supports the load of 2.5 k on the trolley.

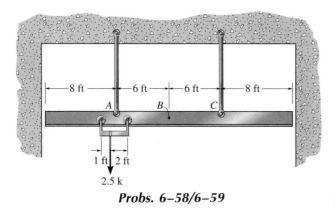

Probs. 6–58/6–59

6–61.   Draw the influence line for the force in member BC of the bridge truss. Compute the maximum live force (tension or compression) that can be developed in the member due to a 5-k truck having the wheel loads shown. Assume the truck can travel in *either direction* along the *center* of the deck, so that *half* the load shown is transferred to each of the two side trusses. Also assume the members are pin-connected at the gusset plates.

6–62.   Draw the influence line for the force in member IC of the bridge truss. Compute the maximum live force (tension or compression) that can be developed in the member due to a 5-k truck having the wheel loads shown. Assume the truck can travel in *either direction* along the *center* of the deck, so that half the load shown is transferred to each of the two side trusses. Also assume the members are pin-connected at the gusset plates.

6–63.   Draw the influence line for the force in member IH of the bridge truss. Compute the maximum live force (tension or compression) that can be developed in the member due to a 5-k truck having the wheel loads shown. Assume the truck can travel in *either direction* along the *center* of the deck, so that half the load shown is transferred to each of the two side trusses. Also assume the members are pin-connected at the gusset plates.

*6–60.   The car has a weight of 4200 lb and a center of gravity at G. Determine the maximum live moment created in the side girder at C as it crosses the bridge. Assume the car can travel in either direction along the *center* of the deck, so that *half* its load is transferred to each of the two side girders.

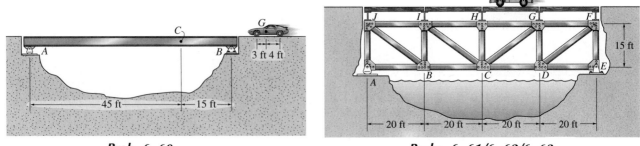

Prob. 6–60

Probs. 6–61/6–62/6–63

*6–64. The machine has a weight of 5000 lb and a mass center at G. Determine the maximum moment it produces in the single girder.

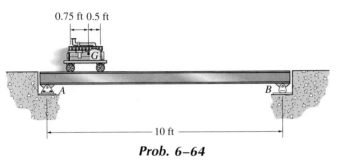

0.75 ft 0.5 ft

10 ft

Prob. 6–64

6–65. Draw the influence line for the force in member *IH* of the bridge truss. Compute the maximum live force (tension or compression) that can be developed in this member due to a 72-k truck having the wheel loads shown. Assume the truck can travel in *either direction* along the *center* of the deck, so that half its load is transferred to each of the two side trusses. Also assume the members are pin-connected at the gusset plates.

6–66. Draw the influence line for the force in member *CD* of the bridge truss. Compute the maximum live force (tension or compression) that can be developed in this member due to a 72-k truck having the wheel loads shown. Assume the truck can travel in *either direction* along the *center* of the deck, so that *half* its load is transferred to each of the two side trusses. Also assume the members are pin-connected at the gusset plates.

6–67. Draw the influence line for the force in member *BC* for the bridge truss. Compute the maximum live force (tension or compression) that can be developed in this member due to the 72-k truck having the wheel loads shown. Assume the truck can travel in *either direction* along the *center* of the deck, so that *half* its load is transferred to each of the two side trusses. Also assume the members are pin-connected at the gusset plates.

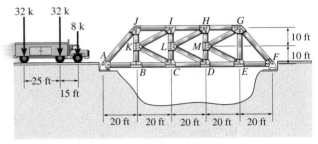

32 k   32 k

8 k

J   I   H   G

K   L   M

A

B   C   D   E

F

10 ft

10 ft

25 ft

15 ft

20 ft   20 ft   20 ft   20 ft   20 ft

Probs. 6–65/6–66/6–67

*6–68. The truck has a mass of 4 Mg and mass center at G_1, and the trailer has a mass of 1 Mg and mass center at G_2. Determine the absolute maximum live moment developed in the bridge.

6–69. Determine the absolute maximum live moment in the bridge in Problem 6–68 if the trailer is removed.

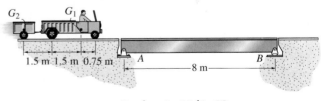

G_2 G_1

1.5 m 1.5 m 0.75 m

A   8 m   B

Probs. 6–68/6–69

6–70. Determine the absolute maximum live shear and absolute maximum live moment in the jib beam AB due to the 10-kN loading. The end constraints require $0.1 \text{ m} \leq x \leq 3.9 \text{ m}$.

6–71. Determine the absolute maximum live moment in the bridge due to the loading shown.

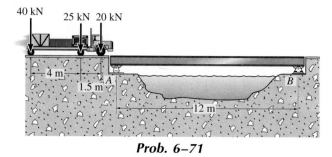

Prob. 6–71

*6–72. Determine the absolute maximum live moment in the bridge due to the loading shown.

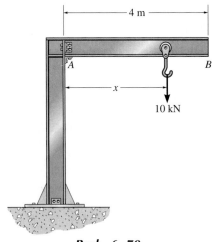

Prob. 6–70

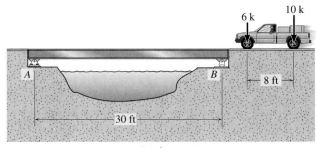

Prob. 6–72

6–73. Determine the absolute maximum live moment in the bridge due to the truck loading shown.

6–74. The trolley rolls at C and D along the bottom and top flange of beam AB. Determine the absolute maximum live moment developed in the beam if the load supported by the trolley is 2 k. Assume the support at A is a pin and at B a roller.

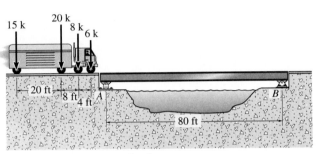

Prob. 6–73

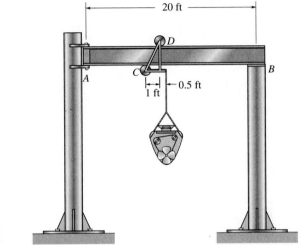

Prob. 6–74

The members of this concrete building are all fixed-connected, so the framework is statically indeterminate. Approximations were made for the preliminary design of the members before a more exact structural analysis was done.

Approximate Analysis of Statically Indeterminate Structures

In this chapter we will present some of the approximate methods used to analyze statically indeterminate trusses and frames. These methods were developed on the basis of structural behavior and their accuracy in most cases compares favorably with more exact methods of analysis. Although not all types of structural forms will be discussed here, it is hoped that enough insight is gained from the study of these methods so that one can judge what would be the best approximations to make when performing an approximate force analysis of a statically indeterminate structure.

7.1 Use of Approximate Methods

If a *model* is used to represent any structure, then the analysis of it must satisfy both the conditions of equilibrium and compatibility of displacement at the joints. As will be shown in later chapters of this text, the compatibility conditions for a *statically indeterminate* structure can be related to the loads, provided we know the material's modulus of elasticity and the size and shape of the members. For design, however, we will *not* know a member's size, and so an analysis that provides further simplifying assumptions for modeling the structure must be made. This analysis is called an *approximate analysis*, and throughout this chapter, we will use it to simplify the model of a statically indeterminate structure to one that is statically determinate. By performing an approximate analysis, a preliminary design of the members of a structure can be made, and once this is complete, the more exact indeterminate analysis can then be performed and the design refined. An approximate analysis also provides insight as to a structure's behavior under load and is beneficial when checking a more exact analysis, or when time, money, or capability are not available for performing the more exact analysis.

Realize that, in a general sense, all methods of structural analysis are approximate, simply because the conditions of loading, geometry, material behavior, and joint resistance at the supports are never known *exactly*. In this text, however, the statically indeterminate analysis of a structure will be referred to as an *exact analysis*, and the simpler statically determinate analysis will be referred to as the *approximate analysis*.

7.2 Trusses

A common type of truss often used for lateral bracing of a building or for the top and bottom cords of a bridge is shown in Fig. 7–1a. (Also see Fig. 3–4.) When used for this purpose, this truss is not considered a primary element for the support of the structure, and as a result it is often analyzed by approximate methods. In the case shown, it will be noticed that if a diagonal is removed from each of the three panels, it

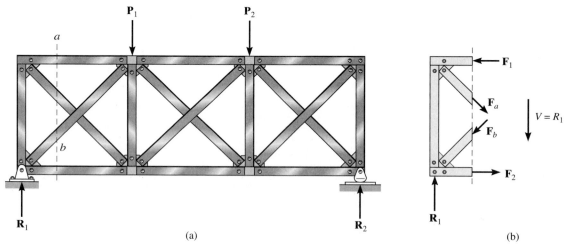

(a)

(b)

Fig. 7–1

will render the truss statically determinate. Hence, the truss is statically indeterminate to the third degree,* and therefore we must make three assumptions regarding the bar forces in order to reduce the truss to one that is statically determinate. These assumptions can be made with regard to the cross-diagonals, realizing that when one diagonal in a panel is in tension the corresponding cross-diagonal will be in compression. This is evident from Fig. 7–1b, where the "panel shear" $\mathbf{V}$ is carried by the *vertical component* of tensile force in member a and the *vertical component* of compressive force in member b. Two methods of analysis are generally acceptable.

Method 1:   If the diagonals are intentionally designed to be *long and slender*, it is reasonable to assume that they *cannot* support a compressive force; otherwise, they may easily buckle. Hence the panel shear is resisted entirely by the *tension diagonal*, whereas the *compressive diagonal is assumed to be a zero-force member*.

Method 2:   If the diagonal members are intended to be constructed from large rolled sections such as angles or channels, they may be equally capable of supporting a tensile and compressive force. Hence, we can assume that the tension and compression diagonals each carry *half* the panel shear.

Both of these methods of approximate analysis are illustrated numerically in the following examples.

---

*Using Eq. 3–1, $b + r = 2j$, or $16 + 3 > 8(2)$.

Example 7–1

Determine (approximately) the forces in the members of the truss shown in Fig. 7–2a. Assume the diagonals are slender and therefore will not support a compressive force. The support reactions have been computed.

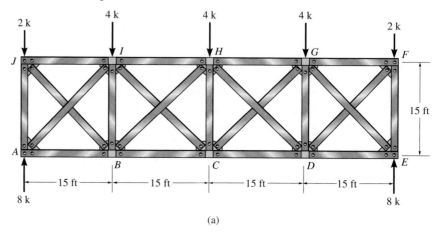

Fig. 7–2a,b,c

(a)

Solution

By inspection the truss is statically indeterminate to the fourth degree. Thus the four assumptions to be used require that each compression diagonal sustain zero force. Hence, from a vertical section through the left panel, Fig. 7–2b, we have

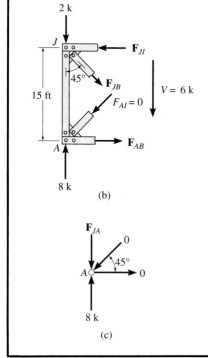

(b)

(c)

$$F_{AI} = 0 \qquad \textbf{Ans.}$$
$$+\uparrow \Sigma F_y = 0; \quad 8 - 2 - F_{JB} \cos 45° = 0$$
$$F_{JB} = 8.49 \text{ k (T)} \qquad \textbf{Ans.}$$
$$\zeta +\Sigma M_A = 0; \quad 8.49 \sin 45°(15) - F_{JI}(15) = 0$$
$$F_{JI} = 6 \text{ k (C)} \qquad \textbf{Ans.}$$
$$\zeta +\Sigma M_J = 0; \quad F_{AB}(15) = 0$$
$$F_{AB} = 0 \qquad \textbf{Ans.}$$

From joint A, Fig. 7–2c,

$$F_{JA} = 8 \text{ k (C)} \qquad \textbf{Ans.}$$

A vertical section of the truss through members IH, IC, BH, and BC is shown in Fig. 7–2d. The panel shear is $V = \Sigma F_y = 8 - 2 - 4 = 2$ k. We require

$$F_{BH} = 0 \qquad \textbf{Ans.}$$
$$+\uparrow\Sigma F_y = 0; \qquad 8 - 2 - 4 - F_{IC}\cos 45° = 0$$
$$F_{IC} = 2.83 \text{ k (T)} \qquad \textbf{Ans.}$$
$$\text{₹}+\Sigma M_B = 0; \qquad 8(15) - 2(15) + 2.83\sin 45°(15) - F_{IH}(15) = 0$$
$$F_{IH} = 8 \text{ k (C)} \qquad \textbf{Ans.}$$
$$\text{₹}+\Sigma M_I = 0; \qquad 8(15) - 2(15) - F_{BC}(15) = 0$$
$$F_{BC} = 6 \text{ k (T)} \qquad \textbf{Ans.}$$

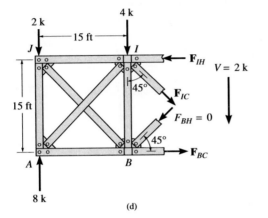

(d)

Fig. 7–2d,e,f

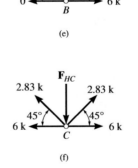

(e)

From joint B, Fig. 7–2e,

$$+\uparrow\Sigma F_y = 0; \qquad 8.49\sin 45° - F_{IB} = 0 \qquad F_{IB} = 6 \text{ k (C)} \qquad \textbf{Ans.}$$

The forces in the other members can be determined by symmetry, except F_{HC}; however, from joint C, Fig. 7–2f, we have

$$+\uparrow\Sigma F_y = 0; \qquad 2(2.83\sin 45°) - F_{HC} = 0 \qquad F_{HC} = 4 \text{ k (C)} \quad \textbf{Ans.}$$

(f)

Example 7–2

Determine (approximately) the forces in the members of the truss shown in Fig. 7–3a. The diagonals are to be designed to support both tensile and compressive forces, and therefore each is assumed to carry half the panel shear. The support reactions have been computed.

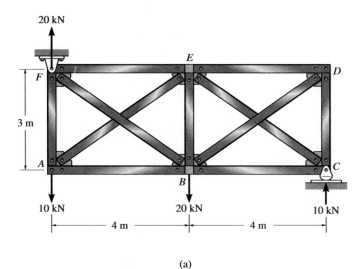

(a)

Fig. 7–3a,b

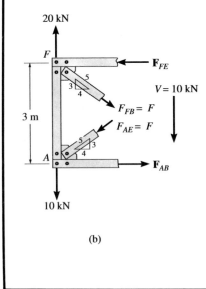

(b)

Solution

By inspection the truss is statically indeterminate to the second degree. The two assumptions require the tensile and compressive diagonals to carry equal forces, that is, $F_{FB} = F_{AE} = F$. For a vertical section through the left panel, Fig. 7–3b, we have

$$+\uparrow\Sigma F_y = 0; \qquad 20 - 10 - 2(\tfrac{3}{5}F) = 0 \qquad F = 8.33 \text{ kN}$$

so that

$$F_{FB} = 8.33 \text{ kN (T)}$$
$$F_{AE} = 8.33 \text{ kN (C)} \qquad\qquad\qquad \textbf{Ans.}$$

$$\curvearrowright +\Sigma M_A = 0; \quad 8.33(\tfrac{4}{5})(3) - F_{FE}(3) = 0$$
$$F_{FE} = 6.67 \text{ kN (C)} \qquad\qquad \textbf{Ans.}$$

$$\curvearrowright +\Sigma M_F = 0; \quad 8.33(\tfrac{4}{5})(3) - F_{AB}(3) = 0$$
$$F_{AB} = 6.67 \text{ kN (T)} \qquad\qquad \textbf{Ans.}$$

From joint A, Fig. 7–3c,

$$+\uparrow\Sigma F_y = 0; \quad F_{AF} - 8.33(\tfrac{3}{5}) - 10 = 0 \qquad F_{AF} = 15 \text{ kN (T)} \textbf{ Ans.}$$

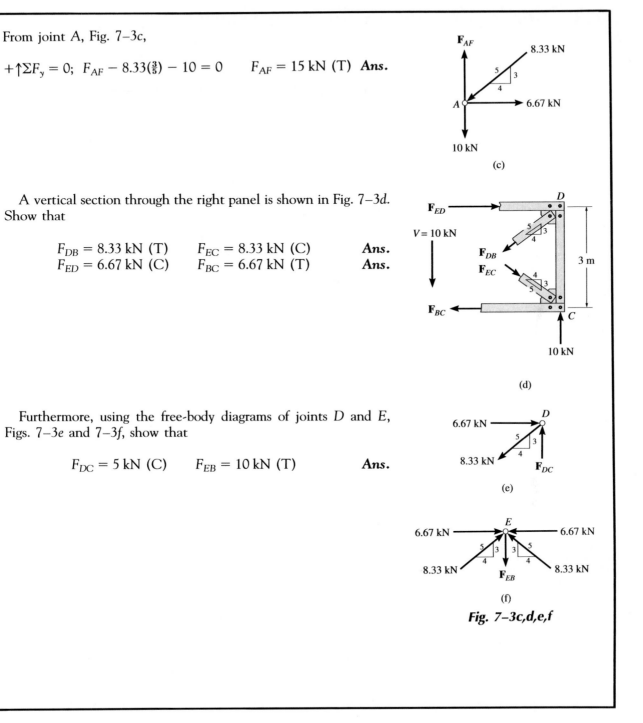

A vertical section through the right panel is shown in Fig. 7–3d. Show that

$$F_{DB} = 8.33 \text{ kN (T)} \qquad F_{EC} = 8.33 \text{ kN (C)} \qquad \textbf{Ans.}$$
$$F_{ED} = 6.67 \text{ kN (C)} \qquad F_{BC} = 6.67 \text{ kN (T)} \qquad \textbf{Ans.}$$

Furthermore, using the free-body diagrams of joints D and E, Figs. 7–3e and 7–3f, show that

$$F_{DC} = 5 \text{ kN (C)} \qquad F_{EB} = 10 \text{ kN (T)} \qquad \textbf{Ans.}$$

Fig. 7–3c,d,e,f

309

7.3   Vertical Loads on Building Frames

Building frames often consist of girders that are *rigidly connected* to columns so that the entire structure is better able to resist the effects of lateral forces due to wind and earthquake. An example of such a rigid framework, often called a building bent, is shown in Fig. 7–4. In this section we will establish a method for analyzing (approximately) the forces in building frames due to vertical loads, and in Secs. 7.5 and 7.6 an approximate analysis for frames subjected to lateral loads will be presented. In all these cases it should be noticed that most of the simplifying assumptions made to reduce a frame from a statically indeterminate structure to one that is statically determinate are based on *the way the structure deforms under load.*

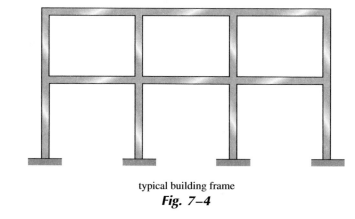

typical building frame
Fig. 7–4

Assumptions for Approximate Analysis

Consider a typical girder located within a building bent and subjected to a uniform vertical load, as shown in Fig. 7–5a. The column supports at A and B will each exert three reactions on the girder, and therefore the girder will be statically indeterminate to the third degree (6 reactions − 3 equations of equilibrium). To make the girder statically determinate, an approximate analysis will therefore require three assumptions. If the columns are extremely stiff, no rotation at A and B will occur, and the deflection curve for the girder will look like that shown in Fig. 7–5b. Using one of the methods presented in Chapters 9 through 11, an exact analysis reveals that for this case inflection points, or zero moment, occur at 0.21l from each support. If, however, the column connections at A and B are very flexible, then like a simply supported beam, zero moment will occur at the supports, Fig. 7–5c. In reality, however, the columns will provide some flexibility at the sup-

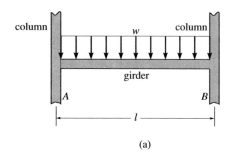

(a)

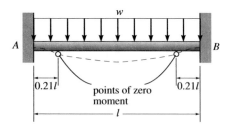

points of zero moment

fixed supported
(b)

Fig. 7–5a,b

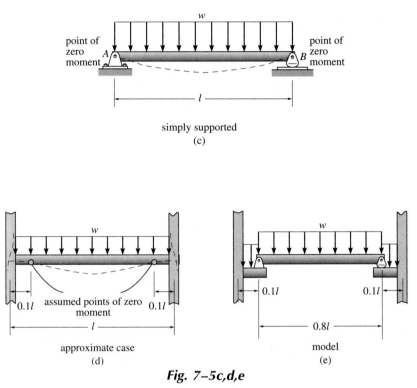

Fig. 7–5c,d,e

ports, and therefore we will assume that zero moment occurs at the *average point* between the two extremes, i.e., at $(0.21l + 0)/2 \approx 0.1l$ from each support, Fig. 7–5d. Furthermore, an exact analysis of frames supporting vertical loads indicates that the axial forces in the girder are negligible.

As a result of the above discussion, each girder of length l may be modeled by a simply supported span of length $0.8l$ resting on two cantilevered ends, each having a length of $0.1l$, Fig. 7–5e. The following three assumptions are incorporated in this model:

1. There is zero moment in the girder, $0.1l$ from the left support.

2. There is zero moment in the girder, $0.1l$ from the right support.

3. The girder does not support an axial force.

By using statics, the internal loadings in the girders can now be obtained and a preliminary design of their cross section can be made. The following example illustrates this numerically.

Example 7–3

Determine (approximately) the moments at the joints E and C caused by members EF and CD of the building bent in Fig. 7–6a.

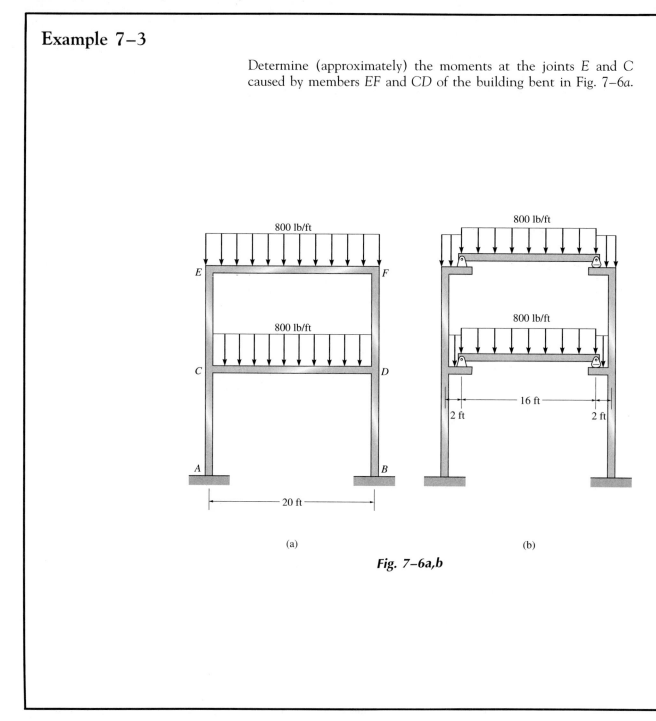

(a)

(b)

Fig. 7–6a,b

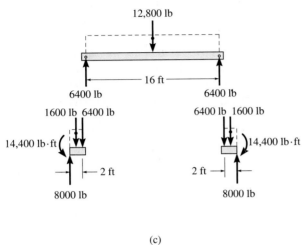

(c)

Fig. 7–6c

Solution

For an approximate analysis the frame is modeled as shown in Fig. 7–6b. Note that the cantilevered spans supporting the center portion of the girder have a length of $0.1l = 0.1(20) = 2$ ft. Equilibrium requires the end reactions for the center portion of the girder to be 6400 lb, Fig. 7–6c. The cantilevered spans are then subjected to a reaction moment of

$$M = 14{,}400 \text{ lb} \cdot \text{ft} = 14.4 \text{ k} \cdot \text{ft} \qquad \textbf{Ans.}$$

This moment, with opposite direction, acts on the joints E and C.

7.4 Portal Frames and Trusses

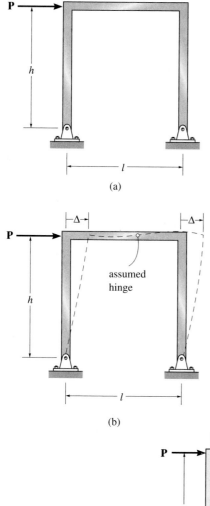

(a)

(b)

Frames

Portal frames are frequently used over the entrance of a bridge* and as a main element in building design in order to transfer horizontal forces applied at their top joints to the foundation. On bridges, these forces are caused by wind, earthquake, and unbalanced traffic loading on the bridge deck. Portals can be pin-supported, fixed-supported, or supported by partial fixity. The approximate analysis of each case will now be discussed for a simple three-member portal.

Pin-Supported. A typical pin-supported portal consists of pin-supported vertical columns having equal length and size and a rigidly connected horizontal girder, Fig. 7–7a. Since four unknowns exist at the supports but only three equilibrium equations are available for solution, this structure is statically indeterminate to the first degree. Consequently, only one assumption must be made to reduce the frame to one that is statically determinate.

The elastic deflection of the portal is shown in Fig. 7–7b. This diagram indicates that a point of inflection, that is, where the moment changes from positive bending to negative bending, is located *approximately* at the girder's midpoint. Since the moment in the girder is zero at this point, we can assume a hinge exists there and then proceed to

*See Fig. 3–4.

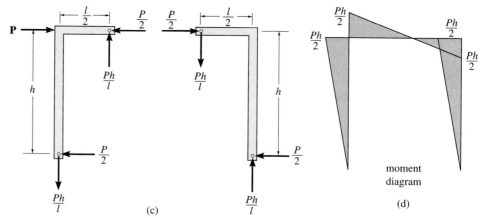

Fig. 7–7

(c)

(d)

determine the reactions at the supports using statics. If this is done, it is found that the horizontal reactions at the base of each column are *equal* and the other reactions are those indicated in Fig. 7–7c. Furthermore, the moment diagrams for this frame are indicated in Fig. 7–7d.

Fixed-Supported. Portals with two fixed supports, Fig. 7–8a, are statically indeterminate to the third degree since there is a total of six unknowns at the supports. If the vertical members have equal lengths and cross-sectional areas, the frame will deflect as shown in Fig. 7–8b. For this case we will assume points of inflection occur at the midpoints of all three members, and therefore hinges are placed at these points. The reactions and moment diagrams for each member can therefore be determined by dismembering the frame at the hinges and applying the equations of equilibrium to each of the four parts. The results are shown in Fig. 7–8c. Note that, as in the case of the pin-connected portal, the horizontal reactions at the base of each column are *equal*. The moment diagram for this frame is indicated in Fig. 7–8d.

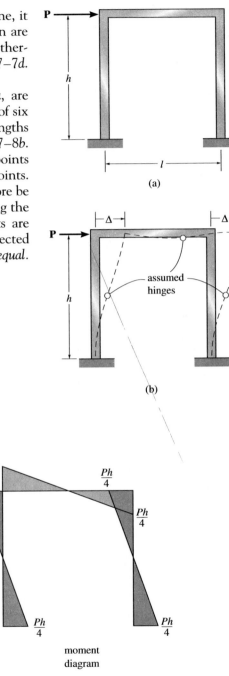

(a)

(b)

(c)

(d)

Fig. 7–8

Partial Fixity.   Since it is both difficult and costly to construct a perfectly fixed support or foundation for a portal frame, it is conservative and somewhat realistic to assume a slight rotation occurs at the supports, Fig. 7–9a. As a result, the points of inflection on the columns lie somewhere between the case of having a pin-supported portal, Fig. 7–7a, where the "inflection points" are at the supports (base of columns), and a fixed-supported portal, Fig. 7–8a, where the inflection points are at the center of the columns. Many engineers arbitrarily define the location at $h/3$, Fig. 7–9b, and therefore place hinges at these points, and also at the center of the girder.†

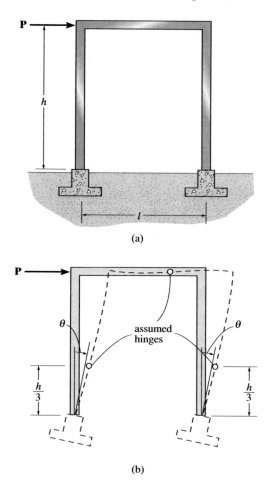

(a)

(b)

Fig. 7–9

†See Prob. 7–12.

Trusses

When a portal is used to span large distances, a truss may be used in place of the horizontal girder. Such a structure is used on large bridges and as transverse bents for large auditoriums and mill buildings. A typical example is shown in Fig. 7–10a. In all cases, the suspended truss is assumed to be pin-connected at its points of attachment to the columns. Furthermore, the truss keeps the columns straight within the region of attachment when the portal is subjected to the sidesway Δ, Fig. 7–10b. Consequently, we can analyze trussed portals using the same assumptions as those used for portal frames. For pin-supported columns, assume the horizontal reactions are equal, as in Fig. 7–7c. For fixed-supported columns, assume the horizontal reactions are equal and an inflection point (or hinge) occurs on each column, measured midway between the base of the column and the *lowest point* of truss member connection to the column. See Fig. 7–8c and Fig. 7–10b.

The following example illustrates how to determine the forces in the members of a trussed portal using the approximate method of analysis described above.

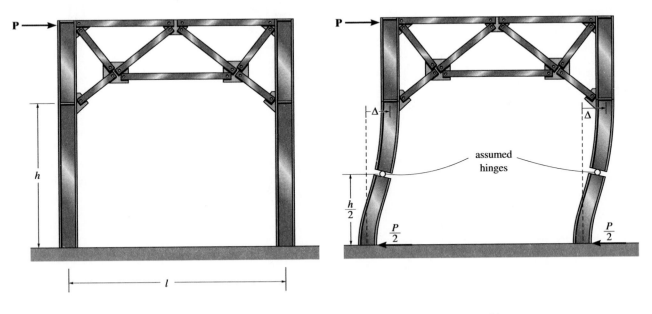

(a)   (b)

Fig. 7–10

Example 7–4

Determine by approximate methods the forces acting in the members of the Warren portal shown in Fig. 7–11a.

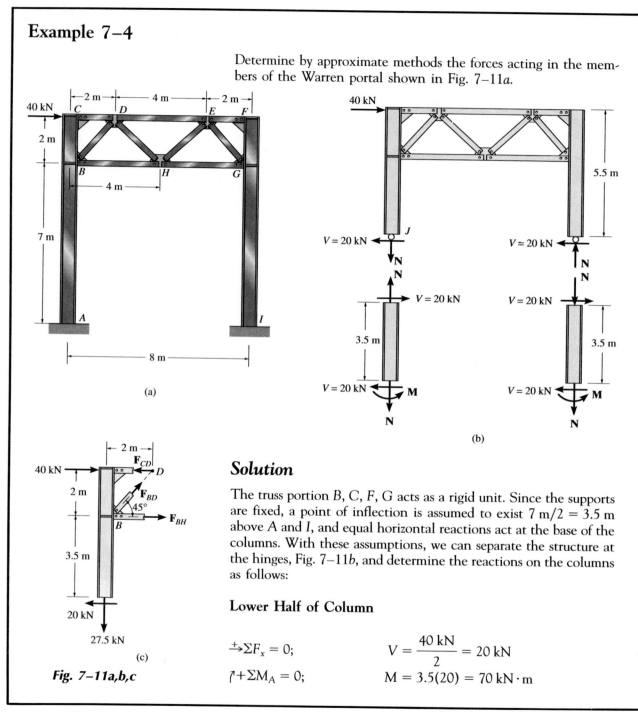

(a)

(b)

(c)

Fig. 7–11a,b,c

Solution

The truss portion B, C, F, G acts as a rigid unit. Since the supports are fixed, a point of inflection is assumed to exist $7\ m/2 = 3.5\ m$ above A and I, and equal horizontal reactions act at the base of the columns. With these assumptions, we can separate the structure at the hinges, Fig. 7–11b, and determine the reactions on the columns as follows:

Lower Half of Column

$$\xrightarrow{+}\Sigma F_x = 0; \qquad V = \frac{40\ kN}{2} = 20\ kN$$

$$\wideparen{+}\Sigma M_A = 0; \qquad M = 3.5(20) = 70\ kN \cdot m$$

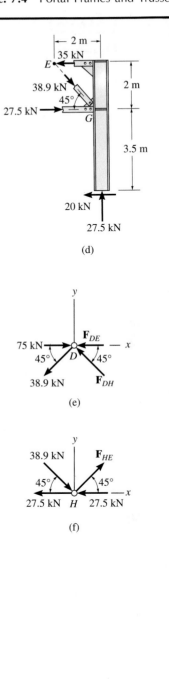

(d)

Upper Portion of Column

$$\uparrow + \Sigma M_J = 0; \qquad 40(5.5) - A(8) = 0 \qquad A = 27.5 \text{ kN}$$

Using the method of sections, Fig. 7–11c, we can now proceed to obtain the forces in members *CD*, *BD*, and *BH*. Show that $\Sigma F_y = 0$ yields $F_{BD} = 38.9 \text{ kN (T)}$, $\Sigma M_B = 0$ yields $F_{CD} = 75 \text{ kN (C)}$, and $\Sigma M_D = 0$ yields $F_{BH} = 27.5 \text{ kN (T)}$. Also, in a similar manner, show that one obtains the results on the free-body diagram of column *FGI* in Fig. 7–11d. Using these results, we can now find the force in each of the other truss members of the portal using the method of joints.

Joint *D*, Fig. 7–11e

$$+\uparrow\Sigma F_y = 0; \quad F_{DH} \sin 45° - 38.9 \sin 45° = 0 \quad F_{DH} = 38.9 \text{ kN (C)}$$
$$\overset{+}{\rightarrow}\Sigma F_x = 0; \quad 75 - 2(38.9 \cos 45°) - F_{DE} = 0 \quad F_{DE} = 20 \text{ kN (C)}$$

Joint *H*, Fig. 7–11f

$$+\uparrow\Sigma F_y = 0; \quad F_{HE} \sin 45° - 38.9 \sin 45° = 0 \quad F_{HE} = 38.9 \text{ kN (T)}$$

These results are summarized in Fig. 7–11g.

(e)

(f)

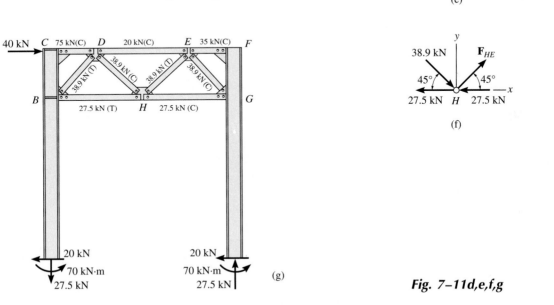

(g)

Fig. 7–11d,e,f,g

7.5 Lateral Loads on Building Frames: Portal Method

In Sec. 7.4 we discussed the action of lateral loads on portal frames and found that for a frame fixed-supported at its base, points of inflection occur at approximately the center of each girder and column and the columns carry equal shear loads, Fig. 7–8. A building bent deflects in the same way as a portal frame, Fig. 7–12a, and therefore it would be appropriate to assume inflection points occur at the center of the columns and girders. If we consider each bay of the frame to be composed of a series of portals, Fig. 7–12b, then as a further assumption, the *interior columns* would represent the effect of *two portal columns* and would therefore carry twice the shear V as the two exterior columns.

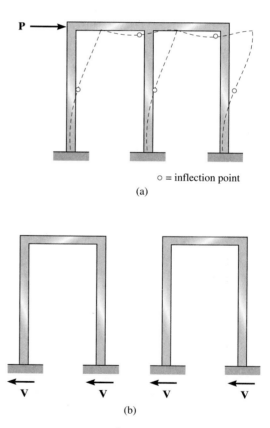

○ = inflection point

(a)

(b)

Fig. 7–12

In summary, then, the portal method for analyzing fixed-supported building frames requires the following assumptions:

1. A hinge is placed at the center of each girder since this is assumed to be a point of zero moment.
2. A hinge is placed at the center of each column since this is assumed to be a point of zero moment.
3. At a given floor level the shear at the interior column hinges is twice that at the exterior column hinges since the frame is considered to be a superposition of portals.

These assumptions provide an adequate reduction of the frame to one that is statically determinate yet stable under loading.

By comparison with the more exact statically indeterminate analysis, *the portal method is most suitable for buildings having low elevation* and *uniform framing*. The reason for this has to do with the structure's action under load. In this regard, *consider the frame as acting like a cantilevered beam* that is fixed to the ground. Recall from mechanics of materials that *shear resistance* becomes more important in the design of *short* beams, whereas *bending* is more important if the *beam is long*. (See Sec. 7.6.) The portal method is based on the assumption related to shear as stated in item 3 above.

The following examples illustrate how to apply the portal method to analyze a building bent.

The portal method of analysis can be used to (approximately) perform a lateral-load analysis of this single-story frame.

Example 7–5

Determine (approximately) the reactions at the base of the columns of the frame shown in Fig. 7–13a. Use the portal method of analysis.

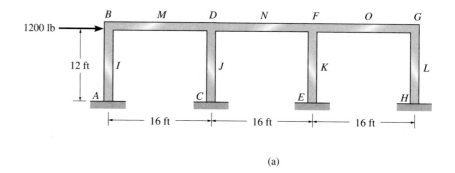

Fig. 7–13a

(a)

Solution

Applying the first two assumptions of the portal method, we place hinges at the centers of the girders and columns of the frame. The locations of these points are indicated by the letters I through O in Fig. 7–13a. A section through the column hinges at I, J, K, L yields the free-body diagram shown in Fig. 7–13b. Here the third assumption regarding the column shears applies. We require

$$\xrightarrow{+} \Sigma F_x = 0; \qquad 1200 - 6V = 0 \qquad V = 200 \text{ lb}$$

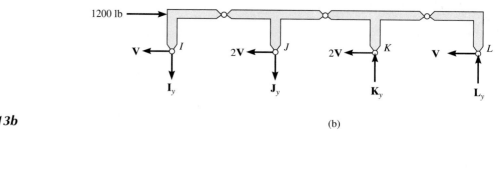

Fig. 7–13b

(b)

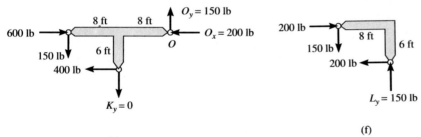

(c)   (d)   ***Fig. 7–13c,d***

Using this result, we can now proceed to dismember the frame at the hinges and determine their reactions. *As a general rule, always start this analysis at the corner where the horizontal load is applied.* Hence, the free-body diagram of segment *IBM* is shown in Fig. 7–13c. The three reaction components at the hinges I_y, M_x, and M_y are determined by applying $\Sigma M_M = 0$, $\Sigma F_x = 0$, $\Sigma F_y = 0$. The adjacent segment *MJN* is analyzed next, Fig. 7–13d, followed by segment *NKO*, Fig. 7–13e, and finally segment *OL*, Fig. 7–13f. Using these results, the free-body diagrams of the columns with their support reactions are shown in Fig. 7–13g.

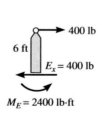

(e)

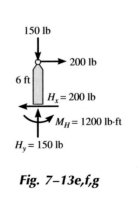

(f)

(g)

Fig. 7–13e,f,g

323

Example 7–6

Determine (approximately) the reactions at the base of the columns of the frame shown in Fig. 7–14a. Use the portal method of analysis.

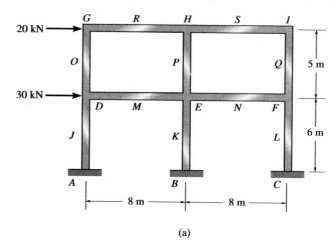

(a)

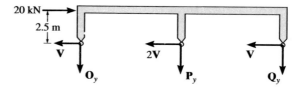

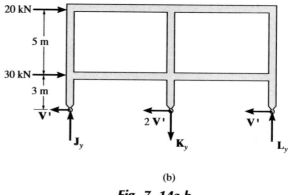

(b)

Fig. 7–14a,b

Solution

First hinges are placed at the *centers* of the girders and columns of the frame. The locations of these points are indicated by the letters J through S in Fig. 7–14a. A section through the hinges at O, P, Q and J, K, L yields the free-body diagrams shown in Fig. 7–14b. The column shears are calculated as follows:

$$\xrightarrow{+}\Sigma F_x = 0; \qquad 20 - 4V = 0 \qquad V = 5 \text{ kN}$$
$$\xrightarrow{+}\Sigma F_x = 0; \qquad 20 + 30 - 4V' = 0 \qquad V' = 12.5 \text{ kN}$$

Using these results, we can now proceed to analyze each part of the frame. The analysis starts with the *corner* segment OGR, Fig. 7–14c. The three unknowns O_y, R_x, and R_y have been calculated using the equations of equilibrium. With these results segment OJM is analyzed next, Fig. 7–14d; then segment JA, Fig. 7–14e; RPS, Fig. 7–14f; PMNK, Fig. 7–14g; and KB, Fig. 7–14h. Complete this example and analyze segments SIQ, then QNL, and finally LC, and show that $C_x = 12.5$ kN, $C_y = 15.625$ kN, and $M_C = 37.5$ kN · m.

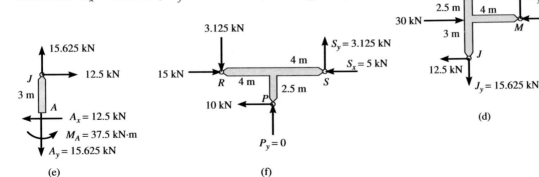

(c)

(d)

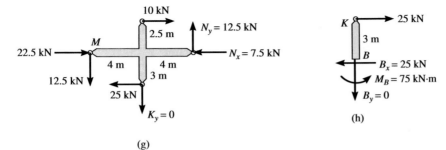

(e)

(f)

(g)

(h)

Fig. 7–14c,d,e,f,g,h

325

7.6 Lateral Loads on Building Frames: Cantilever Method

The cantilever method assumes that the *axial stress* in a column is proportional to its distance from the centroid of all the column areas at a given floor level. This assumption is based on the same action as a cantilevered beam subjected to a transverse load. It may be recalled from mechanics of materials that such a loading causes a bending stress in the beam that varies linearly from the beam's neutral axis, Fig. 7–15a. In a similar manner, the lateral loads on a frame tend to tip the frame over, or cause a rotation of the frame about a "neutral axis" lying in a horizontal plane that passes through the columns at each floor level. To counteract this tipping, the axial forces (or stress) in the columns will be tensile on one side of the neutral axis and compressive on the other side, Fig. 7–15b. Like the cantilevered beam, it therefore seems reasonable to assume this axial stress has a linear variation from the centroid of the column areas or neutral axis. *The cantilever method is therefore appropriate if the frame is tall and slender, or has columns with different cross-sectional areas.*

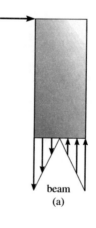

P

beam
(a)

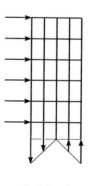

building frame
(b)

Fig. 7–15

In summary, then, using the cantilever method, the following assumptions apply to a fixed-supported frame.

1. A hinge is placed at the center of each girder since this is assumed to be a point of zero moment.
2. A hinge is placed at the center of each column since this is assumed to be a point of zero moment.
3. The axial *stress* in a column is proportional to its distance from the centroid of the cross-sectional areas of the columns at a given floor level. Since stress equals force per area, then in the special case of the *columns having equal cross-sectional areas*, the *force* in a column is also proportional to its distance from the centroid of the column areas.

These three assumptions reduce the frame to one that is both stable and statically determinate.

The following examples illustrate how to apply the cantilever method to analyze a building frame.

The building framework has rigid connections, and yet the designer has spliced the girders at their midpoints. Lateral-load analysis can be performed (approximately) by using the cantilever method of analysis. *(Peter Vandermark/Stock, Boston)*

Example 7–7

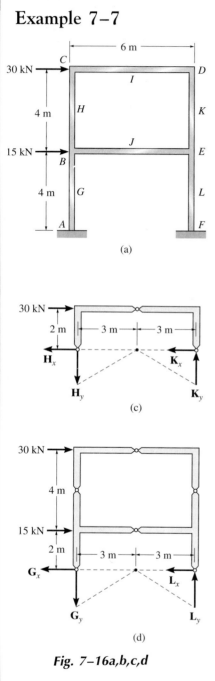

(a)

(c)

(d)

Fig. 7–16a,b,c,d

Determine (approximately) the reactions at the base of the columns of the frame shown in Fig. 7–16a. The columns are assumed to have equal cross-sectional areas. Use the cantilever method of analysis.

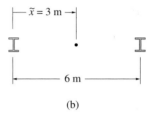

(b)

Solution

First hinges are placed at the midpoints of the columns and girders. The locations of these points are indicated by the letters G through L in Fig. 7–16a. The centroid of the columns' cross-sectional areas can be determined by inspection, Fig. 7–16b, or analytically as follows:

$$\bar{x} = \frac{\Sigma \tilde{x} A}{\Sigma A} = \frac{0(A) + 6(A)}{A + A} = 3 \text{ m}$$

The axial force in each column is thus proportional to its distance from this point. Hence, a section through the hinges H and K at the top story yields the free-body diagram shown in Fig. 7–16c. Note how the column to the left of the centroid must be subjected to tension and the one on the right is subjected to compression. This is necessary in order to counteract the tipping caused by the 30-kN force. Summing moments about the neutral axis (NA), we have

$$\zeta + \Sigma M_{NA} = 0; \qquad 30(2) - 3H_y - 3K_y = 0$$

The unknowns can be related by proportional triangles, Fig. 7–16c, that is,

$$\frac{H_y}{3} = \frac{K_y}{3} \qquad \text{or} \qquad H_y = K_y$$

Thus,

$$H_y = K_y = 10 \text{ kN}$$

In a similar manner, using a section of the frame through the hinges at G and L, Fig. 7–16d, we have

$$\overset{\curvearrowleft}{+}\Sigma M_{NA} = 0; \qquad 30(6) + 15(2) - 3G_y - 3L_y = 0$$

Since

$$\frac{G_y}{3} = \frac{L_y}{3} \qquad \text{or} \qquad G_y = L_y$$

Then

$$G_y = L_y = 35 \text{ kN}$$

Each part of the frame can now be analyzed using the above results. As in Examples 7–5 and 7–6, we begin at the upper corner where the applied loading occurs, i.e., segment HCI, Fig. 7–16a. Applying the three equations of equilibrium, $\Sigma M_I = 0$, $\Sigma F_x = 0$, $\Sigma F_y = 0$, yields the results for H_x, I_x, and I_y shown on the free-body diagram in Fig. 7–16e. Using these results, segment IDK is analyzed next, Fig. 7–16f; followed by HJG, Fig. 7–16g; then KJL, Fig. 7–16h; and finally the bottom portions of the columns, Fig. 7–16i.

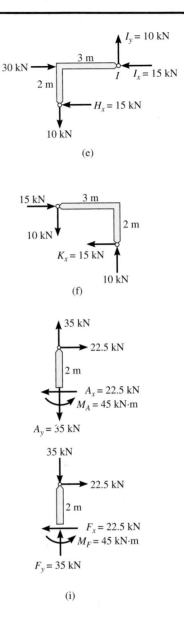

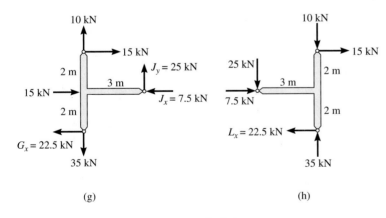

Fig. 7–16e,f,g,h,i

Example 7–8

Show how to determine (approximately) the reactions at the base of the columns of the frame shown in Fig. 7–17a. The columns have the cross-sectional areas shown in Fig. 7–17b. Use the cantilever method of analysis.

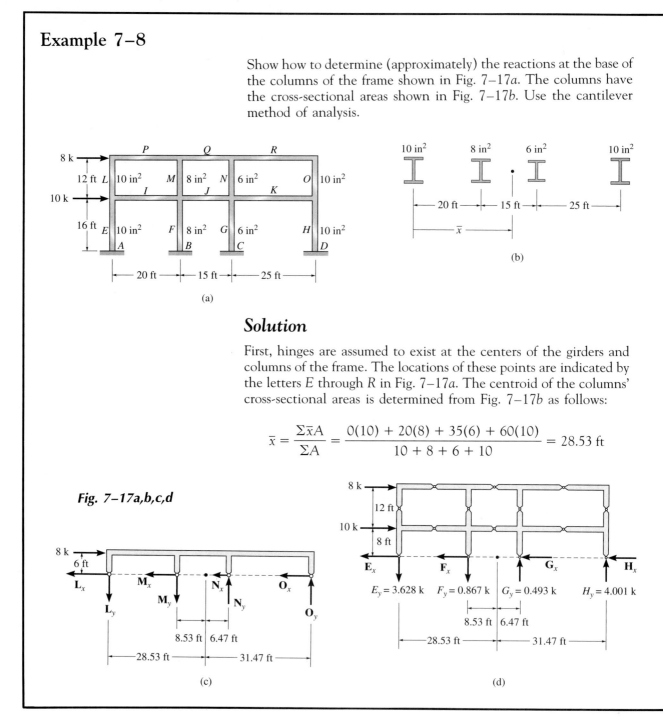

(a)

(b)

Solution

First, hinges are assumed to exist at the centers of the girders and columns of the frame. The locations of these points are indicated by the letters E through R in Fig. 7–17a. The centroid of the columns' cross-sectional areas is determined from Fig. 7–17b as follows:

$$\bar{x} = \frac{\Sigma \bar{x} A}{\Sigma A} = \frac{0(10) + 20(8) + 35(6) + 60(10)}{10 + 8 + 6 + 10} = 28.53 \text{ ft}$$

Fig. 7–17a,b,c,d

(c)

(d)

Here the columns have different cross-sectional areas, so the *axial stress* in a column is proportional to its distance from the neutral axis, located at $\bar{x} = 28.53$ ft. Hence, a section through the hinges at L, M, N, O yields the free-body diagram shown in Fig. 7–17c. Note how the columns to the left of the centroid are subjected to tension and those on the right are subjected to compression. Why? Summing moments about the neutral axis, we have

$$\zeta+\Sigma M_{NA} = 0; \quad 8(6) - L_y(28.53) - M_y(8.53) -$$
$$N_y(6.47) - O_y(31.47) = 0 \quad (1)$$

Since any column stress σ is proportional to its distance from the neutral axis, we can relate the column stresses by proportional triangles. Expressing the relations in terms of the force L_y, we have

$$\sigma_M = \frac{8.53}{28.53}\sigma_L; \quad \frac{M_y}{(8)} = \frac{8.53}{28.53}\frac{L_y}{(10)} \quad M_y = 0.239L_y \quad (2)$$

$$\sigma_N = \frac{6.47}{28.53}\sigma_L; \quad \frac{N_y}{(6)} = \frac{6.47}{28.53}\frac{L_y}{(10)} \quad N_y = 0.136L_y \quad (3)$$

$$\sigma_O = \frac{31.47}{28.53}\sigma_L; \quad \frac{O_y}{(10)} = \frac{31.47}{28.53}\frac{L_y}{(10)} \quad O_y = 1.103L_y \quad (4)$$

Solving Eqs. (1)–(4) yields

$$L_y = 0.726 \text{ k} \quad M_y = 0.173 \text{ k} \quad N_y = 0.0987 \text{ k} \quad O_y = 0.800 \text{ k}$$

Using this same method, show that one obtains the results in Fig. 7–17d for the columns at E, F, G, and H.

We can now proceed to analyze each part of the frame. As in the previous examples, we begin with the upper corner segment LP, Fig. 7–17e. Using the calculated results, segment LEI is analyzed next, Fig. 7–17f, followed by segment EA, Fig. 7–17g. One can continue to analyze the other segments in sequence, i.e., PQM, then MJFI, then FB, and so on.

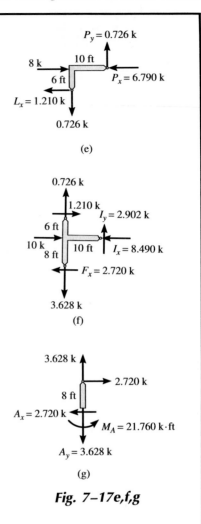

$P_y = 0.726$ k

8 k 10 ft

6 ft $P_x = 6.790$ k

$L_x = 1.210$ k

0.726 k

(e)

0.726 k

1.210 k

$I_y = 2.902$ k

6 ft

10 k 10 ft $I_x = 8.490$ k

8 ft

$F_x = 2.720$ k

3.628 k

(f)

3.628 k

2.720 k

8 ft

$A_x = 2.720$ k

$M_A = 21.760$ k·ft

$A_y = 3.628$ k

(g)

Fig. 7–17e,f,g

REFERENCE

Smith, A., "Wind Stresses in the Frames of Office Buildings," *Journal of the Western Society of Engineers*, Vol. 20, No. 4, 1915, p. 341.

PROBLEMS

7–1. Determine (approximately) the force in each member of the truss. Assume the diagonals can support either a tensile or compressive force.

7–2. Solve Prob. 7–1 assuming that the diagonals cannot support a compressive force.

7–5. Determine (approximately) the force in each member of the truss. Assume the diagonals can support both tensile and compressive forces.

7–6. Solve Prob. 7–5 assuming that the diagonals cannot support a compressive force.

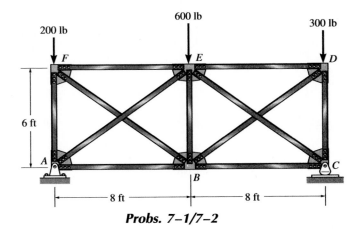

Probs. 7–1/7–2

7–3. Determine (approximately) the force in each member of the truss. Assume the diagonals can support either a tensile or a compressive force.

*7–4. Solve Prob. 7–3 assuming that the diagonals cannot support a compressive force.

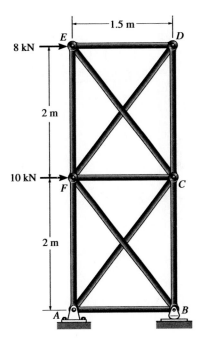

Probs. 7–5/7–6

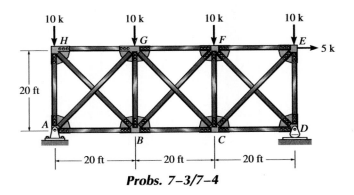

Probs. 7–3/7–4

333

7–7. Determine (approximately) the support reactions at A and B due to the uniform vertical load acting on the frame.

7–9. Determine (approximately) the support reactions at A, B, and C of the frame.

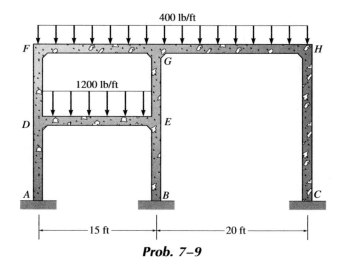

Prob. 7–9

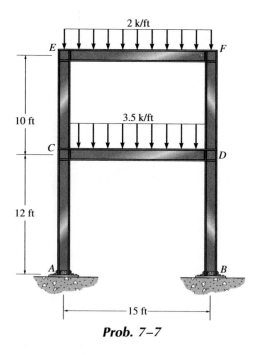

Prob. 7–7

7–10. Determine (approximately) the internal moment and shear at the ends of each member of the portal frame. Assume the supports at A and D are partially fixed, such that an inflection point is located at $h/3$ from the bottom of each column.

*7–8. Determine (approximately) the support reactions at A, B, and C of the frame.

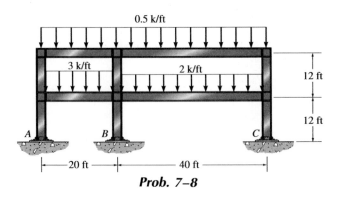

Prob. 7–8

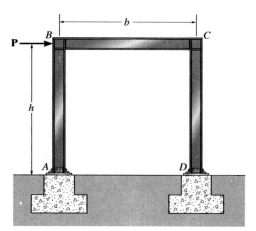

Prob. 7–10

7–11. Determine (approximately) the internal moment and shear at the ends of each member of the portal frame. Assume the supports at A and D are (a) pinned, (b) fixed, and (c) partially fixed such that the inflection point for the columns is located $h/3 = 6$ ft up from A and D.

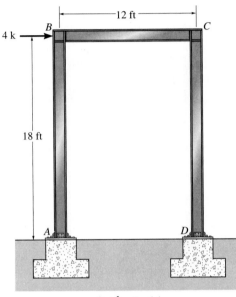

Prob. 7–11

*7–12. Determine (approximately) the internal moment at joints D and C. Assume the supports at A and B are pins.

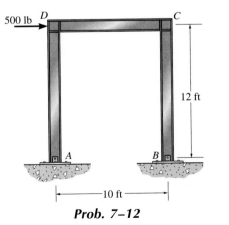

Prob. 7–12

7–13. Determine (approximately) the internal moment at joints C and D. Assume the supports at A and B are fixed.

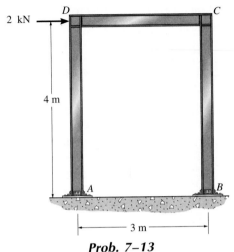

Prob. 7–13

7–14. Draw (approximately) the moment diagram for member ACE of the portal constructed with a *rigid* member EG and knee braces CF and DH. Assume that all points of connection are pins. Also determine the force in the knee brace CF.

7–15. Solve Prob. 7–14 if the supports at A and B are fixed instead of pinned.

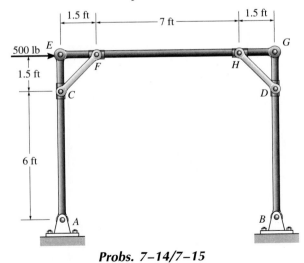

Probs. 7–14/7–15

335

*7–16. Determine (approximately) the force in each truss member of the portal frame. Also compute the reactions at the fixed column supports *A* and *B*. Assume all members of the truss to be pin-connected at their ends.

7–19. Determine (approximately) the force in each of the truss members of the portal frame. Also, determine the reactions at the column supports *A* and *B*. Assume all members of the truss and the columns to be pin-connected at their ends.

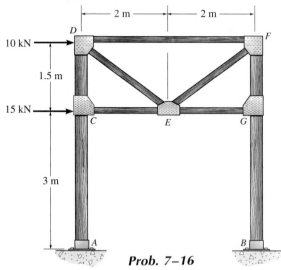

Prob. 7–16

7–17. Draw (approximately) the moment diagram for column *ACE* of the portal. Assume that all points of connection are pins. Also determine the force in the truss members *EG*, *CG*, and *CD*.

7–18. Solve Prob. 7–17 if the supports at *A* and *B* are fixed instead of pinned.

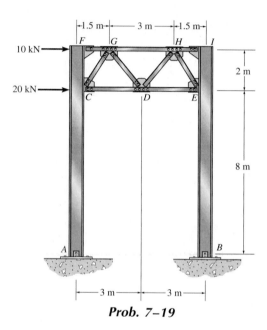

Prob. 7–19

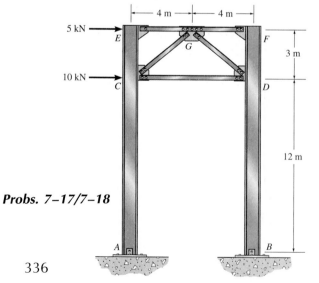

Probs. 7–17/7–18

*7–20. Determine (approximately) the force in each of the truss members of the portal frame. Also, determine the reactions at the column supports A and B. Assume all truss members and the columns to be pin-connected at their ends.

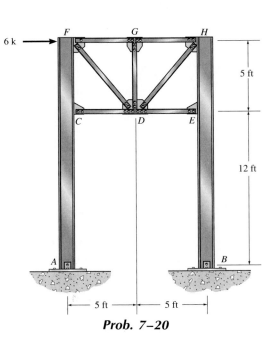

Prob. 7–20

7–21. Determine (approximately) the force in each truss member of the portal frame. Also compute the reactions at the fixed column supports A and B. Assume all members of the truss to be pin-connected at their ends.

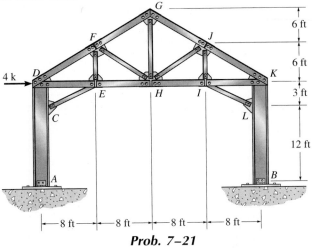

Prob. 7–21

7–22. Determine (approximately) the force in each truss member of the portal frame. Also compute the reactions at the fixed column supports A and B. Assume all members of the truss to be pin-connected at their ends.

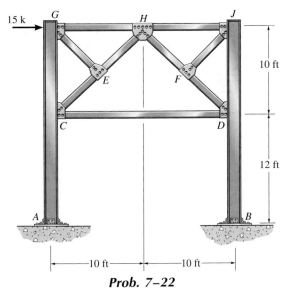

Prob. 7–22

337

7–23. Determine (approximately) the force in each truss member of the Warren truss portal frame. Also, determine the reactions at the column supports *A* and *B*. Assume all truss members and the column to be pin-connected at their ends.

*7–24. Solve Prob. 7–23 if the supports at *A* and *B* are fixed instead of pinned.

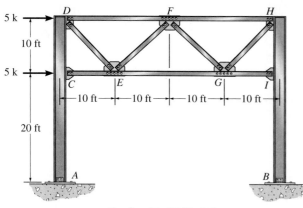

Probs. 7–23/7–24

7–25. Use the portal method of analysis and draw the moment diagram for girder *EFGH*.

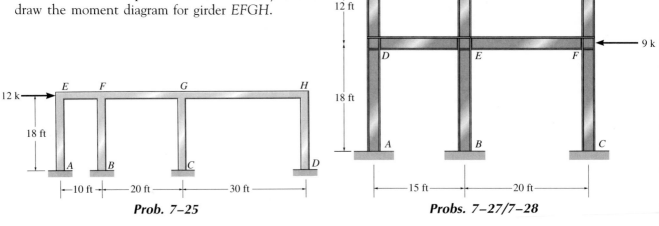

Prob. 7–25

7–26. Use the portal method of analysis and draw the moment diagram for girder *EFGH*.

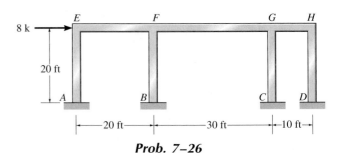

Prob. 7–26

7–27. Use the portal method of analysis and draw the moment diagram for column *ADG*.

*7–28. Solve Prob. 7–27 using the cantilever method of analysis. All the columns have the same cross-sectional area.

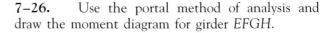

Probs. 7–27/7–28

7–29. Draw the moment diagram for girder *EFGH*. Use the portal method of analysis.

7–30. Solve Prob. 7–29 using the cantilever method of analysis. Each column has the same cross-sectional area.

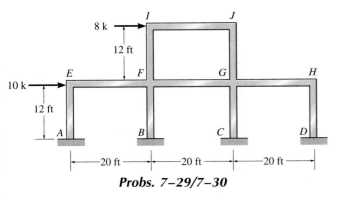

Probs. 7–29/7–30

7–31. Draw the moment diagram for girder *IJKL* of the building frame. Use the portal method of analysis.

***7–32.** Solve Prob. 7–31 using the cantilever method of analysis. Each column has the cross-sectional area indicated.

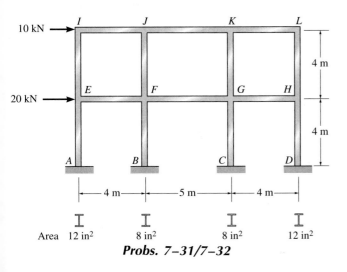

Probs. 7–31/7–32

7–33. Draw the moment diagram for girders *PQRST* and column *BGLQ* for the building frame. Use the portal method of analysis.

7–34. Solve Prob. 7–33 using the cantilever method of analysis. All columns have the same cross-sectional area.

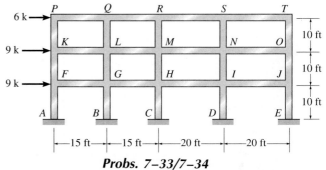

Probs. 7–33/7–34

A structure, such as this truss, may be subjected to unusual deflections when it is moved. Calculations for deflections must be performed to ensure proper support during transit. *(Photo courtesy American Institute of Steel Construction)*

Deflections

In this chapter we will determine the elastic deflections of a structure using both geometrical and energy methods. Also, the method of double integration will be discussed. The geometrical methods we will consider include the moment-area theorems and the conjugate-beam method, and the energy methods to be considered are based on virtual work and Castigliano's theorem. Each of these methods has particular advantages or disadvantages, which will be discussed at the time each is presented.

8.1 Deflection Diagrams and the Elastic Curve

Deflections of structures can occur from various sources, such as loads, temperature, fabrication errors, or settlement. In design, deflections must be limited in order to prevent cracking of attached brittle materials such as concrete or plaster. Furthermore, a structure must not vibrate or deflect severely in order to "appear" safe for its occupants. More important, though, deflections at specified points in a structure must be computed if one is to analyze statically indeterminate structures.

Table 8–1

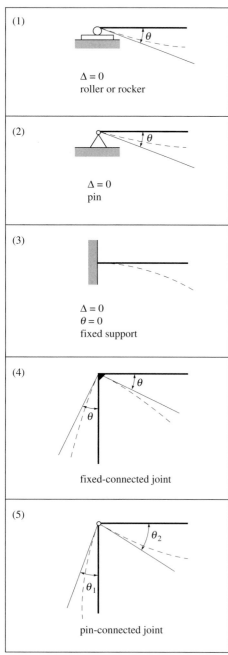

(1) $\Delta = 0$
roller or rocker

(2) $\Delta = 0$
pin

(3) $\Delta = 0$
$\theta = 0$
fixed support

(4) fixed-connected joint

(5) pin-connected joint

The analysis of deflections to be considered in this text applies only to structures having *linear elastic* material response. As a result, a structure subjected to a load will return to its original undeformed position after the load is removed. Most often the deflection of a structure is caused by internal loadings such as normal force, shear force, or bending moment. For *beams* and *frames* the greatest deflections are caused by *internal bending,* whereas *internal axial forces* cause the deflections of a *truss*.

Before the slope or displacement of a point on a beam or frame is computed, it is often helpful to sketch the shape of the structure when it is loaded in order to visualize the computed results and thereby partially check the results. This deflection diagram represents the *elastic curve* for the points representing the centroids of the cross-sectional areas along each of the members. For most problems the elastic curve can be sketched without much difficulty. When doing so, however, it is necessary to know the restrictions as to slope or displacement that often occur at a support or a connection. With reference to Table 8–1, supports that *resist a force,* such as a pin, *restrict displacement;* and those that *resist moment,* such as the fixed wall, *restrict rotation.* Note also that frame members that are fixed-connected (4) cause the joint to rotate the members by the same amount, θ, when the frame is loaded. On the other hand, if a pin connection is used at the joint, the members will each have a different slope or rotation at the pin, since a pin cannot support a moment (5). Using these restrictions, typical examples of deflected beams and frames, sketched to a greatly exaggerated scale, are shown in Fig. 8–1.

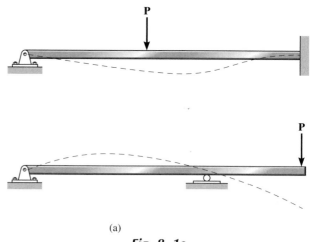

(a)

Fig. 8–1a

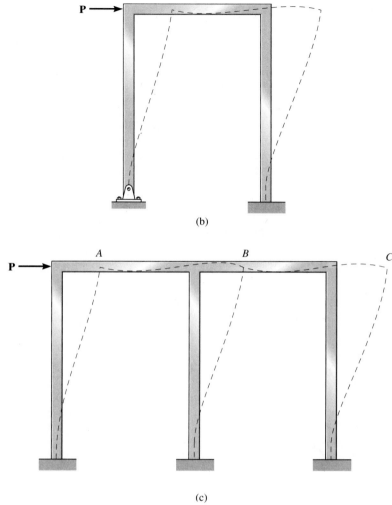

(b)

(c)

Fig. 8–1b,c

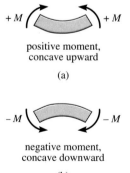

+ M positive moment, concave upward

(a)

− M negative moment, concave downward

(b)

Fig. 8–2

If the elastic curve seems difficult to establish, it is suggested that the moment diagram for the beam or frame be drawn first. By our sign convention for positive moments established in Chapter 4, a *positive moment* tends to bend a beam or horizontal member *concave upward*, Fig. 8–2a. Likewise, a *negative moment* tends to bend the beam or member *concave downward*, Fig. 8–2b. Therefore, if the moment diagram is

343

known, it will be easy to construct the elastic curve. For example, consider the beam in Fig. 8–3a with its associated moment diagram shown in Fig. 8–3b. Due to the pin-and-roller support, the displacement at A and D must be zero, Fig. 8–3c. Within the region of negative moment, Fig. 8–3b, the elastic curve is concave downward; and within the region of positive moment, the elastic curve is concave upward. In particular, there must be an *inflection point* at the point where the curve changes from concave down to concave up, since this is a point of zero moment. Using these same principles, note how the elastic curve for the beam in Fig. 8–4a was drawn in Fig. 8–4c, realizing that it is fixed at its end. Its moment diagram is shown in Fig. 8–4b.

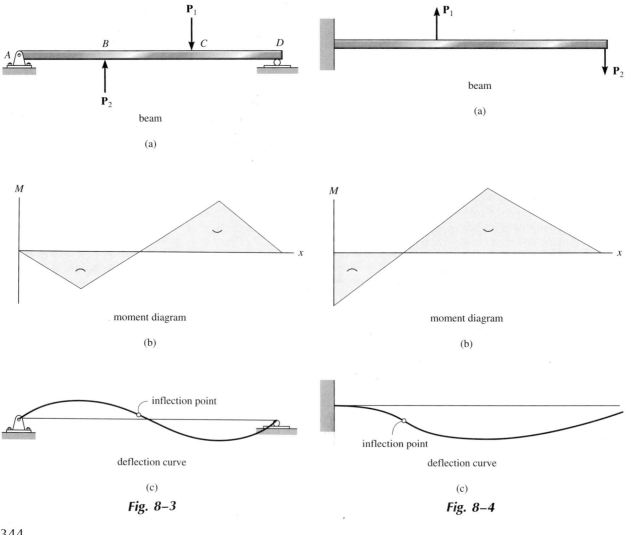

beam

(a)

moment diagram

(b)

deflection curve

(c)

Fig. 8–3

beam

(a)

moment diagram

(b)

deflection curve

(c)

Fig. 8–4

8.2   Elastic-Beam Theory

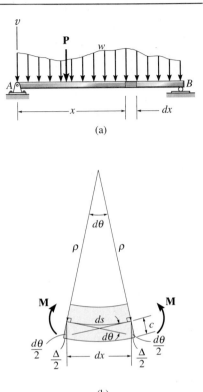

(a)

In this section we will develop two important differential equations that relate the internal moment in a beam to the displacement and slope of its elastic curve. These equations form the basis for the deflection methods presented in this chapter, and for this reason the assumptions and limitations used in their development should be fully understood.

To derive these relationships we will consider an initially *straight beam, elastically deformed* by loads that are applied perpendicular to the beam's axis and that lie in the longitudinal plane of symmetry for the beam's cross-sectional area. During deflection of the beam, *plane cross sections of the beam will remain plane.** With this in mind, consider the differential element dx, located a distance x from the left support of the beam in Fig. 8–5a. Isolating this element, Fig. 8–5b, it is seen that the beam's internal moment M distorts the element such that the tangents on each side of the element intersect at an angle $d\theta$. We can relate M to $d\theta$ by realizing that M causes the bottom fiber of the beam to be elongated Δ ($\Delta/2$ on each side) so that the strain in this fiber is $\epsilon = \Delta/dx$. Since linear-elastic distortions occur, Hooke's law applies, $\epsilon = \sigma/E$. Furthermore, the stress in the fiber occurs due to bending and therefore, from mechanics of materials, $\sigma = Mc/I$. Combining these equations $\epsilon = Mc/EI$ or $\Delta/dx = Mc/EI$. From the shaded triangle in Fig. 8–5b, $\Delta/2 = (d\theta/2)c$ or $d\theta = \Delta/c$.† Thus, the final result is

(b)

Fig. 8–5

$$d\theta = \frac{M}{EI}\,dx \qquad (8\text{–}1)$$

where

E = the material's modulus of elasticity
I = the moment of inertia of the beam's cross-sectional area computed about the neutral axis

The product EI is often called the *flexural rigidity*, and it is always a positive quantity.

---

*This important concept is based strictly on the geometry of deformation and is often referred to as the Euler–Bernoulli hypothesis.

†Since the distortion is small, here we have applied $s = \theta\,r$, where s is the arc of a circle inscribed by the sweep θ of the radius r.

345

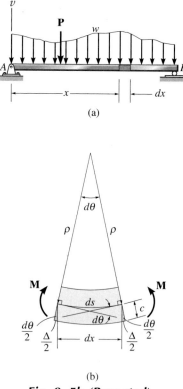

(a)

(b)

Fig. 8–5b (Repeated)

If the radius of curvature ρ (rho) of the elastic curve at x is to be determined, then from Fig. 8–5b, $\rho\, d\theta = dx$, and therefore

$$\frac{1}{\rho} = \frac{M}{EI} \tag{8–2}$$

If we choose the v axis positive upward, Fig. 8–5a, and if we can express the curvature $(1/\rho)$ in terms of x and v, we can then determine the elastic curve for the beam. In most calculus books it is shown that the curvature relationship is

$$\frac{1}{\rho} = \frac{d^2v/dx^2}{[1 + (dv/dx)^2]^{3/2}}$$

Therefore,

$$\frac{M}{EI} = \frac{d^2v/dx^2}{[1 + (dv/dx)^2]^{3/2}} \tag{8–3}$$

This equation represents a nonlinear second-order differential equation. Its solution, $v = f(x)$, gives the exact shape of the elastic curve—assuming, of course, that beam deflections occur only due to bending. In order to facilitate the solution of a greater number of problems, Eq. 8–3 can be modified by making an important simplification. Since elastic response has been assumed, the slope of the elastic curve, dv/dx, will be *very small*, and consequently its square will be negligible compared to unity. Therefore the curvature can be approximated by $1/\rho = d^2v/dx^2$. Using this simplification, Eq. 8–3 can be written as

$$\frac{d^2v}{dx^2} = \frac{M}{EI} \tag{8–4}$$

It should also be pointed out that by assuming $dv/dx \approx 0$, the original length of the beam's axis and the *arc* of its elastic curve will be approximately the same. In other words, ds in Fig. 8–5b is approximately equal to dx, since

$$ds = \sqrt{dx^2 + dv^2} = \sqrt{1 + (dv/dx)^2}\, dx \approx dx$$

This result implies that points on the elastic curve will only be displaced vertically and not horizontally.

Once M is expressed as a function of position x, then successive integrations of Eq. 8–4 will yield the beam's slope, $\theta = dv/dx$ (Eq. 8–1), and the equation of the elastic curve, $v = f(x)$, respectively. In the next section, we will show how this integration method is applied. However, the results from such an analysis for some common beam loadings often encountered in structural analysis are given in the table on the inside front cover of this book. Also listed are the slope and displacement at critical points on the beam. Obviously, no single table can account for the many different cases of loading and geometry that are encountered in structural analysis. When a table is not available or is incomplete, the displacement or slope of a specific point on a beam or frame can be determined by using the integration method or one of the other methods discussed in this chapter. All these methods, and any others like them, however, are formulated from Eqs. 8–1 and 8–4. These methods are therefore limited to problems involving small deflections caused only by elastic bending, as noted by the assumptions and limitations given above.

8.3   The Double Integration Method

As stated in the previous section, we can determine the equation of the elastic curve by direct integration of Eq. 8–4 ($d^2v/dx^2 = M/EI$). Solution of this equation requires two successive integrations to obtain the deflection v of the elastic curve. For each integration, it is necessary to introduce a "constant of integration," and then solve for the constants to obtain a unique solution for a particular problem. Recall from Sec. 4–2 that if the loading on a beam is discontinuous—that is, it consists of a series of several distributed and concentrated loads—then several functions must be written for the internal moment, each valid within the region between the discontinuities. For example, consider the beam shown in Fig. 8–6. The internal moment in regions AB, BC, and CD must be written in terms of the x_1, x_2, and x_3 coordinates. Once these functions are integrated through the application of Eq. 8–4 and the constants of integration determined, the functions will give the slope and deflection (elastic curve) for each region of the beam for which they are valid.

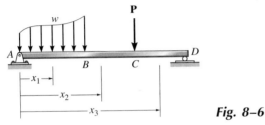

Fig. 8–6

347

Sign Convention

When applying Eq. 8–4, it is important to use the proper sign for M as established by the sign convention that was used in the derivation of this equation, Fig. 8–7a. Furthermore, recall that positive deflection, v, is upward, and as a result, the positive slope angle θ will be measured counterclockwise from the x axis. The reason for this is shown in Fig. 8–7b. Here, positive increases dx and dv in x and v create an increase $d\theta$ that is counterclockwise. Also, since the slope angle θ will be very small, its value in radians can be determined directly from $\theta \approx \tan \theta = dv/dx$.

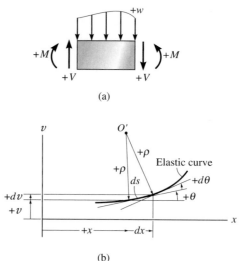

(a)

(b)

Fig. 8–7

Boundary and Continuity Conditions

The constants of integration are determined by evaluating the functions for slope or displacement at a particular point on the beam where the value of the function is known. These values are called *boundary conditions*. For example, if the beam is supported by a roller or pin, then it is required that the displacement be zero at these points. Also, at a fixed support the slope and displacement are both zero.

If a single x coordinate cannot be used to express the equation for the beam's slope or the elastic curve, then continuity conditions must be used to evaluate some of the integration constants. Consider the beam in Fig. 8–8. Here the x_1 and x_2 coordinates are valid only within the regions AB, and BC, respectively. Once the functions for the slope and deflection are obtained, they must give the same values for the slope and deflection at point B so that the elastic curve is physically continuous. Expressed mathematically, this requires that $\theta_1(a) = \theta_2(a)$ and $v_1(a) = v_2(a)$. These equations can then be used to evaluate two constants of integration.

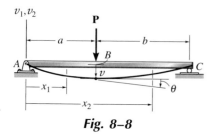

Fig. 8–8

Procedure for Analysis

The following procedure provides a method for determining the slope and deflection of a beam (or shaft) using the method of double integration. It should be realized that this method is suitable only for *elastic deflections* such that the beam's slope is very small. Furthermore, the method considers *only deflections due to bending*. Additional deflection due to shear generally represents only a few percent of the bending deflection, and so it is usually neglected in engineering practice.

Elastic Curve.   Draw an exaggerated view of the beam's elastic curve. Recall that points of zero slope and zero displacement occur at a fixed support, and zero displacement occurs at all pin and roller supports.

   Establish the x and v coordinate axes. The x axis must be parallel to the undeflected beam and has its origin at the left side of the beam, with a positive direction to the right. If several discontinuous loads are pesent, establish x coordinates that are valid for each region of the beam between the discontinuities. In all cases, the associated positive v axis should be directed upward.

Load or Moment Function.   For each region in which there is an x coordinate, express the internal moment M as a function of x. In particular, *always* assume that M acts in the positive direction when applying the equation of moment equilibrium to determine $M = f(x)$.

Slope and Elastic Curve.   Provided EI is constant, apply the moment equation $EI\, d^2v/dx^2 = M(x)$, which requires two integrations. For each integration it is important to include a constant of integration. The constants are evaluated using the boundary conditions for the supports and the continuity conditions that apply to slope and displacement at points where two functions meet. Once the constants are evaluated and substituted back into the slope and deflection equations, the slope and displacement at *specific points* on the elastic curve can then be determined. The numerical values obtained can be checked graphically by comparing them with the sketch of the elastic curve. Realize that *positive* values for *slope* are *counterclockwise and positive displacement* is *upward*.

The following examples illustrate application of this procedure.

Example 8–1

The cantilevered beam shown in Fig. 8–9a is subjected to a couple moment M_0 at its end. Determine the equation of the elastic curve. EI is constant.

Solution

Elastic Curve. The load tends to deflect the beam as shown in Fig. 8–9a. By inspection, the internal moment can be represented throughout the beam using a single x coordinate.

Moment Function. From the free-body diagram, with **M** acting in the *positive direction*, Fig. 8–9b, we have

$$M = M_0$$

Slope and Elastic Curve. Applying Eq. 8–4 and integrating twice yields

$$EI\frac{d^2v}{dx^2} = M_0 \tag{1}$$

$$EI\frac{dv}{dx} = M_0x + C_1 \tag{2}$$

$$EI\,v = \frac{M_0x^2}{2} + C_1x + C_2 \tag{3}$$

Using the boundary conditions $dv/dx = 0$ at $x = 0$ and $v = 0$ at $x = 0$, then $C_1 = C_2 = 0$. Substituting these results into Eqs. (2) and (3) with $\theta = dv/dx$, we get

$$\theta = \frac{M_0x}{EI}$$

$$v = \frac{M_0x^2}{2EI} \qquad\qquad \textbf{Ans.}$$

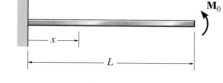

(a)

(b)

Fig. 8–9

Maximum slope and displacement occur at $A(x = 0)$, for which

$$\theta_A = \frac{M_0 L}{EI} \tag{4}$$

$$v_A = \frac{M_0 L^2}{2EI} \tag{5}$$

The *positive* result for θ_A indicates *counterclockwise* rotation and the *positive* result for v_A indicates that v_A is *upward*. This agrees with the results sketched in Fig. 8–9a.

In order to obtain some idea as to the actual *magnitude* of the slope and displacement at the end A, consider the beam in Fig. 8–9a to have a length of 12 ft, support a couple moment of 15 k · ft, and be made of steel having $E_{st} = 29(10^3)$ ksi. If this beam was designed without a factor of safety by assuming the allowable normal stress is equal to the yield stress $\sigma_{allow} = 36$ ksi, then a W18 × 35 would be found to be adequate ($I = 510$ in^4). From Eqs. (4) and (5) we get

$$\theta_A = \frac{15\ \text{k} \cdot \text{ft}(12\ \text{in./ft})(12\ \text{ft})(12\ \text{in./ft})}{29(10^3)\ \text{k/in}^2(510\ \text{in}^4)} = 0.00175\ \text{rad}$$

$$v_A = \frac{15\ \text{k} \cdot \text{ft}(12\ \text{in./ft})(12\ \text{ft})^2(12\ \text{in./1 ft})^2}{2(29(10^3)\ \text{k/in}^2)(510\ \text{in}^4)} = 0.126\ \text{in.}$$

Since $\theta_A^2 = 3.07(10^{-6})\ \text{rad}^2 \ll 1$, this justifies the use of Eq. 8–4, rather than applying the more exact Eq. 8–3, for computing the deflection of beams. Also, since this numerical application is for a *cantilevered beam*, we have obtained *larger values* for θ and v than would have been obtained if the beam was supported using pins, rollers, or other fixed supports.

Example 8–2

The beam in Fig. 8–10a is subjected to a load **P** at its end. Determine the displacement at C. EI is constant.

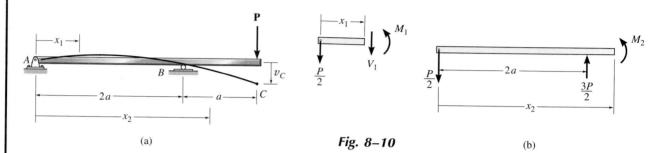

Fig. 8–10

(a)   (b)

Solution

Elastic Curve. The beam deflects into the shape shown in Fig. 8–10a. Due to the loading, two x-coordinates must be considered. x_2 is directed to the left from C, since the internal moment is easy to formulate.

Moment Functions. Using the free-body diagrams shown in Fig. 8–10b, we have

$$M_1 = -\frac{P}{2}x_1$$

$$M_2 = -\frac{P}{2}x_2 + \frac{3P}{2}(x_2 - 2a)$$

$$= Px_2 - 3Pa$$

Slope and Elastic Curve. Applying Eq. 8–4,

for x_1, $$EI\frac{d^2v_1}{dx_1^2} = -\frac{P}{2}x_1$$

$$EI\frac{dv_1}{dx_1} = -\frac{P}{4}x_1^2 + C_1 \tag{1}$$

$$EIv_1 = -\frac{P}{12}x_1^3 + C_1x + C_2 \tag{2}$$

For x_2,
$$EI \frac{d^2 v_2}{dx_2^2} = Px_2 - 3Pa$$

$$EI \frac{dv_2}{dx_2} = \frac{P}{2}x_2^2 - 3Pax_2 + C_3 \tag{3}$$

$$EIv_2 = \frac{P}{6}x_2^3 - \frac{3}{2}Pax_2^2 + C_3 x_2 + C_4 \tag{4}$$

The *four* constants of integration are determined using *three* boundary conditions, namely, $v_1 = 0$ at $x_1 = 0$, $v_1 = 0$ at $x_1 = 2a$, and $v_2 = 0$ at $x_2 = 2a$ and *one* continuity equation. Here the continuity of slope at the roller requires $dv_1/dx_1 = dv_2/dx_2$ at $x_1 = x_2 = 2a$. Note that continuity of displacement at B has been indirectly considered in the boundary conditions, since $v_1 = v_2 = 0$ at $x_1 = x_2 = 2a$. Applying these four conditions yields

$v_1 = 0$ at $x_1 = 0$; $0 = 0 + 0 + C_2$

$v_1 = 0$ at $x_1 = 2a$; $0 = -\dfrac{P}{12}(2a)^3 + C_1(2a) + C_2$

$v_2 = 0$ at $x_2 = 2a$; $0 = \dfrac{P}{6}(2a)^3 - \dfrac{3}{2}Pa(2a)^2 + C_3(2a) + C_4$

$\dfrac{dv_1(2a)}{dx_1} = \dfrac{dv_2(2a)}{dx^2}; -\dfrac{P}{4}(2a)^2 + C_1 = \dfrac{P}{2}(2a)^2 - 3Pa(2a) + C_3$

Solving, we obtain

$$C_1 = \frac{Pa^2}{3} \qquad C_2 = 0 \qquad C_3 = \frac{10}{3}Pa^2 \qquad C_4 = -2Pa^3$$

Substituting C_3 and C_4 into Eq. (4) gives

$$v_2 = \frac{P}{6EI}x_2^3 - \frac{3}{2}\frac{Pa}{EI}x_2^2 + \frac{10Pa^2}{3EI}x^2 - \frac{2Pa^3}{EI}$$

The displacement at C is determined by setting $x_2 = 3a$. We get

$$v_C = -\frac{Pa^3}{EI} \qquad\qquad \textbf{Ans.}$$

8.4 Moment-Area Theorems

The initial ideas for the two moment-area theorems were developed by Otto Mohr and later stated formally by Charles E. Greene in 1872. These theorems provide a semigraphical technique for determining the slope of the elastic curve and its deflection due to bending. They are particularly advantageous when used to solve problems involving beams, especially those subjected to a series of concentrated loadings or having segments with different moments of inertia. To develop these theorems, reference is made to the beam in Fig. 8–11a. If we draw the moment diagram for the beam and then divide it by the flexural rigidity, EI, the "M/EI diagram" shown in Fig. 8–11b results. Since Eq. 8–1 states that

$$d\theta = \left(\frac{M}{EI}\right) dx$$

it can be seen that the change $d\theta$ in the slope of the tangents on either side of the element dx is equal to the area under the M/EI diagram. Integrating from point A on the elastic curve to point B, Fig. 8–11c, we have

$$\theta_{AB} = \int_A^B \frac{M}{EI} dx \tag{8–5}$$

This equation forms the basis for the first moment-area theorem.

> **Theorem 1: The change in slope between any two points on the elastic curve equals the area of the M/EI diagram between these two points.**

From the proof it should be evident that the slope from A to B *increases* if the area of the M/EI diagram is *positive*, Fig. 8–11c. Conversely, if this area is *negative*, or below the x axis, the slope *decreases*. Furthermore, from a dimensional analysis of Eq. 8–5, θ_{AB} (θ "A to B") is measured in radians.

The second moment-area theorem is based on the relative deviation of *tangents* to the elastic curve. Shown in Fig. 8–12c is a greatly exaggerated view of the *vertical deviation dt* of the tangents on each side of the differential element dx. Using the circular arc formula $s = \theta r$, where r is of length x, we can write $dt = x\, d\theta$. This approximation is actually quite accurate since the slope of the elastic curve and its de-

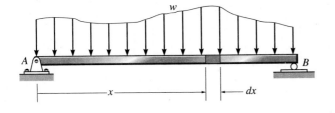

(a)

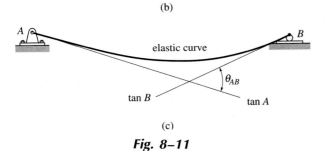

(b)

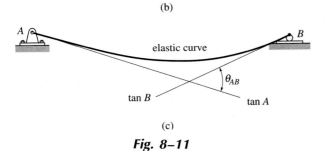

elastic curve

θ_{AB}

tan B   tan A

(c)

Fig. 8–11

flection are very small. Using Eq. 8–1, the deviation of the tangent at
B from that at A can be found by integration, in which case*

$$t_{A/B} = \int_A^B x \, \frac{M}{EI} \, dx \qquad (8\text{--}6)$$

Recall from statics that the centroid of an area is determined from
$\bar{x} \int dA = \int x \, dA$. Since $\int_A^B M/EI \, dx$ represents an area of the M/EI

---

*The notation $t_{A/B}$ refers to the deviation of the tangent at A with respect to
that at B.

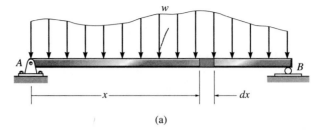

(a)

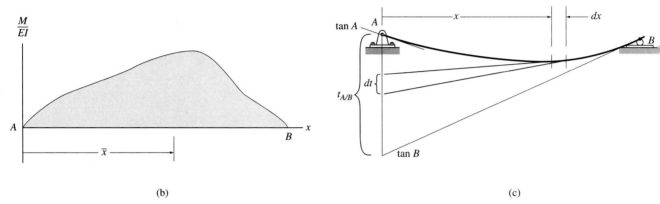

(b)

(c)

Fig. 8–12

diagram, we can also write

$$t_{A/B} = \bar{x} \int_A^B \frac{M}{EI}\, dx \qquad (8\text{–}7)$$

where $\bar{x}$ is the distance from point A to the *centroid* of the area, Fig. 8–12*b*. The second moment-area theorem can now be stated as follows:

> **Theorem 2: The deviation of the tangent at point B on the elastic curve with respect to the tangent at point A equals the "moment" of the M/EI diagram between the two points A and B computed about point A (the point on the elastic curve), where the deviation $t_{A/B}$ is to be determined.**

Provided the moment of a *positive* M/EI area from A to B is computed, as in Fig. 8–12*b*, it indicates that the tangent at point A is *above* the tangent to the curve extended from point B, Fig. 8–12*c*. Similarly, *negative* M/EI areas indicate that the tangent at A is *below* the tangent extended from B.

Procedure for Analysis

The following procedure provides a method that may be used to determine the displacement and slope at a point on the elastic curve of a beam using the moment-area theorems.

M/EI *Diagram*   Determine the support reactions and draw the beam's M/EI diagram. If the beam is loaded with concentrated forces, the M/EI diagram will consist of a series of straight line segments, and the areas and their moments required for the moment-area theorems will be relatively easy to compute. If the loading consists of a series of concentrated forces and distributed loads, it may be simpler to compute the required M/EI areas and their moments by drawing the M/EI diagram in parts, using the method of superposition as indicated in Sec. 4.5. In any case, the M/EI diagram will consist of parabolic or perhaps higher-order curves, and it is suggested that the table on the inside back cover be used to locate the area and centroid under each curve.

Elastic Curve   Draw an exaggerated view of the beam's elastic curve. Recall that points of zero slope occur at fixed supports and zero displacement occurs at all fixed, pin, and roller supports. If it becomes difficult to draw the general shape of the elastic curve, use the moment (or M/EI) diagram. Realize that when the beam is subjected to a *positive moment* the beam bends *concave up*, whereas *negative moment* bends the beam *concave down*. Furthermore, an inflection point or change in curvature occurs where the moment in the beam (or M/EI) is zero.

The displacement and slope to be determined should be sketched on the curve. Draw extended (or long) tangents on the curve at these points and often times at the supports. Since the moment-area theorems apply only between two tangents, attention should be given as to which tangents will be used to determine the slope or displacement.

Moment-Area Theorems   Apply Theorem 1 to determine the angle only between two tangents, and Theorem 2 to determine tangential deviations between these tangents. Keep in mind that Theorem 2 in general *will not* yield the displacement of a point on the elastic curve. When applied properly, it will only give the vertical distance or deviation of a tangent for point B from a point A (or tangent at A) *located on the elastic curve*. After applying either Theorem 1 or Theorem 2, the algebraic sign of the answer can be verified from the angle or deviation as indicated on the elastic curve.

The following examples illustrate this method of solution.

Example 8–3

Determine the slope at points B and C of the beam shown in Fig. 8–13a. Take $E = 29(10^3)$ ksi and $I = 600$ in^4.

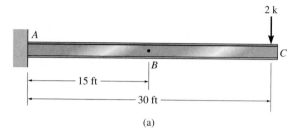

Fig. 8–13a

(a)

Solution

M/EI Diagram. This diagram is shown in Fig. 8–13b. It is easier to solve the problem in terms of EI and substitute the numerical data as a last step.

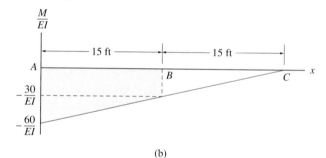

Fig. 8–13b

(b)

Elastic Curve. The 2-k load causes the beam to deflect as shown in Fig. 8–13c. (The beam is deflected concave down, since M/EI is negative.) Note that the tangent at A is *always horizontal*. The tangents at B and C are also indicated. We are required to find θ_B and θ_C. By the construction, note that the angle between tan A and tan B, that is, θ_{AB}, is equivalent to θ_B.

$$\theta_B = \theta_{AB}$$

Also,

$$\theta_C = \theta_{AC}$$

358

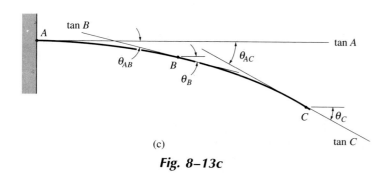

(c)

Fig. 8–13c

Moment-Area Theorem.   Applying Theorem 1, θ_{AB} is equal to the area under the M/EI diagram between points A and B; that is,

$$\theta_B = \theta_{AB} = -\left(\frac{30 \text{ k} \cdot \text{ft}}{EI}\right)(15 \text{ ft}) - \frac{1}{2}\left(\frac{60 \text{ k} \cdot \text{ft}}{EI} - \frac{30 \text{ k} \cdot \text{ft}}{EI}\right)(15 \text{ ft})$$
$$= -\frac{675 \text{ k} \cdot \text{ft}^2}{EI}$$

Substituting numerical data for E and I, we have

$$\theta_B = \frac{-675 \text{ k} \cdot \text{ft}^2(144 \text{ in}^2/1 \text{ ft}^2)}{(29(10^3) \text{ k/in}^2)(600 \text{ in}^4)}$$
$$= -0.00559 \text{ rad} \qquad \textbf{Ans.}$$

The *negative sign* indicates that the slope at B *decreases* from that at A.

In a similar manner, the area under the M/EI diagram between points A and C equals θ_{AC}. We have

$$\theta_C = \theta_{AC} = \frac{1}{2}\left(-\frac{60 \text{ k} \cdot \text{ft}}{EI}\right)(30 \text{ ft}) = -\frac{900 \text{ k} \cdot \text{ft}^2}{EI}$$

Substituting numerical values for EI, and converting feet to inches, we have

$$\theta_C = \frac{-900 \text{ k} \cdot \text{ft}^2(144 \text{ in}^2/\text{ft}^2)}{29(10^3) \text{ k/in}^2(600 \text{ in}^4)}$$
$$= -0.00745 \text{ rad} \qquad \textbf{Ans.}$$

Example 8–4

Determine the deflection at points B and C of the beam shown in Fig. 8–14a. Values for the moment of inertia of each segment are indicated in the figure. Take $E = 200$ GPa.

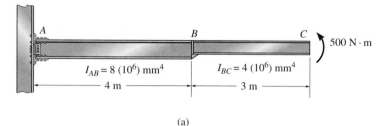

Fig. 8–14a

(a)

Solution

M/EI Diagram.   By inspection, the moment diagram for the beam is a rectangle. (See Fig. 4–12.) Here we will construct the M/EI diagram relative to I_{BC}, realizing that $I_{AB} = 2I_{BC}$, Fig. 8–14b. Numerical data for EI_{BC} will be substituted as a last step.

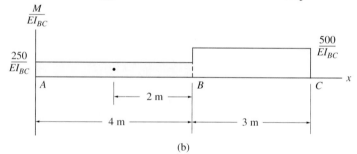

Fig. 8–14b

(b)

Elastic Curve.   The couple moment at C causes the beam to deflect as shown in Fig. 8–14c. The tangents at A, B, and C are indicated. We are required to find Δ_B and Δ_C. These displacements can be related directly to the deviations between the tangents, whereby from the construction Δ_B is equal to the deviation of tan B from tan A; that is,

$$\Delta_B = t_{B/A}$$

Also,

$$\Delta_C = t_{C/A}$$

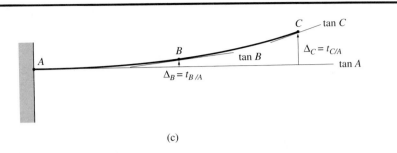

(c)

Fig. 8–14c

Moment-Area Theorem. Applying Theorem 2, $t_{B/A}$ is equal to the moment of the M/EI_{BC} diagram between A and B computed about point B, since this is the point where the tangential deviation is to be determined. Hence, from Fig. 8–14b,

$$\Delta_B = t_{B/A} = \left[\frac{250\,\text{N}\cdot\text{m}}{EI_{BC}}(4\,\text{m})\right](2\,\text{m}) = \frac{2000\,\text{N}\cdot\text{m}^3}{EI_{BC}}$$

Substituting the numerical data yields

$$\Delta_B = \frac{2000\,\text{N}\cdot\text{m}^3}{[200(10^9)\,\text{N/m}^2][4(10^6)\,\text{mm}^4(1\,\text{m}^4/(10^3)^4\,\text{mm}^4)]}$$
$$= 0.0025\,\text{m} = 2.5\,\text{mm} \qquad \textbf{Ans.}$$

Likewise, for $t_{C/A}$ we must compute the moment of the entire M/EI_{BC} diagram from A to C about point C. We have

$$\Delta_C = t_{C/A} = \left[\frac{250\,\text{N}\cdot\text{m}}{EI_{BC}}(4\,\text{m})\right](5\,\text{m}) + \left[\frac{500\,\text{N}\cdot\text{m}}{EI_{BC}}(3\,\text{m})\right](1.5\,\text{m})$$
$$= \frac{7250\,\text{N}\cdot\text{m}^3}{EI_{BC}}$$

or

$$\Delta_C = \frac{7250\,\text{N}\cdot\text{m}^3}{[200(10^9)\,\text{N/m}^2][4(10^6)(10^{-12})\,\text{m}^4]}$$
$$= 0.00906\,\text{m} = 9.06\,\text{mm} \qquad \textbf{Ans.}$$

Since both answers are *positive*, they indicate that points B and C lie *above* the tangent at A.

Example 8–5

Determine the slope at point C of the beam in Fig. 8–15a. $E = 200$ GPa, $I = 6(10^6)$ mm^4.

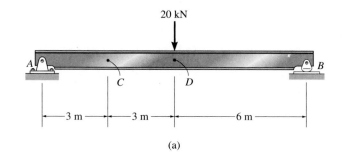

Fig. 8–15a

(a)

Solution

M/EI Diagram. Fig. 8–15b.

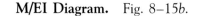

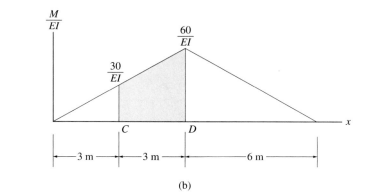

Fig. 8–15b

(b)

Elastic Curve. Since the loading is applied symmetrically to the beam, the elastic curve is symmetric, as shown in Fig. 8–15c. We are required to find θ_C. This can easily be done, realizing that the tangent at D is *horizontal*, and therefore, by the construction, the angle θ_{CD} between tan C and tan D is equal to θ_C; that is,

$$\theta_C = \theta_{CD}$$

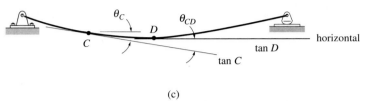

(c)

Fig. 8–15c

Moment-Area Theorem.   Using Theorem 1, θ_{CD} is equal to the shaded area under the M/EI diagram between points C and D. We have

$$\theta_C = \theta_{CD} = 3 \text{ m}\left(\frac{30 \text{ kN} \cdot \text{m}}{EI}\right) + \frac{1}{2}(3 \text{ m})\left(\frac{60 \text{ kN} \cdot \text{m}}{EI} - \frac{30 \text{ kN} \cdot \text{m}}{EI}\right)$$

$$= \frac{135 \text{ kN} \cdot \text{m}^2}{EI}$$

Thus,

$$\theta_C = \frac{135 \text{ kN} \cdot \text{m}^2}{[200(10^6) \text{ kN/m}^2][6(10^6)(10^{-12}) \text{ m}^4]} = 0.112 \text{ rad} \quad \textbf{Ans.}$$

Example 8–6

Determine the slope at point C of the beam in Fig. 8–16a. $E = 29(10^3)$ ksi, $I = 600$ in⁴.

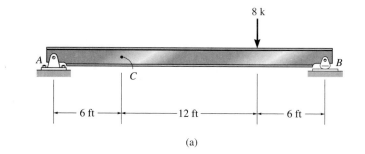

Fig. 8–16a

(a)

Solution

M/EI Diagram. Fig. 8–16b.

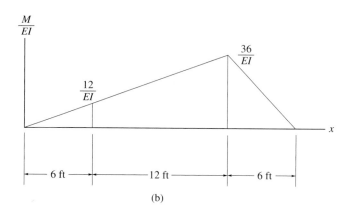

(b)

Fig. 8–16b

Elastic Curve. The elastic curve is shown in Fig. 8–16c. We are required to find θ_C. Angle θ_{AC} is the angle between the tangents at A and C. The angle ϕ in Fig. 8–16c can be found using $\phi = t_{B/A}/L_{AB}$. This equation is valid since $t_{B/A}$ is very small, so that $t_{B/A}$ can be approximated by the length of a circular arc defined by a radius of $L_{AB} = 24$ ft and sweep of ϕ. (Recall that $s = \theta r$.) From the geometry of Fig. 8–16c, we have

$$\theta_C = \phi - \theta_{AC} = \frac{t_{B/A}}{24} - \theta_{AC} \qquad (1)$$

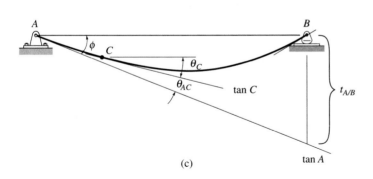

(c)

tan A

Fig. 8–16c

Moment-Area Theorems.   Using Theorem 1, θ_{AC} is equivalent to the area under the M/EI diagram between points A and C; that is,

$$\theta_{AC} = \frac{1}{2}(6 \text{ ft})\left(\frac{12 \text{ k} \cdot \text{ft}}{EI}\right) = \frac{36 \text{ k} \cdot \text{ft}^2}{EI}$$

Applying Theorem 2, $t_{B/A}$ is equivalent to the moment of the area under the M/EI diagram between B and A about point B, since this is the point where the tangential deviation is to be determined. We have

$$t_{B/A} = \left[6 \text{ ft} + \frac{1}{3}(18 \text{ ft})\right]\left[\frac{1}{2}(18 \text{ ft})\left(\frac{36 \text{ k} \cdot \text{ft}}{EI}\right)\right]$$
$$+ \frac{2}{3}(6 \text{ ft})\left[\frac{1}{2}(6 \text{ ft})\left(\frac{36 \text{ k} \cdot \text{ft}}{EI}\right)\right]$$
$$= \frac{4320 \text{ k} \cdot \text{ft}^3}{EI}$$

Substituting these results into Eq. (1), we have

$$\theta_C = \frac{4320 \text{ k} \cdot \text{ft}^3}{(24 \text{ ft}) EI} - \frac{36 \text{ k} \cdot \text{ft}^2}{EI} = \frac{144 \text{ k} \cdot \text{ft}^2}{EI}$$

so that

$$\theta_C = \frac{144 \text{ k} \cdot \text{ft}^2}{29(10^3) \text{ k/in}^2(144 \text{ in}^2/\text{ft}^2)600 \text{ in}^4(\text{ft}^4/(12)^4 \text{ in}^4)}$$
$$= 0.00119 \text{ rad} \qquad\qquad \textbf{Ans.}$$

Example 8–7

Determine the deflection at C of the beam shown in Fig. 8–17a. Take $E = 29(10^3)$ ksi, $I = 21$ in⁴.

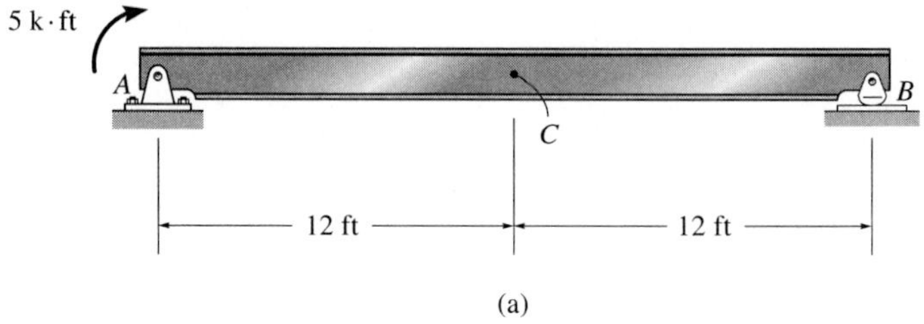

Fig. 8–17a

(a)

Solution

M/EI Diagram.   Fig. 8–17b.

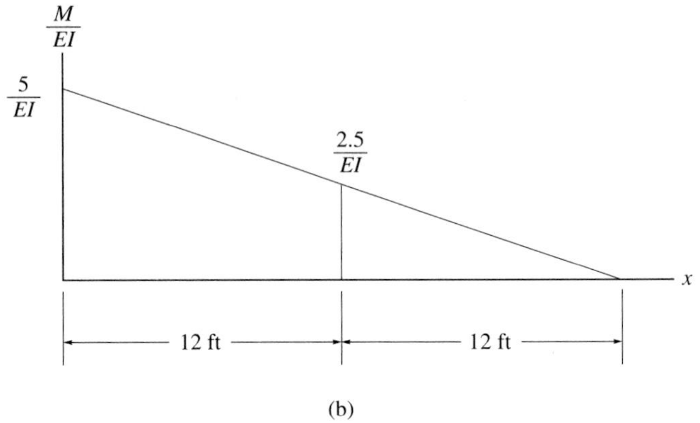

Fig. 8–17b

(b)

Elastic Curve.   Here we are required to find Δ_C, Fig. 8–17c. Note that this is *not* the maximum deflection of the beam, since the loading and hence the elastic curve are *not symmetric*. Also indicated in Fig. 8–17c are the tangents at A, B, and C. Notice that if $t_{A/B}$ is determined, then Δ' can be found from proportional triangles, that is, $\Delta'/12 = t_{A/B}/24$ or $\Delta' = t_{A/B}/2$. From the construction in Fig. 8–17c, we have

$$\Delta_C = \frac{t_{A/B}}{2} - t_{C/B} \qquad (1)$$

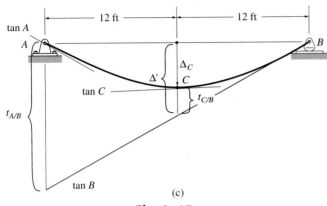

(c)

Fig. 8–17c

Moment-Area Theorem. Applying Theorem 2 to determine $t_{A/B}$ and $t_{C/B}$, we have

$$t_{A/B} = \left[\frac{1}{3}(24 \text{ ft})\right]\left[\frac{1}{2}(24 \text{ ft})\left(\frac{5 \text{ k} \cdot \text{ft}}{EI}\right)\right] = \frac{480 \text{ k} \cdot \text{ft}^3}{EI}$$

$$t_{C/B} = \left[\frac{1}{3}(12 \text{ ft})\right]\left[\frac{1}{2}(12 \text{ ft})\left(\frac{2.5 \text{ k} \cdot \text{ft}}{EI}\right)\right] = \frac{60 \text{ k} \cdot \text{ft}^3}{EI}$$

Substituting these results into Eq. (1) yields

$$\Delta_C = \frac{1}{2}\left(\frac{480 \text{ k} \cdot \text{ft}^3}{EI}\right) - \frac{60 \text{ k} \cdot \text{ft}^3}{EI} = \frac{180 \text{ k} \cdot \text{ft}^3}{EI}$$

Working in units of kips and inches, we have

$$\Delta_C = \frac{180 \text{ k} \cdot \text{ft}^3(1728 \text{ in}^3/\text{ft}^3)}{(29(10^3) \text{ k/in}^2)(21 \text{ in}^4)}$$

$$= 0.511 \text{ in.} \qquad\qquad \textbf{\textit{Ans.}}$$

Example 8–8

Determine the deflection at point C of the beam shown in Fig. 8–18a. $E = 200$ GPa, $I = 250(10^6)$ mm^4.

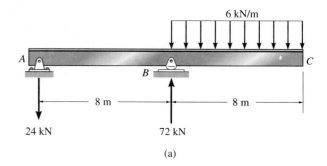

Fig. 8–18a

(a)

Solution

M/EI Diagram. As shown in Fig. 8–18b, this diagram consists of triangular and parabolic segments.

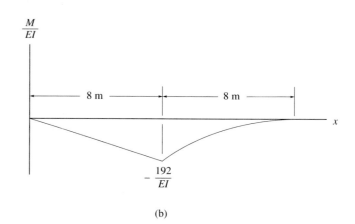

Fig. 8–18b

(b)

Elastic Curve. The loading causes the beam to deform as shown in Fig. 8–18c. We are required to find Δ_C. By constructing tangents at A, B, and C, it is seen that $\Delta_C = t_{C/A} - \Delta'$. However, Δ' can be related to $t_{B/A}$ by proportional triangles, that is, $\Delta'/16 = t_{B/A}/8$ or $\Delta' = 2t_{B/A}$. Hence

$$\Delta_C = t_{C/A} - 2t_{B/A} \tag{1}$$

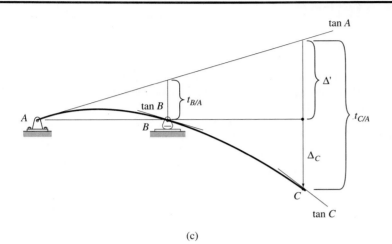

(c)

Fig. 8–18c

Moment-Area Theorem.   Applying Theorem 2 to determine $t_{C/A}$ and $t_{B/A}$, we have

$$t_{C/A} = \left[\frac{3}{4}(8\text{ m})\right]\left[\frac{1}{3}(8\text{ m})\left(-\frac{192\text{ kN}\cdot\text{m}}{EI}\right)\right]$$
$$+ \left[\frac{1}{3}(8\text{ m}) + 8\text{ m}\right]\left[\frac{1}{2}(8\text{ m})\left(-\frac{192\text{ kN}\cdot\text{m}}{EI}\right)\right]$$
$$= -\frac{11264\text{ kN}\cdot\text{m}^3}{EI}$$

$$t_{B/A} = \left[\frac{1}{3}(8\text{ m})\right]\left[\frac{1}{2}(8\text{ m})\left(-\frac{192\text{ kN}\cdot\text{m}}{EI}\right)\right] = -\frac{2048\text{ kN}\cdot\text{m}^3}{EI}$$

Why are these terms negative? Substituting the results into Eq. (1) yields

$$\Delta_C = -\frac{11264\text{ kN}\cdot\text{m}^3}{EI} - 2\left(-\frac{2048\text{ kN}\cdot\text{m}^3}{EI}\right)$$
$$= -\frac{7168\text{ kN}\cdot\text{m}^3}{EI}$$

Thus,

$$\Delta_C = \frac{-7168\text{ kN}\cdot\text{m}^3}{[200(10^6)\text{ kN/m}^2][250(10^6)(10^{-12})\text{ m}^4]}$$
$$= 0.143\text{ m} \qquad\qquad\qquad \textbf{\textit{Ans.}}$$

8.5   Conjugate-Beam Method

The conjugate-beam method was first presented by Otto Mohr in 1860. Essentially, it requires the same amount of computation as the moment-area theorems to determine a beam's slope or deflection; however, this method relies only on the principles of statics and hence its application will be more familiar.

The basis for the method comes from the similarity between both Eq. 4–1 ($dV/dx = -w$) and Eq. 4–2 ($dM/dx = V$ or $d^2M/dx^2 = -w$), which relate a beam's internal shear and moment to its applied loading, and Eq. 8–1 ($d\theta/dx = M/EI$) and Eq. 8–4 ($d^2y/dx^2 = M/EI$), which relate the slope and deflection of its elastic curve to the internal moment. To show this similarity, we can write these equations as follows:

$$\frac{dV}{dx} = -w \qquad\qquad \frac{d^2M}{dx^2} = -w$$

$$\frac{d\theta}{dx} = \frac{M}{EI} \qquad\qquad \frac{d^2y}{dx^2} = \frac{M}{EI}$$

Or integrating,

$$V = -\int w\, dx \qquad\qquad M = \int \left[-\int w\, dx \right] dx$$

$$\theta = \int \left(\frac{M}{EI} \right) dx \qquad\qquad y = \int \left[\int \left(\frac{M}{EI} \right) dx \right] dx$$

Note that the *shear* V compares with the *slope* θ, the *moment* M compares with the *displacement* y, and the intensity of the *external load* w compares with the *area* under the M/EI diagram. To make use of this comparison we will now consider a beam having the same length as the real beam, but referred to here as the "conjugate beam." Examples of

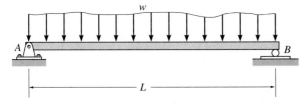

real beam

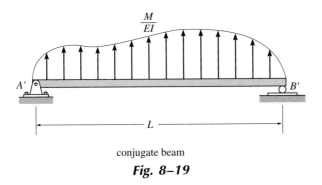

conjugate beam

Fig. 8–19

both a real beam and a conjugate beam are shown in Fig. 8–19. Note that the conjugate beam is "loaded" with the M/EI diagram, in order to conform with the load w on the real beam. As noted above, we can therefore state two theorems related to the conjugate beam, namely,

Theorem 1: The slope at a point in the real beam is equal to the shear at the corresponding point in the conjugate beam.

Theorem 2: The displacement of a point in the real beam is equal to the moment at the corresponding point in the conjugate beam.

371

Conjugate-Beam Supports

Since each of the previous equations requires integration, it is important that the proper boundary conditions be used when they are applied. Likewise, when the conjugate beam is drawn it is important that the shear and moment developed at its supports account for the corresponding slope and displacement of the real beam at its supports, a consequence of Theorems 1 and 2. For example, as shown in Table 8–2, a pin or roller support at the end of the real beam provides *zero*

Table 8–2

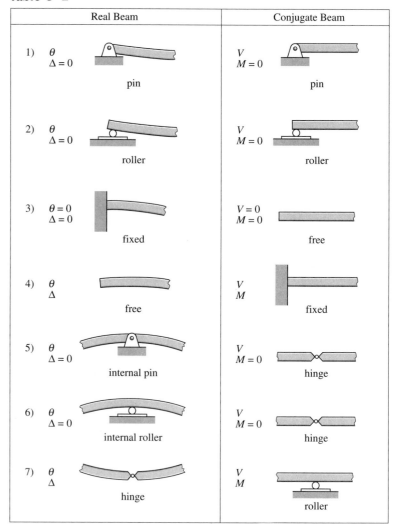

displacement, but the beam has a slope. Consequently, from Theorems 1 and 2, the conjugate beam must be supported by a pin, since this support has *zero moment* but has a shear or end reaction. When the real beam is fixed-supported (3), both the slope and displacement at the support are zero. Here the conjugate beam has a free end, since at this end there is zero shear and zero moment. Corresponding real and conjugate-beam supports for other cases are listed in the table. Examples of real and conjugate beams are shown in Fig. 8–20. Note that, as a rule, neglecting axial force, statically determinate real beams have statically determinate conjugate beams; and statically indeterminate real beams, as in the last case in Fig. 8–20, become unstable conjugate beams. Although this occurs, the M/EI loading will provide the necessary "equilibrium" to hold the conjugate beam stable.

Real Beam   Conjugate Beam

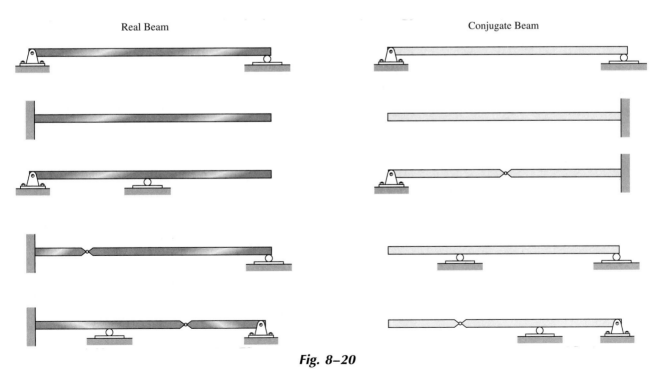

Fig. 8–20

373

Procedure for Analysis

The following procedure provides a method that may be used to determine the displacement and slope at a point on the elastic curve of a beam using the conjugate-beam method.

Conjugate Beam Draw the conjugate beam for the real beam. This beam has the same length as the real beam and has corresponding supports as listed in Table 8–2. In general, though, remember that if the real support allows a *slope*, the conjugate support must develop a *shear*; and if the real support allows a *displacement*, the conjugate support must develop a *moment*. The conjugate beam is loaded with the real beam's M/EI diagram. This loading is assumed to be *distributed* over the conjugate beam and is directed *upward* when M/EI is *positive* and *downward* when M/EI is *negative*.

Equilibrium Using the equations of equilibrium, determine the reactions at the conjugate beam's supports.* Then section the conjugate beam at the point where the slope θ and displacement Δ of the real beam are to be determined. At the section show the unknown shear V' and moment M' acting in their positive sense. (See Fig. 4–1.) Determine the shear and moment using the equations of equilibrium. V' and M' equal θ and Δ, respectively, for the real beam. In particular, if these values are positive, the slope is counterclockwise and the displacement is upward.

The following examples illustrate this method of solution.

---

*The computations involving areas under the M/EI diagram can be facilitated using the table on the inside back cover of the book.

Example 8–9

Determine the slope and deflection at point B of the steel beam shown in Fig. 8–21a. The reactions have been computed. $E = 29(10^3)$ ksi, $I = 800$ in^4.

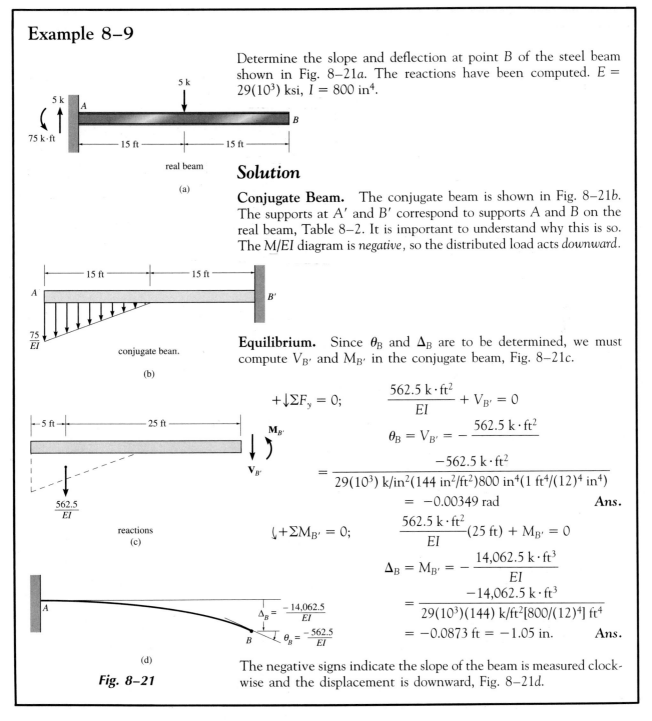

real beam

(a)

Solution

Conjugate Beam. The conjugate beam is shown in Fig. 8–21b. The supports at A' and B' correspond to supports A and B on the real beam, Table 8–2. It is important to understand why this is so. The M/EI diagram is *negative*, so the distributed load acts *downward*.

conjugate beam.

(b)

Equilibrium. Since θ_B and Δ_B are to be determined, we must compute $V_{B'}$ and $M_{B'}$ in the conjugate beam, Fig. 8–21c.

$$+\downarrow\Sigma F_y = 0; \qquad \frac{562.5 \text{ k}\cdot\text{ft}^2}{EI} + V_{B'} = 0$$

$$\theta_B = V_{B'} = -\frac{562.5 \text{ k}\cdot\text{ft}^2}{EI}$$

$$= \frac{-562.5 \text{ k}\cdot\text{ft}^2}{29(10^3) \text{ k/in}^2(144 \text{ in}^2/\text{ft}^2)800 \text{ in}^4(1 \text{ ft}^4/(12)^4 \text{ in}^4)}$$

$$= -0.00349 \text{ rad} \qquad \textbf{Ans.}$$

reactions

(c)

$$\zeta+\Sigma M_{B'} = 0; \qquad \frac{562.5 \text{ k}\cdot\text{ft}^2}{EI}(25 \text{ ft}) + M_{B'} = 0$$

$$\Delta_B = M_{B'} = -\frac{14{,}062.5 \text{ k}\cdot\text{ft}^3}{EI}$$

$$= \frac{-14{,}062.5 \text{ k}\cdot\text{ft}^3}{29(10^3)(144) \text{ k/ft}^2[800/(12)^4] \text{ ft}^4}$$

$$= -0.0873 \text{ ft} = -1.05 \text{ in.} \qquad \textbf{Ans.}$$

(d)

Fig. 8–21

The negative signs indicate the slope of the beam is measured clockwise and the displacement is downward, Fig. 8–21d.

Example 8–10

Determine the maximum deflection of the steel beam shown in Fig. 8–22a. The reactions have been computed. $E = 200$ GPa, $I = 60(10^6)$ mm^4.

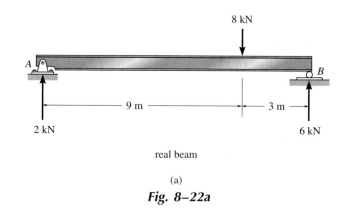

real beam

(a)

Fig. 8–22a

Solution

Conjugate Beam.   The conjugate beam loaded with the M/EI diagram is shown in Fig. 8–22b. Here the distributed load acts upward, since the M/EI diagram is positive.

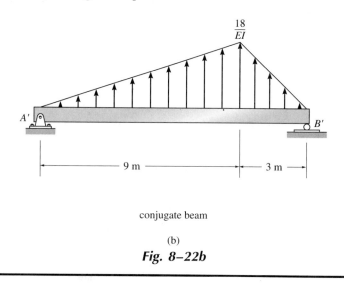

conjugate beam

(b)

Fig. 8–22b

Equilibrium.   The external reactions on the conjugate beam are determined first and are indicated on the free-body diagram in Fig. 8–22c. *Maximum deflection* of the real beam occurs at the point where the *slope* of the beam is *zero*. This corresponds to the same point in the conjugate beam where the *shear* is *zero*. Assuming this point acts within the region $0 \leq x < 9$ m from A', we can isolate the section shown in Fig. 8–22d. Note that the peak of the distributed loading was determined from proportional triangles, that is, $w/x = (18/EI)/9$. We require $V' = 0$ so that

$$+\uparrow \Sigma F_y = 0; \qquad -\frac{45}{EI} + \frac{1}{2}\left(\frac{2x}{EI}\right)x = 0$$

$$x = 6.71 \text{ m} \qquad (0 \leq x < 9 \text{ m OK})$$

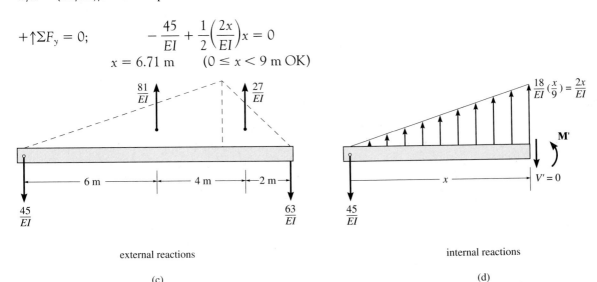

external reactions

(c)

internal reactions

(d)

Fig. 8–22c,d

Using this value for x, the maximum deflection corresponds to the moment M'. Hence,

$$\curvearrowright + \Sigma M = 0; \quad -\frac{45}{EI}(6.71) + \left[\frac{1}{2}\left(\frac{2(6.71)}{EI}\right)6.71\right]\frac{1}{3}(6.71) - M' = 0$$

$$\Delta_{max} = M' = -\frac{201.2 \text{ kN} \cdot \text{m}^3}{EI}$$

$$= \frac{-201.2 \text{ kN} \cdot \text{m}^3}{[200(10^6) \text{ kN/m}^2][60(10^6) \text{ mm}^4(1 \text{ m}^4/(10^3)^4 \text{ mm}^4)]}$$

$$= -0.0168 \text{ m} = -16.8 \text{ mm} \qquad \qquad \textbf{\textit{Ans.}}$$

The negative sign indicates the deflection is downward.

Example 8–11

Determine the displacement of the pin at B and the slope of each beam segment connected to the pin for the steel beam shown in Fig. 8–23a. $E = 29(10^3)$ ksi, $I = 30$ in^4.

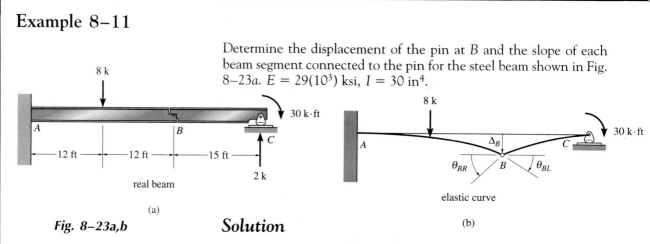

real beam

(a)

elastic curve

(b)

Fig. 8–23a,b

Solution

Conjugate Beam.   The elastic curve for the beam is shown in Fig. 8–23b in order to identify the unknown displacement Δ_B and the slopes $(\theta_B)_L$ and $(\theta_B)_R$ to the left and right of the pin. Using Table 8–2, the conjugate beam is shown in Fig. 8–23c. For simplicity in calculation, the M/EI diagram has been drawn in *parts* using the principle of superposition as described in Sec. 4.5. In this regard, the real beam is thought of as cantilevered from the left support. The moment diagram for the 8-k load, the 2-k and 30-k · ft loadings are given. Notice that negative regions of this diagram develop a downward distributed load and the positive regions have a distributed load that acts upward.

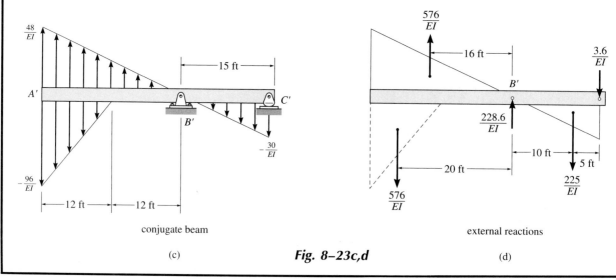

conjugate beam

(c)

external reactions

(d)

Fig. 8–23c,d

Equilibrium. The external reactions at B' and C' are calculated first and the results are indicated in Fig. 8–23d. In order to determine $(\theta_B)_R$, the conjugate beam is sectioned just to the *right* of B' and the shear force $(V_{B'})_R$ is computed, Fig. 8–23e. Thus,

$$+\uparrow\Sigma F_y = 0; \qquad (V_{B'})_R - \frac{225}{EI} - \frac{3.6}{EI} = 0$$

$$(\theta_B)_R = (V_{B'})_R = \frac{228.6 \text{ k} \cdot \text{ft}^2}{EI}$$

$$= \frac{228.6 \text{ k} \cdot \text{ft}^2}{[29(10^3)(144) \text{ k/ft}^2][30/(12)^4] \text{ ft}^4}$$

$$= 0.0378 \text{ rad} \qquad\qquad \textbf{Ans.}$$

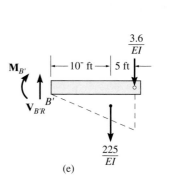

(e)

The internal moment at B' yields the displacement of the pin. Thus,

$$\gamma + \Sigma M_{B'} = 0; \qquad M_{B'} + \frac{225}{EI}(10) + \frac{3.6}{EI}(15) = 0$$

$$\Delta_B = M_{B'} = -\frac{2304 \text{ k} \cdot \text{ft}^3}{EI}$$

$$= \frac{-2304 \text{ k} \cdot \text{ft}^3}{[29(10^3)(144) \text{ k/ft}^2][30/(12)^4] \text{ ft}^4}$$

$$= -0.381 \text{ ft} = -4.58 \text{ in.} \qquad \textbf{Ans.}$$

The slope $(\theta_B)_L$ can be found from a section of beam just to the *left* of B', Fig. 8–23f. Thus,

$$+\uparrow\Sigma F_y = 0; \qquad (V_{B'})_L + \frac{228.6}{EI} - \frac{225}{EI} - \frac{3.6}{EI} = 0$$

$$(\theta_B)_L = (V_{B'})_L = 0 \qquad \textbf{Ans.}$$

Obviously, $\Delta_B = M_{B'}$ for this segment is the *same* as previously calculated, since the moment arms are only slightly different in Fig. 8–23e and f.

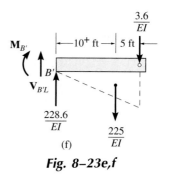

(f)

Fig. 8–23e,f

Example 8–12

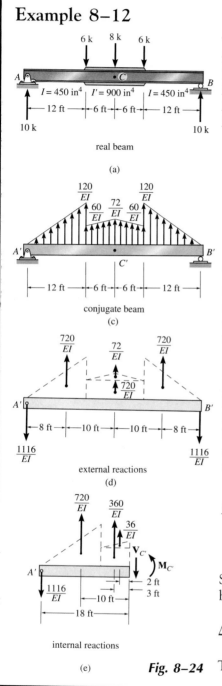

real beam

(a)

conjugate beam

(c)

external reactions

(d)

internal reactions

(e)

Fig. 8–24

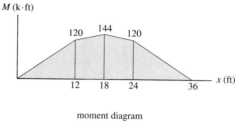

moment diagram

(b)

The girder in Fig. 8–24a is made from a continuous beam and reinforced at its center with cover plates where its moment of inertia is larger. The 12-ft end segments have a moment of inertia of $I = 450$ in^4, and the center portion has a moment of inertia of $I' = 900$ in^4. Determine its deflection at the center C. Take $E = 29(10^3)$ ksi. The reactions have been calculated.

Solution

Conjugate Beam. The moment diagram for the beam is determined first, Fig. 8–24b. Since $I' = 2I$, for simplicity, we can express the load on the conjugate beam in terms of the constant EI as shown in Fig. 8–24c.

Equilibrium. The reactions on the conjugate beam can be calculated by the symmetry of the loading or using the equations of equilibrium. The results are shown in Fig. 8–24d. Since the deflection at C is to be determined, we must compute the internal moment at C'. Using the method of sections, segment A'C' is isolated and the resultants of the distributed loads and their locations are determined, Fig. 8–24e. Thus,

$$\zeta + \Sigma M_{C'} = 0; \frac{1116}{EI}(18) - \frac{720}{EI}(10) - \frac{360}{EI}(3) - \frac{36}{EI}(2) + M_{C'} = 0$$

$$M_{C'} = -\frac{11{,}736 \text{ k} \cdot \text{ft}^3}{EI}$$

Substituting the numerical data for EI and converting units, we have

$$\Delta_C = M_{C'} = -\frac{11{,}736 \text{ k} \cdot \text{ft}^3(1728 \text{ in}^3/\text{ft}^3)}{29(10^3) \text{ k/in}^2(450 \text{ in}^4)} = -1.55 \text{ in.} \qquad \textbf{Ans.}$$

The negative sign indicates that the deflection is downward.

8.6   External Work and Strain Energy

The semigraphical methods presented in the previous sections are very effective for finding the displacements and slopes at points in *beams* subjected to rather simple loadings. For more complicated loadings or for structures such as trusses and frames, it is suggested that energy methods be used for the computations. All energy methods are based on the *conservation of energy principle*, which states that the work done by all the external forces acting on a structure, U_e, is transformed into internal work or strain energy U_i, which is developed when the structure deforms. If the material's elastic limit is not exceeded, the *elastic strain energy* will return the structure to its undeformed state when the loads are removed. The conservation of energy principle can therefore be stated mathematically as

$$U_e = U_i$$

(8–8)

Before developing any of the energy methods based on this principle, however, we will first determine the external work and strain energy caused by a force and a moment. The formulations to be presented will provide a basis for understanding the work and energy methods that follow throughout the rest of this chapter.

External Work—Force

When a force **F** undergoes a displacement dx in the *same direction* as the force, the work done is $dU_e = F\,dx$. If the total displacement is x, the work becomes

$$U_e = \int_0^x F\,dx$$

(8–9)

381

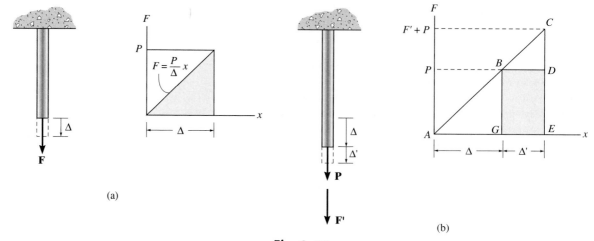

Fig. 8–25

Consider now the effect caused by an axial force applied to the end of a bar as shown in Fig. 8–25a. As the magnitude of **F** is *gradually* increased from zero to some limiting value $F = P$, the final deflection of the bar becomes **Δ**. If the material has a linear elastic response, then $F = (P/\Delta)x$. Substituting into Eq. 8–9, and integrating from 0 to **Δ**, we get

$$U_e = \tfrac{1}{2}P\Delta \qquad (8\text{–}10)$$

which represents the shaded *triangular area*.

We may also conclude from this equation that as a force is gradually applied to the bar, its magnitude builds linearly from zero to some value P. Consequently, the work done is equal to the *average force magnitude* $(P/2)$ times the total displacement **(Δ)**.

Suppose now that **P** is already applied to the bar and that *another force* **F′** is now applied, so the bar deflects further by an amount **Δ′**, Fig. 8–25b. The work done by **P** (not **F′**) when the bar undergoes the further deflection **Δ′** is then

$$U_e = P\Delta' \qquad (8\text{–}11)$$

Here the work represents the shaded *rectangular area* in Fig. 8–25b. In this case **P** does not change its magnitude since **Δ′** is caused only by **F′**. Therefore, work is simply the force magnitude (P) times the displacement **(Δ′)**.

In summary, then, when a force **P** is applied to the bar, followed by application of the force **F'**, the total work done by both forces is represented by the triangular area ACE in Fig. 8–25b. The triangular area ABG represents the work of **P** that is caused by its displacement **Δ**, the triangular area BCD represents the work of **F'** since this force causes a displacement **Δ'**, and lastly, the shaded rectangular area BDEG represents the additional work done by **P** when displaced **Δ'** as caused by **F'**.

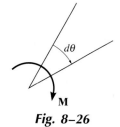

Fig. 8–26

External Work—Moment

The work of a moment is defined by the product of the magnitude of the moment **M** and the angle $d\theta$ through which it rotates, that is, $dU = M\, d\theta$, Fig. 8–26. If the total angle of rotation is θ radians, the work becomes

$$U_e = \int_0^\theta M\, d\theta \qquad (8\text{--}12)$$

As in the case of force, if the moment is applied to a structure *gradually*, from zero to M, the work is then

$$U_e = \tfrac{1}{2}M\theta \qquad (8\text{--}13)$$

However, if the moment is already applied to the structure and other loadings further distort the structure by an amount θ', then **M** rotates θ', and the work is

$$U_e' = M\theta' \qquad (8\text{--}14)$$

Strain Energy—Axial Force

When an axial force **S** is applied gradually to the bar in Fig. 8–27, it will strain the material such that the *external work* done by **S** will be converted into *strain energy*, which is stored in the bar (Eq. 8–8). Provided the material is *linear elastic*, Hooke's law applies, $\sigma = E\epsilon$, and if the bar has a constant cross-sectional area A and length L, the final stress is $\sigma = S/A$ and the final strain is $\epsilon = \Delta/L$. Consequently, $S/A = E(\Delta/L)$, and the final deflection is

$$\Delta = \frac{SL}{AE} \qquad (8\text{--}15)$$

Substituting into Eq. 8–10, with $P = S$, the strain energy in the bar is therefore

$$U_i = \frac{S^2 L}{2AE} \qquad (8\text{--}16)$$

Fig. 8–27

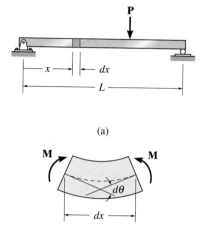

(a)

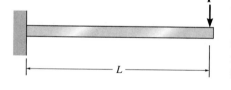

(b)

Fig. 8–28

Strain Energy—Bending

Consider the beam shown in Fig. 8–28a, which is distorted by the *gradually* applied load **P**. This force creates an internal moment **M** in the beam at a section located a distance x from the left support. The resulting rotation of the differential element dx, Fig. 8–28b, can be found from Eq. 8–1, that is, $d\theta = (M/EI)\, dx$. Consequently, based on the conservation of energy principle, the strain energy, or work stored in the element, is determined from Eq. 8–13 since the internal moment is gradually developed. Hence,

$$dU_i = \frac{M^2\, dx}{2EI} \qquad (8\text{–}17)$$

The strain energy for the beam is determined by integrating this result over the beam's entire length L. The result is

$$U_i = \int_0^L \frac{M^2 dx}{2EI} \qquad (8\text{–}18)$$

8.7   Principle of Work and Energy

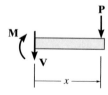

(a)

(b)

Fig. 8–29

Now that the work and strain energy for a force and a moment have been formulated, we will illustrate how the conservation of energy or the principle of work and energy can be applied to determine the displacement at a point on a structure. To do this, consider finding the displacement Δ at the point where the force **P** is applied to the cantilever beam in Fig. 8–29a. We require $U_e = U_i$. From Eq. 8–10, the external work is $U_e = \frac{1}{2}P\Delta$. To obtain the resulting strain energy we must first determine the internal moment as a function of position x in the beam, and then apply Eq. 8–18. In this case $M = -Px$, Fig. 8–29b, so that

$$U_i = \int_0^L \frac{M^2\, dx}{2EI} = \int_0^L \frac{(-Px)^2\, dx}{2EI} = \frac{1}{6}\frac{P^2L^3}{EI}$$

Equating the external work to internal strain energy and solving for the unknown displacement Δ, we have

$$U_e = U_i$$
$$\frac{1}{2}P\Delta = \frac{1}{6}\frac{P^2L^3}{EI}$$
$$\Delta = \frac{PL^3}{3EI}$$

Although the solution here is quite direct, application of this method is limited to only a few select problems. It will be noted that only *one load* may be applied to the structure. If more than one load was applied, there would be an unknown displacement under each load, and yet it is possible to write only *one* "work" equation for the structure. Furthermore, *only the displacement under the load can be obtained,* since the external work depends upon both the force and its corresponding displacement. One way to circumvent these limitations is to use the method of virtual work or Castigliano's theorem, both of which are explained in the following sections.

8.8   Principle of Virtual Work

The principle of virtual work, like other energy methods of analysis, is based on the conservation of energy for the structure, $U_e = U_i$ (Eq. 8–8). This method was developed by John Bernoulli in 1717 and is sometimes referred to as the unit-load method. It provides a general means of obtaining the displacement and slope at a point on a structure, be it a beam, frame, or truss.

Before developing the principle of virtual work, it is necessary to make some general statements regarding the principle of work and energy, which was discussed in the previous section. If we take a deformable structure of any shape or size and apply a series of *external loads* **P** to it, it will cause *internal* loads **u** at points throughout the structure. *It is necessary that the external and internal loads be related by the equations of equilibrium.* As a consequence of these loadings, external displacements Δ will occur at the **P** loads and internal displacements δ will occur at each point of internal load **u**. In general, *these displacements do not have to be elastic,* and they may not be related to the loads; however, *the external and internal displacements must be related by the compatibility of the displacements.* In other words, if the external displacements are known, the corresponding internal displacements are uniquely defined. In general, then, the principle of work and energy states:

$$\underset{\substack{\text{Work of} \\ \text{External Loads}}}{\Sigma P\Delta} = \underset{\substack{\text{Work of} \\ \text{Internal Loads}}}{\Sigma u\delta} \qquad (8\text{–}19)$$

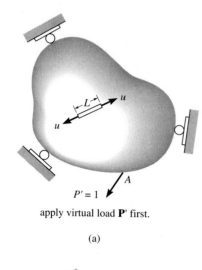

apply virtual load **P'** first.

(a)

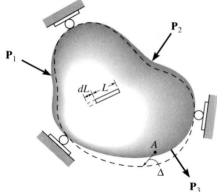

then apply real loads P_1, P_2, P_3.

(b)

Fig. 8–30

Based on this concept, we will now develop the principle of virtual work so that it can be used to determine the *displacement* of a point on a structure. To do this, we will consider the structure (or body) to be of arbitrary shape as shown in Fig. 8–30b.* Suppose it is necessary to determine the displacement Δ of point A on the body caused by the "real loads" P_1, P_2, and P_3. It is to be understood that these loads cause no movement of the supports; in general, however, they can strain the material *beyond the elastic limit*. Since no external load acts on the body at A and in the direction of Δ, the displacement Δ can be determined by *first* placing on the body a "*virtual*" load such that this force P' acts in the *same direction* as Δ, Fig. 8–30a. For convenience, which will be apparent later, we will choose P' to have a "unit" magnitude, that is, $P' = 1$. The term "virtual" is used to describe the load, since *it is imaginary and does not actually exist as part of the real loading*. The unit load (P') does, however, create an internal virtual load u in a representative element or fiber of the body, as shown in Fig. 8–30a. Here it is required that P' and u be related by the equations of equilibrium. As a result of these loadings, the body and the element will each undergo a virtual displacement due to the load P', although we will not be concerned with its magnitude. Once the virtual loadings are applied and *then* the body is subjected to the *real loads* P_1, P_2, and P_3, Fig. 8–30b, point A is displaced an amount Δ, causing the element to deform dL. As a result, the external virtual force P' and internal virtual load u "ride along" by Δ and dL, respectively, and therefore perform *external virtual work* of $1 \cdot \Delta$ on the body and internal virtual work of $u \cdot dL$ on the element. Realizing that the external virtual work is equal to the internal virtual work done on all the elements of the body, we can write the virtual-work equation as

$$1 \cdot \Delta = \Sigma u \cdot dL \qquad (8\text{--}20)$$

where

$P' = 1 =$ external virtual unit load acting in the direction of Δ
$u =$ internal virtual load acting on the element in the direction of dL
$\Delta =$ external displacement caused by the real loads
$dL =$ internal deformation of the element caused by the real loads

---

*This arbitrary shape will later represent a specific truss, beam, or frame.

By choosing $P' = 1$, it can be seen that the solution for Δ follows directly, since $\Delta = \Sigma u \, dL$.

In a similar manner, if the rotational displacement or slope of the tangent at a point on a structure is to be determined, a virtual *couple moment* $\mathbf{M'}$ having a "unit" magnitude is applied at the point. As a consequence, this couple moment causes a virtual load $\mathbf{u}_\theta$ in one of the elements of the body. Assuming that the real loads deform the element an amount dL, the rotation θ can be found from the virtual-work equation

$$
\underset{\text{real displacements}}{\overset{\text{virtual loadings}}{1 \cdot \theta = \Sigma u_\theta \cdot dL}}
\tag{8-21}
$$

where

$M' = 1$ = external virtual unit couple moment acting in the direction of θ

u_θ = internal virtual load acting on an element in the direction of dL

θ = external rotational displacement or slope in radians caused by the real loads

dL = internal deformation of the element caused by the real loads

This method for applying the principle of virtual work is often referred to as the *method of virtual forces*, since a virtual force is applied resulting in the calculation of a *real displacement*. The equation of virtual work in this case represents a *compatibility requirement* for the structure. Although not important here, realize that we can also apply the principle of virtual work as a *method of virtual displacements*. In this case virtual displacements are imposed on the structure, while the structure is subjected to *real loadings*. This method can be used to determine a force on or in a structure,* so that the equation of virtual work is then expressed as an *equilibrium requirement*.

---

*It was used in this manner in Sec. 5.3 with reference to the Müller–Breslau principle.

8.9 Method of Virtual Work: Trusses _____

External Loads

We can use Eq. 8–20 to determine the displacement Δ of a truss joint when the truss is subjected to an applied loading. For example, consider joint A in Fig. 8–31. For application, a typical element of the "body" would represent a *truss member* having a length L. If the applied loading causes a *linear elastic* material response, dL in Eq. 8–20 is determined from $\Delta L = NL/AE$, where N is the normal or axial force in the member, caused by the loads. The virtual work equation for a truss is therefore

$$1 \cdot \Delta = \sum \frac{nNL}{AE} \qquad (8\text{–}22)$$

where

1 = external virtual unit load acting on the truss joint in the stated direction of Δ

n = internal virtual normal force in a truss member caused by the external virtual unit load

Δ = external joint displacement caused by the real loads on the truss

N = internal normal force in a truss member caused by the real loads

L = length of a member

A = cross-sectional area of a member

E = modulus of elasticity of a member

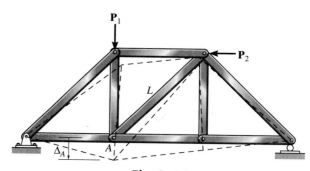

Fig. 8–31

The formulation of this equation follows naturally from the development in Sec. 8.8. Here the external virtual unit load creates internal virtual forces **n** in each of the truss members. The real loads then cause the truss joint to be displaced Δ in the same direction as the virtual unit load, and each member is displaced NL/AE in the same direction as its respective **n** force. Consequently, the external virtual work $1 \cdot \Delta$ equals the internal virtual work or the internal (virtual) strain energy stored in *all* the truss members, that is, $\Sigma nNL/AE$.

Temperature

In some cases, truss members may change their length due to temperature. If α is the coefficient of thermal expansion for a member and ΔT is the change in its temperature, the change in length of a member is $\Delta L = \alpha \, \Delta T \, L$. Hence, we can determine the displacement of a selected truss joint due to this temperature change from Eq. 8–20, written as

$$1 \cdot \Delta = \Sigma n \, \alpha \, \Delta T \, L \qquad\qquad (8\text{--}23)$$

where

1 = external virtual unit load acting on the truss joint in the stated direction of Δ

n = internal virtual normal force in a truss member caused by the external virtual unit load

Δ = external joint displacement caused by the temperature change

α = coefficient of thermal expansion of member

ΔT = change in temperature of member

L = length of member

Fabrication Errors and Camber

Occasionally, errors in fabricating the lengths of the members of a truss may occur. Also, in some cases truss members must be made slightly longer or shorter in order to give the truss a camber. Camber is often built into a bridge truss so that the bottom cord will curve upward by an amount equivalent to the downward deflection of the cord when subjected to the bridge's full dead weight. If a truss member is shorter or longer than intended, the displacement of a truss joint from its expected position can be determined from direct application of Eq. 8–20, written as

$$1 \cdot \Delta = \Sigma n \, \Delta L \qquad (8\text{–}24)$$

where

1 = external virtual unit load acting on the truss joint in the stated direction of Δ

n = internal virtual normal force in a truss member caused by the external virtual unit load

Δ = external joint displacement caused by the fabricating errors

ΔL = difference in length of the member from its intended size as caused by a fabricating error

A combination of the right sides of Eqs. 8–22 through 8–24 will be necessary if both external loads act on the truss and some of the members undergo a thermal change or have been fabricated with the wrong dimensions.

Procedure for Analysis

The following procedure may be used to determine the displacement of any joint on a truss using the method of virtual work.

Virtual Forces u   Place the unit load on the truss at the joint where the desired displacement is to be determined. The load should be directed along the line of action of the displacement. With the unit load so placed, and all the real loads *removed* from the truss, use the method of joints or the method of sections and calculate the internal **u** force in each truss member. Assume that tensile forces are positive and compressive forces are negative.

Real Forces N   Use the method of sections or the method of joints to determine the **N** forces in each member. These forces are caused only by the real loads acting on the truss. Again, assume tensile forces are positive and compressive forces are negative.

Virtual-Work Equation   Apply the equation of virtual work, to determine the desired displacement. It is important to retain the algebraic sign for each of the corresponding **n** and **N** forces when substituting these terms into the equation. If the resultant sum $\Sigma nNL/AE$ is positive, the displacement Δ is in the same direction as the unit load. If a negative value results, Δ is opposite to the unit load.

In a similar manner, when applying $1 \cdot \Delta = \Sigma n\alpha \, \Delta T \, L$, realize that if any of the members undergoes an *increase in temperature*, ΔT will be *positive*, whereas a *decrease in temperature* results in a *negative* value for ΔT. Likewise, for $1 \cdot \Delta = \Sigma n \, \Delta L$, when a fabricating error *increases the length* of a member, ΔL is *positive*, whereas a *decrease in length* is *negative*.

When applying any formula, attention should be paid to the units of each numerical quantity. In particular, the virtual unit load can be assigned any arbitrary unit (lb, kip, N, etc.), since the **n** forces will have these *same units*, and as a result the units for both the virtual unit load and the **n** forces will cancel from both sides of the equation.

The following examples illustrate this method of solution.

Example 8–13

Determine the vertical displacement of joint C of the steel truss shown in Fig. 8–32a. The cross-sectional area of each member is $A = 0.5$ in^2 and $E = 29(10^3)$ ksi.

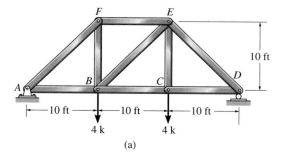

(a)

Solution

Virtual Forces n.   Only a vertical 1-k load is placed at joint C, and the force in each member is calculated using the method of joints. The results of this analysis are shown in Fig. 8–32b. As in the text discussion, positive numbers indicate tensile forces and negative numbers indicate compressive forces.

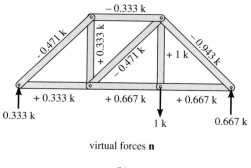

virtual forces **n**

Fig. 8–32a,b

(b)

Real Forces N.   The real forces in the members are calculated using the method of joints. The results of this analysis are shown in Fig. 8–32c.

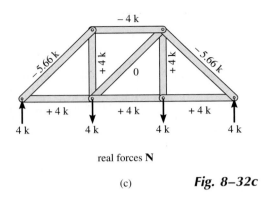

real forces **N**

(c)   ***Fig. 8–32c***

Virtual-Work Equation.   Arranging the data in tabular form, we have

Member	n (k)	N (k)	L (ft)	nNL (k²·ft)
AB	0.333	4	10	13.33
BC	0.667	4	10	26.67
CD	0.667	4	10	26.67
DE	−0.943	−5.66	14.14	75.47
FE	−0.333	−4	10	13.33
EB	−0.471	0	14.14	0
BF	0.333	4	10	13.33
AF	−0.471	−5.66	14.14	37.70
CE	1	4	10	40
				Σ246.50

Thus,

$$1\,k \cdot \Delta_{C_v} = \sum \frac{nNL}{AE} = \frac{246.50\ k^2 \cdot ft}{AE}$$

Converting the units of member length to inches and substituting the numerical values for A and E, we have

$$1\,k \cdot \Delta_{C_v} = \frac{(246.50\ k^2 \cdot ft)(12\ in./ft)}{(0.5\ in^2)(29(10^3)\ k/in^2)}$$

$$\Delta_{C_v} = 0.204\ in. \qquad\qquad \textbf{Ans.}$$

Example 8–14

Consider the truss shown in Fig. 8–33a. The cross-sectional area of each member is $A = 400 \text{ mm}^2$ and $E = 200 \text{ GPa}$. (a) Determine the vertical displacement of joint C if a 4-kN force is applied to the truss at C. (b) If no loads act on the truss, what would be the vertical displacement of joint C if member AB was 5 mm too short?

Solution

Part (a)

Virtual Forces n. Since the *vertical displacement* of joint C is to be determined, a virtual force of 1 kN is applied at C in the vertical direction. Note that the units of this force are the *same* as those of the real loading. The support reactions at A and B have been calculated and are shown on the free-body diagram of the truss, Fig. 8–33b. The **n** force in each member is determined by the method of joints as shown on the free-body diagrams of joints A and B. The results are indicated on the truss free-body diagram—a positive number for a tensile force and negative for a compressive force.

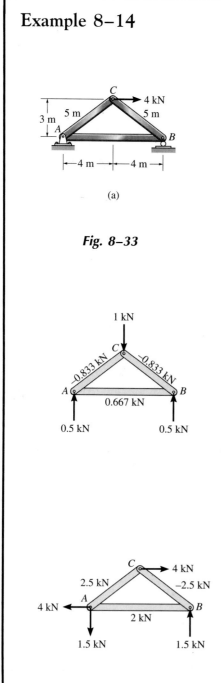

(a)

Fig. 8–33

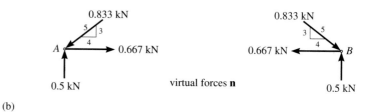

(b)

Real Forces N. The support reactions at A and B caused by the real load of 4 kN are shown on the free-body diagram in Fig. 8–33c. Also shown is the joint analysis of A and B with the results shown on the truss free-body diagram.

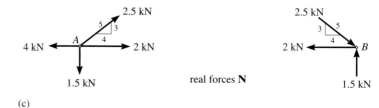

(c)

Virtual-Work Equation.   Since AE is constant, each of the terms ΣnNL can be arranged in tabular form and computed.

Member	n (kN)	N (kN)	L (m)	nNL (kN$^2 \cdot$ m)
AB	0.667	2	8	10.67
AC	−0.833	2.5	5	−10.41
CB	−0.833	−2.5	5	10.41
				$\Sigma 10.67$

Thus,

$$1 \text{ kN} \cdot \Delta_{C_v} = \sum \frac{nNL}{AE} = \frac{10.67 \text{ kN}^2 \cdot \text{m}}{AE}$$

Substituting the values $A = 400 \text{ mm}^2 = 400(10^{-6}) \text{ m}^2$, $E = 200 \text{ GPa} = 200(10^6) \text{ kN/m}^2$, we have

$$1 \text{ kN} \cdot \Delta_{C_v} = \frac{10.67 \text{ kN}^2 \cdot \text{m}}{400(10^{-6}) \text{ m}^2 (200(10^6) \text{ kN/m}^2)}$$

$$\Delta_{C_v} = 0.000133 \text{ m} = 0.133 \text{ mm} \qquad \textbf{Ans.}$$

Part (b).   Here we must apply Eq. 8–33. Since the vertical displacement of C is to be determined, we can use the results of Fig. 8–33b. Only member AB undergoes a deformation of $\Delta L = -0.005$ m. Thus,

$$1 \cdot \Delta = \Sigma n \, \Delta L$$
$$1 \text{ kN} \cdot \Delta_{C_v} = (0.667 \text{ kN})(-0.005 \text{ m})$$
$$\Delta_{C_v} = -0.00333 \text{ m} = -3.33 \text{ mm} \qquad \textbf{Ans.}$$

The negative sign indicates joint C is displaced *upward*, opposite to the 1-kN vertical load. Note that if the 4-kN load and fabrication error are both accounted for, the resultant displacement is then $\Delta_{C_v} = 0.13 - 3.33 = -3.20$ mm (upward).

Example 8–15

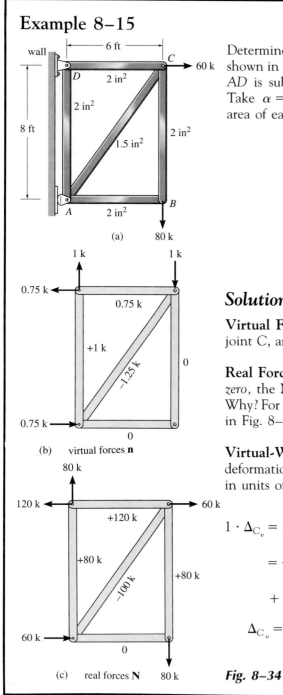

(a)

(b)   virtual forces **n**

(c)   real forces **N**

Fig. 8–34

Determine the vertical displacement of joint C of the steel truss shown in Fig. 8–34a. Due to radiant heating from the wall, member AD is subjected to an *increase* in temperature of $\Delta T = +120°F$. Take $\alpha = 0.6(10^{-5})/°F$ and $E = 29(10^3)$ ksi. The cross-sectional area of each member is indicated in the figure.

Solution

Virtual Forces n.   A *vertical* 1-k load is applied to the truss at joint C, and the forces in each member are computed, Fig. 8–34b.

Real Forces N.   Since the **n** forces in members AB and BC are *zero*, the **N** forces in these members do *not* have to be computed. Why? For completion though, the entire real force analysis is shown in Fig. 8–29c.

Virtual-Work Equation.   Both loads and temperature affect the deformation; therefore, Eqs. 8–22 and 8–23 are combined. Working in units of kips and inches, we have

$$1 \cdot \Delta_{C_v} = \sum \frac{nNL}{AE} + \Sigma n\alpha\, \Delta T\, L$$

$$= \frac{(0.75)(120)(6)(12)}{2[29(10^3)]} + \frac{(1)(80)(8)(12)}{2[29(10^3)]}$$

$$+ \frac{(-1.25)(-100)(10)(12)}{1.5[29(10^3)]} + (1)[0.6(10^{-5})](120)(8)(12)$$

$$\Delta_{C_v} = 0.658 \text{ in.} \qquad\qquad \textbf{Ans.}$$

8.10   Method of Virtual Work: Beams and Frames

The equations of virtual work developed in Sec. 8.8 can also be applied to deflection problems involving beams and frames. Since strains due to *bending* are the *primary cause* of beam or frame deflections, we will discuss their effects first. Deflections due to shear, axial and torsional loadings, and temperature will be considered in Sec. 8.11.

The principle of virtual work, or more exactly, the method of virtual force, may be formulated for beam and frame deflections by considering the beam shown in Fig. 8–35a. Here the displacement Δ of point A is to be determined. To compute Δ a virtual unit load acting in the direction of Δ is placed on the beam at A, and the *internal virtual moment m* is determined by the method of sections at an arbitrary location x from the left support, Fig. 8–35b. When the real loads act on the beam, Fig. 8–35a, point A is displaced Δ. Provided these loads cause *linear elastic* material response, then from Eq. 8–1, the element dx deforms or rotates $d\theta = (M/EI)\,dx$.* Here M is the internal moment at x caused by the real loads. Consequently, the *external virtual work* done

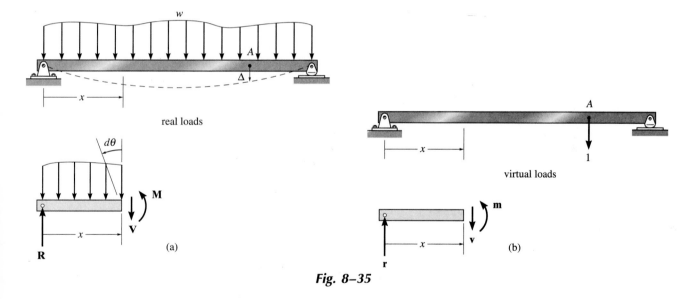

real loads

(a)

virtual loads

(b)

Fig. 8–35

---

*Recall that if the material is strained beyond its elastic limit, the principle of virtual work can still be applied, although in this case a nonlinear or plastic analysis must be used.

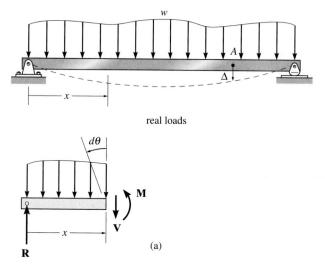

real loads

(a)

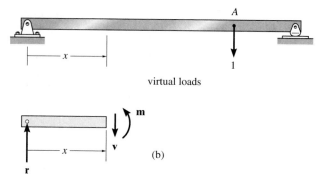

virtual loads

(b)

Fig. 8–35 (Repeated)

by the unit load is $1 \cdot \Delta$, and the *internal virtual work* done by the moment m is $m \, d\theta = m(M/EI) \, dx$. Summing the effects on all the elements dx along the beam requires an integration, and therefore Eq. 8–20 becomes

$$1 \cdot \Delta = \int_0^L \frac{mM}{EI} \, dx \qquad (8\text{–}25)$$

where

1 = external virtual unit load acting on the beam or frame in the direction of Δ

m = internal virtual moment in the beam or frame, expressed as a function of x and caused by the external virtual unit load

Δ = external displacement of the point caused by the real loads acting on the beam or frame

M = internal moment in the beam or frame, expressed as a function of x and caused by the real loads

E = modulus of elasticity of the material

I = moment of inertia of cross-sectional area, computed about the neutral axis

In a similar manner, if the tangent rotation or slope angle θ at a point on the beam's elastic curve is to be determined, a unit couple moment is applied at the point, and the corresponding internal moments m_θ have to be determined. Since the work of the unit couple is $1 \cdot \theta$, then

$$1 \cdot \theta = \int_0^L \frac{m_\theta M}{EI} dx \qquad (8\text{--}26)$$

When applying Eqs. 8–25 and 8–26, it is important to realize that the definite integrals on the right side actually represent the amount of virtual strain energy that is *stored* in the beam. If concentrated forces or couple moments act on the beam or the distributed load is discontinuous, a single integration cannot be performed across the beam's entire length. Instead, separate x coordinates will have to be chosen within regions that have no discontinuity of loading. Also, it is not necessary that each x have the same origin; however, the x selected for determining the real moment M in a particular region must be the *same x* as that

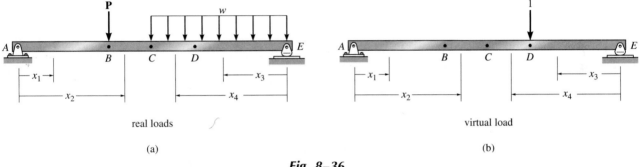

real loads

virtual load

(a)

(b)

Fig. 8–36

selected for determining the virtual moment m or m_θ within the same region. For example, consider the beam shown in Fig. 8–36. In order to determine the displacement of D, four regions of the beam must be considered and therefore four integrals having the form $\int(mM/EI)\,dx$ must be evaluated. We can use x_1 to determine the strain energy in region AB, x_2 for region BC, x_3 for region DE, and x_4 for region DC. In any case, each x coordinate should be selected so that both **M** and **m** (or $\mathbf{m}_\theta$) can be easily formulated.

When the structure is subjected to a relatively simple loading, and yet the solution for a displacement requires several integrations, a *tabular method* may be used to perform these integrations. To do so the moment diagrams for each member are drawn first for both the real and virtual loadings. By matching these diagrams for m and M with those given in the table on the inside front cover, the integral $\int mM\,dx$ can be determined from the appropriate formula. Examples 8–17 and 8–20 illustrate the application of this method.

Procedure for Analysis

The following procedure may be used to determine the displacement and/or the slope at a point on the elastic curve of a beam or frame using the method of virtual work.

Virtual Moments m or m_θ Place a *unit load* on the beam or frame at the point and in the direction of the desired *displacement*. If the *slope* is to be determined, place a *unit couple moment* at the point. Establish appropriate *x* coordinates that are valid within regions of the beam or frame where there is no discontinuity of real or virtual load. With the virtual loads so placed, and all the real loads *removed* from the beam or frame, calculate the internal moment m or m_θ as a function of each *x* coordinate. In each case, assume positive m or m_θ acting in the conventional direction as indicated in Fig. 4–1.

Real Moments Using the *same x* coordinates as those established for m or m_θ, determine the internal moments M caused only by the real loads. Since *positive m or m_θ* was assumed to act in the conventional direction, *it is important that positive M acts in this same* direction. This is necessary since positive or negative internal work depends upon the directional sense of load (defined by $\pm m$ or $\pm m_\theta$) and displacement (defined by $\pm M \, d\theta/EI$).

Virtual-Work Equation Apply the equation of virtual work to determine the desired displacement Δ or rotation θ. It is important to retain the algebraic sign of each integral calculated within its specified region. If the algebraic sum of all the integrals for the entire beam or frame is positive, Δ or θ is in the same direction as the unit load or unit couple moment, respectively. If a negative value results, the direction of Δ or θ is opposite to that of the unit load or couple moment.

The following examples illustrate this method of solution.

Example 8–16

Determine the displacement of point B of the steel beam shown in Fig. 8–37a. Take $E = 200$ GPa, $I = 500(10^6)$ mm^4.

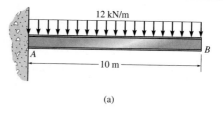

(a)

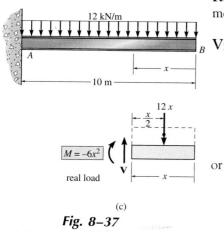

(b)

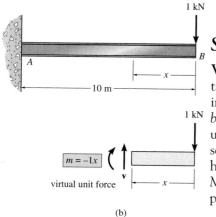

(c)

Fig. 8–37

Solution

Virtual Moment m. The vertical displacement of point B is obtained by placing a virtual unit load of 1 kN at B, Fig. 8–37b. By inspection there are no discontinuities of loading on the beam for *both* the real and virtual loads. Thus, a *single* x coordinate can be used to determine the virtual strain energy. This coordinate will be selected with its origin at B, since then the reactions at A do not have to be determined in order to find the internal moments m and M. Using the method of sections, the internal moment m is computed as shown in Fig. 8–37b.

Real Moment M. Using the *same* x coordinate, the internal moment M is computed as shown in Fig. 8–37c.

Virtual-Work Equation. The vertical displacement of B is thus

$$1 \cdot \Delta_B = \int \frac{mM}{EI} \, dx = \int_0^{10} \frac{(-1x)(-6x^2) \, dx}{EI}$$

$$1 \, kN \cdot \Delta_B = \frac{15(10^3) \, \text{kN}^2 \cdot \text{m}^3}{EI}$$

or

$$\Delta_B = \frac{15(10^3) \, \text{kN} \cdot \text{m}^3}{200(10^6) \, \text{kN/m}^2 (500(10^6) \, \text{mm}^4)(10^{-12} \, \text{m}^4/\text{mm}^4)}$$

$$= 0.150 \, \text{m} = 150 \, \text{mm} \qquad \textbf{Ans.}$$

Example 8–17

Determine the slope θ at point B of the steel beam shown in Fig. 8–38a. Take $E = 200$ GPa, $I = 60(10^6)$ mm⁴.

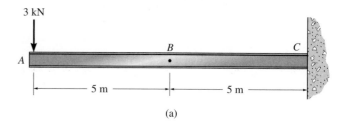

(a)

Solution

Virtual Moments m_θ. The slope at B is determined by placing a virtual unit couple moment of 1 kN · m at B, Fig. 8–38b. Here two x coordinates must be selected in order to determine the total virtual strain energy in the beam. Coordinate x_1 accounts for the strain energy within segment AB and coordinate x_2 accounts for that in segment BC. The internal moments m_θ within each of these segments are computed using the method of sections as shown in Fig. 8–38b.

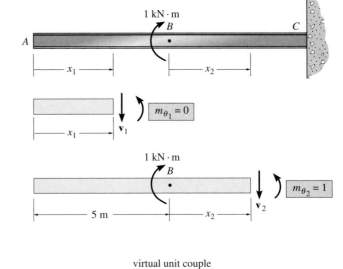

Fig. 8–38a,b

virtual unit couple

(b)

Real Moments M. Using the *same* coordinates x_1 and x_2, the internal moments M are computed as shown in Fig. 8–38c.

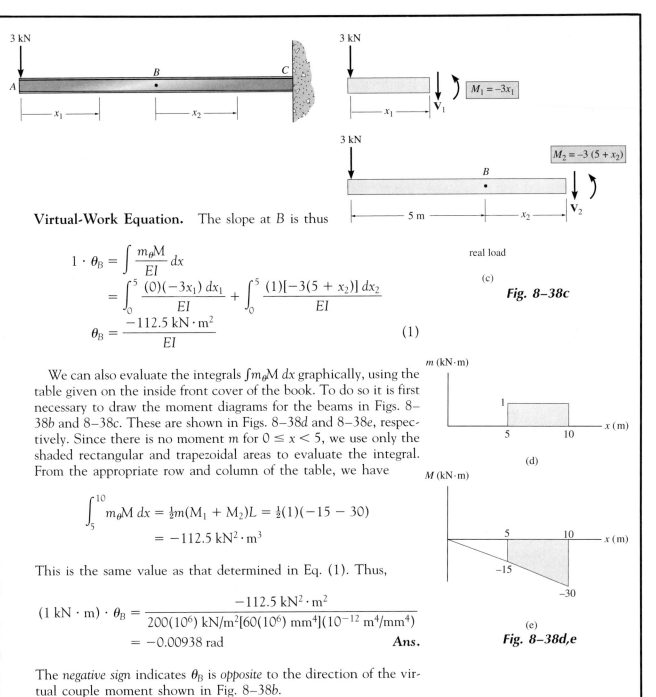

Virtual-Work Equation. The slope at B is thus

$$1 \cdot \theta_B = \int \frac{m_\theta M}{EI}\, dx$$

$$= \int_0^5 \frac{(0)(-3x_1)\, dx_1}{EI} + \int_0^5 \frac{(1)[-3(5 + x_2)]\, dx_2}{EI}$$

$$\theta_B = \frac{-112.5\ \text{kN} \cdot \text{m}^2}{EI} \qquad (1)$$

real load

(c)

Fig. 8–38c

We can also evaluate the integrals $\int m_\theta M\, dx$ graphically, using the table given on the inside front cover of the book. To do so it is first necessary to draw the moment diagrams for the beams in Figs. 8–38b and 8–38c. These are shown in Figs. 8–38d and 8–38e, respectively. Since there is no moment m for $0 \le x < 5$, we use only the shaded rectangular and trapezoidal areas to evaluate the integral. From the appropriate row and column of the table, we have

$$\int_5^{10} m_\theta M\, dx = \tfrac{1}{2} m(M_1 + M_2)L = \tfrac{1}{2}(1)(-15 - 30)$$

$$= -112.5\ \text{kN}^2 \cdot \text{m}^3$$

This is the same value as that determined in Eq. (1). Thus,

$$(1\ \text{kN} \cdot \text{m}) \cdot \theta_B = \frac{-112.5\ \text{kN}^2 \cdot \text{m}^2}{200(10^6)\ \text{kN/m}^2[60(10^6)\ \text{mm}^4](10^{-12}\ \text{m}^4/\text{mm}^4)}$$

$$= -0.00938\ \text{rad} \qquad \textbf{Ans.}$$

(e)

Fig. 8–38d,e

The *negative sign* indicates θ_B is *opposite* to the direction of the virtual couple moment shown in Fig. 8–38b.

Example 8–18

Determine the displacement at D of the beam in Fig. 8–39a. Take $E = 29(10^3)$ ksi, $I = 800$ in^4.

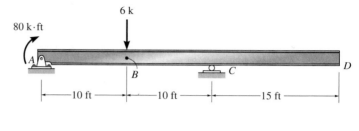

Fig. 8–39a

(a)

Solution

Virtual Moments m. The beam is subjected to a virtual unit load at D as shown in Fig. 8–39b. By inspection, *three coordinates*, such as x_1, x_2, and x_3, must be used to cover all the regions of the beam. Notice that these coordinates cover regions where no discontinuities in either real or virtual load occur. The internal moments m have been computed in Fig. 8–39b using the method of sections.

Fig. 8–39b

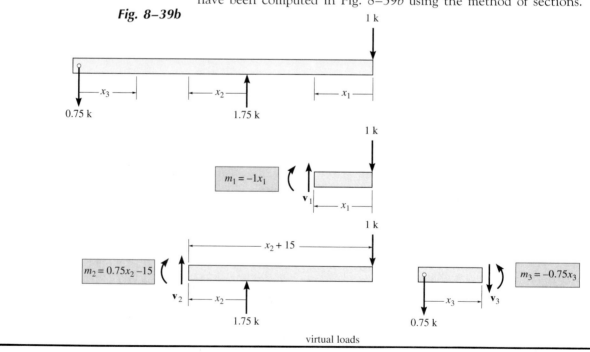

virtual loads

(b)

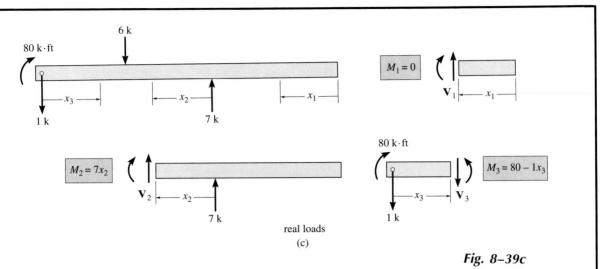

real loads
(c)

Fig. 8–39c

Real Moments M.   The reactions on the beam are computed first, then, using the *same* x coordinates as those used for m, the internal moments M are determined as shown in Fig. 8–39c.

Virtual-Work Equation.   Applying the equation of virtual work to the beam using the data in Figs. 8–39b and 8–39c, we have

$$1 \cdot \Delta_D = \int \frac{mM}{EI} \, dx$$

$$= \int_0^{15} \frac{(-1x_1)(0) \, dx_1}{EI} + \int_0^{10} \frac{(0.75x_2 - 15)(7x_2) \, dx_2}{EI}$$

$$+ \int_0^{10} \frac{(-0.75x_3)(80 - 1x_3) \, dx_3}{EI}$$

$$\Delta_D = \frac{0}{EI} - \frac{3500}{EI} - \frac{2750}{EI} = -\frac{6250 \text{ k} \cdot \text{ft}^3}{EI}$$

or

$$\Delta_D = \frac{-6250 \text{ k} \cdot \text{ft}^3 (12)^3 \text{ in}^3/\text{ft}^3}{29(10^3) \text{ k/in}^2 (800 \text{ in}^4)}$$

$$= -0.466 \text{ in.} \qquad \qquad \textbf{Ans.}$$

The negative sign indicates the displacement is upward, opposite to the downward unit load, Fig. 8–39b. Also note that m_1 did not have to be calculated since $M_1 = 0$.

Example 8–19

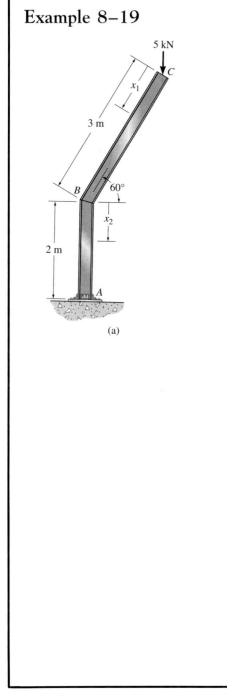

5 kN

x_1

3 m

B 60°

x_2

2 m

A

(a)

Determine the tangential rotation at point C of the frame shown in Fig. 8–40a. Take $E = 200$ GPa, $I = 15(10^6)$ mm^4.

Solution

Virtual Moments m_θ.   The coordinates x_1 and x_2 shown in Fig. 8–40a will be used. A unit couple moment is applied at C and the internal moments m_θ are calculated, Fig. 8–40b.

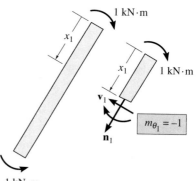

1 kN·m

x_1

x_1 1 kN·m

$\mathbf{v}_1$

$m_{\theta_1} = -1$

$\mathbf{n}_1$

1 kN·m

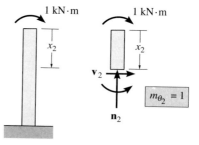

1 kN·m   1 kN·m

x_2 x_2

$\mathbf{v}_2$

$m_{\theta_2} = 1$

$\mathbf{n}_2$

virtual loads

(b)

Fig. 8–40a,b

406

Real Moments M.   In a similar manner, the real moments M are calculated as shown in Fig. 8–40c.

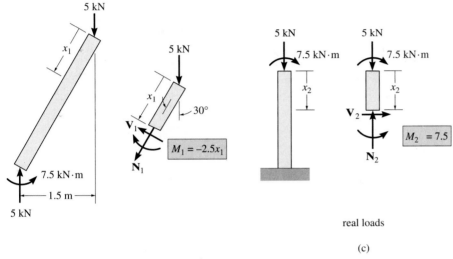

real loads

(c)

Fig. 8–40c

Virtual-Work Equation.   Using the data in Figs. 8–40b and 8–40c, we have

$$1 \cdot \theta_C = \int \frac{m_\theta M}{EI} \, dx = \int_0^3 \frac{(-1)(-2.5x_1) \, dx_1}{EI} + \int_0^2 \frac{(1)(7.5) \, dx_2}{EI}$$

$$\theta_C = \frac{11.25}{EI} + \frac{15}{EI} = \frac{26.25 \text{ kN} \cdot \text{m}^2}{EI}$$

or

$$\theta_C = \frac{26.25 \text{ kN} \cdot \text{m}^2}{200(10^6) \text{ kN/m}^2[15(10^6) \text{ mm}^4](10^{-12} \text{ m}^4/\text{mm}^4)}$$

$$= 0.00875 \text{ rad} \qquad\qquad \textbf{\textit{Ans.}}$$

Example 8–20

Determine the horizontal displacement of point C on the frame shown in Fig. 8–41a. Take $E = 29(10^3)$ ksi and $I = 600$ in⁴ for both members.

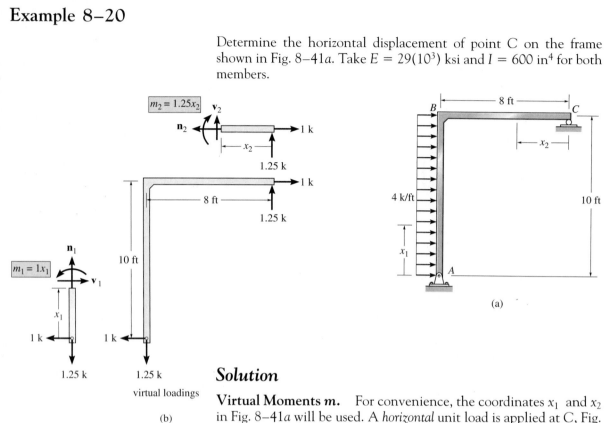

virtual loadings

(b)

Fig. 8–41a,b

Solution

Virtual Moments m.   For convenience, the coordinates x_1 and x_2 in Fig. 8–41a will be used. A *horizontal* unit load is applied at C, Fig. 8–41b. Why? The support reactions and internal virtual moments are computed and shown.

Real Moments M.   In a similar manner the support reactions and real moments are computed as shown in Fig. 8–41c.

Virtual-Work Equation.   Using the data in Figs. 8–41b and 8–41c, we have

$$1 \cdot \Delta_{C_h} = \int \frac{mM}{EI}\, dx = \int_0^{10} \frac{(1x_1)(40x_1 - 2x_1^2)\, dx_1}{EI}$$
$$+ \int_0^8 \frac{(1.25x_2)(25x_2)\, dx_2}{EI}$$

$$\Delta_{C_h} = \frac{8333.3}{EI} + \frac{5333.3}{EI} = \frac{13{,}666.7 \text{ k} \cdot \text{ft}^3}{EI} \qquad (1)$$

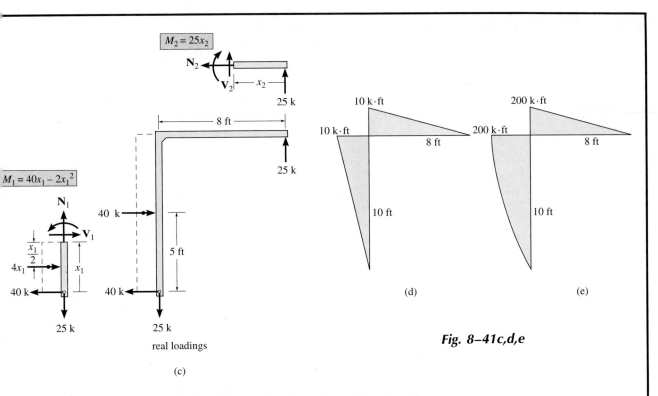

real loadings

(c)

Fig. 8–41c,d,e

If desired, the integrals $\int mM/dx$ can also be evaluated graphically using the table on the inside front cover. The moment diagrams for the frame in Fig. 8–41b and 8–41c are shown in Figs. 8–41d and 8–41e, respectively. Thus,

$$\int mM \, dx = \tfrac{5}{12}(10)(200)(10) + \tfrac{1}{3}(10)(200)(8)$$
$$= 8333.3 + 5333.3 = 13,666.7 \text{ k}^2 \cdot \text{ft}^3$$

Which is the same as that calculated in Eq. (1). Thus

$$\Delta_{C_h} = \frac{13,666.7 \text{ k} \cdot \text{ft}^3}{[29(10^3)\text{k/in}^2((12)^2 \text{ in}^2/\text{ft}^2)][600 \text{ in}^4(\text{ft}^4/(12)^4 \text{ in}^4)]}$$
$$= 0.113 \text{ ft} = 1.36 \text{ in.} \hspace{2cm} \textbf{\textit{Ans.}}$$

8.11 Virtual Strain Energy Caused by Axial Load, Shear, Torsion, and Temperature _____

Although deflections of beams and frames are caused primarily by bending strain energy, in some structures the additional strain energy of axial load, shear, torsion, and perhaps temperature may become important. Each of these effects will now be considered.

Axial Load

Frame members can be subjected to axial loads, and the virtual strain energy caused by these loadings has been established in Sec. 8.9. For members having a constant cross-sectional area, we have

$$U_a = \frac{nNL}{AE} \tag{8-27}$$

where

n = internal virtual axial load caused by the external virtual unit load
N = internal axial force in the member caused by the real loads
E = modulus of elasticity for the material
A = cross-sectional area of the member
L = member's length

Shear

In order to determine the virtual strain energy in a beam due to shear, we will consider the beam element dx shown in Fig. 8–42. The shearing distortion dy of the element as caused by the *real loads* is $dy = \gamma\, dx$. If the shearing strain γ is caused by *linear-elastic* material response, then Hooke's law applies, $\gamma = \tau/G$. Therefore, $dy = (\tau/G)\, dx$. We can express the shear stress as $\tau = K(V/A)$, where K is a *form factor* that depends upon the shape of the beam's cross-sectional area A. Hence, we can write $dy = K(V/GA)\, dx$. The internal virtual work done by a virtual shear force v, acting on the element *while* it is deformed dy, is therefore $dU_i = v\, dy = v(KV/GA)\, dx$. For the entire beam, the virtual strain energy is determined by integration.

$$U_s = \int_0^L K\left(\frac{vV}{GA}\right) dx \tag{8-28}$$

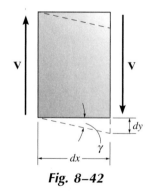

Fig. 8–42

where

v = internal virtual shear in the member, expressed as a function of x
and caused by the external virtual unit load

V = internal shear in the member, expressed as a function of x and
caused by the real loads

A = cross-sectional area of the member

K = form factor for the cross-sectional area:

$K = 1.2$ for rectangular cross sections

$K = 10/9$ for circular cross sections

$K \approx 1$ for wide-flange and I-beams, when A is the area of the web

G = shear modulus of elasticity for the material

Torsion

Often three-dimensional frameworks are subjected to torsional load-ings. If the member has a *circular* cross-sectional area, no warping of its cross section will occur when it is loaded. As a result, the virtual strain energy in the member can easily be derived. To do so consider an element dx of the member that is subjected to an applied torque T, Fig. 8–43. This torque causes a shear strain of $\gamma = (c\,d\theta)/dx$. Provided linear elastic material response occurs, then $\gamma = \tau/G$, where $\tau = Tc/J$. Thus, the angle of twist, $d\theta = (\gamma\,dx)/c = (\tau/Gc)\,dx = (T/GJ)\,dx$. If a virtual unit load is applied to the structure that causes an internal virtual torque t in the member, then after applying the real loads, the virtual strain energy in the member of length dx will be $dU = t\,d\theta = tT\,dx/GJ$. Integrating over the length L of the member yields

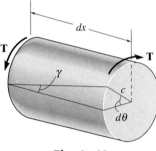

Fig. 8–43

$$U_t = \frac{tTL}{GJ} \qquad (8\text{–}29)$$

where

t = internal virtual torque caused by the external virtual unit load

T = internal torque in the member caused by the real loads

G = shear modulus of elasticity for the material

J = polar moment of inertia for the cross section, $J = \pi c^4/2$, where c is the radius of the cross-sectional area

L = member's length

The virtual strain energy due to torsion for members having noncircular cross-sectional areas is determined using a more rigorous analysis than that presented here.*

Temperature

In Sec. 8.9 we considered the effect of a *uniform temperature change* ΔT on a truss member and indicated that the member will elongate or shorten by an amount $\Delta L = \alpha \Delta TL$. The result of this change in length may be accounted for by using the virtual work Eq. 8–23 ($1 \cdot \Delta = u\alpha \Delta TL$). In some cases, however, a structural member could be subjected to a *temperature difference across its depth*, as in the case of the beam shown in Fig. 8–44a. If this occurs, it is also possible to determine the displacement of points along the elastic curve of the beam by using the principle of virtual work. To do so we must first compute the amount of *rotation* of a differential element dx of the beam as caused by the thermal gradient that acts over the beam's cross section. For the sake of discussion we will choose the most common case

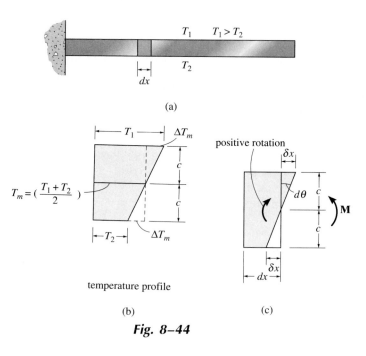

(a)

temperature profile

(b)   (c)

Fig. 8–44

*See L. Bairstow and A. J. S. Pippard, "The Determination of the Stresses in a Shaft of Any Cross-Section," *Proceedings, Institute of Civil Engineers*, vol. 214, 1921–22.

of a beam having a neutral axis located at the mid-depth (c) of the beam. If we plot the temperature profile, Fig. 8–44b, it will be noted that the mean temperature $T_m = (T_1 + T_2)/2$ will simply cause the *entire beam to elongate* without causing the element to rotate. If $T_1 > T_2$, the temperature difference at the top of the element causes strain elongation, while that at the bottom causes strain contraction. In both cases the difference in temperature is $\Delta T_m = T_1 - T_m = T_m - T_2$. Since the thermal change of length at these fibers is $\delta x = \alpha \Delta T\, dx$, Fig. 8–44c, then the rotation of the element is

$$d\theta = \frac{\alpha\, \Delta T_m\, dx}{c}$$

If we apply a virtual unit load at the point where a displacement is to be determined, or apply a virtual unit couple moment at the point where a rotational displacement of the tangent is to be determined, then this creates a virtual moment **m** in the beam at the point where the element dx is located. When the temperature gradient is imposed, the virtual strain energy in the beam is then

$$U_{temp} = \int_0^L \frac{m\alpha\, \Delta T_m\, dx}{c} \qquad (8\text{–}30)$$

where

m = internal virtual moment in the beam expressed as a function of x and caused by the external virtual unit load

α = coefficient of thermal expansion

ΔT_m = temperature difference between the mean temperature and the temperature at the top or bottom of the beam

c = mid-depth of the beam

Unless otherwise stated, *this text will consider only beam and frame deflections due to bending.* In general, though, beam and frame members may be subjected to several of the other loadings discussed in this section. However, as previously mentioned, the additional deflections caused by shear and axial force alter the deflection of beams by only a few percent and are therefore generally ignored for even "small" two- or three-member frame analysis of one-story height. If these and the other effects of torsion and temperature are to be considered for the analysis, then one simply adds their virtual strain energy as defined by Eqs. 8–27 through 8–30 to the equation of virtual work defined by Eq. 8–25 or 8–26. The following examples illustrate application of these equations.

Example 8–21

Determine the horizontal displacement of point C on the frame shown in Fig. 8–45a. Take $E = 29(10^3)$ ksi, $G = 12(10^3)$ ksi, $I = 600$ in^4, and $A = 80$ in^2 for both members. The cross-sectional area is rectangular. Include the internal strain energy due to axial load and shear.

Solution

Here we must apply a horizontal unit load at C. The necessary free-body diagrams for the real and virtual loadings are shown in Figs. 8–45b and 8–45c.

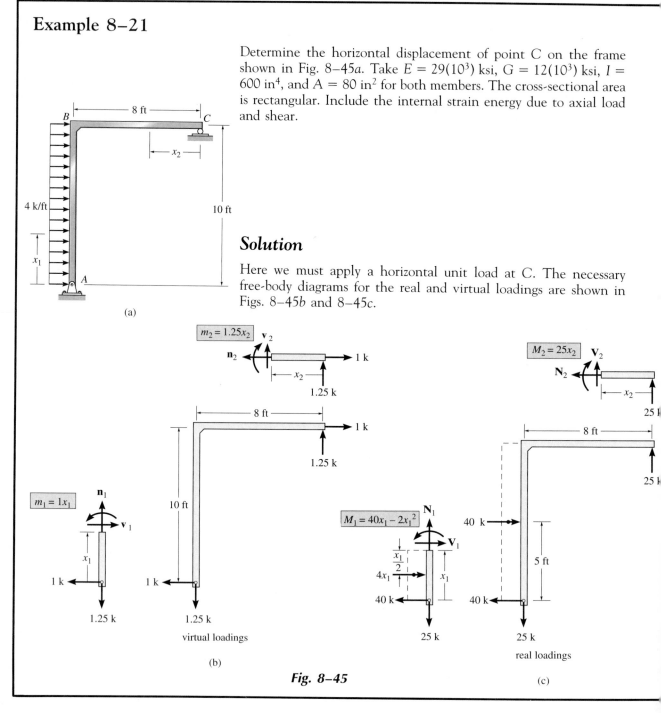

(a)

(b)

virtual loadings

real loadings

(c)

Fig. 8–45

Bending. The virtual strain energy due to bending has been cal-culated in Example 8–18. There it was shown that

$$U_b = \int \frac{mM\,dx}{EI} = \frac{13{,}666.7 \text{ k}^2 \cdot \text{ft}^3 (12^3 \text{ in}^3/1 \text{ ft}^3)}{[29(10^3)\text{ k/in}^2](600 \text{ in}^4)} = 1.357 \text{ in.} \cdot \text{k}$$

Axial load. From the data in Fig. 8–45b and 8–45c, we have

$$U_a = \sum \frac{nNL}{AE}$$

$$= \frac{1.25 \text{ k}(25 \text{ k})(120 \text{ in.})}{80 \text{ in}^2[29(10^3)\text{ k/in}^2]} + \frac{(1 \text{ k})(0)(96 \text{ in.})}{(80 \text{ in}^2)[29(10^3)\text{ k/in}^2]}$$

$$= 0.001616 \text{ in.} \cdot \text{k}$$

Shear. Applying Eq. 8–28 with $K = 1.2$ for rectangular cross sec-tions, and using the shear functions shown in Fig. 8–45b and 8–45c, we have

$$U_s = \int_0^L K\left(\frac{vV}{GA}\right) dx$$

$$= \int_0^{10} \frac{1.2(1)(40 - 4x_1)\,dx_1}{GA} + \int_0^8 \frac{1.2(-1.25)(-25)\,dx_2}{GA}$$

$$= \frac{540 \text{ k}^2 \cdot \text{ft}(12 \text{ in./ft})}{[12(10^3)\text{ k/in}^2](80 \text{ in}^2)} = 0.00675 \text{ in.} \cdot \text{k}$$

Applying the equation of virtual work, we have

$$1 \text{ k} \cdot \Delta_{C_h} = 1.357 \text{ in.} \cdot \text{k} + 0.001616 \text{ in.} \cdot \text{k} + 0.00675 \text{ in.} \cdot \text{k}$$

$$\Delta_{C_h} = 1.37 \text{ in.} \qquad\qquad \textbf{\textit{Ans.}}$$

Including the effects of shear and axial load contributed only a 0.6% increase in the answer to that determined only from bending.

Example 8–22

The beam shown in Fig. 8–46a is used in a building subjected to two different thermal environments. If the temperature at the top surface of the beam is 80°F and at the bottom surface 160°F, determine the vertical deflection of the beam at its midpoint due to the temperature gradient. Take $\alpha = 6.5(10^{-6})/°F$.

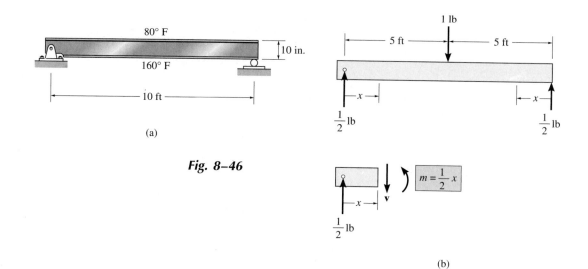

Fig. 8–46

Solution

Since the deflection at the center of the beam is to be determined, a virtual unit load is placed there and the internal virtual moment in the beam is calculated, Fig. 8–46b.

The mean temperature at the center of the beam is (160° + 80°)/2 = 120°F, so that for application of Eq. 8–30, ΔT_m = 120°F − 80°F = 40°F. Also, c = 10 in./2 = 5 in. Applying the principle of virtual work, we have

$$1 \text{ lb} \cdot \Delta_{C_v} = \int_0^L \frac{m\alpha\,\Delta T_m\,dx}{c}$$

$$= 2\int_0^{60 \text{ in.}} \frac{(\tfrac{1}{2}x)6.5(10^{-6})/°F(40°F)}{5 \text{ in.}}\,dx$$

$$\Delta_{C_v} = 0.0936 \text{ in.} \qquad\qquad \textbf{Ans.}$$

The result indicates a very negligible deflection.

8.12   Castigliano's Theorem

In 1879 Alberto Castigliano, an Italian railroad engineer, published a book in which he outlined a method for determining the deflection or slope at a point in a structure, be it a truss, beam, or frame. This method, which is referred to as *Castigliano's second theorem*, or the *method of least work*, applies only to structures that have constant temperature, unyielding supports, and *linear elastic* material response. If the displacement of a point is to be determined, the theorem states that it is equal to the first partial derivative of the strain energy in the structure with respect to a force acting at the point and in the direction of displacement. In a similar manner, the slope at a point in a structure is equal to the first partial derivative of the strain energy in the structure with respect to a couple moment acting at the point and in the direction of rotation.

To derive Castigliano's second theorem, consider a body (structure) of any arbitrary shape, which is subjected to a series of n forces P_1, P_2. $\ldots$, P_n. Since the external work done by these loads is equal to the internal strain energy stored in the body, we can write

$$U_i = U_e$$

However, as indicated by Eq. 8–9 ($U_e = \Sigma \int P \, dx$), the external work is a function of the external loads. Thus,

$$U_i = U_e = f(P_1, P_2, \ldots, P_n)$$

Now, if any one of the forces, say P_i, is increased by a differential amount dP_i, the internal work is also increased such that the new strain energy becomes

$$U_i + dU_i = U_i + \frac{\partial U_i}{\partial P_i} \, dP_i \qquad (8-31)$$

This value, however, should not depend on the sequence in which the n forces are applied to the body. For example, if we apply dP_i to the body *first*, then this causes the body to be displaced a differential amount $d\Delta_i$ in the direction of dP_i. By Eq. 8–10 ($U_e = \frac{1}{2}P\Delta$), the increment of strain energy would be $\frac{1}{2}dP_i \, d\Delta_i$. This quantity, however, is a second-order differential and may be neglected. Further application of the loads P_1, P_2, $\ldots$, P_n, which displace the body Δ_1, Δ_2, $\ldots$, Δ_n, yields the strain energy.

$$U_i + dU_i = U_i + dP_i\Delta_i \qquad (8\text{--}32)$$

Here, as before, U_i is the internal strain energy in the body, caused by the loads $P_1, P_2, \ldots, P_n$, and $dU_i = dP_i\Delta_i$ is the *additional* strain energy caused by dP_i. (See Eq. 8–11, $U_e = P\Delta'$.)

In summary, then, Eq. 8–31 represents the strain energy in the body determined by first applying the loads $P_1, P_2, \ldots, P_n$, *then* dP_i, and Eq. 8–32 represents the strain energy determined by first applying dP_i and *then* the loads $P_1, P_2, \ldots, P_n$. Since these two equations must be equal, we require

$$\Delta_i = \frac{\partial U_i}{\partial P_i} \qquad (8\text{--}33)$$

which proves the theorem; i.e., the displacement Δ_i in the direction of P_i is equal to the first partial derivative of the strain energy with respect to P_i.*

It should be noted that Eq. 8–33 is a statement regarding the *structure's compatibility*. Also, the above derivation requires that *only conservative forces* be considered for the analysis. These forces do work that is independent of the path and therefore create no energy loss. Since forces causing a linear elastic response are conservative, the theorem is restricted to *linear elastic behavior* of the material. This is unlike the method of virtual force discussed in the previous section, which applied to *both* elastic and inelastic behavior.

---

*Castigliano's first theorem is similar to his second theorem; however, it relates the load P_i to the partial derivative of the strain energy with respect to the corresponding displacement, that is, $P_i = \partial U_i/\partial\Delta_i$. The proof is similar to that given above and, like the method of virtual displacement, Castigliano's first theorem applies to both elastic and inelastic material behavior. This theorem is another way of expressing the *equilibrium requirements* for a structure, and since it has very limited use in structural analysis, it will not be discussed in this book.

8.13 Castigliano's Theorem for Trusses _____

The strain energy for a member of a truss is given by Eq. 8–16, $U_i = N^2L/2AE$. Substituting this equation into Eq. 8–33 and omitting the subscript i, we have

$$\Delta = \frac{\partial}{\partial P} \sum \frac{N^2L}{2AE}$$

It is generally easier to perform the differentiation prior to summation. In the general case L, A, and E are constant for a given member, and therefore we may write

$$\Delta = \sum N\left(\frac{\partial N}{\partial P}\right)\frac{L}{AE} \qquad (8\text{–}34)$$

where

Δ = external joint displacement of the truss
P = external force applied to the truss joint in the direction of Δ
N = internal force in a member caused by *both* the force P and the loads on the truss
L = length of a member
A = cross-sectional area of a member
E = modulus of elasticity of a member

This equation is similar to that used for the method of virtual work, Eq. 8–22 $(1 \cdot \Delta = \sum nNL/AE)$, except u is replaced by $\partial N/\partial P$. Notice that in order to determine this partial derivative it will be necessary to treat P as a *variable* (not a specific numerical quantity), and furthermore, each bar force N must be expressed as a function of P. As a result, computing $\partial N/\partial P$ generally requires slightly more calculation than that required to compute each n force directly. These terms will of course be the same, since n or $\partial N/\partial P$ is simply the rate of change of the internal bar force with respect to the load P, or the change in bar force per unit load.

Procedure for Analysis

The following procedure provides a method that may be used to determine the displacement of any joint on a truss using Castigliano's theorem.

External Force P Place a force P on the truss at the joint where the desired displacement is to be determined. This force is assumed to have a *variable magnitude* and should be directed along the line of action of the displacement.

Internal Forces N Determine the force N in each member caused by both the real (numerical) loads and the variable force P. Assume tensile forces are positive and compressive forces are negative. Also compute the respective partial derivative $\partial N/\partial P$ for each member. *After N and $\partial N/\partial P$ have been determined*, assign P its numerical value if it has replaced a real force on the truss. Otherwise, set P equal to zero.

Castigliano's Theorem Apply Eq. 8–34 to determine the desired displacement Δ. It is important to retain the algebraic signs for corresponding values of N and $\partial N/\partial P$ when substituting these terms into the equation. If the resultant sum $\Sigma S(\partial N/\partial P)L/AE$ is positive, Δ is in the same direction as P. If a negative value results, Δ is opposite to P.

The following examples numerically illustrate this method of solution.

Example 8–23

Determine the horizontal displacement of joint D of the truss shown in Fig. 8–47a. Take $E = 29(10^3)$ ksi. The cross-sectional area of each member is indicated in the figure.

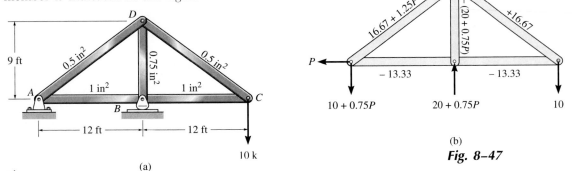

(a)

10 k

(b)

Fig. 8–47

Solution

External Force P. Since the horizontal displacement of D is to be determined, a horizontal variable force P is applied to joint D, Fig. 8–47b.

Internal Forces N. Using the method of joints, the force N in each member is computed.* The results are shown in Fig. 8–47b. Arranging the data in tabular form, we have

Member	N	$\dfrac{\partial N}{\partial P}$	$N\,(P=0)$	L	$N\!\left(\dfrac{\partial N}{\partial P}\right)L$
AB	-13.33	0	-13.33	12	0
BC	-13.33	0	-13.33	12	0
CD	$+16.67$	0	$+16.67$	15	0
DA	$16.67 + 1.25P$	$+1.25$	$+16.67$	15	312.50
BD	$-(20 + 0.75P)$	-0.75	-20	9	135.00

Castigliano's Theorem. Applying Eq. 8–34, we have

$$\Delta_{D_h} = \sum N\!\left(\frac{\partial N}{\partial P}\right)\frac{L}{AE} = 0 + 0 + 0 + \frac{312.5 \text{ k} \cdot \text{ft}(12 \text{ in./ft})}{(0.5 \text{ in}^2)(29(10^3) \text{ k/in}^2)} + \frac{135.00 \text{ k} \cdot \text{ft}(12 \text{ in./ft})}{(0.75 \text{ in.})[29(10^3) \text{ k/in}^2]}$$

$$= 0.333 \text{ in.} \qquad\qquad \textbf{Ans.}$$

*As in the preceding example, it may be preferable to perform a separate analysis of the truss loaded with 10 k and loaded with P and then superimpose the results.

Example 8–24

Determine the vertical displacement of joint C of the truss shown in Fig. 8–48a. The cross-sectional area of each member is $A = 400 \text{ mm}^2$ and $E = 200 \text{ GPa}$.

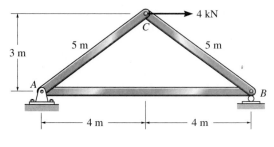

(a)

Solution

External Force P. A vertical force **P** is applied to the truss at joint C, since this is where the vertical displacement is to be determined, Fig. 8–48b.

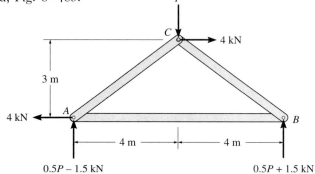

(b)

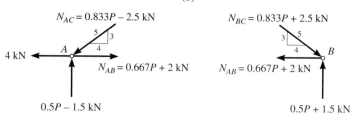

(c)

Fig. 8–48

Internal Forces N.   The reactions at the truss supports at A and B are determined and the results are shown in Fig. 8–48b. Using the method of joints, the N forces in each member are determined, Fig. 8–48c.* For convenience, these results along with the partial derivatives $\partial N/\partial P$ are listed in tabular form as follows:

Member	N	$\dfrac{\partial N}{\partial P}$	$N\ (P = 0)$	L	$N\left(\dfrac{\partial N}{\partial P}\right)L$
AB	$0.667P + 2$	0.667	2	8	10.67
AC	$-(0.833P - 2.5)$	-0.833	2.5	5	-10.41
BC	$-(0.833P + 2.5)$	-0.833	-2.5	5	10.41
					$\Sigma 10.67\ \text{kN} \cdot \text{m}$

Since P does not actually exist as a real load on the truss, we require $P = 0$ in the table above.

Castigliano's Theorem.   Applying Eq. 8–34, we have

$$\Delta_{C_v} = \sum N\left(\frac{\partial N}{\partial P}\right)\frac{L}{AE} = \frac{10.67\ \text{kN} \cdot \text{m}}{AE}$$

Substituting $A = 400\ \text{mm}^2 = 400(10^{-6})\ \text{m}^2$, $E = 200$ GPa $= 200(10^9)$ Pa, and converting the units of S from kN to N, we have

$$\Delta_{C_v} = \frac{10.67(10^3)}{400(10^{-6})200(10^9)} = 0.000133\ \text{m} = 0.133\ \text{mm} \quad \textbf{Ans.}$$

This solution should be compared with the virtual-work method of Example 8–12a.

*It may be more convenient to analyze the truss with just the 4-kN load on it, then analyze the truss with the P load on it. The results can then be added together to give the N forces.

Example 8–25

Determine the vertical displacement of joint C of the truss shown in Fig. 8–49a. Assume that $A = 0.5$ in^2 and $E = 29(10^3)$ ksi.

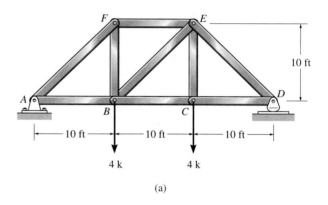

(a)

Solution

External Force P.   The 4-k force at C is replaced with a *variable force P* at joint C, Fig. 8–49b.

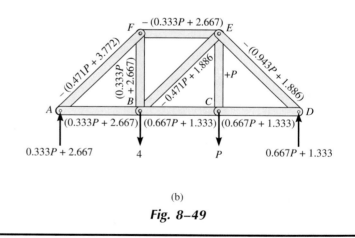

(b)

Fig. 8–49

Internal Forces N. The method of joints is used to determine the force N in each member of the truss. The results are summarized in Fig. 8–49*b*. The required data can be arranged in tabulated form as follows:

Member	N	$\dfrac{\partial N}{\partial P}$	N $(P = 4\text{ k})$	L	$N\left(\dfrac{\partial N}{\partial P}\right)L$
AB	$(0.333P + 2.667)$	$+0.333$	$+4$	10	13.33
BC	$(0.667P + 1.333)$	$+0.667$	$+4$	10	26.67
CD	$(0.667P + 1.333)$	$+0.667$	$+4$	10	26.67
DE	$-(0.943P + 1.886)$	-0.943	-5.66	14.14	75.47
EF	$-(0.333P + 2.667)$	-0.333	-4	10	13.33
FA	$-(0.471P + 3.772)$	-0.471	-5.66	14.14	37.70
BF	$(0.333P + 2.667)$	$+0.333$	$+4$	10	13.33
BE	$(-0.471P + 1.886)$	-0.471	0	14.14	0
CE	P	1	4	10	40

$$\Sigma = 246.50\text{ k}\cdot\text{ft}$$

Castigliano's Theorem. Substituting the data into Eq. 8–34, we have

$$\Delta_{C_v} = N\left(\frac{\partial N}{\partial P}\right)\frac{L}{AE} = \frac{246.50\text{ k}\cdot\text{ft}}{AE}$$

Converting the units of member length to inches and substituting the numerical value for AE, we have

$$\Delta_{C_v} = \frac{(246.50\text{ k}\cdot\text{ft})(12\text{ in./ft})}{(0.5\text{ in}^2)(29(10^3)\text{ k/in}^2)} = 0.204\text{ in.} \qquad \textbf{Ans.}$$

The similarity between this solution and that of the virtual-work method, Example 8–11, should be noted.

8.14 Castigliano's Theorem for Beams and Frames

The internal strain energy for a beam or frame is given by Eq. 8–18 ($U_i = \int M^2\, dx/2EI$). Substituting this equation into Eq. 8–33 ($\Delta_i = \partial U_i/\partial P_i$) and omitting the subscript i, we have

$$\Delta = \frac{\partial}{\partial P}\int_0^L \frac{M^2\, dx}{2EI}$$

Rather than squaring the expression for internal moment M, integrating, and then taking the partial derivative, it is generally easier to differentiate prior to integration. Provided E and I are constant, we have

$$\Delta = \int_0^L M\left(\frac{\partial M}{\partial P}\right)\frac{dx}{EI} \qquad (8\text{–}35)$$

where

Δ = external displacement of the point caused by the real loads acting on the beam or frame

P = external force applied to the beam or frame in the direction of Δ

M = internal moment in the beam or frame, expressed as a function of x and caused by both the force P and the real loads on the beam

E = modulus of elasticity of beam material

I = moment of inertia of cross-sectional area computed about the neutral axis

If the slope θ at a point is to be determined, the partial derivative of the internal moment M with respect to an *external couple moment M'* acting at the point must be computed, i.e.,

$$\theta = \int_0^L M\left(\frac{\partial M}{\partial M'}\right)\frac{dx}{EI} \qquad (8\text{–}36)$$

The above equations are similar to those used for the method of virtual work, Eqs. 8–25 and 8–26, except m and m_θ replace $\partial M/\partial P$ and $\partial M/\partial M'$, respectively. Like the case for trusses, slightly more calculation is generally required to determine the partial derivatives and apply Castigliano's theorem rather than use the method of virtual work. Also, recall that this theorem applies only to material having a linear elastic response. If a more complete accountability of strain energy in the structure is desired, the strain energy due to shear, axial force, and torsion must be included. The derivations for shear and torsion follow the same development as Eqs. 8–28 and 8–29. The strain energies and their derivatives are, respectively,

$$U_s = K\int_0^L \frac{V^2\,dx}{2AG} \qquad \frac{\partial U_s}{\partial P} = \int_0^L \frac{V}{AG}\left(\frac{\partial V}{\partial P}\right) dx$$

$$U_t = \int_0^L \frac{T^2\,dx}{2JG} \qquad \frac{\partial U_t}{\partial P} = \int_0^L \frac{T}{JG}\left(\frac{\partial T}{\partial P}\right) dx$$

These effects, however, will not be included in the analysis of the problems in this text since beam and frame deflections are caused mainly by bending strain energy. Larger frames, or those with unusual geometry, can be analyzed by computer, where these effects can readily be incorporated into the analysis.

Procedure for Analysis

The following procedure provides a method that may be used to determine the deflection and/or slope at a point in a beam or frame using Castigliano's theorem.

External Force P or Couple Moment M′   Place a force **P** on the beam or frame at the point and in the direction of the desired displacement. If the slope is to be determined, place a couple moment **M′** at the point. It is assumed that both P and M' have a variable magnitude.

Internal Moments M   Establish appropriate x coordinates that are valid within regions of the beam or frame where there is no discontinuity of force, distributed load, or couple moment. Calculate the internal moment M as a function of P or M' and each x coordinate. Also, compute the partial derivative $\partial M/\partial P$ or $\partial M/\partial M'$ for each coordinate x. After M and $\partial M/\partial P$ or $\partial M/\partial M'$ have been determined, assign P or M' its numerical value if it has replaced a real force or couple moment. Otherwise, set P or M' equal to zero.

Castigliano's Theorem   Apply Eq. 8–35 or 8–36 to determine the desired displacement Δ or slope θ. It is important to retain the algebraic signs for corresponding values of M and $\partial M/\partial P$ or $\partial M/\partial M'$. If the resultant sum of all the definite integrals is positive, Δ or θ are in the same direction as **P** or **M′**.

The following examples illustrate this method of solution.

Example 8–26

Determine the displacement of point B of the beam shown in Fig. 8–50a. Take $E = 200$ GPa, $I = 500(10^6)$ mm^4.

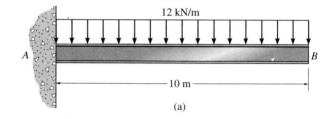

(a)

Solution

External Force P. A vertical force $\mathbf{P}$ is placed on the beam at B as shown in Fig. 8–50b.

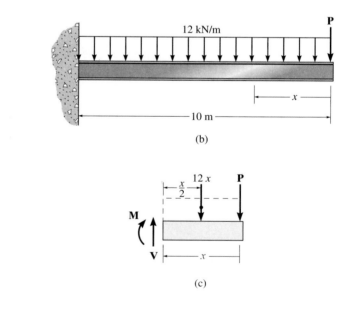

(b)

(c)

Fig. 8–50

Internal Moments M. A single x coordinate is needed for the solution, since there are no discontinuities of loading between A and B. Using the method of sections, Fig. 8–50c, we have

$$\gamma+\Sigma M = 0; \qquad M + (12x)\left(\frac{x}{2}\right) + P(x) = 0$$

$$M = -6x^2 - Px \qquad \frac{\partial M}{\partial P} = -x$$

Setting $P = 0$ yields

$$M = -6x^2 \qquad \frac{\partial M}{\partial P} = -x$$

Castigliano's Theorem. Applying Eq. 8–35, we have

$$\Delta_B = \int M\left(\frac{\partial M}{\partial P}\right)\frac{dx}{EI} = \int_0^{10} \frac{(-6x^2)(-x)\,dx}{EI}$$

$$= \frac{15(10^3)\ \text{kN}\cdot\text{m}^3}{EI}$$

or

$$\Delta_B = \frac{15(10^3)\ \text{kN}\cdot\text{m}^3}{200(10^6)\ \text{kN/m}^2[500(10^6)\ \text{mm}^4](10^{-12}\ \text{m}^4/\text{mm}^4)}$$

$$= 0.150\ \text{m} = 150\ \text{mm} \qquad\qquad \textbf{Ans.}$$

The similarity between this solution and that of the virtual-work method, Example 8–14, should be noted.

Example 8–27

Determine the slope at point B of the beam shown in Fig. 8–51a. Take $E = 200$ GPa, $I = 60(10^6)$ mm^4.

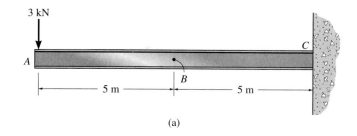

(a)

Solution

External Couple Moment M′.   Since the slope at point B is to be determined, an external couple $\mathbf{M}'$ is placed on the beam at this point, Fig. 8–51b.

Fig. 8–51

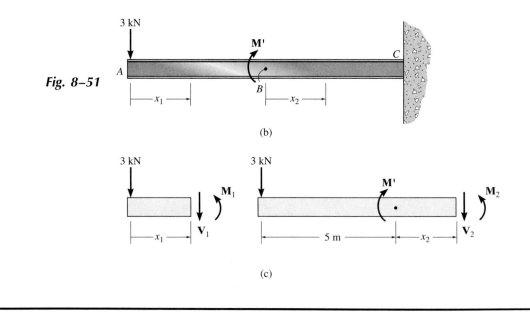

(b)

(c)

Internal Moments M.   Two coordinates, x_1 and x_2, must be used to determine the internal moments within the beam since there is a discontinuity, M', at B. As shown in Fig. 8–51b, x_1 ranges from A to B and x_2 ranges from B to C. Using the method of sections, Fig. 8–51c, the internal moments and the partial derivatives are computed as follows:

For x_1:

$\zeta + \Sigma M = 0;$ $\qquad$ $M_1 + 3x_1 = 0$
$$M_1 = -3x_1$$
$$\frac{\partial M_1}{\partial M'} = 0$$

For x_2:

$\zeta + \Sigma M = 0;$ $\qquad$ $M_2 - M' + 3(5 + x_2) = 0$
$$M_2 = M' - 3(5 + x_2)$$
$$\frac{\partial M_2}{\partial M'} = 1$$

Castigliano's Theorem.   Setting $M' = 0$ and applying Eq. 8–36, we have

$$\theta_B = \int M\left(\frac{\partial M}{\partial M'}\right)\frac{dx}{EI}$$
$$= \int_0^5 \frac{(-3x_1)(0)\,dx_1}{EI} + \int_0^5 \frac{-3(5 + x_2)(1)\,dx_2}{EI}$$
$$= -\frac{112.5\ \text{kN}\cdot\text{m}^2}{EI}$$

or

$$\theta_B = \frac{-112.5\ \text{kN}\cdot\text{m}^2}{200(10^6)\ \text{kN/m}^2[60(10^6)\ \text{mm}^4](10^{-12}\ \text{m}^4/\text{mm}^4)}$$
$$= -0.00938\ \text{rad} \qquad\qquad\qquad \textbf{Ans.}$$

The negative sign indicates that θ_B is opposite to the direction of the couple moment **M'**. Note the similarity between this solution and that of Example 8–17.

Example 8–28

Determine the vertical displacement of point C of the beam shown in Fig. 8–52a. Take $E = 200$ GPa, $I = 150(10^6)$ mm^4.

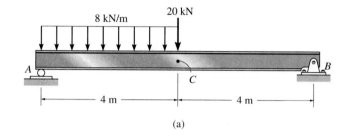

(a)

Solution

External Force P. A vertical force P is applied at point C, Fig. 8–52b. Later this force will be set equal to a fixed value of 20 kN.

Fig. 8–52

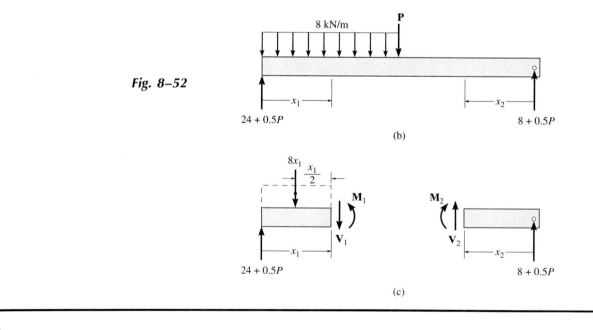

(b)

(c)

Internal Moments M.   In this case two x coordinates are needed for the integration, Fig. 8–52b, since the load is discontinuous at C. Using the method of sections, Fig. 8–52c, we have

For x_1:

$$\gamma+\Sigma M = 0; \quad (24 + 0.5P)x_1 - 8x_1\left(\frac{x_1}{2}\right) - M_1 = 0$$

$$M_1 = (24 + 0.5P)x_1 - 4x_1^2$$

$$\frac{\partial M_1}{\partial P} = 0.5x_1$$

For x_2:

$$\gamma+\Sigma M = 0; \qquad M_2 - (8 + 0.5P)x_2 = 0$$

$$M_2 = (8 + 0.5P)x_2$$

$$\frac{\partial M_2}{\partial P} = 0.5x_2$$

Castigliano's Theorem.   Setting $P = 20\,\text{kN}$ and applying Eq. 8–35 yields

$$\Delta_{C_v} = \int M\left(\frac{\partial M}{\partial P}\right)\frac{dx}{EI}$$

$$= \int_0^4 \frac{(34x_1 - 4x_1^2)(0.5x_1)\,dx_1}{EI} + \int_0^4 \frac{(18x_2)(0.5x_2)\,dx_2}{EI}$$

$$= \frac{234.7\,\text{kN}\cdot\text{m}^3}{EI} + \frac{192\,\text{kN}\cdot\text{m}^3}{EI} = \frac{426.7\,\text{kN}\cdot\text{m}^3}{EI}$$

or

$$\Delta_{C_v} = \frac{426.7\,\text{kN}\cdot\text{m}^3}{(200)(10^6)\,\text{kN/m}^2[150(10^6)\,\text{mm}^4](10^{-12}\,\text{m}^4/\text{mm}^4)}$$

$$= 0.0142\,\text{m} = 14.2\,\text{mm} \hspace{2cm} \textbf{Ans.}$$

Example 8–29

Determine the slope at point C of the two-member frame shown in Fig. 8–53a. The support at A is fixed. Take $E = 29(10^3)$ ksi, $I = 600$ in^4.

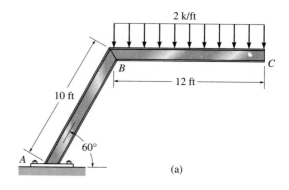

(a)

Solution

External Couple Moment M′.   A variable moment **M′** is applied to the frame at point C, since the slope at this point is to be determined, Fig. 8–53b. Later this moment will be set equal to zero.

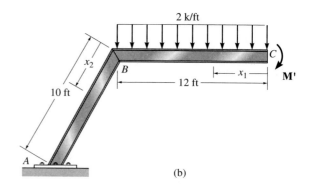

(b)

Fig. 8–53a,b

Internal Moments M.   Due to the discontinuity of internal loading at B, two coordinates, x_1 and x_2, are chosen as shown in Fig. 8–53b. Using the method of sections, Fig. 8–53c, we have

For x_1:

$$\curvearrowleft +\Sigma M = 0; \qquad M_1 + 2x_1\left(\frac{x_1}{2}\right) + M' = 0$$

$$M_1 = -(x_1^2 + M')$$

$$\frac{\partial M_1}{\partial M'} = -1$$

For x_2:

$$\curvearrowleft +\Sigma M = 0; \qquad M_2 + 24(x_2 \cos 60° + 6) + M' = 0$$

$$M_2 = -24(x_2 \cos 60° + 6) - M'$$

$$\frac{\partial M_2}{\partial M'} = -1$$

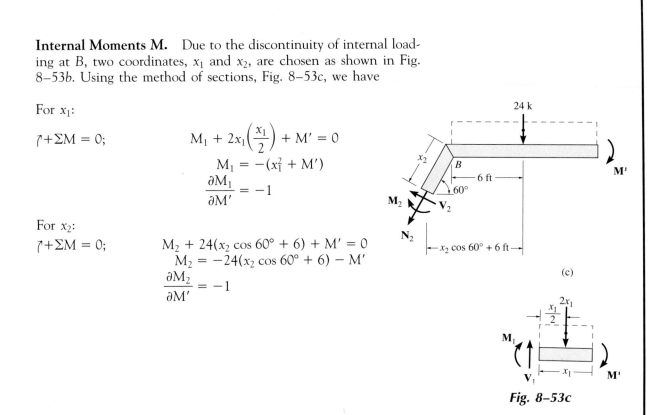

Fig. 8–53c

Castigliano's Theorem.   Setting $M' = 0$ and applying Eq. 8–36 yields

$$\theta_C = \int M\left(\frac{\partial M}{\partial M'}\right)\frac{dx}{EI}$$

$$= \int_0^{12} \frac{(-x_1^2)(-1)\,dx_1}{EI} + \int_0^{10} \frac{-24(x_2 \cos 60° + 6)(-1)\,dx_2}{EI}$$

$$= \frac{576\ \text{k}\cdot\text{ft}^2}{EI} + \frac{2040\ \text{k}\cdot\text{ft}^2}{EI} = \frac{2616\ \text{k}\cdot\text{ft}^2}{EI}$$

or

$$\theta_C = \frac{2616\ \text{k}\cdot\text{ft}^2(144\ \text{in}^2/\text{ft}^2)}{29(10^3)\ \text{k/in}^2(600\ \text{in}^4)} = 0.0216\ \text{rad} \qquad \textbf{Ans.}$$

435

REFERENCES

Betti, E. *Il Nuovo Cimento*, ser. 2, Vols. 7 and 8, 1872.

Castigliano, A. *Théorie de l'équilibre des systèmes élastiques et ses applications*, Paris, 1879. (See translation E. S. Andrews. *The Theory of Equilibrium of Elastic Systems and its Applications.* Dover Publications, Inc., New York, 1966.)

Mohr, O. *Beitrag zur Theorie der Holz-und Eisenkonstruktionen.* Zeitschrift des Architekten-und Ingenieuren-Vereins Zu Hannover, 6 (1860), 323–346.

PROBLEMS

8–1. Determine the slope and deflection at *B*. Use the method of integration. *EI* is constant.

8–2. Solve Prob. 8–1 using the moment-area theorems.

8–3. Solve Prob. 8–1 using the conjugate-beam method.

*8–4. Solve Prob. 8–1 using the method of virtual work.

8–5. Solve Prob. 8–1 using Castigliano's theorem.

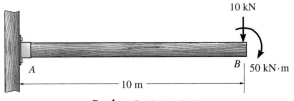

Probs. 8–1 to 8–5

8–6. Determine the deflection at the center *B* of the beam and the slope at *A*. *EI* is constant. Use the method of integration. Assume the support at *C* is a roller.

8–7. Solve Prob. 8–6 using the moment-area theorems.

*8–8. Solve Prob. 8–6 using the method of virtual work.

8–9. Determine the deflection at the center *B* of the beam and the slope at *A*. *EI* is constant. Use the method of integration. Assume the support at *A* is a pin and *C* is a roller.

8–10. Solve Prob. 8–9 using the moment-area theorems.

8–11. Solve Prob. 8–9 using the conjugate-beam method.

*8–12. Solve Prob. 8–9 using the method of virtual work.

8–13. Solve Prob. 8–9 using Castigliano's theorem.

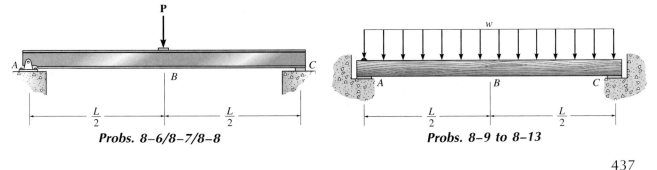

Probs. 8–6/8–7/8–8   **Probs. 8–9 to 8–13**

8–14.   Determine the deflection at the center B of the beam and the slope at A. *EI* is constant. Use the method of integration.

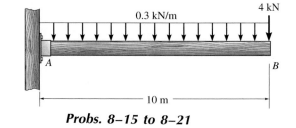

Prob. 8–14

8–15.   Determine the slope and deflection at B. Use the method of integration. The beam has a rectangular cross section, where $E = 13\,\text{GPa}$, $A = 0.08\,\text{m}^2$, $I = 0.500(10^{-3})\,\text{m}^4$.

*8–16.   Solve Prob. 8–15 using the moment-area theorems.

8–17.   Solve Prob. 8–15 using the conjugate-beam method.

8–18.   Solve Prob. 8–15 using the method of virtual work.

8–19.   Solve Prob. 8–15 using the method of virtual work, including the effect of shear.

*8–20.   Solve Prob. 8–15 using Castigliano's theorem.

8–21.   Solve Prob. 8–15 using Castigliano's theorem, including the effect of shear.

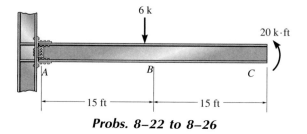

Probs. 8–15 to 8–21

8–22.   Determine the slope and deflection at C. Use the method of integration. $E = 29(10^3)$ ksi, $I = 800\,\text{in}^4$.

8–23.   Solve Prob. 8–22 using the moment-area theorems.

*8–24.   Solve Prob. 8–22 using the conjugate-beam method.

8–25.   Solve Prob. 8–22 using the method of virtual work.

8–26.   Solve Prob. 8–22 using Castigliano's theorem.

Probs. 8–22 to 8–26

438

8–27. Determine the deflection at C. Use the method of integration. Assume the support at A is a pin and B is a roller. EI is constant.

*8–28. Solve Prob. 8–27 using the moment-area theorems.

8–29. Solve Prob. 8–27 using the conjugate-beam method.

8–30. Solve Prob. 8–27 using the method of virtual work.

8–31. Solve Prob. 8–27 using Castigliano's theorem.

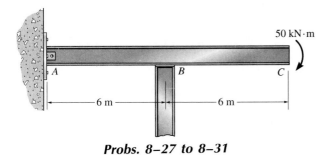

Probs. 8–27 to 8–31

*8–32. Determine the slope and the deflection of the beam at C. Use the moment-area theorems. EI is constant. Assume the support at A is a roller and B is a pin.

8–33. Solve Prob. 8–32 using the conjugate-beam method.

8–34. Solve Prob. 8–32 using the method of virtual work.

8–35. Solve Prob. 8–32 using Castigliano's theorem.

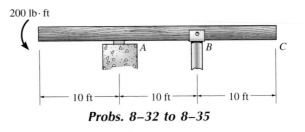

Probs. 8–32 to 8–35

*8–36. Determine the deflection at C. Assume B is a roller and A is a pin. Use the method of integration. $E = 200$ GPa, and $I = 80(10^{-6})$ m^4.

8–37. Solve Prob. 8–36 using the moment-area theorems.

8–38. Solve Prob. 8–36 using the conjugate-beam method.

8–39. Solve Prob. 8–36 using the method of virtual work.

*8–40. Solve Prob. 8–36 using Castigliano's theorem.

Probs. 8–36 to 8–40

8–41. Determine the deflection and slope of the end C of the cantilever beam. $E = 29(10^3)$ ksi. The moment of inertia of each segment is indicated in the figure. Use the moment-area theorems.

8–42. Solve Prob. 8–41 using the conjugate-beam method.

8–43. Solve Prob. 8–41 using the method of virtual work.

*8–44. Solve Prob. 8–41 using Castigliano's theorem.

8–49. Determine the slope at B and the deflection at C. Use the moment-area theorems. EI is constant. Assume A is a roller and B is a pin.

8–50. Solve Prob. 8–49 using the conjugate-beam method.

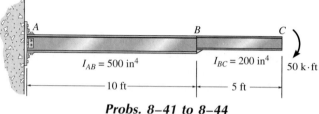

Probs. 8–41 to 8–44

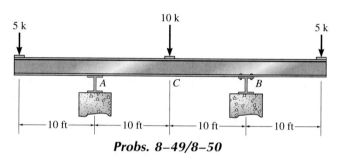

Probs. 8–49/8–50

8–45. Determine the slope and deflection at C. Assume A is a pin and B is a roller. EI is constant. Use the moment-area theorems.

8–46. Solve Prob. 8–45 using the conjugate-beam method.

8–47. Solve Prob. 8–45 using the method of virtual work.

*8–48. Solve Prob. 8–45 using Castigliano's theorem.

8–51. Determine the slope at the end C of the beam. Use the moment-area theorems. $E = 200$ GPa, $I = 70(10^6)$ mm^4. Assume B is a roller and A is a pin.

*8–52. Solve Prob. 8–51 using the conjugate-beam method.

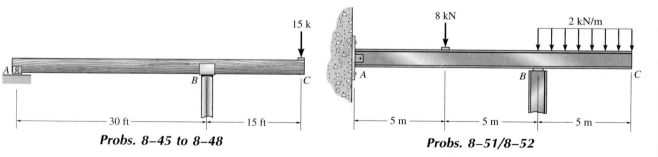

Probs. 8–45 to 8–48   **Probs. 8–51/8–52**

8–53. Determine the slope just to the left and just to the right of the pin at B. Also, determine the deflection at D. Assume the beam is fixed-supported at A, and that C is a roller. Use the conjugate-beam method. EI is constant.

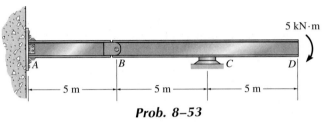

5 kN·m

5 m — 5 m — 5 m

Prob. 8–53

8–54. Determine the deflection at B and the slope at C. Assume A is a fixed support, B is a pin, and C is a roller. Use the moment-area theorems. EI is constant.

8–55. Solve Prob. 8–54 using the conjugate-beam method.

*8–56. Determine the deflection at the pin B of the beam in Prob. 8–54. Use the method of virtual work.

8–57. Solve Prob. 8–54 for the deflection at B using Castigliano's theorem.

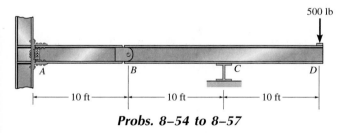

500 lb

10 ft — 10 ft — 10 ft

Probs. 8–54 to 8–57

8–58. Determine the deflection of the center C of the wide-flange beam. Use the method of virtual work. $E = 200$ GPa and $I = 75(10^6)$ mm⁴.

8–59. Solve Prob. 8–58 using Castigliano's theorem.

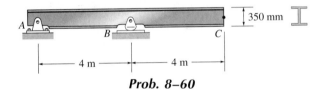

2 kN/m

6 m — 6 m

Probs. 8–58/8–59

*8–60. The top of the beam is subjected to a temperature of $T_t = 200°C$, while the temperature of its bottom is $T_b = 30°C$. If $\alpha = 12(10^{-6})/°C$, determine the vertical displacement of its end C due to the temperature gradient. The beam has a depth of 350 mm.

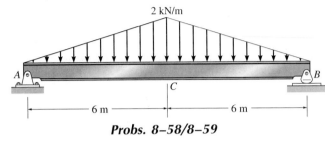

350 mm

4 m — 4 m

Prob. 8–60

8–61. The bottom of the beam is subjected to a temperature of $T_b = 250°F$, while the temperature of its top is $T_t = 50°F$. If $\alpha = 6.5(10^{-6})/°F$, determine the vertical displacement of its end B due to the temperature gradient. The beam has a rectangular cross section with a depth of 14 in.

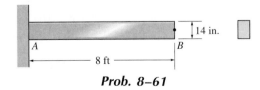

14 in.

8 ft

Prob. 8–61

441

8–62.   Determine the horizontal deflection at C. The cross-sectional area of each member is indicated in the figure. Assume the members are pin-connected at their end points. $E = 29\,(10^3)$ ksi. Use the method of virtual work.

8–63.   Solve Prob. 8–62 using Castigliano's theorem.

8–66.   Remove the loads on the truss in Prob. 8–64 and determine the vertical displacement of point B if member EB is fabricated 0.75 in. too long.

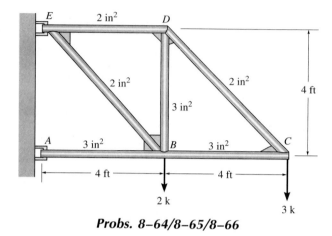

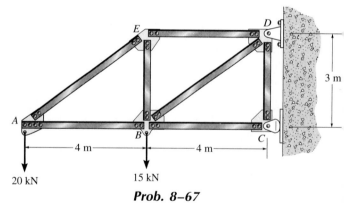

Probs. 8–64/8–65/8–66

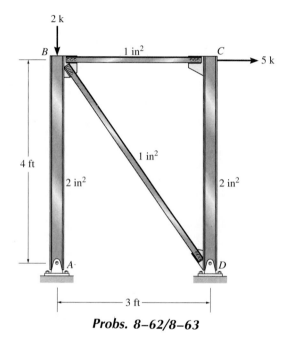

Probs. 8–62/8–63

*8–64.**   Determine the vertical deflection at C. Member EB has been fabricated 0.2 in. too short. The cross-sectional area of each member is indicated in the figure. $E = 29\,(10^6)$ psi. Assume all members are pin-connected at their end points. Use the method of virtual work.

8–65.   Remove the loads on the truss in Prob. 8–64 and determine the vertical displacement of point B if members AB and BC experience a temperature increase of $\Delta T = 200°F$. Take $A = 1.5$ in^2 and $E = 29(10^3)$ ksi. Also, $\alpha = 10^{-6}/°F$.

8–67.   Determine the vertical deflection at A. Use the method of virtual work. AE is constant. Assume the members are pin-connected at their ends.

Prob. 8–67

*8–68.**   Solve Prob. 8–66 using Castigliano's theorem.

442

8–69. Determine the horizontal deflection at C. Use the method of virtual work. Assume the members are pin-connected at their end points. AE is constant.

8–70. Remove the loads on the truss in Prob. 8–69 and determine the horizontal displacement of point C if members AB and BC experience a temperature increase of $\Delta T = 200°F$. Take $A = 2$ in^2 and $E = 29(10^3)$ ksi. Also, $\alpha = 10^{-6}/°F$.

8–71. Remove the loads on the truss in Prob. 8–69 and determine the horizontal displacement of point C if member CD is fabricated 0.5 in. too short.

*8–72. Determine the horizontal deflection of the truss at B. AE is the same for all members. Assume all members are pin-connected at their end points. Use the method of virtual work.

8–73. Determine the vertical deflection of the truss at F. AE is the same for all members. Assume all members are pin-connected at their end points. Use the method of virtual work.

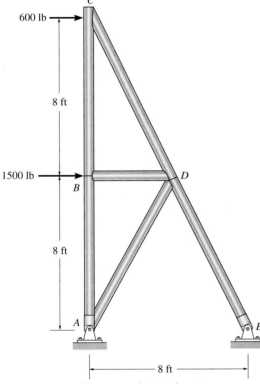

Probs. 8–69/8–70/8–71

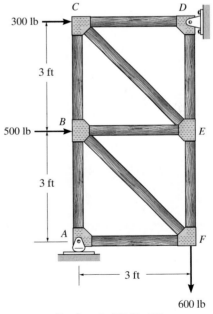

Probs. 8–72/8–73

443

8–74. Determine the vertical deflection at C. The cross-sectional area of each member is indicated in the figure. Assume the members are pin-connected at their end points. $E = 29 (10^3)$ ksi. Use the method of virtual work.

8–75. Solve Prob. 8–74 using Castigliano's theorem.

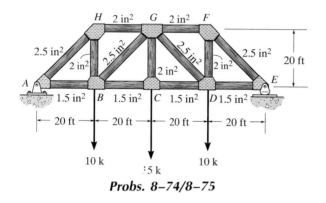

Probs. 8–74/8–75

*8–76. Determine the vertical deflection at C. Use the method of virtual work. AE is constant. Assume the members are pin-connected at their end points and the supports at A and E only exert vertical reactions on the truss.

8–77. Solve Prob. 8–76 using Castigliano's theorem.

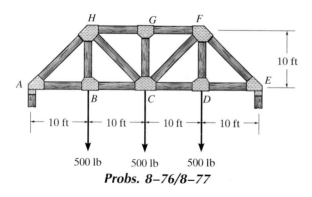

Probs. 8–76/8–77

8–78. Determine the vertical deflection at C. AE is constant. Assume each member is pin-connected at its ends. Use the method of virtual work.

8–79. Solve Prob. 8–78 using Castigliano's theorem.

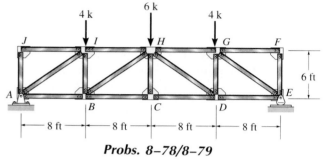

Probs. 8–78/8–79

*8–80. Determine the horizontal and vertical deflection components at C. The support at A is fixed. Each wide-flange member has $E = 29(10^3)$ ksi, and $I = 250$ in⁴. Use the method of virtual work.

8–81. Solve Prob. 8–80, including the effect of axial strain energy. Take $A = 14$ in².

8–82. Solve Prob. 8–80 using Castigliano's theorem.

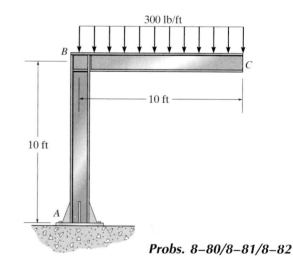

Probs. 8–80/8–81/8–82

8–83. Determine the vertical deflection at A. Use the method of virtual work. $E = 200\,\text{GPa}$, $I = 90(10^6)\,\text{mm}^4$.

*8–84. Solve Prob. 8–83 using Castigliano's theorem.

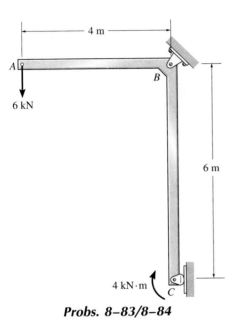

Probs. 8–83/8–84

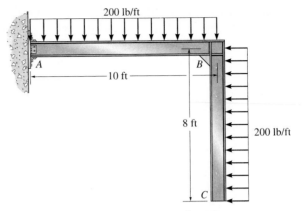

Probs. 8–85/8–86/8–87

*8–88. Determine the vertical deflection at C. The cross-sectional area and moment of inertia of each segment is shown in the figure. Take $E = 200\,\text{GPa}$. Assume A is a fixed support. Use the method of virtual work.

8–89. Solve Prob. 8–88, including the effect of shear and axial strain energy.

8–90. Solve Prob. 8–88 using Castigliano's theorem.

8–85. Determine the horizontal and vertical components of deflection at C. Each wide-flange member has $E = 29(10^3)\,\text{ksi}$, and $I = 80\,\text{in}^4$. Use the method of virtual work.

8–86. Solve Prob. 8–85, including the effect of axial strain energy. Take $A = 8\,\text{in}^2$.

8–87. Solve Prob. 8–85 using Castigliano's theorem.

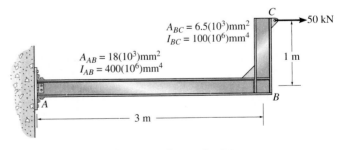
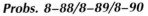

Probs. 8–88/8–89/8–90

445

8–91. Determine the slope at A. $E = 29(10^3)$ ksi. The moment of inertia of each segment of the frame is indicated in the figure. Assume D is a pin support. Use the method of virtual work.

*8–92. Solve Prob. 8–91 using Castigliano's theorem.

Probs. 8–91/ 8–92

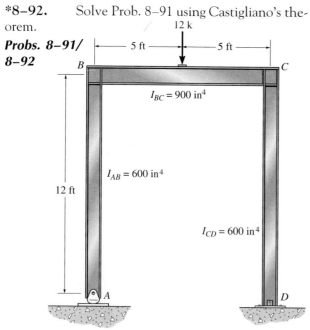

8–93. Determine the vertical deflection at the rocker support D. EI is constant. Use the method of virtual work.

8–94. Solve Prob. 8–93 using Castigliano's theorem.

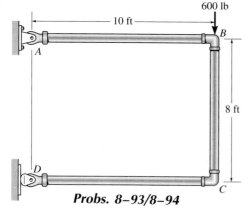

Probs. 8–93/8–94

8–95. Determine the horizontal deflection at C. EI is constant. Use the method of virtual work. There is a pin at A and assume C is a roller, and B is a fixed joint.

*8–96. Solve Prob. 8–95 using Castigliano's theorem.

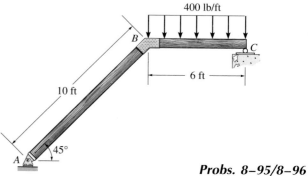

Probs. 8–95/8–96

8–97. The frame is subjected to the load of 5 k. Determine the vertical displacement at C. The members are pin-connected at A, C, and E; and fixed-connected at the knee joints B and D. EI is constant. Use the method of virtual work.

8–98. Solve Prob. 8–97 using Castigliano's theorem.

Probs. 8–97/8–98

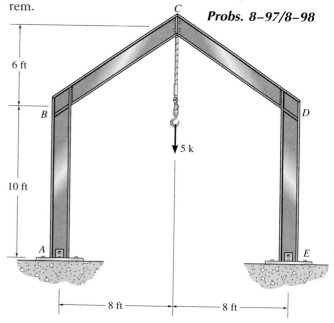

8–99. The bent rod has an $E = 200\,\text{GPa}$, $G = 75\,\text{GPa}$, and a radius of 30 mm. Determine the vertical deflection at C, include the effects of bending, shear, and torsional strain energy. Use the method of virtual work.

*8–100. Solve Prob. 8–99 using Castigliano's theorem.

8–101. The bent rod has an $E = 29(10^3)\,\text{ksi}$, $G = 11(10^3)\,\text{ksi}$, and a radius of 0.75 in. Determine the deflection in the direction of the load at C. Include the effects of bending, axial, shear, and torsional strain energy. Use the method of virtual work.

8–102. Solve Prob. 8–101 using Castigliano's theorem.

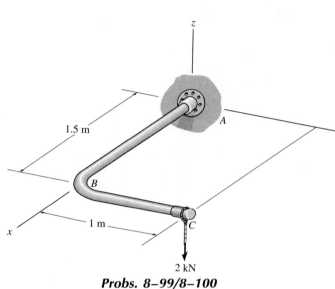

Probs. 8–99/8–100

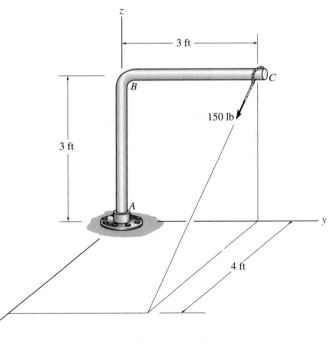

Probs. 8–101/8–102

447

This building frame is statically indeterminate since the connections between the beams and columns form welded and stiffened joints. *(Photo courtesy American Institute of Steel Construction)*

9

Analysis of Statically Indeterminate Structures by the Force Method

In this chapter we will apply the *force* or *flexibility* method to analyze statically indeterminate trusses, beams, and frames. Also, we will discuss application of the three-moment equation, a force method of analysis used to analyze indeterminate beams. At the end of the chapter we will present a method for drawing the influence line for a statically indeterminate beam or frame.

9.1 Statically Indeterminate Structures

Recall from Sec. 2.4 that a structure of any type is classified as *statically indeterminate* when the number of unknown reactions or internal forces exceeds the number of equilibrium equations available for its analysis. In this section we will discuss the merits of using indeterminate structures and two fundamental ways in which they may be analyzed. Realize that most of the structures designed today are statically indeterminate. This indeterminacy may arise as a result of added supports or members, or by the general form of the structure. For example, reinforced concrete buildings are almost always statically indeterminate since the columns and beams are poured as continuous members through the joints and over supports.

Advantages and Disadvantages

Although the analysis of a statically indeterminate structure is more involved than that of one that is statically determinate, there are usually several very important reasons for choosing this type of structure for design. Most important, for a given loading the maximum stress and deflection of an indeterminate structure are generally *smaller* than those of its statically determinate counterpart. For example, the statically indeterminate, fixed-supported beam in Fig. 9–1a will be subjected to a maximum moment of $M_{max} = PL/8$, whereas the same beam, when simply-supported, Fig. 9–1b, will be subjected to twice the moment, that is, $M_{max} = PL/4$. As a result, the fixed-supported beam has one fourth the deflection and one half the stress at its center of the one that is simply supported.

Another important reason for selecting a statically indeterminate structure is because it has a tendency to redistribute its load to its redundant supports in cases where faulty design or overloading occurs. In these cases, the structure maintains its stability and collapse is prevented. This is particularly important when *sudden* lateral loads, such as wind or earthquake, are imposed on the structure. To illustrate, consider again the fixed-end beam loaded at its center, Fig. 9–1a. As **P** is increased, the beam's material at the walls and at the center of the beam begins to *yield* and forms a localized "plastic hinge," which causes

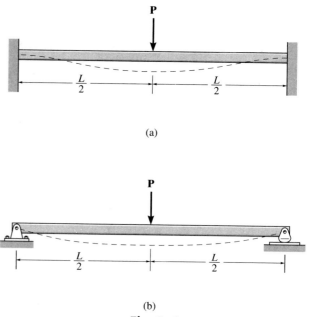

(a)

(b)

Fig. 9–1

the beam to deflect as if it were hinged or pin-commuted at these points. Although the deflection becomes large, the walls will develop horizontal force and moment reactions that will hold the beam and thus prevent it from totally collapsing. In the case of the simply-supported beam, Fig. 9–1b, an excessive load **P** will cause the "plastic hinge" to form only at the center of the beam, and due to the large vertical deflection, the supports will not develop the horizontal force and moment reactions that may be necessary to prevent total collapse.

Although statically indeterminate structures can support a loading with thinner members and with increased stability compared to their statically determinate counterparts, there are cases when these advantages may instead become disadvantages. The cost savings in material must be compared with the added cost necessary to fabricate the structure, since oftentimes it becomes more costly to construct the supports and joints of an indeterminate structure compared to one that is determinate. More important, though, because statically indeterminate structures have redundant support reactions, one has to be very careful to prevent differential displacement of the supports, since this effect will introduce internal stress in the structure. For example, if the wall at one end of the fixed-end beam in Fig. 9–1a were to settle, stress would be developed in the beam because of its "forced" deformation. On the other hand, if the beam was simply-supported or statically determinate, Fig. 9–1b, then any settlement of its end would not cause the beam to deform, and therefore no stress would be developed in the beam. In general, then, any deformation, such as that caused by relative support displacement, or changes in member lengths caused by temperature or fabrication errors, will introduce additional stresses in the structure, which must be considered when designing indeterminate structures.

Methods of Analysis

When analyzing any indeterminate structure, it is necessary to satisfy *equilibrium*, *compatibility*, and *force-displacement* requirements for the structure. *Equilibrium* is satisfied when the reactive forces hold the structure at rest, and *compatibility* is satisfied when the various segments of the structure fit together without intentional breaks or overlaps. The *force-displacement* requirements depend upon the way the material responds, which in this text we have assumed linear-elastic response. In general there are two different ways to satisfy these requirements when analyzing a statically indeterminate structure: the *force* or *flexibility method*, and the *stiffness* or *displacement method*.

451

Force Method

The force method was originally developed by James Clerk Maxwell in 1864 and later refined by Otto Mohr and Heinrich Müller-Breslau. This method was one of the first available for the analysis of statically indeterminate structures. As suggested by the name, the force method consists of writing equations that satisfy the *compatibility* and *force-displacement requirements* for the structure and involve redundant *forces* as the *unknowns*. The coefficients of these unknowns are called *flexibility coefficients*. Since compatibility forms the basis for this method, it has sometimes been referred to as the *compatibility* method or the *method of consistent displacements*. Once the redundant forces have been determined, the remaining reactive forces on the structure are determined by satisfying the equilibrium requirements for the structure. The fundamental principles involved in applying this method are easy to understand and develop, and they will be discussed in this chapter.

Displacement Method

The displacement method of analysis is based on first writing force-displacement relations for the members and then satisfying the *equilibrium requirements* for the structure. In this case the *unknowns* in the equations are *displacements* and their coefficients are called *stiffness coefficients*. Once the displacements are obtained, the forces are determined from the compatibility and force-displacement equations. We will study some of the classical techniques used to apply it in Chapters 9 and 10. A matrix formulation of the method is given in Chapters 13 and 14.

Each of these two methods of analysis, which are outlined in Fig. 9–2, has particular advantages and disadvantages, depending upon the geometry of the structure and its degree of indeterminacy. A discussion of the usefulness of each method will be given after each has been presented.

	Unknowns	Equations Used for Solution
Force Method	Forces	Compatibility and Force Displacement
Displacement Method	Displacements	Equilibrium and Force Displacement

Fig. 9–2

9.2   Force Method of Analysis: General Procedures

Perhaps the best way to illustrate the principles involved in the force method of analysis is to consider the beam shown in Fig. 9–3a. If its free-body diagram were drawn, there would be four unknown support reactions; and since three equilibrium equations are available for solution, the beam is indeterminate to the first degree. Consequently, one additional equation is necessary for solution. To obtain this equation, we will use the principle of superposition and consider the *compatibility of displacement* at one of the supports. This is done by choosing one of the support reactions as "redundant" and temporarily removing its effect on the beam so that the beam then becomes statically determinate and stable. This beam is referred to as the *primary structure*. Here we will remove the restraining action of the rocker at B. As a result, the load **P** will cause B to be displaced downward by an amount Δ_B as shown in Fig. 9–3b. By superposition, however, the unknown reaction at B, i.e., **B**$_y$, causes the beam at B to be displaced Δ'_{BB} upward, Fig. 9–3c. Here the first letter in this double-subscript notation refers to the point (B) where the deflection is specified, and the second letter refers to the point (B) where the unknown reaction acts. Assuming positive displacements act downward, then from Fig. 9–3a through 9–3c we can write the necessary compatibility equation at the rocker as

$$(+\downarrow) \qquad\qquad 0 = \Delta_B - \Delta'_{BB}$$

Let us now denote the displacement at B caused by a *unit load* acting in the direction of **B**$_y$ as the *linear flexibility coefficient* f_{BB}, Fig. 9–3d. Using the same scheme for this double subscript notation as above, f_{BB} is the deflection at B caused by a unit load at B. Since the material behaves in a linear-elastic manner, a force of **B**$_y$ acting at B, instead of the unit load, will cause a proportionate increase in f_{BB}. Thus we can write

$$\Delta'_{BB} = B_y f_{BB}$$

When written in this format, it can be seen that the linear flexibility coefficient f_{BB} is a *measure of the deflection per unit force*, and so its units are m/N, ft/lb, etc. The compatibility equation above can therefore be written in terms of the unknown B_y as

$$0 = \Delta_B - B_y f_{BB}$$

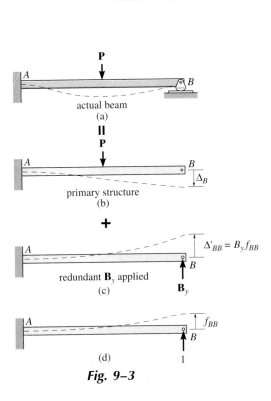

actual beam
(a)

‖

primary structure
(b)

Δ_B

+

$\Delta'_{BB} = B_y f_{BB}$

redundant **B**$_y$ applied
(c)

B$_y$

f_{BB}

(d)

1

Fig. 9–3

453

Using the methods of Chapter 8, or the deflection table on the inside front cover of the book, the appropriate load-displacement relations for the deflection Δ_B, Fig. 9–3b, and the flexibility coefficient f_{BB}, Fig. 9–3d, can be obtained and the solution for B_y determined, that is, $B_y = \Delta_B/f_{BB}$. Once this is accomplished, the three reactions at the wall A can then be found from the equations of equilibrium.

As stated previously, the choice of the redundant is *arbitrary*. For example, the moment at A can be determined *directly* by removing the capacity of the beam to support a moment at A, that is, by replacing the fixed support by a pin. As shown in Fig. 9–4b, the rotation at A caused by the load $\mathbf{P}$ is θ_A, and the rotation at A caused by the redundant $\mathbf{M}_A$ at A is θ'_{AA}, Fig. 9–4c. If we denote an *angular flexibility coefficient* α_{AA} as the angular displacement at A caused by a unit couple moment applied to A, Fig. 9–4d, then

$$\theta'_{AA} = M_A\,\alpha_{AA}$$

Thus, the angular flexibility coefficient measures the angular displacement per unit couple moment, and therefore it has units of $\text{rad}/N \cdot m$ or $\text{rad}/\text{lb} \cdot \text{ft}$, etc. The compatibility equation for rotation at A therefore requires

$$(\curvearrowright+) \qquad\qquad 0 = \theta_A + M_A\,\alpha_{AA}$$

In this case, $M_A = -\theta_A/\alpha_{AA}$, a negative value, which simply means that $\mathbf{M}_A$ acts in the opposite direction to the unit couple moment.

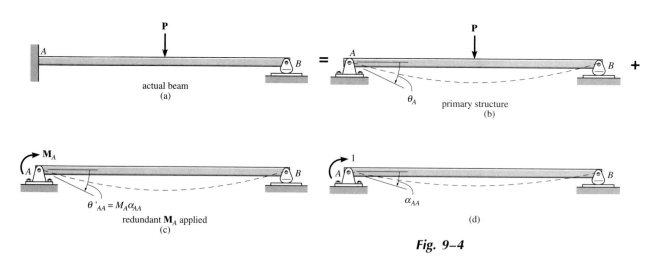

actual beam
(a)

primary structure
(b)

$\theta'_{AA} = M_A\alpha_{AA}$
redundant $\mathbf{M}_A$ applied
(c)

α_{AA}
(d)

Fig. 9–4

A third example that illustrates application of the force method is given in Fig. 9–5a. Here the beam is indeterminate to the second degree and therefore two compatibility equations will be necessary for the solution. We will choose the vertical forces at the roller supports, B and C, as redundants. The resultant statically determinate beam deflects as shown in Fig. 9–5b when the redundants are removed. Each redundant force deflects this beam as shown in Figs. 9–5c and 9–5d, respectively. Here the flexibility coefficients* f_{BB} and f_{CB} are found from a unit load acting at B, Fig. 9–5e; and f_{CC} and f_{BC} are found from a unit load acting at C, Fig. 9–5f. By superposition, the compatibility equations for the deflection at B and C, respectively, are

$$(+\downarrow) \qquad 0 = \Delta_B + B_y f_{BB} + C_y f_{BC}$$
$$(+\downarrow) \qquad 0 = \Delta_C + B_y f_{CB} + C_y f_{CC} \qquad (9\text{–}1)$$

Once the load-displacement relations are established using the methods of Chapter 8, these equations may be solved simultaneously for the two unknown forces B_y and C_y.

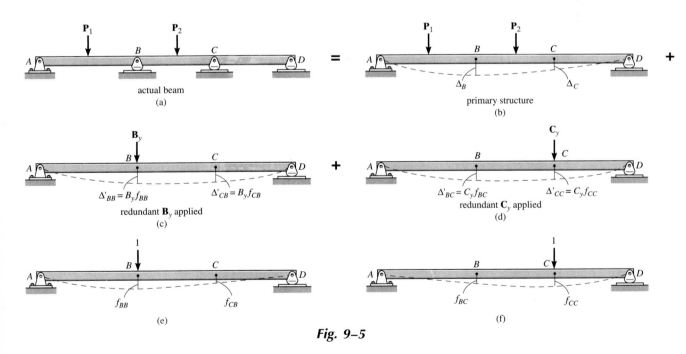

Fig. 9–5

*f_{BB} is the deflection at B caused by a unit load at B; f_{CB} the deflection at C caused by a unit load at B.

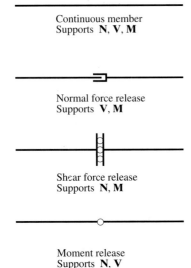

Continuous member
Supports **N, V, M**

Normal force release
Supports **V, M**

Shear force release
Supports **N, M**

Moment release
Supports **N, V**

Fig. 9–6

Having illustrated the application of the force method of analysis by example, we will now discuss its application in general terms and then we will use it as a basis for solving problems involving trusses, beams, and frames. For all these cases, however, realize that since the method depends on superposition of displacements, it is necessary that *the material remain linear-elastic when loaded*. Also, recognize that *any* external reaction or internal loading at a point in the structure can be directly determined by first releasing the capacity of the structure to support the loading and then writing a compatibility equation at the point. Connections used to symbolically represent releases of internal normal force, shear, and moment in a continuous member are shown in Fig. 9–6. Note that each connection will support internal loadings *except* the one being released. A *compatibility equation* written for each of these connections will then require that zero relative displacement occur at the connection when the displacement on the primary structure is superimposed with that caused by the redundant loading. This equation will then allow a direct solution for the redundant loading at the connection.

Procedure for Analysis

The following procedure provides a general method for determining the reactions or internal loadings of statically indeterminate structures using the force or flexibility method of analysis.

Principle of Superposition   Determine the number of degrees n to which the structure is indeterminate. Then specify the n unknown redundant forces or moments that must be removed from the structure in order to make it statically determinate and stable. Using the principle of superposition, draw the statically indeterminate structure and show it to be equal to a sequence of corresponding statically *determinate* structures. The primary structure supports the same external loads as the statically indeterminate structure, and each of the other structures added to the primary structure shows the structure loaded with a separate redundant force or moment. Also, sketch the elastic curve on each structure and indicate symbolically the displacement or rotation at the point of each redundant force or moment. (See Figs. 9–3 through 9–5.)

Compatibility Equations Write a compatibility equation for the displacement or rotation at each point where there is a redundant force or moment. These equations should be expressed in terms of the unknown redundants and their corresponding flexibility coefficients obtained from unit loads or unit couple moments that are collinear with the redundant forces or moments.

Determine all the flexibility coefficients using the table on the inside front cover or the methods of Chapter 8.* Substitute these load-displacement relations into the compatibility equations and solve for the unknown redundants. If the numerical value for a redundant is positive, it acts in the same direction as its corresponding unit force or unit couple moment. Likewise, a "negative" numerical value indicates the redundant acts opposite to its corresponding unit force or unit couple moment.

Equilibrium Equations Draw a free-body diagram of the structure. Since the redundant forces and/or moments have been calculated, the remaining unknown reactions can be determined from the equations of equilibrium.

It should be realized that once all the support reactions have been obtained, the shear and moment diagrams can then be drawn, and the deflection at any point on the structure can be determined using the same methods outlined previously for statically determinate structures.

---

*It is suggested that if the M/EI diagram for a beam consists of simple segments, the moment-area theorems or the conjugate-beam method be used. Beams with complicated M/EI diagrams, that is those with many curved segments (parabolic, cubic, etc.), can be readily analyzed using the method of virtual work or by Castigliano's second theorem.

9.3 Maxwell's Theorem of Reciprocal Displacements; Betti's Law

When Maxwell developed the force method of analysis, he also published a theorem that relates the flexibility coefficients of any two points on an elastic structure—be it a truss, a beam, or a frame. This theorem is referred to as the theorem of reciprocal displacements and may be stated as follows: *The displacement of a point B on a structure due to a unit load acting at point A is equal to the displacement of point A when the unit load is acting at point B, that is,* $f_{BA} = f_{AB}$.

Proof of this theorem is easily demonstrated using the principle of virtual work. For example, consider the beam in Fig. 9–7. When a real unit load acts at A, Fig. 9–7a, assume that the internal moments in the beam are represented by m_A. To determine the flexibility coefficient at B, that is, f_{BA}, a virtual unit load is placed at B, Fig. 9–7b, and the internal moments m_B are computed. Then applying Eq. 7–25 yields

$$f_{BA} = \int \frac{m_B m_A}{EI} \, dx$$

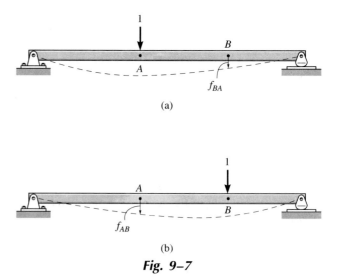

(a)

(b)

Fig. 9–7

Likewise, if the flexibility coefficient f_{AB} is to be determined when a real unit load acts at B, Fig. 9–7b, then m_B represents the internal moments in the beam due to a real unit load. Furthermore, m_A represents the internal moments due to a virtual unit load at A, Fig. 9–7a. Hence,

$$ f_{AB} = \int \frac{m_A m_B}{EI} \, dx $$

Both integrals obviously give the same result, which proves the theorem. The theorem also applies for reciprocal rotations, and may be stated as follows: *The rotation at point B on a structure due to a unit couple moment acting at point A is equal to the rotation at point A when the unit couple moment is acting at point B.* Furthermore, using a unit force and unit couple moment, applied at separate points on the structure, we may also state: *The rotation in radians at point B on a structure due to a unit load acting at point A is equal to the displacement at point A when a unit couple moment is acting at point B.*

As a consequence of this theorem, some work can be saved when applying the force method to problems that are statically indeterminate to the second degree or higher. For example, only the flexibility coefficient, f_{BC} or f_{CB}, has to be calculated in Eqs. 9–1 since $f_{BC} = f_{CB}$. Furthermore, the theorem of reciprocal displacements has applications in structural model analysis and for constructing influence lines using the Müller-Breslau principle (see Sec. 9.11).

When the theorem of reciprocal displacements is formalized in a more general sense, it is referred to as *Betti's law*. Briefly stated: The virtual work U_{AB} done by a system of forces $\Sigma \mathbf{P}_B$ that undergo a displacement caused by a system of forces $\Sigma \mathbf{P}_A$ is equal to the virtual work U_{BA} caused by the forces $\Sigma \mathbf{P}_A$ when the structure deforms due to the system of forces $\Sigma \mathbf{P}_B$. In other words, $U_{AB} = U_{BA}$. The proof of this statement is similar to that given above for the reciprocal displacement theorem.

9.4   Force Method of Analysis: Beams

The force method applied to beams was outlined in Sec. 9.2. Using the "procedure for analysis" also given in Sec. 9.2, we will now present several examples that illustrate the application of this technique.

Example 9–1

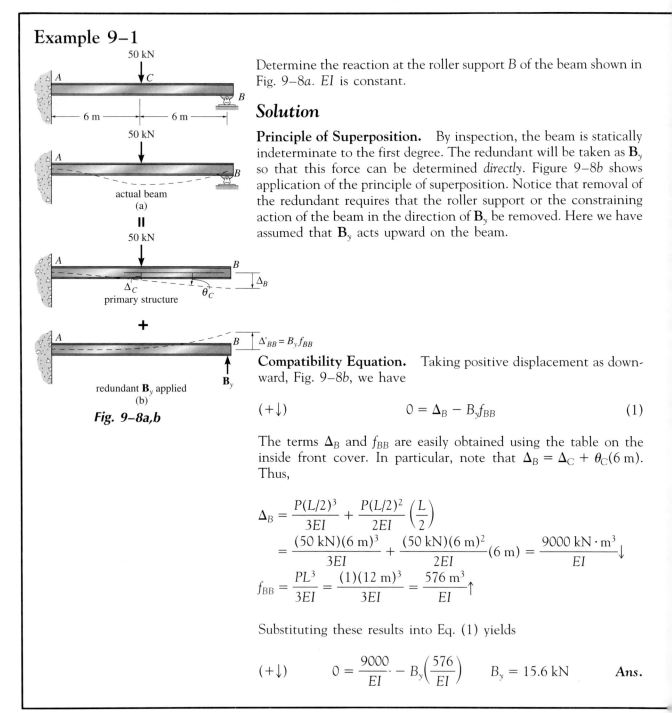

Fig. 9–8a,b

Determine the reaction at the roller support B of the beam shown in Fig. 9–8a. EI is constant.

Solution

Principle of Superposition.   By inspection, the beam is statically indeterminate to the first degree. The redundant will be taken as $\mathbf{B}_y$ so that this force can be determined *directly*. Figure 9–8b shows application of the principle of superposition. Notice that removal of the redundant requires that the roller support or the constraining action of the beam in the direction of $\mathbf{B}_y$ be removed. Here we have assumed that $\mathbf{B}_y$ acts upward on the beam.

Compatibility Equation.   Taking positive displacement as downward, Fig. 9–8b, we have

$$(+\downarrow) \qquad\qquad 0 = \Delta_B - B_y f_{BB} \qquad\qquad (1)$$

The terms Δ_B and f_{BB} are easily obtained using the table on the inside front cover. In particular, note that $\Delta_B = \Delta_C + \theta_C(6 \text{ m})$. Thus,

$$\Delta_B = \frac{P(L/2)^3}{3EI} + \frac{P(L/2)^2}{2EI}\left(\frac{L}{2}\right)$$

$$= \frac{(50 \text{ kN})(6 \text{ m})^3}{3EI} + \frac{(50 \text{ kN})(6 \text{ m})^2}{2EI}(6 \text{ m}) = \frac{9000 \text{ kN} \cdot \text{m}^3}{EI}\downarrow$$

$$f_{BB} = \frac{PL^3}{3EI} = \frac{(1)(12 \text{ m})^3}{3EI} = \frac{576 \text{ m}^3}{EI}\uparrow$$

Substituting these results into Eq. (1) yields

$$(+\downarrow) \qquad 0 = \frac{9000}{EI} - B_y\left(\frac{576}{EI}\right) \qquad B_y = 15.6 \text{ kN} \qquad \textbf{Ans.}$$

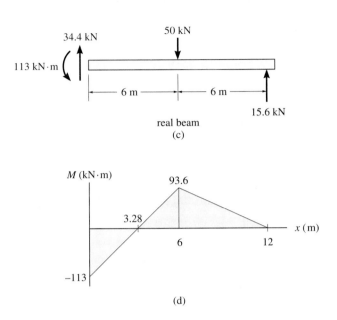

real beam
(c)

(d)

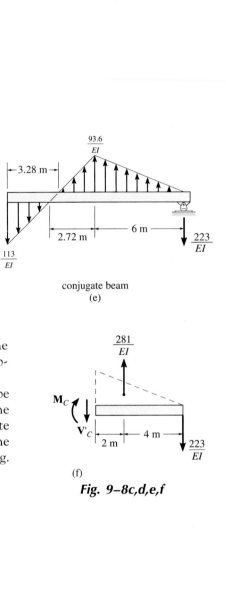

conjugate beam
(e)

If this reaction is placed on the free-body diagram of the beam, the reactions at A can be obtained from the three equations of equilibrium, Fig. 9–8c.

Having determined all the reactions, the moment diagram can be constructed as shown in Fig. 9–8d. Also, we may determine the deflection of the beam at any point. For example, the conjugate beam for the real beam in Fig. 9–8a is shown in Fig. 9–8e.* If the deflection under the 50-kN load is to be determined, then from Fig. 9–8f we have

(f)

Fig. 9–8c,d,e,f

$$M'_C + \frac{223}{EI}(6) - \frac{281}{EI}(2) = 0$$

$$\Delta_C = M'_C = -\frac{776 \text{ kN} \cdot \text{m}^3}{EI}$$

The negative sign indicates the deflection is downward.

*Note that although the conjugate beam will always be unstable for a *statically indeterminate* real beam, it will be held in equilibrium by its loading (M/EI) diagram.

Example 9–2

Determine the moment at the fixed wall for the beam shown in Fig. 9–9a. EI is constant.

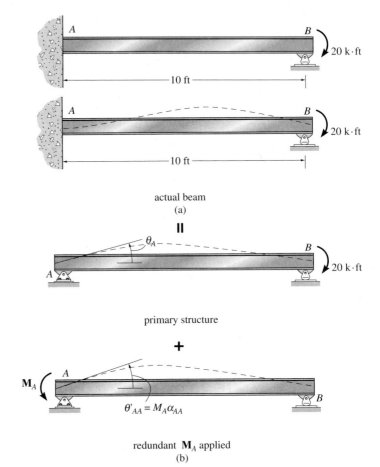

actual beam
(a)

$\parallel$

primary structure

$+$

$\theta'_{AA} = M_A \alpha_{AA}$

redundant $\mathbf{M}_A$ applied
(b)

Fig. 9–9

Solution

Principle of Superposition.   The redundant here will be taken as $\mathbf{M}_A$ since this moment can then be determined directly. Application of the principle of superposition is shown in Fig. 9–9b. Here the capacity of the beam to support a moment at A has been removed. This requires substituting a pin for the fixed support at A. We have assumed $\mathbf{M}_A$ acts counterclockwise.

Compatibility Equation.   Taking positive rotation as counterclockwise, Fig. 9–9b, we have

$$(\zeta+) \qquad\qquad 0 = \theta_A + M_A\,\alpha_{AA} \qquad\qquad (1)$$

The terms θ_A and α_{AA} can be determined from the table on the inside front cover. We have

$$\theta_A = \frac{ML}{6EI} = \frac{20\,\text{k}\cdot\text{ft}(10\,\text{ft})}{6EI} = \frac{33.3\,\text{k}\cdot\text{ft}}{EI}$$

$$\alpha_{AA} = \frac{ML}{3EI} = \frac{1\,(10\,\text{ft})}{3EI} = \frac{3.33\,\text{ft}}{EI}$$

Substituting these results into Eq. (1) yields

$$0 = \frac{33.3}{EI} + M_A\!\left(\frac{3.33}{EI}\right) \qquad M_A = -10\,\text{k}\cdot\text{ft} \qquad \textbf{Ans.}$$

The negative sign indicates that $\mathbf{M}_A$ acts opposite to that shown in Fig. 9–9b.

Example 9-3

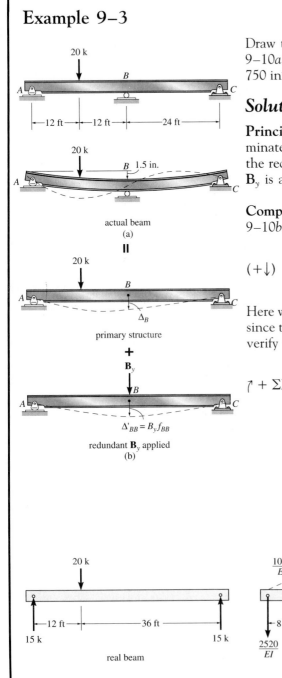

actual beam
(a)

||

primary structure

+

$\Delta'_{BB} = B_y f_{BB}$

redundant **B**$_y$ applied
(b)

Draw the shear and moment diagrams for the beam shown in Fig. 9–10a. The support at B settles 1.5 in. Take $E = 29(10^3)$ ksi, $I = 750$ in^4.

Solution

Principle of Superposition. By inspection, the beam is indeterminate to the first degree. The center support B will be chosen as the redundant, so that the roller at B is removed, Fig. 9–10b. Here **B**$_y$ is assumed to act downward on the beam.

Compatibility Equation. With reference to point B in Fig. 9–10b, using units of ft, we require

$$(+\downarrow) \qquad\qquad \frac{1.5}{12} = \Delta_B + B_y f_{BB} \qquad\qquad (1)$$

Here we will use the conjugate-beam method to compute Δ_B and f_{BB} since the moment diagrams consist of straight line segments. For Δ_B, verify the calculations shown in Fig. 9–10c. Note that

$$\curvearrowleft + \Sigma M_{B'} = 0; \qquad M_{B'} - \frac{1440}{EI}(8) + \frac{1800}{EI}(24) = 0$$

$$M_{B'} = -\frac{31{,}680}{EI} = \frac{31{,}680}{EI}\downarrow$$

Fig. 9–10a,b,c

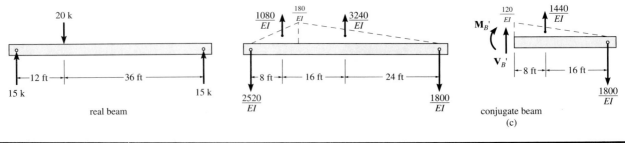

Verify the calculations in Fig. 9–10d for calculating f_{BB}. Note that

$\curvearrowright + \Sigma M_{B'} = 0;$ $m_{B'} - \dfrac{144}{EI}(8) + \dfrac{144}{EI}(24) = 0$

$$m_{B'} = -\dfrac{2304}{EI} = \dfrac{2304}{EI} \downarrow$$

Substituting these results into Eq. (1), we have

$$\dfrac{1.5}{12} = \dfrac{31{,}680}{EI} + B_y\left(\dfrac{2304}{EI}\right)$$

Expressing the units of E and I in terms of k and ft, we have

$$\left(\dfrac{1.5}{12}\ \text{ft}\right)[29(10^3)\ \text{k/in}^2((12)^2\ \text{in}^2/\text{ft}^2)][750\ \text{in}^4(\text{ft}^4/(12)^4\ \text{in}^4)]$$

$$= 31{,}680 + B_y(2304)$$

$$B_y = -5.55\ \text{k}$$

Equilibrium Equations.   The negative sign indicates that $\mathbf{B}_y$ acts *upward* on the beam. From the free-body diagram shown in Fig. 9–10e we have

$\curvearrowright + \Sigma M_A = 0;$ $20(12) - 5.55(24) - C_y(48) = 0$
$$C_y = 2.22\ \text{k}$$
$+\uparrow\Sigma F_y = 0;$ $A_y - 20 + 5.55 + 2.22 = 0$
$$A_y = 12.23\ \text{k}$$

Using these results, verify the shear and moment diagrams shown in Fig. 9–10f.

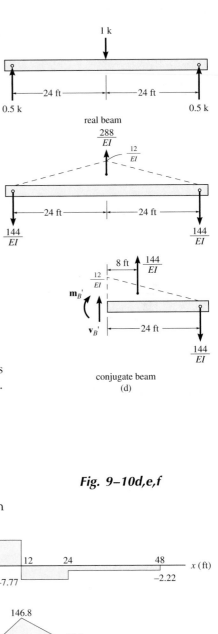

real beam

conjugate beam
(d)

Fig. 9–10d,e,f

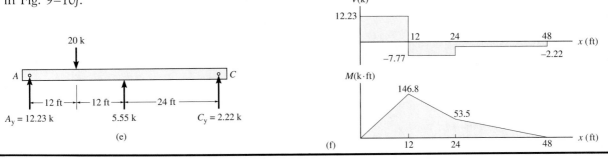

(e)

(f)

465

Example 9–4

Draw the shear and moment diagrams for the beam shown in Figure 9–11a. *EI* is constant. Neglect the effects of axial load.

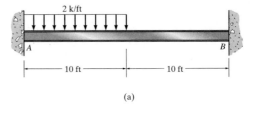

(a)

Solution

Principle of Superposition.   Since axial load is neglected, the beam is indeterminate to the second degree. The two end moments at *A* and *B* will be considered as the redundants. The beam's capac-

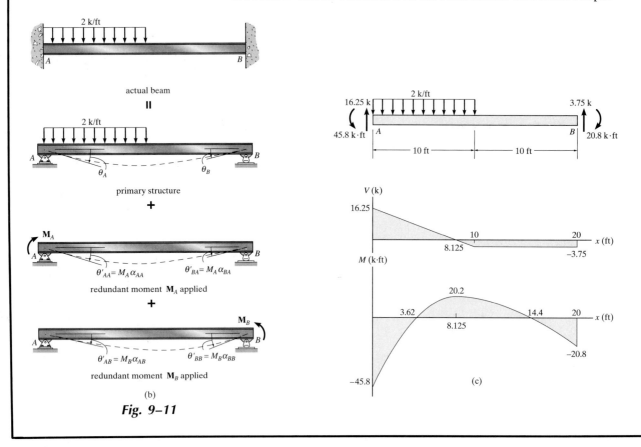

actual beam

‖

primary structure

+

$\theta'_{AA} = M_A \alpha_{AA}$ $\theta'_{BA} = M_A \alpha_{BA}$

redundant moment $\mathbf{M}_A$ applied

+

$\theta'_{AB} = M_B \alpha_{AB}$ $\theta'_{BB} = M_B \alpha_{BB}$

redundant moment $\mathbf{M}_B$ applied

(b)

Fig. 9–11

ity to resist these moments is removed by placing a pin at A and a rocker at B. The principle of superposition applied to the beam is shown in Fig. 9–11b.

Compatibility Equations. Reference to points A and B, Fig. 9–11b, requires

$$(\curvearrowright +) \qquad\qquad 0 = \theta_A + M_A\alpha_{AA} + M_B\alpha_{AB} \qquad\qquad (1)$$

$$(\curvearrowleft +) \qquad\qquad 0 = \theta_B + M_A\alpha_{BA} + M_B\alpha_{BB} \qquad\qquad (2)$$

The required slopes and angular flexibility coefficients can be determined using the table on the inside front cover. We have

$$\theta_A = \frac{3wL^3}{128EI} = \frac{3(2)(20)^3}{128EI} = \frac{375}{EI}$$

$$\theta_B = \frac{7wL^3}{384EI} = \frac{7(2)(20)^3}{384EI} = \frac{291.7}{EI}$$

$$\alpha_{AA} = \frac{ML}{3EI} = \frac{1(20)}{3EI} = \frac{6.67}{EI}$$

$$\alpha_{BB} = \frac{ML}{3EI} = \frac{1(20)}{3EI} = \frac{6.67}{EI}$$

$$\alpha_{AB} = \frac{ML}{6EI} = \frac{1(20)}{6EI} = \frac{3.33}{EI}$$

Note that $f_{BA} = f_{AB}$, a consequence of Maxwell's theorem of reciprocal displacements.

Substituting the data into Eqs. (1) and (2) yields

$$0 = \frac{375}{EI} + M_A\left(\frac{6.67}{EI}\right) + M_B\left(\frac{3.33}{EI}\right)$$

$$0 = \frac{291.7}{EI} + M_A\left(\frac{3.33}{EI}\right) + M_B\left(\frac{6.67}{EI}\right)$$

Canceling EI and solving these equations simultaneously, we have

$$M_A = -45.8 \text{ k} \cdot \text{ft} \qquad M_B = -20.8 \text{ k} \cdot \text{ft}$$

Using these results, the end shears are calculated, Fig. 9–11c, and the shear and moment diagrams plotted.

Example 9–5

Determine the reactions at the supports for the beam shown in Fig. 9–12a. EI is constant.

Solution

Principle of Superposition. By inspection, the beam is indeterminate to the first degree. Here, for the sake of illustration, we will choose the internal moment at support B as the redundant. Consequently, the beam is cut open and end pins are placed at B in order to release *only* the capacity of the beam to resist moment at this point, Fig. 9–12b. The internal moment at B is applied to the beam in Fig. 9–12c.

Compatibility Equations. From Fig. 9–12 we require the relative rotation of one end of the beam with respect to the other end to be zero, that is,

$$(\curvearrowright +) \qquad\qquad \theta_B + M_B\alpha_{BB} = 0$$

where

$$\theta_B = \theta_B' + \theta_B''$$

and

$$\alpha_{BB} = \alpha_{BB_1} + \alpha_{BB_2}$$

The slope θ_B and angular flexibility coefficients can be determined from the table on the inside front cover, that is,

$$\theta_B' = \frac{wL^3}{24EI} = \frac{120(12)^3}{24EI} = \frac{8640 \text{ lb} \cdot \text{ft}^2}{EI}$$

$$\theta_B'' = \frac{PL^2}{16EI} = \frac{500(10)^2}{16EI} = \frac{3125 \text{ lb} \cdot \text{ft}^2}{EI}$$

$$\alpha_{BB_1} = \frac{ML}{3EI} = \frac{1(12)}{3EI} = \frac{4 \text{ ft}}{EI}$$

$$\alpha_{BB_2} = \frac{ML}{3EI} = \frac{1(10)}{3EI} = \frac{3.33 \text{ ft}}{EI}$$

Thus

$$\frac{8640 \text{ lb} \cdot \text{ft}^2}{EI} + \frac{3125 \text{ lb} \cdot \text{ft}^2}{EI} + M_B\left(\frac{4 \text{ ft}}{EI} + \frac{3.33 \text{ ft}}{EI}\right) = 0$$

$$M_B = -1604 \text{ lb} \cdot \text{ft}$$

The negative sign indicates M_B acts in the opposite direction to that shown in Fig. 9–12c. Using this result, the reactions at the supports are calculated as shown in Fig. 9–12d. Furthermore, the shear and moment diagrams are shown in Fig. 9–12e.

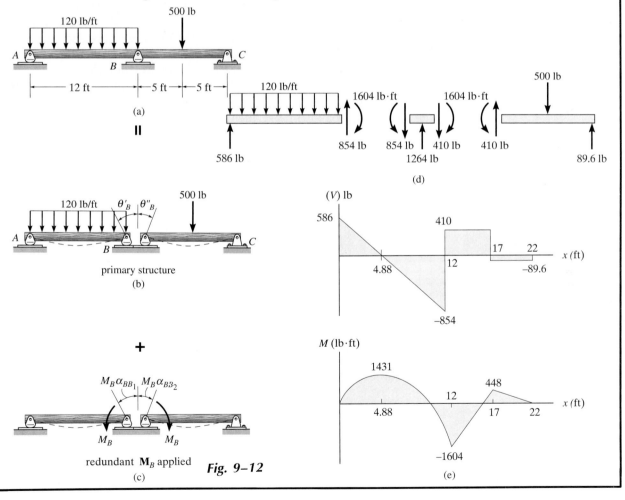

Fig. 9–12

9.5   General Moment Diagrams for Statically Indeterminate Beams

The moment diagram for a beam is of fundamental importance for the design of the beam and its stress analysis. Therefore, this diagram was drawn in most of the preceding example problems after the reactions for the beam were determined. Because continuous beams occur so frequently in structural analysis, engineers should have a "general feeling" as to what this diagram will look like for a variety of different loadings. For this reason, we list in Fig. 9–13 some examples of continuous beams having common supports and loadings. No specific values for the moments are listed on the diagrams since it is expected that numerical values can be obtained using the basic principles outlined in Sec. 4.3. Although peaks for these diagrams will depend upon the relative magnitudes of the loadings, their general shape should be thoroughly understood. When "sketching" moment diagrams, a few general rules apply. Pin supports at the *ends* of a beam and internal hinges are points of zero moment. Fixed-wall and interior pin supports generally develop a moment. Concentrated loads create moment diagrams with straight, sloping line segments, and uniform distributed loads create parabolic-shaped segments.

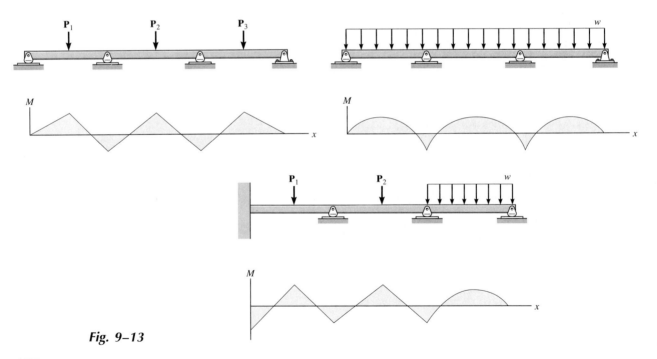

Fig. 9–13

9.6   Force Method of Analysis: Frames _____

The force method is very useful for solving problems involving statically indeterminate frames that have a single story and unusual geometry, such as gabled frames. Problems involving multistory frames, or those with a high degree of indeterminacy, are best solved using the slope-deflection, moment-distribution, or matrix method discussed in later chapters.

The following examples illustrate the application of the force method using the procedure for analysis outlined in Sec. 9.2.

Here is an example of lift-slab construction, where concrete slabs are poured on the ground and jacked up, and fixed to the columns. This technique is often used for buildings supporting light loads, such as in apartment and commercial buildings. *(Photo courtesy of Portland Cement Association)*

Example 9–6

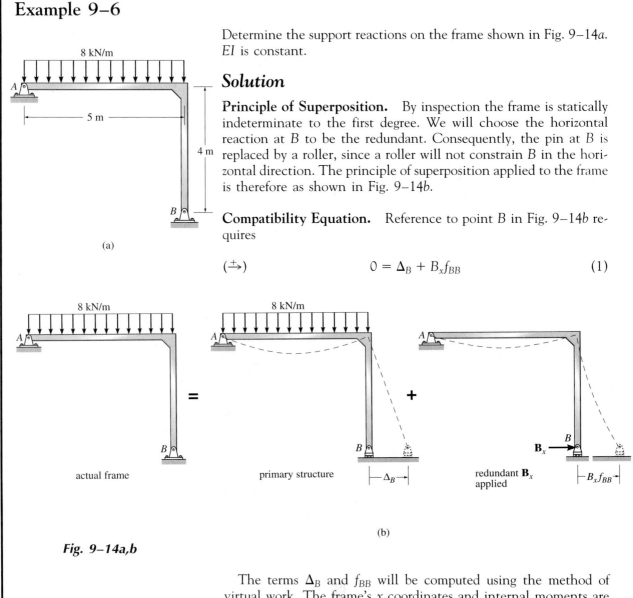

Determine the support reactions on the frame shown in Fig. 9–14a. EI is constant.

Solution

Principle of Superposition. By inspection the frame is statically indeterminate to the first degree. We will choose the horizontal reaction at B to be the redundant. Consequently, the pin at B is replaced by a roller, since a roller will not constrain B in the horizontal direction. The principle of superposition applied to the frame is therefore as shown in Fig. 9–14b.

Compatibility Equation. Reference to point B in Fig. 9–14b requires

$$(\overset{+}{\rightarrow}) \qquad\qquad 0 = \Delta_B + B_x f_{BB} \qquad\qquad (1)$$

(a)

actual frame

=

primary structure

+

redundant B_x applied

(b)

Fig. 9–14a,b

The terms Δ_B and f_{BB} will be computed using the method of virtual work. The frame's x coordinates and internal moments are shown in Fig. 9–14c and 9–14d. It is important that in each case the selected coordinate x_1 or x_2 be the *same* for both the real and virtual loadings. Also, the positive directions for $\mathbf{M}$ and $\mathbf{m}$ must be the *same*.

472

For Δ_B we require application of real loads, Fig. 9–14c, and a virtual unit load at B, Fig. 9–14d. Thus,

$$\Delta_B = \int \frac{Mm}{EI}\, dx = \int_0^5 \frac{(20x_1 - 4x_1^2)(0.8x_1)\, dx_1}{EI} + \int_0^4 \frac{0(1x_2)\, dx_2}{EI}$$

$$= \frac{166.7}{EI} + 0 = \frac{166.7}{EI}$$

For f_{BB} we require application of a real unit load acting at B, Fig. 9–14d, and a virtual unit load acting at B, Fig. 9–14d. Thus,

$$f_{BB} = \int \frac{mm}{EI}\, dx = \int_0^5 \frac{(0.8x_1)^2\, dx_1}{EI} + \int_0^4 \frac{(1x_2)^2\, dx_2}{EI}$$

$$= \frac{26.7}{EI} + \frac{21.3}{EI} = \frac{48.0}{EI}$$

Substituting the data into Eq. (1) and solving yields

$$0 = \frac{166.7}{EI} + B_x\left(\frac{48.0}{EI}\right) \qquad B_x = -3.47\ \text{kN} \qquad \textbf{Ans.}$$

Equilibrium Equations. Showing B_x on the free-body diagram of the frame in the correct direction, Fig. 9–14e, and applying the equations of equilibrium, we have

$$\xrightarrow{+}\Sigma F_x = 0; \qquad\qquad A_x - 3.47 = 0 \quad A_x = 3.47\ \text{kN} \qquad \textbf{Ans.}$$

$$\curvearrowleft +\Sigma M_A = 0;$$
$$40(2.5) - B_y(5) + 3.47(4) = 0 \quad B_y = 22.8\ \text{kN} \qquad \textbf{Ans.}$$

$$+\uparrow\Sigma F_y = 0; \qquad A_y - 40 + 22.8 = 0 \quad A_y = 17.2\ \text{kN} \qquad \textbf{Ans.}$$

Using these results, the moment diagram for the frame is shown in Fig. 9–14f.

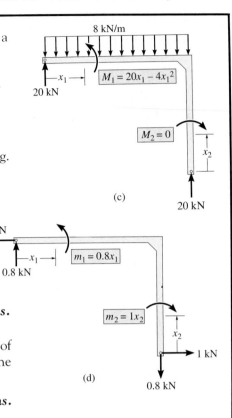

(c)

(d)

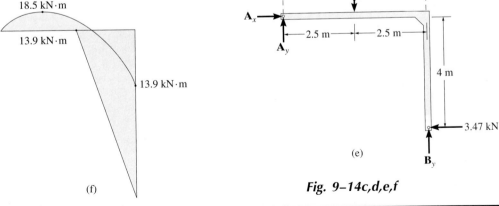

(f)

(e)

Fig. 9–14c,d,e,f

473

Example 9–7

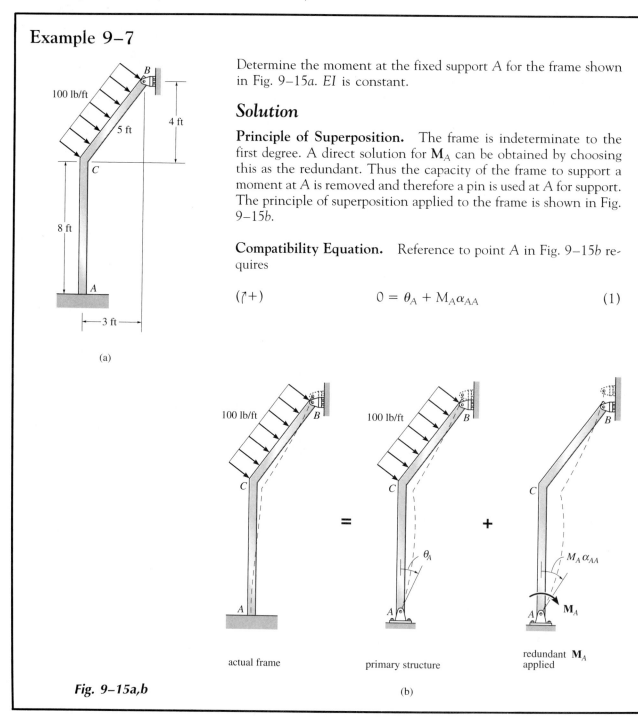

Determine the moment at the fixed support A for the frame shown in Fig. 9–15a. EI is constant.

Solution

Principle of Superposition. The frame is indeterminate to the first degree. A direct solution for $\mathbf{M}_A$ can be obtained by choosing this as the redundant. Thus the capacity of the frame to support a moment at A is removed and therefore a pin is used at A for support. The principle of superposition applied to the frame is shown in Fig. 9–15b.

Compatibility Equation. Reference to point A in Fig. 9–15b requires

$$(\curvearrowright +) \qquad\qquad 0 = \theta_A + M_A \alpha_{AA} \qquad\qquad (1)$$

100 lb/ft

5 ft

4 ft

8 ft

3 ft

(a)

actual frame

primary structure

redundant $\mathbf{M}_A$ applied

(b)

Fig. 9–15a,b

As in the preceding example, θ_A and α_{AA} will be computed using the method of virtual work. The frame's x coordinates and internal moments are shown in Figs. 9–15c and 9–15d.

For θ_A we require application of the real loads, Fig. 9–15c, and a virtual unit couple moment at A, Fig. 9–15d. Thus,

$$
\begin{aligned}
\theta_A &= \sum \int \frac{M m_\theta \, dx}{EI} \\
&= \int_0^8 \frac{(29.17x_1)(1 - 0.0833x_1) \, dx_1}{EI} \\
&\quad + \int_0^5 \frac{(296.7x_2 - 50x_2^2)(0.0667x_2) \, dx_2}{EI} \\
&= \frac{518.7}{EI} + \frac{303.5}{EI} = \frac{822.2}{EI}
\end{aligned}
$$

For α_{AA} we require application of a real unit couple moment acting at A, Fig. 9–15d, and a virtual unit couple moment acting at A, Fig. 9–15d. Thus,

$$
\begin{aligned}
\alpha_{AA} &= \sum \int \frac{m_\theta m_\theta}{EI} \, dx \\
&= \int_0^8 \frac{(1 - 0.0833x_1)^2 \, dx_1}{EI} + \int_0^5 \frac{(0.0667x_2)^2 \, dx_2}{EI} \\
&= \frac{3.85}{EI} + \frac{0.185}{EI} = \frac{4.04}{EI}
\end{aligned}
$$

Substituting these results into Eq. (1) and solving yields

$$
0 = \frac{822.2}{EI} + M_A\left(\frac{4.04}{EI}\right) \qquad M_A = -204 \text{ lb} \cdot \text{ft} \qquad \textbf{Ans.}
$$

The negative sign indicates $\mathbf{M}_A$ acts in the opposite direction to that shown in Fig. 9–15b.

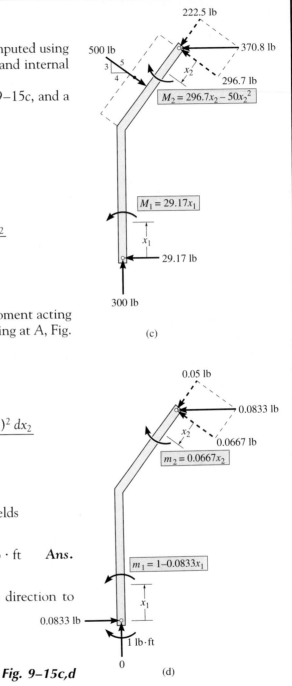

Fig. 9–15c,d

9.7 Force Method of Analysis: Trusses _____

The degree of indeterminacy of a truss can usually be determined by inspection; however, if this becomes difficult, use Eq. 3–1, $b + r > 2j$. Here the unknowns are represented by the number of bar forces (b) plus the support reactions (r), and the number of available equilibrium equations is $2j$ since two equations can be written for each of the (j) joints.

The force method is quite suitable for analyzing trusses that are statically indeterminate to the first or second degree. The following examples illustrate application of this method using the procedure for analysis outlined in Sec. 9.2.

Example 9–8

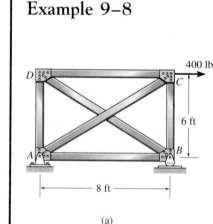

(a)

Fig. 9–16a,b

Determine the force in member AC of the truss shown in Fig. 9–16a. AE is the same for all the members.

Solution

Principle of Superposition. By inspection the truss is indeterminate to the first degree.* Since the force in member AC is to be determined, member AC will be chosen as the redundant. This requires "cutting" this member so that it cannot sustain a force, thereby making the truss statically determinate and stable. The principle of superposition applied to the truss is shown in Fig. 9–16b.

Compatibility Equation. With reference to member AC in Fig. 9–16b, we require the relative displacement Δ_{AC}, which occurs at the ends of the cut member AC due to the 400-lb load, plus the relative displacement $F_{AC}f_{AC\,AC}$ caused by the redundant force acting alone, be equal to zero, that is,

$$0 = \Delta_{AC} + F_{AC}f_{AC\,AC} \tag{1}$$

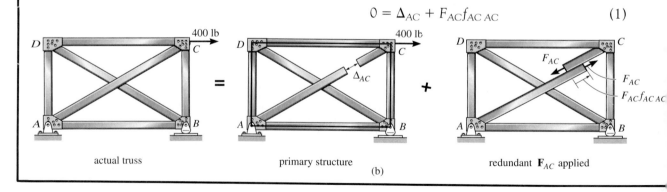

actual truss   primary structure   redundant $\mathbf{F}_{AC}$ applied

(b)

Here the flexibility coefficient $f_{AC\,AC}$ represents the relative displacement of the cut ends of member AC caused by a "real" unit load acting at the cut ends of member AC. This term, $f_{AC\,AC}$, and Δ_{AC} will be computed using the method of virtual work. The force analysis, using the method of joints, is summarized in Fig. 9–16c and 9–16d.

For Δ_{AC} we require application of the real load of 400 lb, Fig. 9–16c, and a virtual unit force acting at the cut ends of member AC, Fig. 9–16d. Thus,

$$\Delta_{AC} = \sum \frac{uSL}{AE}$$

$$= 2\left[\frac{(-0.8)(400)(8)}{AE}\right] + \frac{(-0.6)(0)(6)}{AE} + \frac{(-0.6)(300)(6)}{AE}$$

$$+ \frac{(1)(-500)(10)}{AE} + \frac{(1)(0)(10)}{AE}$$

$$= -\frac{11{,}200}{AE}$$

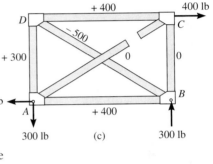

For $f_{AC\,AC}$ we require application of real unit forces acting on the cut ends of member AC, Fig. 9–16d, and virtual unit forces acting on the cut ends of member AC, Fig. 9–16d. Thus,

$$f_{AC\,AC} = \sum \frac{u^2 L}{AE}$$

$$= 2\left[\frac{(-0.8)^2(8)}{AE}\right] + 2\left[\frac{(-0.6)^2(6)}{AE}\right] + 2\left[\frac{(1)^2 10}{AE}\right]$$

$$= \left(\frac{34.56}{AE}\right)$$

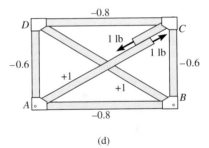

(d)

Fig. 9–16c,d

Substituting the data into Eq. (1) and solving yields

$$0 = -\frac{11{,}200}{AE} + \frac{34.56}{AE}F_{AC}$$

$$F_{AC} = 324 \text{ lb (T)} \qquad\qquad \textbf{Ans.}$$

Since the numerical result is positive, AC is subjected to tension as assumed, Fig. 9–16b. Using this result, the forces in the other members can be found by equilibrium, using the method of joints.

*Applying Eq. 3–1, $b + r > 2j$ or $6 + 3 > 2(4)$, $9 > 8$, $9 - 8 = 1°$.

Example 9–9

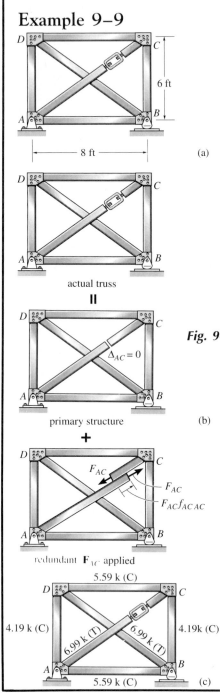

8 ft

6 ft

(a)

actual truss

=

$\Delta_{AC} = 0$

primary structure   (b)

+

F_{AC}

F_{AC}

$F_{AC}f_{AC\,AC}$

redundant $\mathbf{F}_{AC}$ applied

5.59 k (C)

4.19 k (C)   4.19k (C)

6.99 k (T)   6.99 k (T)

5.59 k (C)   (c)

Determine the force in each member of the truss shown in Fig. 9–17a if the turnbuckle on member AC is used to shorten the member by 0.5 in. Each bar has a cross-sectional area of 0.2 in², and $E = 29(10^6)$ psi.

Solution

Principle of Superposition.   This truss has the same geometry as that in Example 9–8. Since AC has been shortened, we will choose it as the redundant, Fig. 9–17b.

Compatibility Equation.   Since no external loads act on the primary structure (truss), there will be no relative displacement between the ends of the sectioned member caused by load; that is, $\Delta_{AC} = 0$. The flexibility coefficient $f_{AC\,AC}$ has been computed in Example 9–8, so

Fig. 9–17

$$f_{AC\,AC} = \frac{34.56}{AE}$$

Assuming the amount by which the bar is shortened is positive, the compatibility equation for the bar is therefore

$$0.5 \text{ in.} = 0 + \frac{34.56}{AE}F_{AC}$$

Realizing that $f_{AC\,AC}$ is a measure of displacement per unit force, we have

$$0.5 \text{ in.} = 0 + \frac{34.56 \text{ ft}(12 \text{ in./ft})}{(0.2 \text{ in}^2)[29(10^6) \text{ lb/in}^2]}F_{AC}$$

Thus,

$$F_{AC} = 6993 \text{ lb} = 6.99 \text{ k (T)} \qquad \textbf{Ans.}$$

Since no external forces act on the truss, the external reactions are zero. Therefore, using F_{AC} and analyzing the truss by the method of joints yields the results shown in Fig. 9–17c.

478

9.8   Composite Structures

Composite structures are composed of some members subjected only to axial force, while other members are subjected to bending. If the structure is statically indeterminate, the force method can conveniently be used for its analysis. The following example illustrates the procedure.

Example 9–10

The beam shown in Fig. 9–18a is supported by a pin at A and two pin-connected bars at B. Determine the forces at the supports for the loading shown. Take $E = 29(10^3)$ ksi, $I = 800$ in⁴ for the beam, and $A = 3$ in² for each bar.

Solution

Principle of Superposition.   By inspection, the beam is indeterminate to the first degree. For solution the force in member *BD* is chosen as the redundant. This member is therefore sectioned to reduce its capacity to sustain a force. The principle of superposition applied to the structure is shown in Fig. 9–18b.

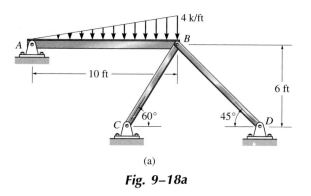

(a)

Fig. 9–18a

Compatibility Equation.   With reference to the relative displacement of the cut ends of member *BD*, Fig. 9–18b, we require

$$0 = \Delta_{BD} + F_{BD}f_{BD\,BD} \qquad (1)$$

(cont'd)

Example 9–10 (continued)

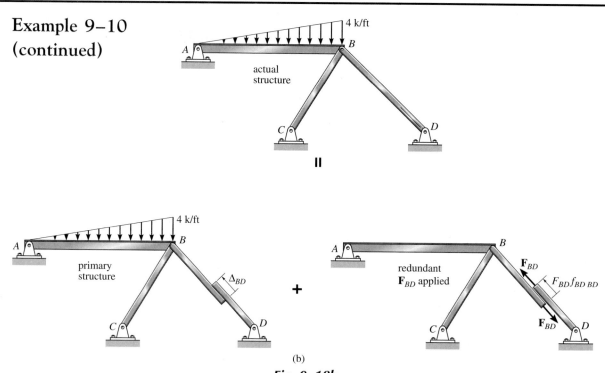

(b)

Fig. 9–18b

The method of virtual work will be used to compute Δ_{BD} and $f_{BD\,BD}$. The necessary force analysis is shown in Fig. 9–18c and 9–18d.

For Δ_{BD} we require application of the real loads, Fig. 9–18c, and a virtual unit load applied to the cut ends of member BD, Fig. 9–18d. Here we will only consider the bending strain energy in the beam and, of course, the axial strain energy in the bars. Thus,

$$
\begin{aligned}
\Delta_{BD} &= \int \frac{Mm}{EI}\,dx + \sum \frac{uSL}{AE} \\
&= \int_0^{10} \frac{(6.67x - 0.0667x^3)(0)\,dx}{EI} \\
&\quad + \frac{(-15.40)(-0.816)(6/\cos 30°)(12)}{AE} \\
&\quad + \frac{(0)(1)(6/\cos 45°)(12)}{AE} \\
&= 0 + \frac{1044.8}{3[29(10^3)]} + 0 = 0.0120 \text{ in.}
\end{aligned}
$$

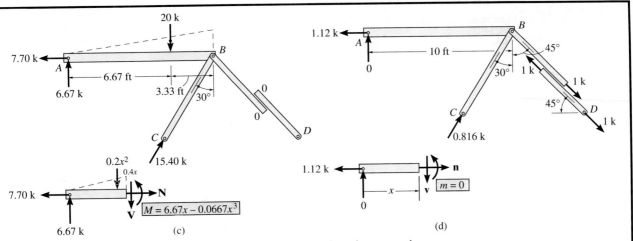

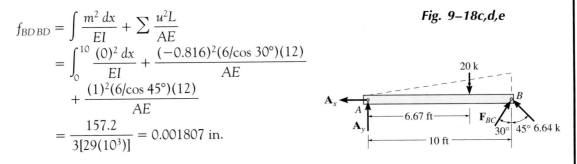

(c)

(d)

For $f_{BD\,BD}$ we require application of a real unit load and a virtual
unit load at the cut ends of member BD, Fig. 9–18d. Thus,

Fig. 9–18c,d,e

$$f_{BD\,BD} = \int \frac{m^2\,dx}{EI} + \sum \frac{u^2 L}{AE}$$

$$= \int_0^{10} \frac{(0)^2\,dx}{EI} + \frac{(-0.816)^2(6/\cos 30°)(12)}{AE}$$

$$+ \frac{(1)^2(6/\cos 45°)(12)}{AE}$$

$$= \frac{157.2}{3[29(10^3)]} = 0.001807 \text{ in.}$$

(e)

Substituting the data into Eq. (1) yields

$$0 = 0.0120 + F_{BD}(0.001807) \qquad F_{BD} = -6.64 \quad k = 6.64 \text{ k (C)}$$
Ans.

Equations of Equilibrium.   Using this result, the free-body diagram of the beam is shown in Fig. 9–18e. We have

$\curvearrowright+\Sigma M_A = 0;$
$$20(6.67) - 6.64 \cos 45°(10) - F_{BC} \cos 30°(10) = 0$$
$$F_{BC} = 9.97 \text{ k (C)} \qquad\qquad \textbf{Ans.}$$
$\xrightarrow{+}\Sigma F_x = 0; \qquad -A_x + 9.97 \sin 30° - 6.64 \sin 45° = 0$
$$A_x = 0.290 \text{ k} \qquad\qquad \textbf{Ans.}$$
$+\uparrow\Sigma F_y = 0; \qquad A_y - 20 + 6.64 \cos 45° + 9.97 \cos 30° = 0$
$$A_y = 6.67 \text{ k} \qquad\qquad \textbf{Ans.}$$

9.9   Additional Remarks on the Force Method of Analysis

Now that the basic ideas regarding the force method have been developed, we will proceed to generalize its application and discuss its usefulness.

When computing the flexibility coefficients, f_{ij} (or α_{ij}), for the structure, it will be noticed that they depend only on the material and geometrical properties of the members and *not* on the loading of the primary structure. Hence these values, once determined, can be used to compute the reactions for any loading.

For a structure having n redundant reactions, $\mathbf{R}_n$, we can write n compatibility equations, namely:

$$\Delta_1 + f_{11}R_1 + f_{12}R_2 + \cdots + f_{1n}R_n = 0$$
$$\Delta_2 + f_{21}R_1 + f_{22}R_2 + \cdots + f_{2n}R_n = 0$$
$$\vdots$$
$$\Delta_n + f_{n1}R_1 + f_{n2}R_2 + \cdots + f_{nn}R_n = 0$$

Here the displacements, $\Delta_1, \ldots, \Delta_n$, are caused by *both* the *real loads* on the primary structure and by *support settlement* or *dimensional changes* due to temperature differences or fabrication errors in the members. To simplify computation for structures having a large degree of indeterminacy, the above equations can be recast into a matrix form,*

$$\begin{bmatrix} f_{11}f_{12}\cdots f_{1n} \\ f_{21}f_{22}\cdots f_{2n} \\ \vdots \\ f_{n1}f_{n2}\cdots f_{nn} \end{bmatrix} \begin{bmatrix} R_1 \\ R_2 \\ \vdots \\ R_n \end{bmatrix} = - \begin{bmatrix} \Delta_1 \\ \Delta_2 \\ \vdots \\ \Delta_n \end{bmatrix} \tag{9–2}$$

or simply

$$\mathbf{fR} = -\mathbf{\Delta}$$

In particular, note that $f_{ij} = f_{ji}$ ($f_{12} = f_{21}$, etc.), a consequence of Maxwell's theorem of reciprocal displacements (or Betti's theorem). Hence the *flexibility matrix* will be *symmetric*, and this feature is beneficial when solving large sets of linear equations, as in the case of a highly indeterminate structure.

---

*Matrix algebra is reviewed in Chapter 13.

A typical highway girder bridge that is statically indeterminate. Notice the two spliced plates located close to (if not at) the points of zero moment on the girder. The shape of the dead-load moment diagram is similar to the uniformly loaded beam shown in Fig. 9–13.

Throughout this chapter we have determined the flexibility coefficients using the method of virtual work as it applies to the *entire structure*. It is possible, however, to obtain these coefficients for *each member* of the structure, and then, using transformation equations, to obtain their values for the entire structure. This approach is covered in books devoted to matrix analysis of structures, and will not be covered in this text.*

Although the details for applying the force method of analysis using computer methods will also be omitted here, we can make some general observations and comments that apply when using this method to solve problems that are highly indeterminate and thus involve large sets of equations. In this regard, numerical accuracy for the solution is improved if the flexibility coefficients located near the main diagonal of the **f** matrix are larger than those located off the diagonal.† To achieve this, some thought should be given to selection of the primary structure. To facilitate computations of f_{ij}, it is also desirable to choose the primary structure so that it is somewhat symmetric. This will tend to yield some flexibility coefficients that are similar or may be zero. Lastly, the deflected shape of the primary structure should be *similar* to that of the actual structure. If this occurs, then the redundants will induce only *small* corrections to the primary structure, which results in a more accurate solution of Eq. 9–2.

---

*See, for example, H. C. Martin, *Introduction to Matrix Methods of Structural Analysis*, McGraw-Hill, New York, 1966.

†A matrix's main diagonal consists of the terms f_{ii}.

9.10   The Three-Moment Equation

The three-moment equation was developed by the French engineer Clapeyron in 1857. This equation relates the internal moments in a continuous beam at three points of support to the loads acting between the supports. By successive application of this equation to segments of the beam, one obtains a set of equations that may be solved simultaneously for the unknown internal moments at the supports.

A general form of the three-moment equation can be developed by considering a segment of a *continuous beam*, Fig. 9–19a, which passes over the left, center, and right supports, L, C, and R. The loads between the supports are arbitrary and the unknown *internal moments* at the supports will be specified as M_L, M_C, and M_R. Furthermore, the left part of the beam has geometrical properties I_L and L_L, and the right part has properties I_R and L_R. The supports are assumed not to settle.* We wish to determine the internal moments at L, C, and R, which are shown acting in the defined positive direction *on the beam* in Fig. 9–19a. The formulation will be based on the conjugate-beam method. Since the "real" beam is continuous over the supports, the conjugate beam has *hinges* at L, C, and R. Using the principle of superposition, the M/EI diagrams for the applied loads and each of the internal moments are separated for clarification as shown in Fig. 9–19b and 9–19c. In particular, A_L/EI and A_R/EI represent the total area under their respective M/EI diagrams, and $\bar{x}_L$ and $\bar{x}_R$ locate their centroids. Since the slope of the real beam is *continuous* over the center support, we require the shear forces $C_{L_1} + C_{L_2} = -(C_{R_1} + C_{R_2})$ for the conjugate beam. Summing moments about point L' for the left segments, we have

$$C_{L_1} + C_{L_2} = \frac{1}{L_L}\left(\frac{A_L}{EI_L}\bar{x}_L\right)$$
$$+ \frac{1}{L_L}\left[\frac{1}{2}\left(\frac{M_L}{EI_L}\right)(L_L)\left(\frac{1}{3}L_L\right) + \frac{1}{2}\left(\frac{M_C}{EI_L}\right)(L_L)\left(\frac{2}{3}L_L\right)\right]$$
$$= \frac{A_L\bar{x}_L}{EI_L L_L} + \frac{M_L L_L}{6EI_L} + \frac{M_C L_L}{3EI_L}$$

*If support settlement is included, the resulting equation becomes more complicated. This situation can, however, be solved using the slope-deflection or moment-distribution techniques presented in Chapters 10 and 11.

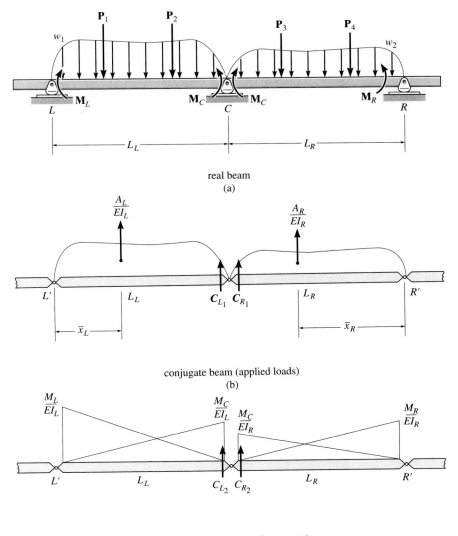

real beam
(a)

conjugate beam (applied loads)
(b)

conjugate beam (internal moments)
(c)

Fig. 9–19

And summing moments about point R' for the right segments yields

$$
C_{R_1} + C_{R_2} = \frac{1}{L_R}\left(\frac{A_R}{EI_R}\,\bar{x}_R\right)
$$

$$
+ \frac{1}{L_R}\left[\frac{1}{2}\left(\frac{M_R}{EI_R}\right)(L_R)\left(\frac{1}{3}L_R\right) + \frac{1}{2}\left(\frac{M_C}{EI_R}\right)(L_R)\left(\frac{2}{3}L_R\right)\right]
$$

$$
= \frac{A_R\bar{x}_R}{EI_R L_R} + \frac{M_R L_R}{6EI_R} + \frac{M_C L_R}{3EI_R}
$$

485

Equating $(C_{L_1} + C_{L_2}) = -(C_{R_1} + C_{R_2})$ and simplifying yields

$$\frac{M_L L_L}{I_L} + 2M_C\left(\frac{L_L}{I_L} + \frac{L_R}{I_R}\right) + \frac{M_R L_R}{I_R} = -\sum \frac{6A_L \bar{x}_L}{I_L L_L} - \sum \frac{6A_R \bar{x}_R}{I_R L_R}$$

(9–3)

Summation signs have been added to the terms on the right so that the M/EI diagrams for each type of applied load can be treated separately. In practice, the most common types of loadings encountered are concentrated and distributed loads, as shown in Fig. 9–20. If the areas and centroidal distances for their M/EI diagrams are computed and substituted into the above equation, we have

$$\frac{M_L L_L}{I_L} + 2M_C\left(\frac{L_L}{I_L} + \frac{L_R}{I_R}\right) + \frac{M_R L_R}{I_R}$$
$$= -\sum \frac{P_L L_L^2}{I_L}(k_L - k_L^3) - \sum \frac{P_R L_R^2}{I_R}(k_R - k_R^3) - \frac{w_L L_L^3}{4I_L} - \frac{w_R L_R^3}{4I_R}$$

(9–4)

where

M_L, M_C, M_R = internal moments at the left, center, and right supports; these moments are assumed to act in the positive sense on the beam, as shown in Fig. 9–19a

$I_L, L_L; I_R, L_R$ = left and right beam moments of inertia and span lengths, Fig. 9–19a

$P_L, w_L; P_R, w_R$ = left and right beam concentrated loads and uniform distributed loads, assumed to act in the positive sense on the beam as shown in Fig. 9–19a

k_L, k_R = fraction of the span length where the concentrated load is acting from the left or right support, Fig. 9–20

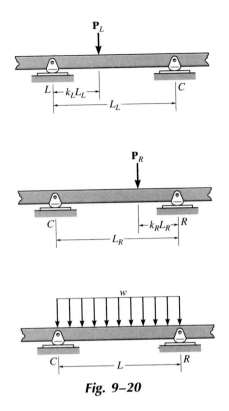

Fig. 9–20

As a special case, if the *moment of inertia is constant for the entire span,* that is, $I_L = I_R$, we have

$$M_L L_L + 2M_C(L_L + L_R) + M_R L_R$$
$$= -\Sigma P_L L_L^2(k_L - k_L^3) - \Sigma P_R L_R^2(k_R - k_R^3) - \frac{w_L L_L^3}{4} - \frac{w_R L_R^3}{4}$$

(9–5)

Application of Eqs. (9–3) through (9–5) is rather straightforward, although care must be taken to follow the positive sign convention for the moment and load terms. Furthermore, a consistent set of units must be used. The following examples illustrate how to apply these equations. Note in particular the technique used to treat a fixed wall support, Example 9–13.

Example 9–11

Determine the reactions at the supports for the beam shown in Fig. 9–21a. The moment of inertia of span AB is one half that of span BC.

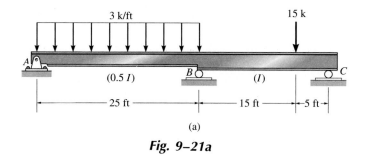

(a)

Fig. 9–21a

Solution

Here we must use Eq. 9–4 for the solution. The data are as follows:

$$
\begin{array}{lll}
M_L = 0 & M_C = M_B & M_R = 0 \\
L_L = 25 \text{ ft} & L_R = 20 \text{ ft} \\
I_L = 0.5I & I_R = I \\
P_L = 0 & P_R = 15 \text{ k} \\
w_L = 3 \text{ k/ft} & w_R = 0 \\
k_L = 0 & k_R = \frac{5}{20} = 0.25
\end{array}
$$

Substituting into Eq. 9–4 yields

$$
0 + 2M_B \left(\frac{25}{0.5I} + \frac{20}{I} \right) + 0
$$

$$
= 0 - \frac{15(20)^2}{I} [0.25 - (0.25)^3] - \left[\frac{3(25)^3}{4(0.5I)} + 0 \right]
$$

Canceling out the common term, I, and solving, we have

$$
M_B = -177.5 \text{ k} \cdot \text{ft}
$$

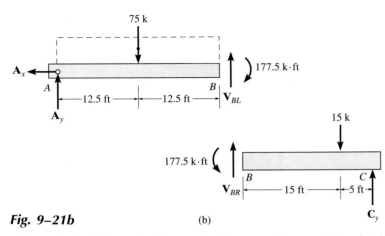

Fig. 9–21b   (b)

Figure 9–21b shows the free-body diagrams of spans AB and BC, with the computed moment at the section.

For span AB:
$\xrightarrow{+}\Sigma F_x = 0;$ $A_x = 0$ **Ans.**
$\curvearrowleft +\Sigma M_B = 0;$ $A_y(25) + 177.5 - 75(12.5) = 0$ $A_y = 30.4$ k **Ans.**
$+\uparrow\Sigma F_y = 0;$ $30.4 - 75 + V_{BL} = 0$ $V_{BL} = 44.6$ k

For span BC:
$\curvearrowleft +\Sigma M_B = 0;$ $-177.5 + 15(15) - C_y(20) = 0$
                    $C_y = 2.38$ k   **Ans.**
$+\uparrow\Sigma F_y = 0;$ $V_{BR} - 15 + 2.38 = 0$
                    $V_{BR} = 12.6$ k

A free-body diagram of the differential segment of the beam that passes over the roller at B, Fig. 9–21c, yields

$+\uparrow\Sigma F_y = 0;$ $B_y - 44.6 - 12.6 = 0$ $B_y = 57.2$ k   **Ans.**

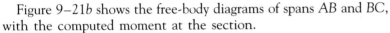

(c)
Fig. 9–21c

489

Example 9–12

Determine the internal moments in the beam at the supports, Fig. 9–22. *EI* is constant.

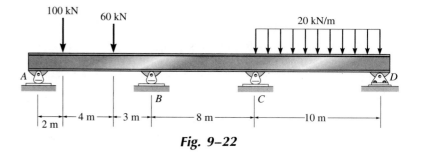

Fig. 9–22

Solution

Since *I* is constant, we can use Eq. 9–5. By inspection,

$$M_A = M_D = 0 \qquad\qquad \textbf{Ans.}$$

There are two unknowns, M_B and M_C; hence, two applications of Eq. 9–5 must be made.

First, we will consider *ABC* in Fig. 9–22, in which case

$$
\begin{array}{lll}
M_L = 0 & M_C = M_B & M_R = M_C \\
L_L = 9 \text{ m} & L_R = 8 \text{ m} & \\
P_{L_1} = 100 \text{ kN} & P_{L_2} = 60 & P_R = 0 \\
w_R = 0 & w_L = 0 & \\
k_{L_1} = \frac{2}{9} = 0.222 & k_{L_2} = \frac{6}{9} = 0.667 &
\end{array}
$$

Substituting the data into Eq. 9–5 yields

$$
\begin{aligned}
0 + 2M_B(9 + 8) &+ M_C(8) \\
&= -100(9)^2[0.222 - (0.222)^3] \\
&\quad - 60(9)^2[0.667 - (0.667)^3] - 0 - 0 - 0 \\
&\qquad 34M_B + 8M_C = -3509.0 \qquad (1)
\end{aligned}
$$

Next consider *BCD* in Fig. 9–22. Here

$$M_L = M_B \qquad M_C = M_C \qquad M_R = 0$$
$$L_L = 8 \text{ m} \qquad L_R = 10 \text{ m}$$
$$P_L = 0 \qquad P_R = 0$$
$$w_L = 0 \qquad w_R = 20 \text{ kN/m}$$
$$k_L = 0 \qquad k_R = 0$$

Substituting the data into Eq. 9–5, we have

$$M_B(8) + 2M_C(8 + 10) + 0 = -0 - 0 - 0 - \frac{20(10)^3}{4}$$
$$8M_B + 36M_C = -5000 \qquad\qquad (2)$$

Solving Eqs. (1) and (2) simultaneously yields

$$M_B = -74.4 \text{ k} \cdot \text{ft} \qquad\qquad \textbf{Ans.}$$
$$M_C = -122.4 \text{ k} \cdot \text{ft} \qquad\qquad \textbf{Ans.}$$

Example 9–13

Determine the internal moments in the beam at the supports, Fig. 9–23a. EI is constant.

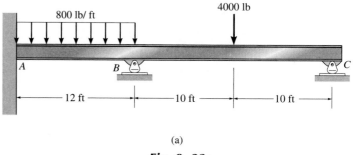

(a)

Fig. 9–23a

Solution

Equation 9–5 can be used for the solution since I is constant. By inspection,

$$M_C = 0 \qquad \textbf{Ans.}$$

The moments at A and B will be determined by two applications of Eq. 9–5. *The fixed support at A can be treated as a roller support provided an imaginary end span of zero length is added to the beam,* Fig. 9–23b.

First we consider $A'AB$, Fig. 9–23b, in which case

$$\begin{array}{lll}
M_L = 0 & M_C = M_A & M_R = M_B \\
L_L = 0 & L_R = 12 & \\
P_L = 0 & P_R = 0 & \\
w_L = 0 & w_R = 800 \text{ lb/ft} & \\
k_L = 0 & k_R = 0 &
\end{array}$$

Substituting the data into Eq. 9–5, we have

$$0 + 2M_A(0 + 12) + M_B(12) = -0 - 0 - 0 - \frac{800(12)^3}{4}$$

$$24M_A + 12M_B = -345{,}600 \qquad (1)$$

Next consider ABC, Fig. 9–23b. Then

$$\begin{array}{lll} M_L = M_A & M_C = M_B & M_R = 0 \\ L_L = 12 \text{ ft} & L_R = 20 \text{ ft} & \\ P_L = 0 & P_R = 4000 \text{ lb} & \\ w_L = 800 \text{ lb/ft} & w_R = 0 & \\ k_L = 0 & k_R = \dfrac{10}{20} = 0.5 & \end{array}$$

Substituting the data into Eq. 9–5, we have

$$M_A(12) + 2M_B(12 + 20) + 0$$
$$= -0 - 4000(20)^2[0.5 - (0.5)^3] - \frac{800(12)^3}{4} - 0$$
$$12M_A + 64M_B = -945,600 \qquad (2)$$

Solving Eqs. (1) and (2) simultaneously yields

$$\begin{array}{ll} M_A = -7.74 \text{ k} \cdot \text{ft} & \textbf{Ans.} \\ M_B = -13.3 \text{ k} \cdot \text{ft} & \textbf{Ans.} \end{array}$$

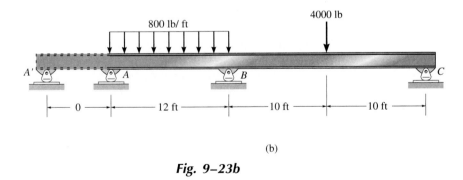

(b)

Fig. 9–23b

9.11 Influence Lines for Statically Indeterminate Beams

In Sec. 5.3 we discussed the use of the Müller-Breslau principle for drawing the influence line for the reaction, shear, and moment at a point in a statically determinate beam. In this section we will extend this method and apply it to statically indeterminate beams.

Recall that, for a beam, *the Müller-Breslau principle states that the influence line for a function (reaction, shear, or moment) is to the same scale as the deflected shape of the beam when the beam is acted upon by the function.* To draw the deflected shape properly, the capacity of the beam to resist the applied function must be *removed* so the beam can deflect when the function is applied. For *statically determinate beams*, the deflected shapes (or the influence lines) will be a series of *straight line segments*. For *statically indeterminate beams*, however, *curves* will result. Construction of each of the three types of influence lines (reaction, shear, and moment) will now be discussed for the statically indeterminate beam in Fig. 9–24a. In each case we will illustrate the validity of the Müller-Breslau principle using Maxwell's theorem of reciprocal displacements.

Reaction at A

To determine the influence line for the reaction at A in Fig. 9–24a, a unit load is placed on the beam at successive points, and at each point the reaction at A must be computed. A plot of these results yields the influence line. For example, when the load is at point D, Fig. 9–24a, the reaction at A, which represents the ordinate of the influence line at D, can be computed by the force method. To do this the principle of superposition is applied as shown in Fig. 9–24a through 9–24c. The compatibility equation for point A is thus $0 = f_{AD} + A_y f_{AA}$ or $A_y = -f_{AD}/f_{AA}$; however, by Maxwell's theorem of reciprocal displacements $f_{AD} = -f_{DA}$, Fig. 9–24d, so that we can also compute A_y (or the ordinate of the influence line at D) using the equation

$$A_y = \left(\frac{1}{f_{AA}}\right) f_{DA}$$

By comparison, the Müller-Breslau principle requires removal of the support at A and application of a vertical unit load. The resulting deflection curve, Fig. 9–24d, is to some scale the shape of the influence line for A_y. From the equation above, however, it is seen that the scale factor is $1/f_{AA}$.

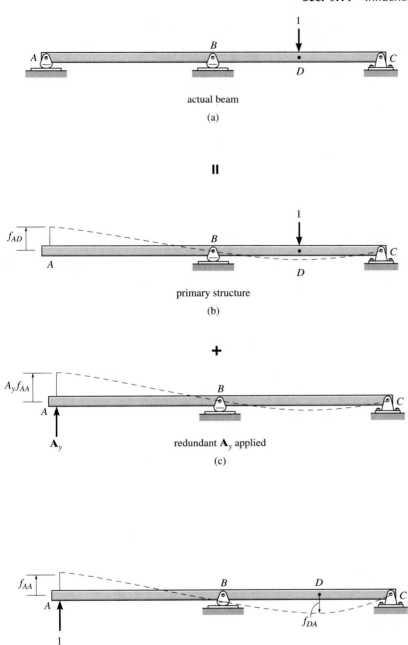

actual beam

(a)

=

primary structure

(b)

+

redundant $\mathbf{A}_y$ applied

(c)

(d)

Fig. 9–24

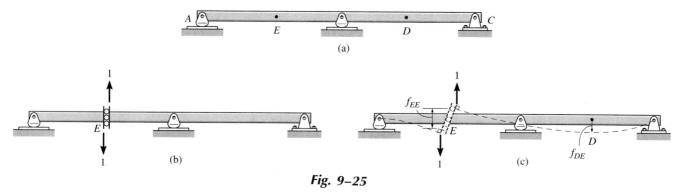

Fig. 9–25

Shear at E

If the influence line for the shear at point E of the beam in Fig. 9–25a is to be determined, then by the Müller-Breslau principle the beam is imagined cut open at this point and a *roller guide* is inserted at E, Fig. 9–25b. This device will transmit a moment and normal force but no shear. When the beam deflects due to positive unit shear loads acting at E, the slope on each side of the guide remains the same, and the deflection curve represents to some scale the influence line for the shear at E, Fig. 9–25c. Had the basic method for establishing the influence line for the shear at E been applied, it would then be necessary to apply a unit load at each point D and compute the internal shear at E, Fig. 9–25a. This value, V_E, would represent the ordinate of the influence line at D. Using the force method and Maxwell's theorem of reciprocal displacements, however, it can be shown that

$$V_E = \left(\frac{1}{f_{EE}}\right) f_{DE}$$

This again establishes the validity of the Müller-Breslau principle, namely, a positive unit shear load applied to the beam at E, Fig. 9–25c, will cause the beam to deflect into the *shape* of the influence line for the shear at E. Here the scale factor is $(1/f_{EE})$.

Moment at E

The influence line for the moment at E in Fig. 9–26a can be determined by placing a *pin* or *hinge* at E, since this connection transmits normal and shear forces but cannot resist a moment, Fig. 9–26b. Applying a positive unit couple moment, the beam then deflects to the dashed position in Fig. 9–26c, which yields to some scale the influence line, again a consequence of the Müller-Breslau principle. Using the

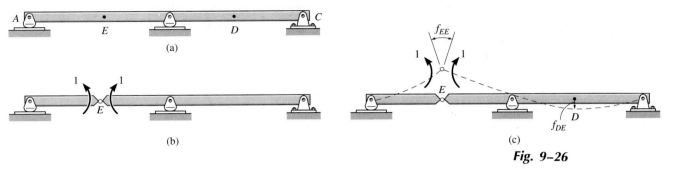

Fig. 9–26

force method and Maxwell's reciprocal theorem, we can show that

$$M_E = \left(\frac{1}{\alpha_{EE}} \right) f_{DE}$$

The scale factor here is $(1/\alpha_{EE})$.

Procedure for Analysis

The following procedure provides a method for determining the influence line for the reaction, shear, or moment at a point on a beam using the Müller-Breslau technique.

Qualitative Influence Line   At the point in the beam for which the influence line is to be determined, place a connection that will remove the capacity of the beam to support the function of the influence line. If the function is a vertical *reaction*, use a vertical *roller guide*; if the function is *shear*, use a *sliding device*; or if the function is *moment*, use a *pin* or *hinge*. Place a unit load at the connection acting on the beam in the "positive direction" of the function. Draw the deflection curve for the beam. This curve represents to some scale the shape of the influence line for the beam.

Quantitative Influence Line   If numerical values of the influence line are to be determined, compute the *displacement* of successive points along the beam when the beam is subjected to the unit load placed at the connection mentioned above. It is generally advisable to use the conjugate-beam method for the computations. Divide each value of displacement by the displacement determined at the point where the unit load acts. By applying this scalar factor, the resulting values are the ordinates of the influence line.

The following examples illustrate these techniques.

Example 9–14

Draw the influence line for the vertical reaction at A for the beam in Fig. 9–27a. EI is constant. Plot numerical values every 6 ft.

Solution

The capacity of the beam to resist the reaction $\mathbf{A}_y$ is removed. This is done using a vertical roller device shown in Fig. 9–27b. Applying a vertical unit load at A yields the shape of the influence line shown in Fig. 9–27c.

The reactions at A and B on the "real beam," when subjected to the unit load at A, are shown in Fig. 9–27b. The corresponding conjugate beam is shown in Fig. 9–27d. Notice that the support at A' remains the *same* as that for A in Fig. 9–27b. This is because a vertical roller device on the conjugate beam supports a moment but no shear, corresponding to a displacement but no slope at A on the real beam, Fig. 9–27c. The reactions at the supports of the conjugate beam have been computed and are shown in Fig. 9–27d. The displacements of points on the real beam, Fig. 9–27b, will now be computed.

Fig. 9–27a,b,c,d

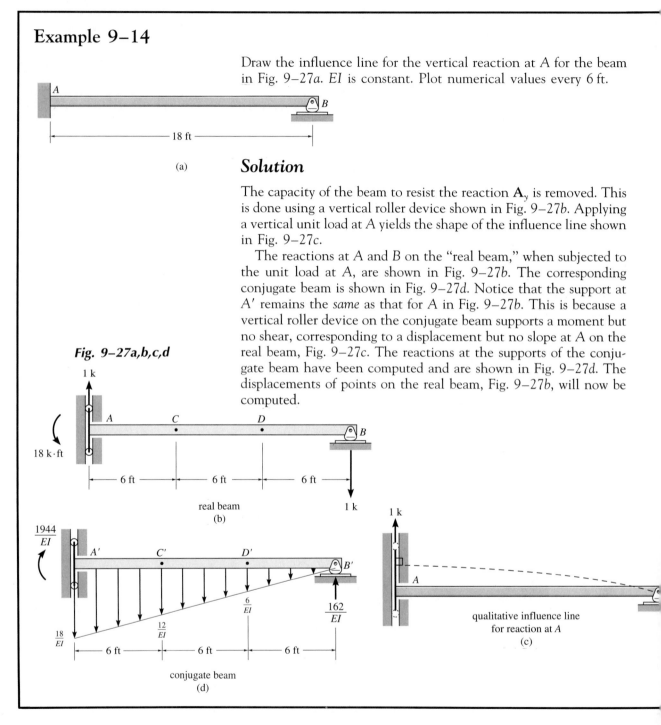

real beam
(b)

conjugate beam
(d)

qualitative influence line
for reaction at A
(c)

498

For B', since no moment exists on the conjugate beam at B', Fig. 9–27d, then

$$\Delta_B = M_{B'} = 0$$

For D', Fig. 9–27e:

$$\Sigma M_{D'} = 0; \quad \Delta_D = M_{D'} = \frac{162}{EI}(6) - \frac{1}{2}\left(\frac{6}{EI}\right)(6)(2) = \frac{936}{EI}$$

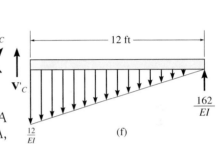

(e)

For C', Fig. 9–27f:

$$\Sigma M_{C'} = 0; \quad \Delta_C = M_{C'} = \frac{162}{EI}(12) - \frac{1}{2}\left(\frac{12}{EI}\right)(12)(4) = \frac{1656}{EI}$$

For A', Fig. 9–27d:

$$\Delta_A = M_{A'} = \frac{1944}{EI}$$

(f)

Since a vertical 1-k load acting on the beam in Fig. 9–27b at A will cause a vertical reaction at A of 1 k, the displacement at A, $\Delta_A = 1944/EI$, should correspond to a numerical value of 1 for the influence-line ordinate at A. Thus, dividing the other computed displacements by this factor, we obtain

x	A_y
A	1
C	0.852
D	0.481
B	0

A plot of these values yields the influence line shown in Fig. 9–27g.

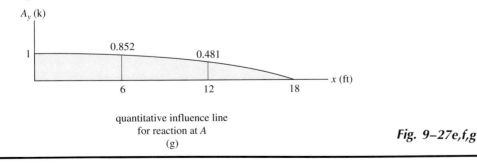

quantitative influence line
for reaction at A
(g)

Fig. 9–27e,f,g

Example 9–15

Draw the influence line for the shear at D for the beam in Fig. 9–28a. EI is constant. Plot numerical values every 9 ft.

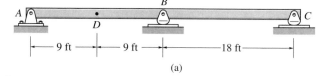

(a)

Solution

The capacity of the beam to resist shear at D is removed. This is done using the roller device shown in Fig. 9–28b. Applying a positive unit shear at D yields the shape of the influence line shown in Fig. 9–28c.

The support reactions at A, B, and C on the "real beam" when subjected to the unit shear at D are shown in Fig. 9–28b. The corresponding conjugate beam is shown in Fig. 9–28d. Here an external couple moment $\mathbf{M}_{D'}$ must be applied at D' in order to cause a different *internal moment* just to the left and just to the right of D'. These

Fig. 9–28a,b,c,d,e

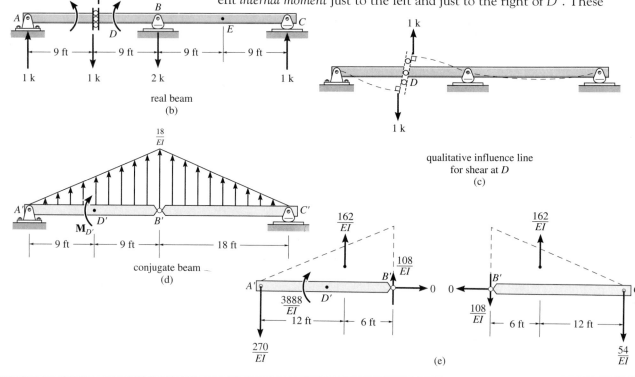

500

internal moments correspond to the displacements just to the left and just to the right of D on the real beam, Fig. 9–28c. The reactions at the supports A', B', C' and the external moment $\mathbf{M}_{D'}$ on the conjugate beam have been computed and are shown in Fig. 9–28e. Verify the calculations.

Since there is a *discontinuity* of moment at D', the internal moment just to the left and right of D' will be computed. Just to the left of D', Fig. 9–28f, we have

$$\Sigma M_{D'_L} = 0; \quad \Delta_{D_L} = M_{D'_L} = \frac{40.5}{EI}(3) - \frac{270}{EI}(9) = -\frac{2309.5}{EI}$$

Just to the right of D', Fig. 9–28g, we have

$$\Sigma M_{D'_R} = 0; \ \Delta_{D_R} = M_{D'_R} = \frac{40.5}{EI}(3) - \frac{270}{EI}(9) + \frac{3888}{EI} = \frac{1579.5}{EI}$$

From Fig. 9–28e,

$$\Delta_A = M_{A'} = 0 \qquad \Delta_B = M_{B'} = 0 \qquad \Delta_C = M_{C'} = 0$$

For point E, Fig. 9–28b, using the method of sections at the corresponding point E' on the conjugate beam, Fig. 9–28h, we have

$$\Sigma M_{E'} = 0; \qquad \Delta_E = M_{E'} = \frac{40.5}{EI}(3) - \frac{54}{EI}(9) = -\frac{364.5}{EI}$$

The ordinates of the influence line are obtained by dividing each of the above values by $M_{D'} = 3888/EI$. We have

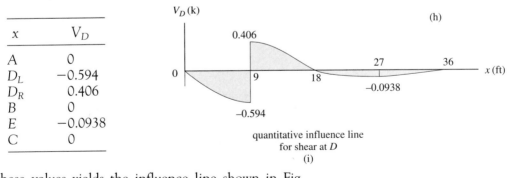

x	V_D
A	0
D_L	-0.594
D_R	0.406
B	0
E	-0.0938
C	0

quantitative influence line
for shear at D
(i)

A plot of these values yields the influence line shown in Fig. 9–28i.

Fig. 9–28f,g,h,i

Example 9–16

Draw the influence line for the moment at D for the beam in Fig. 9–29a. EI is constant. Plot numerical values every 9 ft.

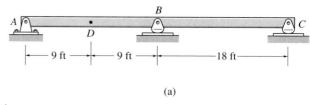

(a)

Solution

A hinge is inserted at D in order to remove the capacity of the beam to resist moment at this point, Fig. 19–29b. Applying positive unit couple moments at D yields the influence line shown in Fig. 9–29c.

The reactions at A, B, and C on the "real beam" when subjected to the unit couple moments at D are shown in Fig. 9–29b. The corresponding conjugate beam and its reactions are shown in Fig. 9–29d. It is suggested that the reactions be verified in both cases. From Fig. 9–29d, note that

$$\Delta_A = M_{A'} = 0 \qquad \Delta_B = M_{B'} = 0 \qquad \Delta_C = M_{C'} = 0$$

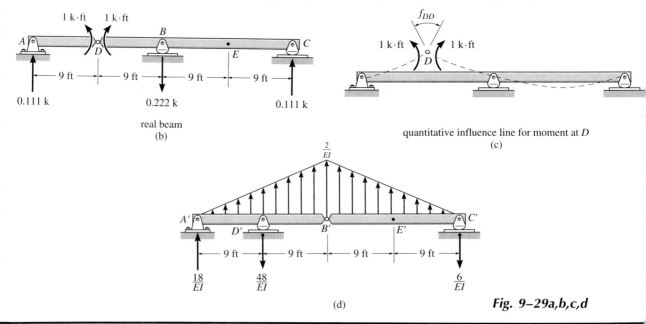

real beam
(b)

quantitative influence line for moment at D
(c)

(d)

Fig. 9–29a,b,c,d

For point D', Fig. 9–29e:

$$\Sigma M_{D'} = 0; \quad \Delta_D = M_{D'} = \frac{4.5}{EI}(3) + \frac{18}{EI}(9) = \frac{175.5}{EI}$$

For point E', Fig. 9–29f:

$$\Sigma M_{E'} = 0; \quad \Delta_E = M_{E'} = \frac{4.5}{EI}(3) - \frac{6}{EI}(9) = -\frac{40.5}{EI}$$

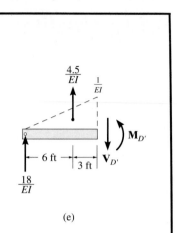

(e)

The angular displacement f_{DD} at D of the "real beam" in Fig. 9–29c is defined by the reaction at D' on the conjugate beam. This factor, $D'_y = 48/EI$, is divided into the above values to give the ordinates of the influence line, that is,

x	M_D
A	0
D	3.656
B	0
E	−0.844
C	0

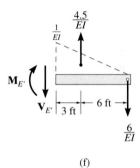

(f)

A plot of these values yields the influence line shown in Fig. 9–29g.

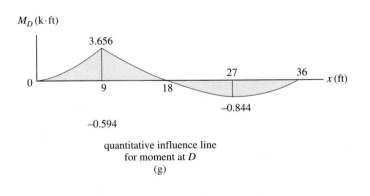

quantitative influence line
for moment at D
(g)

Fig. 9–29e,f,g

9.12   Qualitative Influence Lines for Frames _____

The Müller-Breslau principle provides a quick method and is of great value for establishing the general shape of the influence line for building frames. Once the influence line *shape* is known, one can immediately specify the *location* of the live loads so as to create the greatest influence of the function (reaction, shear, or moment) in the frame. For example, the shape of the influence line for the *positive* moment at the center *I* of girder *FG* of the frame in Fig. 9–30a is shown by the dashed lines. Thus, uniform loads would be placed only on girders *AB*, *CD*, and *FG* in order to create the largest positive moment at *I*. With the frame loaded in this manner, Fig. 9–30b, an indeterminate analysis of the frame could then be performed to determine the critical moment at *I*.

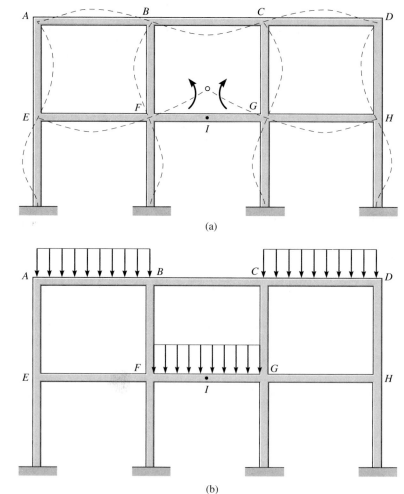

Fig. 9–30

REFERENCES

Clapeyron, B. P. E., "Calcul d'une poutre élastique reposant librement sur des appuis inégalement espacés," *Comptes Rendus des Séances de l'Académie des Sciences,* Paris 45, 1857, 1076–1080.

Maxwell, J. C., "On Reciprocal Figures and Diagrams of Forces," *Philosophical Magazine,* (4), 27, 1864.

Maxwell, J. C. "On the Calculation of the Equilibrium and Stiffness of Frames," *Philosophical Magazine,* (4), 27, 1864.

Mohr, O., "Beitrag zur Theorie des Fachwerks," *Zeitschrift des Architekten und Ingenieur-Vereins zu Hannover,* 20, 1874.

Müller-Breslau, H. F. B., *Die Neueren Methoden der Festigkeitslehre und der Statik der Baukonstruktionen.* Berlin, 1886.

PROBLEMS

9–1. Determine the reactions at the supports and then draw the moment diagram for the beam. Assume the support at A is fixed and B is a roller. EI is constant.

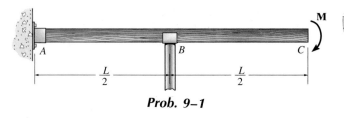

Prob. 9–1

9–2. Determine the reactions at the supports. EI is constant. The support at B is fixed.

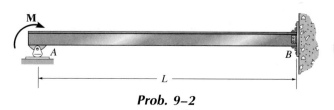

Prob. 9–2

9–3. Determine the reactions at the supports. Assume the support at A is fixed and B is a roller. EI is constant.

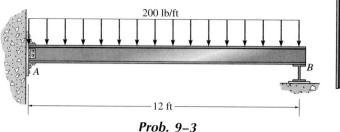

Prob. 9–3

*9–4. Determine the reactions at the supports. Assume the support at A is fixed and B is a roller. Take $E = 29(10^3)$ ksi. The moment of inertia for each segment is shown in the figure.

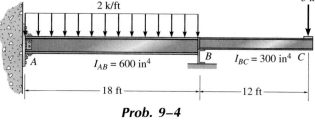

Prob. 9–4

9–5. Determine the reactions at the supports. Assume A is fixed and C is a roller. Take $E = 29(10^3)$ ksi. The moment of inertia for each segment is shown in the figure.

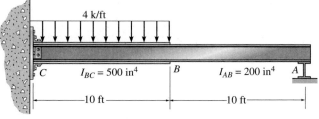

Prob. 9–5

9–6. Determine the reactions at the supports. Assume A is fixed. EI is constant.

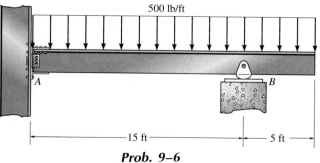

Prob. 9–6

9–7. Determine the reactions at the supports. Assume the support at A is fixed and B is a roller.

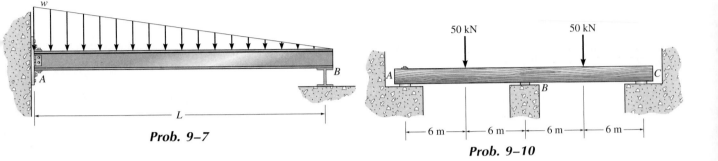

Prob. 9–7

9–10. Determine the reactions at the supports. Assume A is a pin and B and C are rollers. EI is constant.

50 kN   50 kN

|← 6 m →|← 6 m →|← 6 m →|← 6 m →|

Prob. 9–10

*9–8. Determine the reactions at the supports and then draw the bending-moment diagram. Assume A is a pin and B and C are rollers. EI is constant.

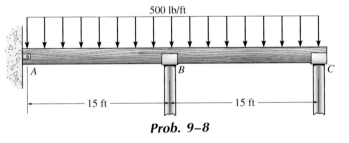

500 lb/ft

|← 15 ft →|← 15 ft →|

Prob. 9–8

9–11. Determine the reactions at the supports. Assume A is pinned and B and C are rollers. The moment of inertia of each span is listed in the figure. Take $E = 29(10^3)$ ksi.

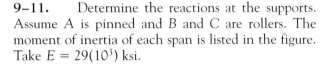

800 lb/ft

$I_{AB} = 600$ in^4 $I_{BC} = 300$ in^4

|← 10 ft →|← 12 ft →|

Prob. 9–11

9–9. Determine the support reactions. Assume B is a pin and A and C are rollers. EI is constant.

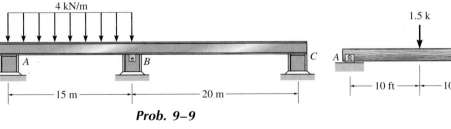

4 kN/m

|← 15 m →|← 20 m →|

Prob. 9–9

*9–12. Determine the reactions at the supports. Assume C is a pin and A and B are rollers. EI is constant.

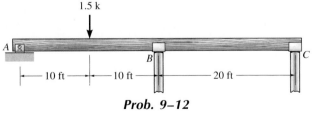

1.5 k

|← 10 ft →|← 10 ft →|← 20 ft →|

Prob. 9–12

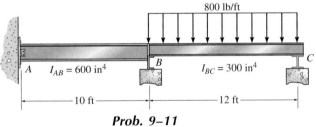

507

9–13. Determine the reactions at the supports, then draw the moment diagram for the beam. The support at B settles downward 0.2 ft. Take $E = 29(10^3)$ ksi, $I = 600$ in⁴.

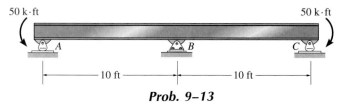

50 k·ft   50 k·ft

Prob. 9–13

9–14. Determine the internal moments acting in the beam at supports B and C. The wall at A moves upward 30 mm. Assume the support at A is a pin. Take $E = 200$ GPa, $I = 90(10^6)$ mm⁴.

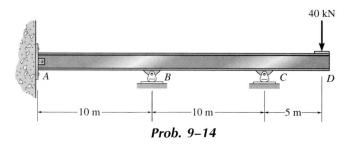

40 kN

Prob. 9–14

9–15. Draw the moment diagram for the beam. EI is constant. Assume the support at C is fixed and A and B are rollers.

*9–16. Two boards each having the same EI and length L are crossed perpendicular to each other as shown. Determine the vertical reactions at the supports. Assume the boards just touch each other before the load **P** is applied.

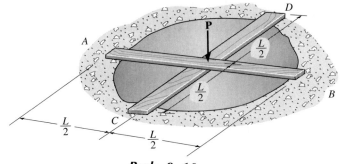

Prob. 9–16

9–17. Determine the reactions at the fixed supports. EI is constant.

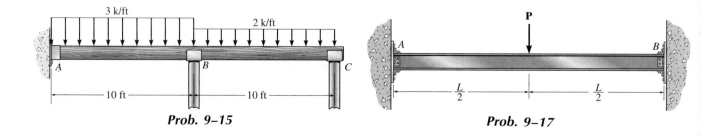

3 k/ft

2 k/ft

Prob. 9–15

P

Prob. 9–17

9–18. Draw the moment diagram for the fixed-end beam. EI is constant.

*9–20. Determine the reactions at the supports. Assume C is fixed. EI is constant.

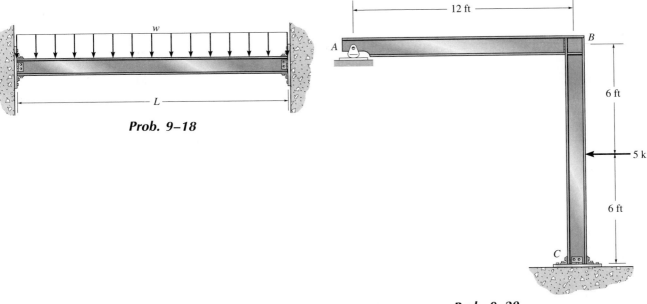

Prob. 9–18

Prob. 9–20

9–19. Determine the reactions at the supports. Assume A is a roller, C is fixed and B is a fixed joint. EI is constant.

9–21. Determine the reactions at the fixed support C and rocker A. The moment of inertia of each segment is listed in the figure. Take $E = 200$ GPa.

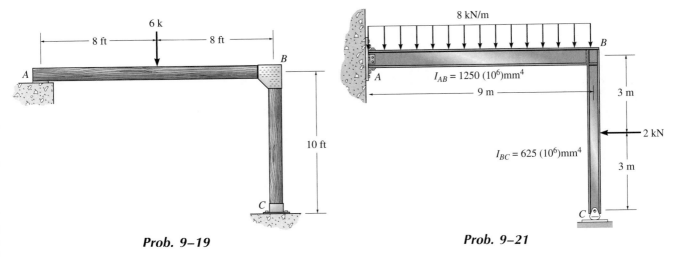

Prob. 9–19

Prob. 9–21

9–22. Determine the reactions at the supports if the support at C moves upward 0.15 in. Assume A is fixed and B is a fixed joint. Take $E = 29(10^3)$ ksi, $I = 600$ in^4.

*9–24. Determine the force in each member of the truss. AE is constant.

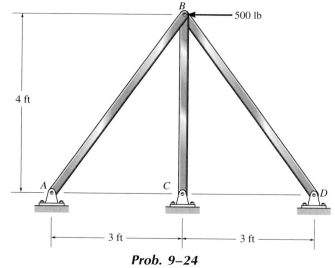

Prob. 9–24

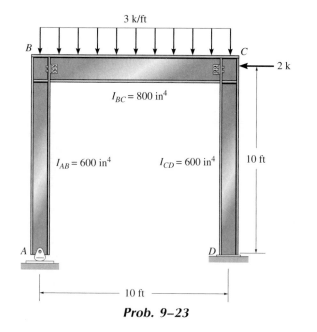

Prob. 9–22

9–23. Determine the reactions at the supports A and D. The moment of inertia of each segment of the frame is listed in the figure. Take $E = 29(10^3)$ ksi.

9–25. Determine the force in bar BC. AE is constant.

9–26. Determine the force in bar BD. AE is constant.

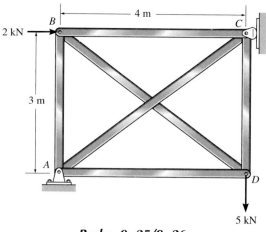

Probs. 9–25/9–26

Prob. 9–23

9–27. Determine the force in each member of the truss. The cross-sectional area of each member is indicated in the figure. $E = 29(10^3)$ ksi. Assume the members are pin-connected at their ends.

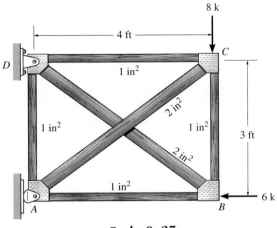

Prob. 9–27

9–29. Determine the force in each member of the truss. AE is constant. Assume the members are pin-connected at their end points.

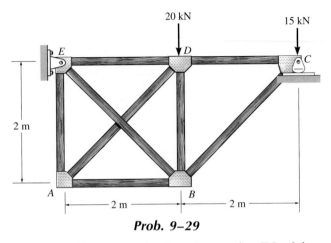

Prob. 9–29

9–30. Determine the force in member EC of the truss. AE is constant.

*9–28. Determine the force in each member of the pin-connected truss. AE is constant.

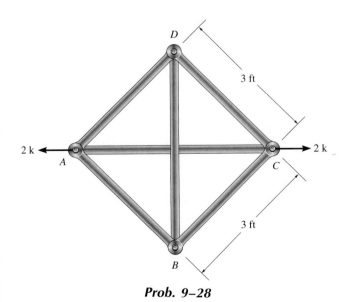

Prob. 9–28

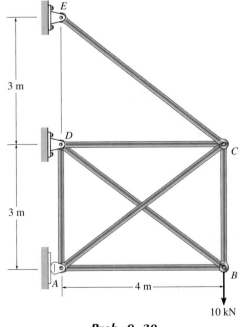

Prob. 9–30

511

9–31.   Determine the force in member AD of the truss. Take $E = 29(10^3)$ ksi. The cross-sectional area of each member is shown in the figure. Assume the members are pin-connected at their end points.

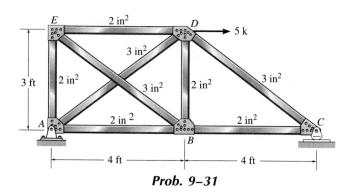

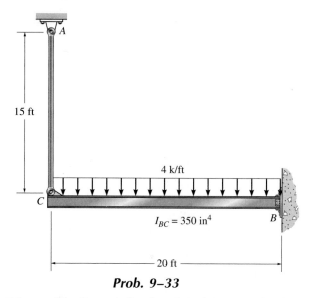

Prob. 9–31

9–32.   Determine the force in member BG. AE is constant. Assume the members are pin-connected at their end points.

9–34.   The beam is fixed at A and supported at its other end by a tie rod BC. Determine the reactions at A, and the force in the tie rod. Take $E = 200$ GPa, $I = 55(10^6)$ mm^4 for the beam, and $A = 600$ mm^2 for the bar. Assume the rod is 5-m long, and neglect the thickness of the beam and its deformation due to axial force.

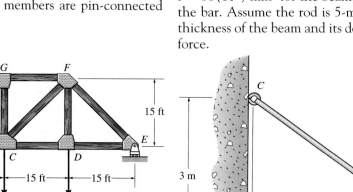

Prob. 9–32

Prob. 9–33

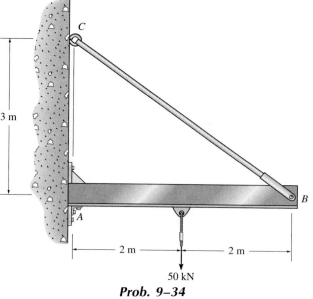

9–33.   The cantilevered beam is supported at one end by a $\frac{1}{2}$-in.-diameter suspender rod AC and fixed at the other end B. Determine the force in the rod due to a uniform loading of 4 k/ft. $E = 29(10^3)$ ksi for both the beam and rod.

Prob. 9–34

9–35. The trussed beam supports a concentrated force of 80 k at its center. Determine the force in each of the three struts and draw the bending-moment diagram for the beam. The struts each have a cross-sectional area of 2 in². Assume they are pin-connected at their end points. Neglect both the depth of the beam and the effect of axial compression in the beam. Take $E = 29(10^3)$ ksi for both the beam and struts. Also, for the beam $I = 400$ in⁴.

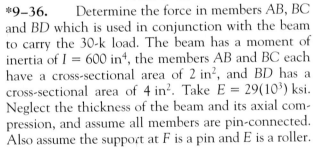

Prob. 9–35

*9–36. Determine the force in members AB, BC and BD which is used in conjunction with the beam to carry the 30-k load. The beam has a moment of inertia of $I = 600$ in⁴, the members AB and BC each have a cross-sectional area of 2 in², and BD has a cross-sectional area of 4 in². Take $E = 29(10^3)$ ksi. Neglect the thickness of the beam and its axial compression, and assume all members are pin-connected. Also assume the support at F is a pin and E is a roller.

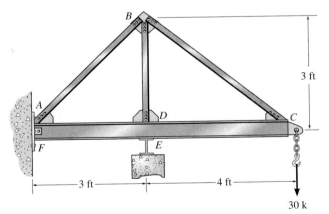

Prob. 9–36

9–37. The trussed beam is used to support a uniform load of 5 k/ft. Determine the forces developed in each of the five struts. The cross-sectional area of each strut is 3 in² and for the beam, $I = 600$ in⁴. Also, $E = 29(10^3)$ ksi. Neglect the thickness of the beam and assume the struts are pin-connected to the beam. Also, neglect the effect of axial compression in the beam.

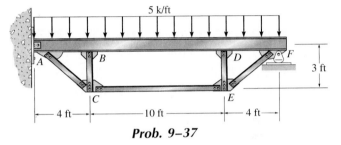

Prob. 9–37

9–38. Use the three-moment equation to determine the reactions at the supports, then draw the bending-moment diagram. EI is constant.

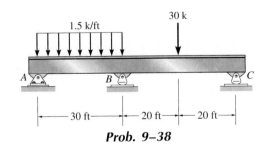

Prob. 9–38

513

9–39.   Use the three-moment equation to determine the reactions at the supports, then draw the bending-moment diagram. Assume A and C are rollers and B is a pin.

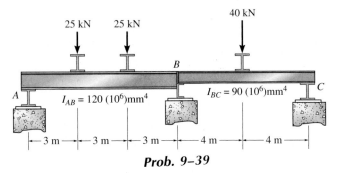

Prob. 9–39

*9–40.   Use the three-moment equation to determine the reactions at the supports, then draw the bending-moment diagram. *EI* is constant. Assume the support at A is a pin and B and C are rollers.

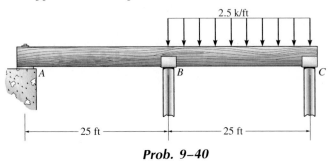

Prob. 9–40

9–41.   Use the three-moment equation to determine the reactions at the supports, then draw the bending-moment diagram. *EI* is constant. Assume the supports at A, B, and C are rollers and D is a pin.

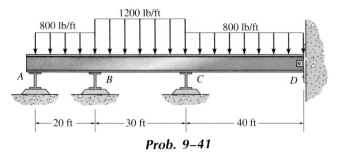

Prob. 9–41

9–42.   Use the three-moment equation to determine the reactions at the supports, then draw the bending-moment diagram. *EI* is constant. Assume A is fixed.

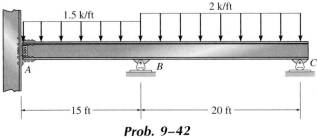

Prob. 9–42

9–43.   Draw the influence line for the moment at point C. Plot numerical values every 5 ft. *EI* is constant. Assume the support at A is fixed and B is a roller.

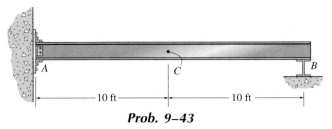

Prob. 9–43

*9–44.   Draw the influence line for the shear at point B. Plot numerical values every 5 ft. Assume the support at A is a roller and C is fixed. *EI* is constant.

9–45.   Draw the influence line for the reaction at point A. Plot numerical values every 5 ft. Assume the support at A is a roller and C is fixed. *EI* is constant.

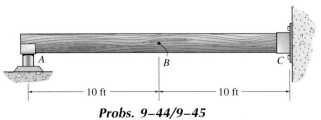

Probs. 9–44/9–45

9–46. Draw the influence line for the reaction at B. Plot numerical values every 5 ft. *EI* is constant.

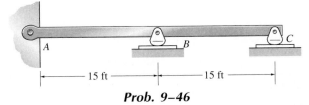

Prob. 9–46

9–47. Sketch the influence line for the moment at D using the Müller-Breslau principle. Determine the maximum positive moment at D due to a uniform live load of 5 kN/m. *EI* is constant. Assume A is a pin and B and C are rollers.

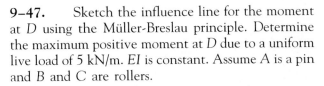

Prob. 9–47

*9–48. Sketch the influence line for the shear at B. If a uniform live load of 6 kN/m is placed on the beam, determine the maximum positive shear at B. Assume the beam is pinned at A. *EI* is constant.

9–49. Sketch the influence line for the moment at B. If a uniform live load of 6 kN/m is placed on the beam, determine the maximum positive moment at B. Assume the beam is pinned at A. *EI* is constant.

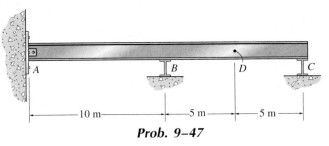

Prob. 9–48/9–49

9–50. Sketch the influence line for (a) the vertical reaction at C, (b) the moment at B, and (c) the shear at E. In each case, indicate on a sketch of the beam where a uniform distributed live load should be placed so as to cause a maximum positive value of these functions. Assume the beam is fixed at F.

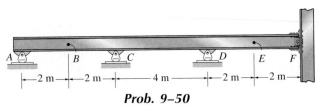

Prob. 9–50

9–51. Use the Müller-Breslau principle to sketch the general shape of the influence line for (a) the moment at A, and (b) the shear at B.

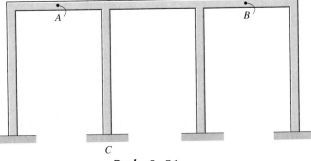

Prob. 9–51

*9–52. Use the Müller-Breslau principle to sketch the general shape of the influence line for (a) the moment at A, and (b) the shear at B.

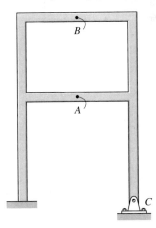

Prob. 9–52

The reinforced-concrete pier framework is statically indeterminate. It can be analyzed using the method of slope deflection.

Displacement Method of Analysis: Slope-Deflection Equations

In this chapter we will briefly outline the basic ideas for analyzing structures using the displacement method of analysis. Once these concepts have been presented, we will develop the general equations of slope deflection and then use them to analyze statically indeterminate beams and frames.

10.1 Displacement Method of Analysis: General Procedures

The *force method* of analysis, discussed in the previous chapter, was based on identifying the unknown redundant loads and then satisfying compatibility equations for the structure. Since these equations actually involved the displacements at points on the structure, it was necessary to use load-displacement relations in order to express the displacements in terms of both the known and unknown loads. The solution of these equations yields the redundant reactions. Once obtained, the equilibrium equations were then used to determine the remaining reactions on the structure.

The *displacement method* works the opposite way. It first requires satisfying equilibrium equations for the structure. Unknown displacements are written in terms of the loads by using the load-displacement relations. These equations are then solved for the unknown displacements. Once obtained, the unknown loads are then determined from the load-displacement relations, which also satisfy the compatibility requirements for the structure. Every displacement method follows this general procedure. In this chapter, the procedure will be generalized to produce the slope-deflection equations. In Chapter 11, the moment-distribution method will be developed. This method sidesteps the calculation of the displacements and instead makes it possible to apply a series of converging corrections that allow direct calculation of the end moments. Finally, in Chapters 14 and 15, we will illustrate how to apply this method using matrix analysis, making it suitable for use on a computer.

In the discussion which follows we will show how to identify the unknown displacements in a structure and we will develop some of the important load-displacement relations for beam and frame members. The results will be used in the next section and in later chapters as the basis for applying the displacement method.

Degrees of Freedom

When a structure is loaded, it deforms into a *unique shape* that can be specified provided we know the displacement of a number of specific points on the structure. These displacements are referred to as the *degrees of freedom* for the structure. In the displacement method of analysis it is important to specify these degrees of freedom since they become the unknowns when the method is applied.

To determine the number of degrees of freedom we can imagine the structure to consist of a series of members connected to *nodes*, which are usually located at *joints, supports*, at the *ends* of a member, or where the members have a sudden *change in cross section*. In three dimensions, each node on a frame or beam can have at most three linear displacements and three rotational displacements; and in two dimensions, each node can have at most two linear displacements and one rotational displacement. If the structure is a truss, the nodes are at the joints and rotational displacement of the nodes is not considered since the members are not subjected to bending. Also, restrictions on nodal displacements may occur at the supports or due to assumptions based on the behavior of the structure. For example, if only deformation due to bending is considered, then there can be no linear displacements along the axis of the members since these displacements are caused by axial-force deformation.

To clarify these concepts we will consider some examples, beginning with the beam in Fig. 10–1a. Here the node at A can only rotate

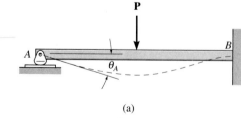

(a)

Fig. 10–1a

(neglecting axial deformation), while node B is completely restricted from moving. Hence the beam has only one degree of freedom, θ_A. The beam in Fig. 10–1b has four degrees of freedom, designated by the rotational displacements θ_A, θ_B, θ_C, and the vertical displacement Δ_C. Consider now the frame in Fig. 10–1c. Again, if we neglect axial deformation of the members, the loading will cause nodes B and C to rotate and these nodes will be displaced horizontally by an *equal* amount. The frame therefore has three degrees of freedom. Lastly, the truss in Fig. 10–1d has a node at each of its joints. Each node can move horizontally and vertically except node A, which cannot move, and node D, which can only move horizontally. As a result, the truss has nine degrees of freedom.

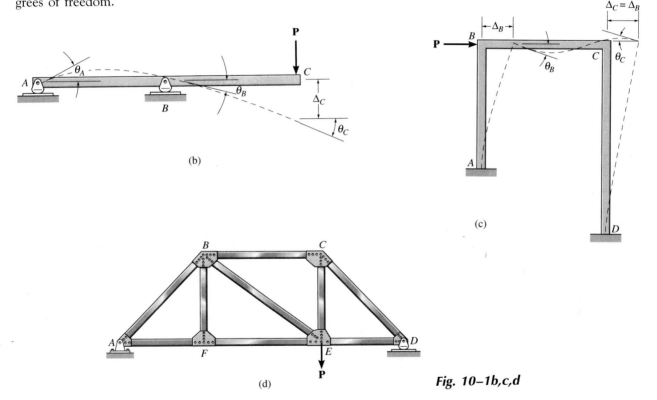

(b)

(c)

(d)

Fig. 10–1b,c,d

In summary, then, specifying the number of unconstrained degrees of freedom for the structure is a necessary first step when applying a displacement method of analysis. It identifies the number of unknowns in the problem, based on the assumptions made regarding the deformation behavior of the structure. Furthermore, once these displacements are known, the deformation of the structure can be completely specified.

519

10.2   Slope-Deflection Equations ⸻

As indicated previously, the method of consistent displacements studied in Chapter 8 is called a force method of analysis, because it requires writing equations that relate the unknown forces or moments in a structure. Unfortunately, its use is limited to structures which are *not* highly indeterminate. This is because much work is required to set up the compatibility equations, and furthermore each equation written involves *all the unknowns,* making it difficult to solve the resulting set of equations unless a computer is available. By comparison, the slope-deflection method is not as involved. As we shall see, it requires less work both to write the necessary equations for the solution of a problem and to solve these equations for the unknowns. Also, the method can be easily programmed on a computer and used to analyze a wide range of indeterminate structures.

The slope-deflection method was originally developed by Heinrich Manderla and Otto Mohr for the purpose of studying secondary stresses in trusses. Later, in 1915, G. A. Maney developed a refined version of this technique and applied it to the analysis of indeterminate beams and framed structures.

General Case

In order to develop the general form of the slope-deflection equations, we will consider the span AB of the continuous beam shown in Fig. 10–2, which is subjected to the arbitrary loading and has a constant EI. We wish to relate the beam's internal end moments M_{AB} and M_{BA} to its three degrees of freedom, namely, the linear displacement Δ and the angular displacements θ_A and θ_B. Since we will be developing a formula, *moments* and *angular displacements* will be considered *positive* when they act *clockwise on the span,* as shown in Fig. 10–2. Furthermore, the *linear displacement* Δ is considered *positive* as shown, since this displacement causes the cord of the span and the span's cord angle ψ to rotate *clockwise.*

We can develop the slope-deflection equations using the principle of

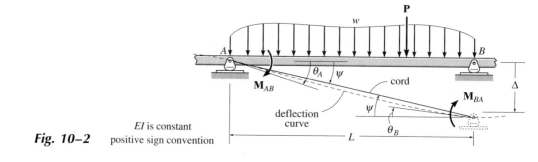

Fig. 10–2 EI is constant
positive sign convention

superposition by considering *separately* the effects caused by each of the displacements and then the loads.

Angular Displacement at A, θ_A.   Consider node A of the member shown in Fig. 10–3a to rotate θ_A while its far-end node B is held fixed. We can determine the moment M_{AB} needed to cause this displacement by using the conjugate-beam method. For this case the conjugate beam is shown in Fig. 10–3b. The end shear at A′ acts downward on the beam, since θ_A is clockwise. The deflection of the "real beam" in Fig. 10–3a is to be zero at A and B, and therefore the corresponding sum of the *moments* at each end A′ and B′ of the conjugate beam must also be zero. This yields

$$\zeta + \Sigma M_{A'} = 0; \qquad \left[\frac{1}{2}\left(\frac{M_{AB}}{EI}\right)L\right]\frac{L}{3} - \left[\frac{1}{2}\left(\frac{M_{BA}}{EI}\right)L\right]\frac{2L}{3} = 0$$

$$\zeta + \Sigma M_{B'} = 0; \qquad \left[\frac{1}{2}\left(\frac{M_{BA}}{EI}\right)L\right]\frac{L}{3} - \left[\frac{1}{2}\left(\frac{M_{AB}}{EI}\right)L\right]\frac{2L}{3} + \theta_A L = 0$$

from which

$$\boxed{M_{AB} = \frac{4EI}{L}\theta_A} \qquad\qquad (10\text{–}1)$$

$$\boxed{M_{BA} = \frac{2EI}{L}\theta_A} \qquad\qquad (10\text{–}2)$$

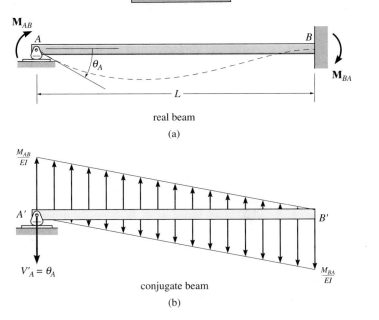

real beam

(a)

conjugate beam

(b)

Fig. 10–3

521

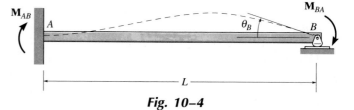

Fig. 10–4

Angular Displacement at B, θ_B.   Finally, if end B of the beam rotates to its final position θ_B, while end A is held fixed, Fig. 10–4, we can relate the applied moment M_{BA} to the angular displacement θ_B and the reaction moment M_{AB} at the wall. The analysis is similar to that discussed above. The results are

$$M_{BA} = \frac{4EI}{L}\theta_B \tag{10–3}$$

$$M_{AB} = \frac{2EI}{L}\theta_B \tag{10–4}$$

Relative Linear Displacement, Δ.   If the far node B of the member is displaced relative to A, so that the cord of the member rotates clockwise (positive displacement) and yet both ends do not rotate, then equal but opposite moment and shear reactions are developed in the member, Fig. 10–5a. As before, the moment $\mathbf{M}$ can be related to the displacement Δ using the conjugate-beam method. In this case, the conjugate beam, Fig. 10–5b, is free at both ends, since the real beam (member) is fixed-supported. However, due to the *displacement* of the real beam at B, the *moment* at the end B' of the conjugate beam must have a magnitude of Δ as indicated.* Summing moments about B', we have

$$\curvearrowleft + \Sigma M_{B'} = 0; \quad \left[\frac{1}{2}\frac{M}{EI}(L)\left(\frac{2}{3}L\right)\right] - \left[\frac{1}{2}\frac{M}{EI}(L)\left(\frac{1}{3}L\right)\right] - \Delta = 0$$

Since $\mathbf{M}$ is counterclockwise,

$$M_{AB} = M_{BA} = -M = \frac{-6EI}{L^2}\Delta \tag{10–5}$$

This induced moment is negative since for equilibrium it acts counterclockwise on the member.

---

*The moment diagrams shown on the conjugate beam were determined by the method of superposition for a simply supported beam, as explained in Sec. 4.5.

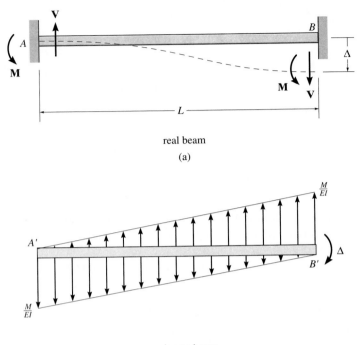

real beam

(a)

conjugate beam

(b)

Fig. 10–5

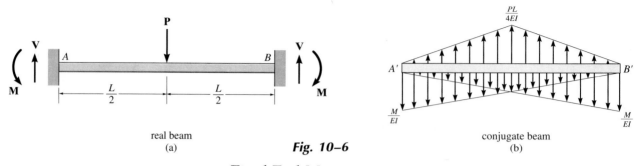

real beam
(a)

conjugate beam
(b)

Fig. 10–6

Fixed-End Moments

In all of the above cases we have considered relationships between the required displacement and an *applied moments* M_{AB} and M_{BA} acting at nodes A and B, respectively. In practice, however, the linear or angular displacements of the nodes are most often caused by loadings acting on the *span* of the member, not by moments acting at its nodes. In order to develop the slope-deflection equations, however, we must transform these *span loadings* into equivalent moments acting at the nodes and then use the load-displacement relationships derived above. This is done simply by finding the reaction moment that each load develops at the nodes. For example, consider the fixed-supported member shown in Fig. 10–6a, which is subjected to a concentrated load $\mathbf{P}$ at its center. The conjugate beam for this case is shown in Fig. 10–6b. Since we require the slope at each end to be zero,

$$+\uparrow\Sigma F_y = 0; \qquad \left[\frac{1}{2}\left(\frac{PL}{4EI}\right)(L)\right] - 2\left[\frac{1}{2}\left(\frac{M}{EI}\right)L\right] = 0$$

$$M = \frac{PL}{8}$$

This moment is called a *fixed-end moment* (FEM). Note that according to our sign convention, it is negative at node A (counterclockwise) and positive at node B (clockwise). For convenience in solving problems, fixed-end moments have been calculated for other loadings and are tabulated on the inside back cover of the book. Assuming these FEMS have been computed for a specific problem (Fig. 10–7), we have

$$M_{AB_1} = (FEM)_{AB} \qquad M_{BA_1} = (FEM)_{BA} \qquad (10\text{–}6)$$

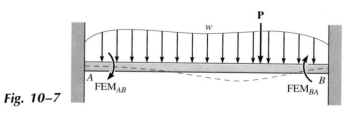

Fig. 10–7

Slope-Deflection Equation.   If the end moments due to each displacement (Eqs. 10–1 through 10–5) and the loading (Eq. 10–6) are added together, the final moments can be written as

$$M_{AB} = 2E\left(\frac{I}{L}\right)\left[2\theta_A + \theta_B - 3\left(\frac{\Delta}{L}\right)\right] + (FEM)_{AB}$$

$$M_{BA} = 2E\left(\frac{I}{L}\right)\left[2\theta_B + \theta_A - 3\left(\frac{\Delta}{L}\right)\right] + (FEM)_{BA} \qquad (10-7)$$

Since these two equations are similar, the result can be expressed as a single equation. Referring to one end of the span as the near end (N) and the other end as the far end (F), and letting the *member stiffness* be represented as $k = I/L$, and the *span's cord rotation* as $\psi = \Delta/L$, we can write

$$M_N = 2Ek(2\theta_N + \theta_F - 3\psi) + (FEM)_N \qquad (10-8)$$
$$\text{For Internal Span or End Span with Far End Fixed}$$

where

M_N = internal moment in the near end of the span; this moment is *positive clockwise* when acting on the span

E, k = modulus of elasticity of material, and span stiffness $k = I/L$

θ_N, θ_F = near- and far-end slopes or angular displacements of the span at the supports; the angles are measured in *radians* and are *positive clockwise*

ψ = span rotation of its cord due to a linear displacement, that is, $\psi = \Delta/L$. This angle is measured in *radians* and is *positive clockwise*

$(FEM)_N$ = fixed-end moment at the near-end support; the moment is *positive clockwise* when acting on the span; refer to the table on the inside back cover for various loading conditions

Equation 10–8 is a load-displacement relationship and is referred to as the general *slope-deflection equation*. When used for the solution of problems, this equation is applied *twice* for each member span (AB); that is, application is from A to B and from B to A for span AB in Fig. 10–2.

Pin-Supported End Span

Occasionally an end span of a beam or frame is supported by a pin or roller at its *far end*, Fig. 10–8. When this occurs, the moment at the roller or pin must be zero; and provided the angular displacement θ_B at this support does not have to be determined, we can modify the general slope-deflection equation so that it has to be applied *only once* to the span. To do this we will apply Eq. 10–8 or Eq. 10–7 to each end of the beam in Fig. 10–8. This results in the following two equations:

$$M_N = 2Ek(2\theta_N + \theta_F - 3\psi) + (FEM)_N$$
$$0 = 2Ek(2\theta_F + \theta_N - 3\psi)$$

(10–9)

Here the $(FEM)_F$ is equal to zero since the far end is pinned, Fig. 10–9. Furthermore, the $(FEM)_N$ can be obtained, for example, using the table in the right-hand column on the inside back cover of this book. Multiplying the first equation by 2 and subtracting the second equation from it eliminates the unknown θ_F and yields

$$M_N = 3Ek(\theta_N - \psi) + (FEM)_N$$

Only for End Span with Far End Pinned or Roller Supported

(10–10)

Since the moment at the far end is zero, only *one* application of this equation is necessary for the end span. This obviously simplifies the analysis since the general equation, Eq. 10–8, requires *two* applications for this span and would therefore involve the (extra) unknown angular displacement θ_B at the end support.

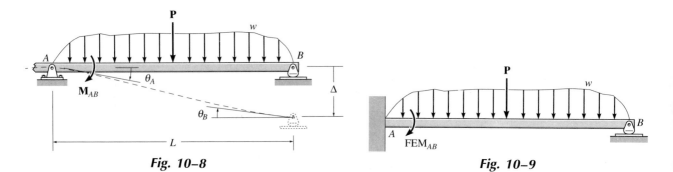

Fig. 10–8   **Fig. 10–9**

10.3   Analysis of Beams

Procedure for Analysis

Degrees of Freedom   Label all the supports and joints (nodes) in order to identify the spans of the beam or frame between the nodes. By drawing the deflected shape of the structure, it will be possible to identify the degrees of freedom for the structure. Here each node can have an angular displacement and a single linear displacement. *Compatibility* at the nodes is maintained provided the members that are fixed-connected to a node undergo the same displacements as the node. If these displacements are unknown, and in general they will be, then for convenience assume they act in the *positive* direction so as to cause *clockwise* rotation of a member or joint, Fig. 10–2.

Slope-Deflection Equations   The slope-deflection equations relate the unknown moments applied to the nodes to the displacements of the nodes for any span of the structure. If a load exists on the span, compute the FEMs using the table given on the inside back cover. Also, if a node has a linear displacement, Δ, compute $\psi = \Delta/L$ for the adjacent spans. Apply Eq. 10–8 to each end of the span, thereby generating *two* slope-deflection equations for each span. However, if the far end of a span at the *end* of a beam or frame is pin-supported, apply Eq. 10–10 only to the restrained end, thereby generating *one* slope-deflection equation for the span.

Equilibrium Equations   Write an equilibrium equation for each unknown degree of freedom for the structure. Each of these equations should be expressed in terms of unknown internal moments as specified by the slope-deflection equations. For beams and frames write the moment equation of equilibrium at each support, and for frames also write joint moment equations of equilibrium. If the frame sidesways or deflects horizontally, column shears should be related to the moments at the ends of the column. This is discussed in Sec. 10.5.

Substitute the slope-deflection equations into the equilibrium equations and solve for the unknown joint rotations and/or displacements. These results are then substituted into the slope-deflection equations to determine the internal moments at the ends of each member. If any of the results are *negative*, they indicate *counterclockwise* rotation; whereas *positive* moments and displacements are applied *clockwise*.

Example 10–1

Draw the shear and moment diagrams for the beam shown in Fig. 10–10a. EI is constant.

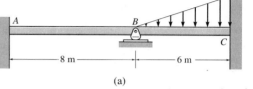

(a)

Solution

Slope-Deflection Equations.   Two spans must be considered in this problem. Since there is *no* span having the far end pinned or roller-supported, Eq. 10–8 applies to the solution. Using the formulas for the FEMs tabulated for the triangular loading given on the inside back cover, we have

$$(\text{FEM})_{BC} = -\frac{wL^2}{30} = -\frac{6(6)^2}{30} = -7.2 \text{ kN} \cdot \text{m}$$

$$(\text{FEM})_{CB} = \frac{wL^2}{20} = \frac{6(6)^2}{20} = 10.8 \text{ kN} \cdot \text{m}$$

Note that $(\text{FEM})_{BC}$ is negative, since it acts counterclockwise *on the beam* at B.

In order to identify the unknowns, the elastic curve for the beam is shown in Fig. 10–10b. As indicated, there are four unknown moments. Only the slope at B, θ_B, is unknown. Since A and C are fixed supports, $\theta_A = \theta_C = 0$. Also, since the supports do not settle, nor are displaced up or down, $\psi_{AB} = \psi_{BC} = 0$. Equation 10–8 will now be applied. For span AB, considering A to be the near end and B to be the far end, we have

$$M_N = 2E\left(\frac{I}{L}\right)(2\theta_N + \theta_F - 3\psi) + (\text{FEM})_N$$

$$M_{AB} = 2E\left(\frac{I}{8}\right)[2(0) + \theta_B - 3(0)] + 0$$

$$M_{AB} = \frac{EI}{4}\theta_B \qquad (1)$$

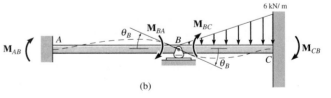

Fig. 10–10a,b

(b)

Now, considering B to be the near end and A to be the far end, we have

$$M_{BA} = 2E\left(\frac{I}{8}\right)[2\theta_B + 0 - 3(0)] + 0$$

$$M_{BA} = \frac{EI}{2}\,\theta_B \tag{2}$$

In a similar manner, for span BC we have

$$M_{BC} = 2E\left(\frac{I}{6}\right)[2\theta_B + 0 - 3(0)] - 7.2$$

$$M_{BC} = \frac{2EI}{3}\,\theta_B - 7.2 \tag{3}$$

$$M_{CB} = 2E\left(\frac{I}{6}\right)[2(0) + \theta_B - 3(0)] + 10.8$$

$$M_{CB} = \frac{EI}{3}\,\theta_B + 10.8 \tag{4}$$

Equilibrium Equations.   The above four equations contain five unknowns. The necessary fifth equation comes from the condition of moment equilibrium at support B. The free-body diagram of a segment of the beam at B is shown in Fig. 10–10c. Here $\mathbf{M}_{BA}$ and $\mathbf{M}_{BC}$ are assumed to act in the positive direction.* The beam shears contribute negligible moment about B, so

$$\zeta + \Sigma M_B = 0; \qquad\qquad M_{BA} + M_{BC} = 0 \tag{5}$$

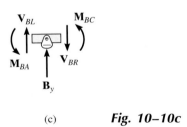

(c)   **Fig. 10–10c**

*Clockwise on the beam segment, but—by the principle of action, equal but opposite reaction—counterclockwise on the support.

(cont'd)

Example 10–1
(continued)

To solve, substitute Eqs. (2) and (3) into Eq. (5), which yields

$$\theta_B = \frac{6.17}{EI}$$

Resubstituting this value into Eqs. (1)–(4) yields

$$M_{AB} = 1.54 \text{ kN} \cdot \text{m}$$
$$M_{BA} = 3.09 \text{ kN} \cdot \text{m}$$
$$M_{BC} = -3.09 \text{ kN} \cdot \text{m}$$
$$M_{CB} = 12.86 \text{ kN} \cdot \text{m}$$

The negative value for M_{BC} indicates that this moment acts counterclockwise on the beam, not clockwise as shown in Fig. 10–10b.

Using these results, the shears at the end spans are determined from the equilibrium equations, Fig. 10–10d. The free-body diagram of the beam and the shear and moment diagrams are shown in Fig. 10–10e.

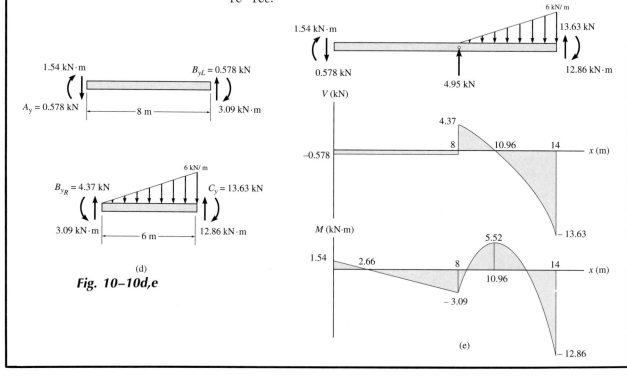

(d)
Fig. 10–10d,e

530

Example 10–2

Draw the shear and moment diagrams for the beam shown in Fig. 10–11a. EI is constant.

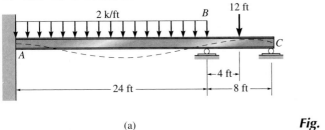

(a)

Fig. 10–11a

Solution

Slope-Deflection Equations.   Two spans must be considered in this problem. Equation 10–8 applies to span AB. We can use Eq. 10–10 for span BC since the *end C is on a roller*. Using the formulas for the FEMs tabulated on the inside back cover, we have

$$(FEM)_{AB} = -\frac{wL^2}{12} = -\frac{1}{12}(2)(24)^2 = -96 \text{ k} \cdot \text{ft}$$

$$(FEM)_{BA} = \frac{wL^2}{12} = \frac{1}{12}(2)(24)^2 = 96 \text{ k} \cdot \text{ft}$$

$$(FEM)_{BC} = -\frac{3PL}{16} = -\frac{3(12)(8)}{16} = -18 \text{ k} \cdot \text{ft}$$

Note that $(FEM)_{AB}$ and $(FEM)_{BC}$ are negative since they act counterclockwise on the beam at A and B, respectively. Also, since the supports do not settle, $\psi_{AB} = \psi_{BC} = 0$. Applying Eq. 10–8 for span AB and realizing that $\theta_A = 0$, we have

$$M_N = 2E\left(\frac{I}{L}\right)(2\theta_N + \theta_F - 3\psi) + (FEM)_N$$

$$M_{AB} = 2E\left(\frac{I}{24}\right)[2(0) + \theta_B - 3(0)] - 96$$

$$M_{AB} = 0.08333EI\theta_B - 96 \qquad (1)$$

$$M_{BA} = 2E\left(\frac{I}{24}\right)[2\theta_B + 0 - 3(0)] + 96$$

$$M_{BA} = 0.1667EI\theta_B + 96 \qquad (2)$$

(cont'd)

531

Example 10–2
(continued)

Applying Eq. 10–10 with B as the near end and C as the far end, we have

$$M_N = 3E\left(\frac{I}{L}\right)(\theta_N - \psi) + (\text{FEM})_N$$

$$M_{BC} = 3E\left(\frac{I}{8}\right)(\theta_B - 0) - 18$$

$$M_{BC} = 0.375EI\theta_B - 18 \qquad (3)$$

Recall that Eq. 10–10 is *not* applied from C (near end) to B (far end).

Equilibrium Equations.   The above three equations contain four unknowns. The necessary fourth equation comes from the conditions of equilibrium at the support B. The free-body diagram is shown in Fig. 10–11b. We have

$$\zeta + \Sigma M_B = 0; \qquad\qquad M_{BA} + M_{BC} = 0 \qquad (4)$$

To solve, substitute Eqs. (2) and (3) into Eq. (4), which yields

$$\theta_B = -\frac{144.0}{EI}$$

Since θ_B is negative (counterclockwise) the elastic curve for the beam has been correctly drawn in Fig. 10-11a. Substituting this value into Eqs. (1)–(3), we get

$$M_{AB} = -108.0 \text{ k} \cdot \text{ft}$$
$$M_{BA} = 72.0 \text{ k} \cdot \text{ft}$$
$$M_{BC} = -72.0 \text{ k} \cdot \text{ft}$$

Using these data for the moments, the shear reactions at the ends of the beam spans have been computed in Fig. 10–11c. The shear and moment diagrams are plotted in Fig. 10–11d and 10–11e.

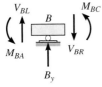

(b)

Fig. 10–11b

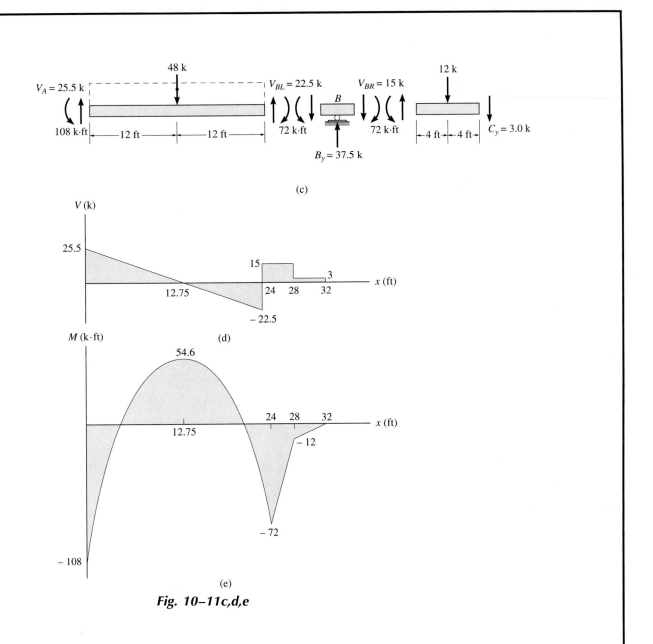

Fig. 10–11c,d,e

533

Example 10–3

Draw the shear and moment diagrams for the beam shown in Fig. 10–12a. The support at B is displaced (settles) 80 mm. Take $E = 200$ GPa, $I = 5(10^6)$ mm^4.

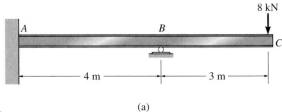

8 kN

A B C

4 m   3 m

(a)

Solution

Slope-Deflection Equations.   Only one span (AB) must be considered in this problem since the moment $\mathbf{M}_{BC}$ due to the overhang can be calculated. Since there is no loading on span AB, the FEMs are zero. As shown in Fig. 10–12b, the downward displacement (settlement) of B causes the cord for span AB to rotate clockwise. Thus,

$$\psi_{AB} = \psi_{BA} = \frac{0.08 \text{ m}}{4} = 0.02 \text{ rad}$$

The stiffness for AB is

$$k = \frac{I}{L} = \frac{5(10^6) \text{ mm}^4(10^{-12}) \text{ m}^4/\text{mm}^4}{4 \text{ m}} = 1.25(10^{-6}) \text{ m}^3$$

Applying the slope-deflection equation, Eq. 10–10, to span AB, with $\theta_A = 0$, we have

$$M_N = 2E\left(\frac{I}{L}\right)(2\theta_N + \theta_F - 3\psi) + (\text{FEM})_N$$

$$M_{AB} = 2(200(10^9) \text{ N/m}^2)[1.25(10^{-6}) \text{ m}^3][2(0) + \theta_B - 3(0.02)] + 0 \tag{1}$$

$$M_{BA} = 2(200(10^9) \text{ N/m}^2)[1.25(10^{-6}) \text{ m}^3][2\theta_B + 0 - 3(0.02)] + 0 \tag{2}$$

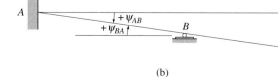

A

$+\psi_{AB}$

$+\psi_{BA}$

B

Fig. 10–12a,b

(b)

Equilibrium Equations. The free-body diagram of the beam at support B is shown in Fig. 10–12c. Moment equilibrium requires

$$\zeta+\Sigma M_B = 0; \qquad M_{BA} - 8000\ N(3\ m) = 0$$

Substituting Eq. (2) into this equation yields

$$1(10^6)\theta_B - 30(10^3) = 24(10^3)$$
$$\theta_B = 0.054\ rad$$

Thus, from Eqs. (1) and (2),

$$M_{AB} = -3.00\ kN \cdot m$$
$$M_{BA} = 24.0\ kN \cdot m$$

Using these results, the equations of equilibrium can be applied to each segment of the beam, and the end shears calculated, Fig. 10–12d. The moment diagram is shown in Fig. 10–12e.

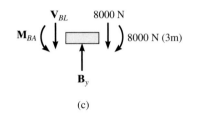

(c)

Fig. 10–12c,d,e

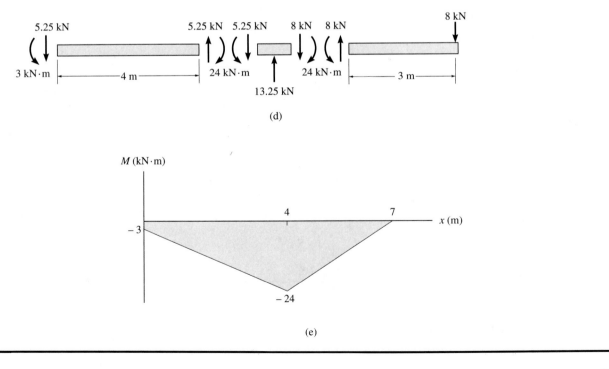

(d)

(e)

Example 10–4

Determine the internal moments at the supports of the beam shown in Fig. 10–13a. The support at C is displaced (settles) 0.1 ft. Take $E = 29(10^3)$ ksi, $I = 1500$ in^4.

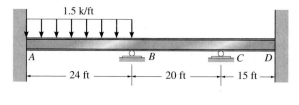

(a)

Solution

Slope-Deflection Equations.   Three spans must be considered in this problem. Equation 10–8 applies since the end supports A and D are fixed. Also, only span AB has FEMs.

$$(\text{FEM})_{AB} = -\frac{wL^2}{12} = -\frac{1}{12}(1.5)(24)^2 = -72.0 \text{ k} \cdot \text{ft}$$

$$(\text{FEM})_{BA} = \frac{wL^2}{12} = \frac{1}{12}(1.5)(24)^2 = 72.0 \text{ k} \cdot \text{ft}$$

As shown in Fig. 10–13b, the displacement (or settlement) of the support C causes ψ_{BC} to be positive, since the cord for span BC rotates clockwise, and ψ_{CD} to be negative, since the cord for span CD rotates counterclockwise. Hence,

$$\psi_{BC} = \frac{0.1 \text{ ft}}{20 \text{ ft}} = 0.005 \text{ rad} \qquad \psi_{CD} = -\frac{0.1 \text{ ft}}{15 \text{ ft}} = -0.00667 \text{ rad}$$

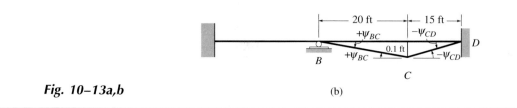

Fig. 10–13a,b

(b)

Also, expressing the units for the stiffness in feet, we have

$$k_{AB} = \frac{1500}{24(12)^4} = 0.003014 \text{ ft}^3 \qquad k_{BC} = \frac{1500}{20(12)^4} = 0.003617 \text{ ft}^3$$

$$k_{CD} = \frac{1500}{15(12)^4} = 0.004823 \text{ ft}^3$$

Noting that $\theta_A = \theta_D = 0$ since A and D are fixed supports, and applying the slope-deflection Eq. 10–8 twice to each span, we have

For span AB:

$$M_{AB} = 2[29(10^3)(12)^2](0.003014)[2(0) + \theta_B - 3(0)] - 72$$
$$M_{AB} = 25{,}172.9\theta_B - 72 \tag{1}$$
$$M_{BA} = 2[29(10^3)(12)^2](0.003014)[2\theta_B + 0 - 3(0)] + 72$$
$$M_{BA} = 50{,}345.9\theta_B + 72 \tag{2}$$

For span BC:

$$M_{BC} = 2[29(10^3)(12)^2](0.003617)[2\theta_B + \theta_C - 3(0.005)] + 0$$
$$M_{BC} = 60{,}418.4\theta_B + 30{,}209.2\theta_C - 453.2 \tag{3}$$
$$M_{CB} = 2[29(10^3)(12)^2](0.003617)[2\theta_C + \theta_B - 3(0.005)] + 0$$
$$M_{CB} = 60{,}418.4\theta_C + 30{,}209.2\theta_B - 453.2 \tag{4}$$

For span CD:

$$M_{CD} = 2[29(10^3)(12)^2](0.004823)[2\theta_C + 0 - 3(-0.00667)] + 0$$
$$M_{CD} = 80{,}563.4\theta_C + 0 + 805.6 \tag{5}$$
$$M_{DC} = 2[29(10^3)(12)^2](0.004823)[2(0) + \theta_C - 3(-0.00667)] + 0$$
$$M_{DC} = 40{,}281.7\theta_C + 805.6 \tag{6}$$

(cont'd)

Example 10–4
(continued)

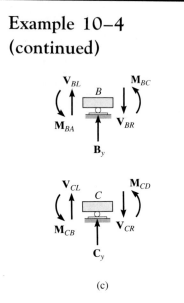

(c)

Fig. 10–13c,d

Equilibrium Equations.   The above six equations contain eight unknowns. Writing the moment equilibrium equations for the supports at B and C, Fig. 10–13c, we have

$$\downarrow + \Sigma M_B = 0; \qquad\qquad M_{BA} + M_{BC} = 0 \qquad\qquad (7)$$
$$\downarrow + \Sigma M_C = 0; \qquad\qquad M_{CB} + M_{CD} = 0 \qquad\qquad (8)$$

In order to solve, substitute Eqs. (2) and (3) into Eq. (7), and Eqs. (4) and (5) into Eq. (8). This yields

$$\theta_C + 3.667\theta_B = 0.01262$$
$$-\theta_C - 0.214\theta_B = 0.00250$$

Solving,

$$\theta_B = 0.00438 \text{ rad} \qquad \theta_C = -0.00344 \text{ rad}$$

The negative value for θ_C indicates counterclockwise rotation of the tangent at C, Fig. 10–13d. Substituting these values into Eqs. (1)–(6) yields

$$\begin{aligned} M_{AB} &= 38.2 \text{ k} \cdot \text{ft} & \textbf{Ans.} \\ M_{BA} &= 292 \text{ k} \cdot \text{ft} & \textbf{Ans.} \\ M_{BC} &= -292 \text{ k} \cdot \text{ft} & \textbf{Ans.} \\ M_{CB} &= -529 \text{ k} \cdot \text{ft} & \textbf{Ans.} \\ M_{CD} &= 529 \text{ k} \cdot \text{ft} & \textbf{Ans.} \\ M_{DC} &= 667 \text{ k} \cdot \text{ft} & \textbf{Ans.} \end{aligned}$$

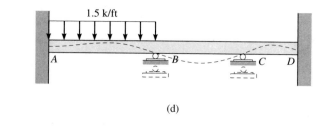

(d)

10.4   Analysis of Frames: No Sidesway

A frame will not sidesway, or be displaced to the left or right, provided it is properly restrained. Examples are shown in Fig. 10–14. Also, no sidesway will occur in an unrestrained frame provided it is symmetric with respect to both loading and geometry, as shown in Fig. 10–15. For both cases the term ψ in the slope-deflection equations is equal to zero, since the joints do not have a linear displacement.

The following examples illustrate application of the slope-deflection equations using the procedure for analysis outlined in Sec. 10.1 for these types of frames.

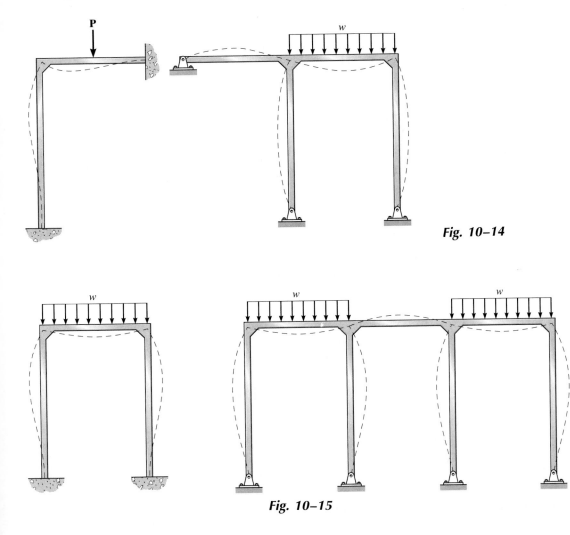

Fig. 10–14

Fig. 10–15

Example 10–5

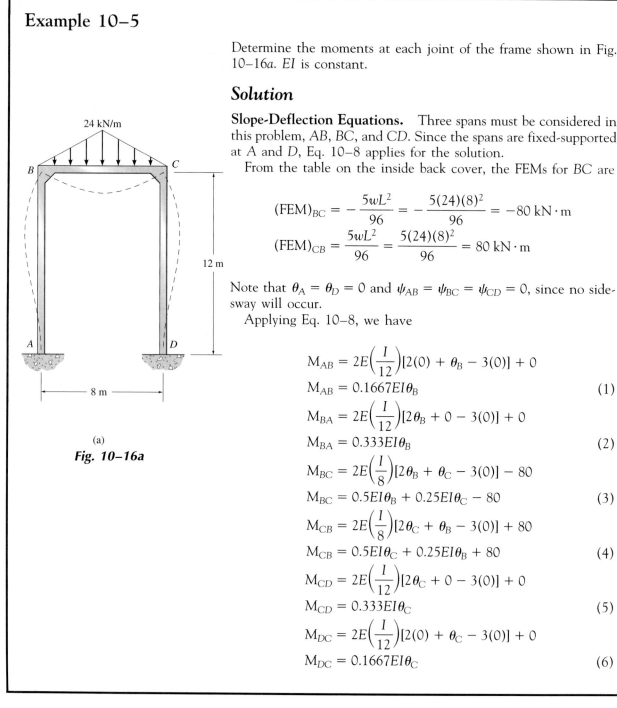

24 kN/m

B

C

12 m

A

D

8 m

(a)

Fig. 10–16a

Determine the moments at each joint of the frame shown in Fig. 10–16a. EI is constant.

Solution

Slope-Deflection Equations.   Three spans must be considered in this problem, AB, BC, and CD. Since the spans are fixed-supported at A and D, Eq. 10–8 applies for the solution.

From the table on the inside back cover, the FEMs for BC are

$$(FEM)_{BC} = -\frac{5wL^2}{96} = -\frac{5(24)(8)^2}{96} = -80 \text{ kN} \cdot \text{m}$$

$$(FEM)_{CB} = \frac{5wL^2}{96} = \frac{5(24)(8)^2}{96} = 80 \text{ kN} \cdot \text{m}$$

Note that $\theta_A = \theta_D = 0$ and $\psi_{AB} = \psi_{BC} = \psi_{CD} = 0$, since no side-sway will occur.

Applying Eq. 10–8, we have

$$M_{AB} = 2E\left(\frac{I}{12}\right)[2(0) + \theta_B - 3(0)] + 0$$

$$M_{AB} = 0.1667EI\theta_B \tag{1}$$

$$M_{BA} = 2E\left(\frac{I}{12}\right)[2\theta_B + 0 - 3(0)] + 0$$

$$M_{BA} = 0.333EI\theta_B \tag{2}$$

$$M_{BC} = 2E\left(\frac{I}{8}\right)[2\theta_B + \theta_C - 3(0)] - 80$$

$$M_{BC} = 0.5EI\theta_B + 0.25EI\theta_C - 80 \tag{3}$$

$$M_{CB} = 2E\left(\frac{I}{8}\right)[2\theta_C + \theta_B - 3(0)] + 80$$

$$M_{CB} = 0.5EI\theta_C + 0.25EI\theta_B + 80 \tag{4}$$

$$M_{CD} = 2E\left(\frac{I}{12}\right)[2\theta_C + 0 - 3(0)] + 0$$

$$M_{CD} = 0.333EI\theta_C \tag{5}$$

$$M_{DC} = 2E\left(\frac{I}{12}\right)[2(0) + \theta_C - 3(0)] + 0$$

$$M_{DC} = 0.1667EI\theta_C \tag{6}$$

Equilibrium Equations. The preceding six equations contain eight unknowns. The remaining two equilibrium equations come from moment equilibrium at joints B and C, Fig. 10–16b. We have

$$M_{BA} + M_{BC} = 0 \qquad\qquad (7)$$
$$M_{CB} + M_{CD} = 0 \qquad\qquad (8)$$

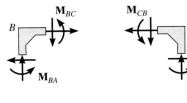

(b)

To solve these eight equations, substitute Eqs. (2) and (3) into Eq. (7) and substitute Eqs. (4) and (5) into Eq. (8). We get

$$0.833EI\theta_B + 0.25EI\theta_C = 80$$
$$0.833EI\theta_C + 0.25EI\theta_B = -80$$

Solving simultaneously yields

$$\theta_B = -\theta_C = \frac{137.1}{EI}$$

which conforms with the way the frame deflects as shown in Fig. 10–16a. Substituting into Eqs. (1)–(6), we get

$M_{AB} = 22.9 \text{ kN} \cdot \text{m}$	**Ans.**
$M_{BA} = 45.7 \text{ kN} \cdot \text{m}$	**Ans.**
$M_{BC} = -45.7 \text{ kN} \cdot \text{m}$	**Ans.**
$M_{CB} = 45.7 \text{ kN} \cdot \text{m}$	**Ans.**
$M_{CD} = -45.7 \text{ kN} \cdot \text{m}$	**Ans.**
$M_{DC} = -22.9 \text{ kN} \cdot \text{m}$	**Ans.**

Using these results, the reactions at the ends of each member can be determined from the equations of equilibrium, and the moment diagram for the frame can be drawn, Fig. 10–16c.

82.3 kN·m

45.7 kN·m

45.7 kN·m   45.7 kN·m

22.9 kN·m   22.9 kN·m

(c)

Fig. 10–16b,c

Example 10–6

Determine the internal moments at each joint of the frame shown in Fig. 10–17a. The moment of inertia for each member is given in the figure. Take $E = 29(10^3)$ ksi.

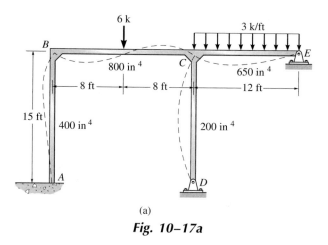

(a)

Fig. 10–17a

Solution

Slope-Deflection Equations.   Four spans must be considered in this problem. Equation 10–8 applies to spans *AB* and *BC*, and Eq. 10–10 applies to *CD* and *CE*, because the ends at *D* and *E* are pinned.

Computing the member stiffnesses, we have

$$k_{AB} = \frac{400}{15(12)^4} = 0.001286 \text{ ft}^3 \qquad k_{CD} = \frac{200}{15(12)^4} = 0.000643 \text{ ft}^3$$

$$k_{BC} = \frac{800}{16(12)^4} = 0.002411 \text{ ft}^3 \qquad k_{CE} = \frac{650}{12(12)^4} = 0.002612 \text{ ft}^3$$

The FEMs due to the loadings are

$$(\text{FEM})_{BC} = -\frac{PL}{8} = -\frac{6(16)}{8} = -12 \text{ k} \cdot \text{ft}$$

$$(\text{FEM})_{CB} = \frac{PL}{8} = \frac{6(16)}{8} = 12 \text{ k} \cdot \text{ft}$$

$$(\text{FEM})_{CE} = -\frac{wL^2}{8} = -\frac{3(12)^2}{8} = -54 \text{ k} \cdot \text{ft}$$

Applying Eqs. 10–8 and 10–10 to the frame and noting that $\theta_A = 0$, $\psi_{AB} = \psi_{BC} = \psi_{CD} = \psi_{CE} = 0$ since no sidesway occurs, we have

$$M_N = 2E\left(\frac{I}{L}\right)(2\theta_N + \theta_F - 3\psi) + (\text{FEM})_N$$

$$M_{AB} = 2[29(10^3)(12)^2](0.001286)[2(0) + \theta_B - 3(0)] + 0$$
$$M_{AB} = 10{,}740.7\theta_B \tag{1}$$

$$M_{BA} = 2[29(10^3)(12)^2](0.001286)[2\theta_B + 0 - 3(0)] + 0$$
$$M_{BA} = 21{,}481.3\theta_B \tag{2}$$

$$M_{BC} = 2[29(10^3)(12)^2](0.002411)[2\theta_B + \theta_C - 3(0)] - 12$$
$$M_{BC} = 40{,}273.3\theta_B + 20{,}136.7\theta_C - 12 \tag{3}$$

$$M_{CB} = 2[29(10^3)(12)^2](0.002411)[2\theta_C + \theta_B - 3(0)] + 12$$
$$M_{CB} = 20{,}136.7\theta_B + 40{,}273.3\theta_C + 12 \tag{4}$$

$$M_N = 3E\left(\frac{I}{L}\right)(\theta_N - \psi) + (\text{FEM})_N$$

$$M_{CD} = 3[29(10^3)(12)^2](0.000643)[\theta_C - 0] + 0$$
$$M_{CD} = 8055.5\theta_C \tag{5}$$

$$M_{CE} = 3[29(10^3)(12)^2](0.002612)[\theta_C - 0] - 54$$
$$M_{CE} = 32{,}723.1\theta_C - 54 \tag{6}$$

(cont'd)

Example 10–6 (continued)

Equations of Equilibrium. These six equations contain eight unknowns. Two moment equilibrium equations can be written for joints B and C, Fig. 10–17b. We have

$$M_{BA} + M_{BC} = 0 \qquad (7)$$
$$M_{CB} + M_{CD} + M_{CE} = 0 \qquad (8)$$

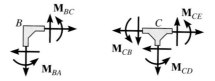

(b)

Fig. 10–17b

In order to solve, substitute Eqs. (2) and (3) into Eq. (7), and Eqs. (4)–(6) into Eq. (8). This gives

$$61{,}754.6\theta_B + 20{,}136.7\theta_C = 12$$
$$20{,}136.7\theta_B + 81{,}051.9\theta_C = 42$$

Solving these equations simultaneously yields

$$\theta_B = 2.758(10^{-5}) \text{ rad} \qquad \theta_C = 5.113(10^{-4}) \text{ rad}$$

These values, being clockwise, tend to distort the frame as shown in Fig. 10–17a. Substituting these values into Eqs. (1)–(6) and solving, we get

$$M_{AB} = 0.296 \text{ k} \cdot \text{ft} \qquad \text{Ans.}$$
$$M_{BA} = 0.592 \text{ k} \cdot \text{ft} \qquad \text{Ans.}$$
$$M_{BC} = -0.592 \text{ k} \cdot \text{ft} \qquad \text{Ans.}$$
$$M_{CB} = 33.2 \text{ k} \cdot \text{ft} \qquad \text{Ans.}$$
$$M_{CD} = 4.12 \text{ k} \cdot \text{ft} \qquad \text{Ans.}$$
$$M_{CE} = -37.3 \text{ k} \cdot \text{ft} \qquad \text{Ans.}$$

10.5 Analysis of Frames: Sidesway

A frame will sidesway, or be displaced to the side, when it or the loading acting on it is nonsymmetric. To illustrate this effect, consider the frame shown in Fig. 10–18. Here the loading **P** causes *unequal* moments M_B and M_C at the joints B and C, respectively. M_B tends to displace joint B to the right, whereas M_C tends to displace joint C to the left. Since M_B is larger than M_C, the net result is a sidesway Δ of both joints B and C to the right, as shown in the figure.* When applying the slope-deflection equation to each column of this frame, we must therefore consider the column rotation ψ (since $\psi = \Delta/L$) as unknown in the equation. As a result an extra equilibrium equation must be included for the solution. In the previous sections it was shown that unknown *angular displacements* θ were related by joint *moment equilibrium equations*. In a similar manner, when unknown joint *linear displacements* Δ (or span rotations ψ) occur, we must write *force equilibrium equations* in order to obtain the complete solution. The unknowns in these equations must, however, involve the internal *moments* acting at the ends of the columns, since the slope-deflection equations involve these moments. The technique for solving problems for frames with sidesway is best explained by examples.

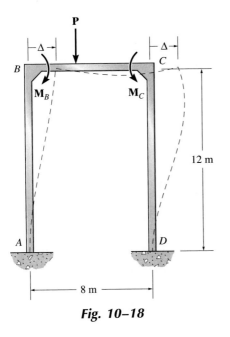

Fig. 10–18

*The deformation of all three members due to axial force is neglected.

545

Example 10–7

Determine the moments at each joint of the frame shown in Fig. 10–19a. EI is constant.

Solution

Slope-Deflection Equations. Since the ends A and D are fixed, Eq. 10–8 applies for all three spans of the frame. Sidesway occurs here since both the applied loading and the geometry of the frame are nonsymmetric. Here the load is applied directly to joint B and therefore no FEMs act at the joints. As shown in Fig. 10–19a, both joints B and C are assumed to be displaced an *equal amount* Δ. Consequently, $\psi_{AB} = \Delta/12$ and $\psi_{DC} = \Delta/18$. Both terms are positive since the cords of members AB and CD "rotate" clockwise. Relating ψ_{AB} to ψ_{DC}, we have $\psi_{AB} = (18/12)\psi_{DC}$. Applying Eq. 10–2 to the frame, we have

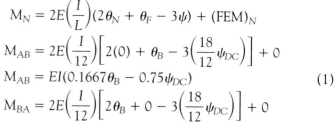

$$M_N = 2E\left(\frac{I}{L}\right)(2\theta_N + \theta_F - 3\psi) + (\text{FEM})_N$$

$$M_{AB} = 2E\left(\frac{I}{12}\right)\left[2(0) + \theta_B - 3\left(\frac{18}{12}\psi_{DC}\right)\right] + 0$$

$$M_{AB} = EI(0.1667\theta_B - 0.75\psi_{DC}) \tag{1}$$

$$M_{BA} = 2E\left(\frac{I}{12}\right)\left[2\theta_B + 0 - 3\left(\frac{18}{12}\psi_{DC}\right)\right] + 0$$

$$M_{BA} = EI(0.333\theta_B - 0.75\psi_{DC}) \tag{2}$$

$$M_{BC} = 2E\left(\frac{I}{15}\right)[2\theta_B + \theta_C - 3(0)] + 0$$

$$M_{BC} = EI(0.267\theta_B + 0.133\theta_C) \tag{3}$$

$$M_{CB} = 2E\left(\frac{I}{15}\right)[2\theta_C + \theta_B - 3(0)] + 0$$

$$M_{CB} = EI(0.267\theta_C + 0.133\theta_B) \tag{4}$$

$$M_{CD} = 2E\left(\frac{I}{18}\right)[2\theta_C + 0 - 3\psi_{DC}] + 0$$

$$M_{CD} = EI(0.222\theta_C - 0.333\psi_{DC}) \tag{5}$$

$$M_{DC} = 2E\left(\frac{I}{18}\right)[2(0) + \theta_C - 3\psi_{DC}] + 0$$

$$M_{DC} = EI(0.111\theta_C - 0.333\psi_{DC}) \tag{6}$$

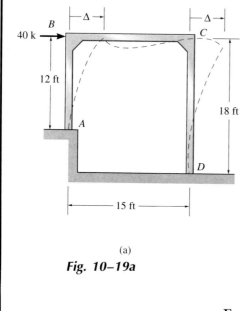

(a)

Fig. 10–19a

Equations of Equilibrium. The above six equations contain nine unknowns. Two moment equilibrium equations for joints B and C,

Fig. 10–19b, can be written, namely,

$$M_{BA} + M_{BC} = 0 \qquad (7)$$
$$M_{CB} + M_{CD} = 0 \qquad (8)$$

Since a horizontal displacement Δ occurs, we will consider summing forces on the *entire frame* in the x direction. This yields

$$\xrightarrow{+}\Sigma F_x = 0; \qquad 40 - V_A - V_D = 0$$

The horizontal reactions or column shears V_A and V_D can be related to the internal moments by considering the free-body diagram of each column separately, Fig. 10–19c. We have

$$\zeta + \Sigma M_B = 0; \qquad V_A = -\frac{M_{AB} + M_{BA}}{12}$$

$$\zeta + \Sigma M_C = 0; \qquad V_D = -\frac{M_{DC} + M_{CD}}{18}$$

Thus,

$$40 + \frac{M_{AB} + M_{BA}}{12} + \frac{M_{DC} + M_{CD}}{18} = 0 \qquad (9)$$

In order to solve, substitute Eqs. (2) and (3) into Eq. (7), Eqs. (4) and (5) into Eq. (8), and Eqs. (1), (2), (5), (6) into Eq. (9). This yields

$$0.6\theta_B + 0.133\theta_C - 0.75\psi_{DC} = 0$$
$$0.133\theta_B + 0.489\theta_C - 0.333\psi_{DC} = 0$$
$$0.5\theta_B + 0.222\theta_C - 1.944\psi_{DC} = -\frac{480}{EI}$$

Solving simultaneously, we have

$$EI\theta_B = 439.05 \qquad EI\theta_C = 136.22 \qquad EI\psi_{DC} = 375.39$$

Now, solving Eqs. (1)–(6) yields

$$
\begin{aligned}
M_{AB} &= -208.36 \text{ k} \cdot \text{ft} & &\textbf{Ans.}\\
M_{BA} &= -135.34 \text{ k} \cdot \text{ft} & &\textbf{Ans.}\\
M_{BC} &= 135.34 \text{ k} \cdot \text{ft} & &\textbf{Ans.}\\
M_{CB} &= 94.8 \text{ k} \cdot \text{ft} & &\textbf{Ans.}\\
M_{CD} &= -94.8 \text{ k} \cdot \text{ft} & &\textbf{Ans.}\\
M_{DC} &= -109.89 \text{ k} \cdot \text{ft} & &\textbf{Ans.}
\end{aligned}
$$

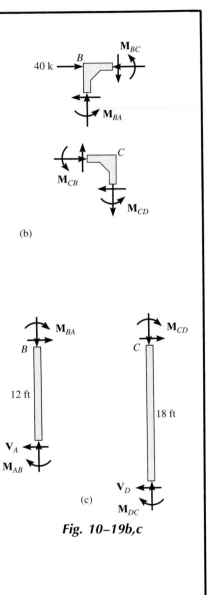

(b)

(c)

Fig. 10–19b,c

Example 10–8

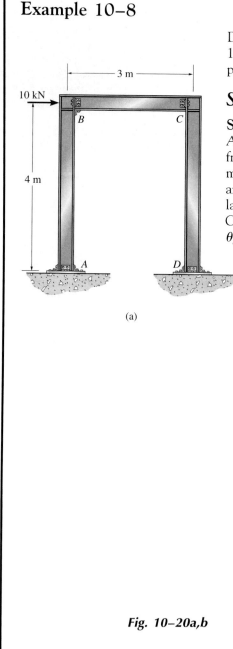

Determine the moments at each joint of the frame shown in Fig. 10–20a. The joints at A and D are fixed and joint C is assumed pin-connected. EI is constant for each member.

Solution

Slope-Deflection Equations. We will apply Eq. 10–8 to member AB since it is fixed at both ends. Equation 10–10 can be applied from B to C and from D to C since the pin at C supports zero moment. As shown by the deflection diagram, Fig. 10–20b, there is an unknown linear displacement Δ of the frame and unknown angular displacement θ_B at joint B.* Due to Δ, the cord members AB and CD rotate clockwise, $\psi = \psi_{AB} = \psi_{DC} = \Delta/4$. Realizing that $\theta_A = \theta_D = 0$ and there are no FEMs for the members, we have,

$$M_{AB} = 2E\left(\frac{I}{4}\right)[2(0) + \theta_B - 3\psi] + 0 \tag{1}$$

$$M_{BA} = 2E\left(\frac{I}{4}\right)(2\theta_B + 0 - 3\psi) + 0 \tag{2}$$

$$M_{BC} = 3E\left(\frac{I}{3}\right)(\theta_B - 0) + 0 \tag{3}$$

$$M_{DC} = 3E\left(\frac{I}{4}\right)(0 - \psi) + 0 \tag{4}$$

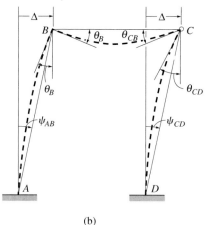

(a)

Fig. 10–20a,b

(b)

*The angular displacements θ_{CB} and θ_{CD} at joint C (pin) are not included in the analysis since Eq. 10–10 is to be used.

Equilibrium Equations. Moment equilibrium of joint B, Fig. 10–20c, requires

$$M_{BA} + M_{BC} = 0 \qquad (5)$$

If forces are summed for the *entire frame* in the horizontal direction, we have

$$\xrightarrow{+} \Sigma F_x = 0; \qquad 10 - V_A - V_D = 0 \qquad (6)$$

As shown on the free-body diagram of each column, Fig. 10–20d, we have

$$\zeta + \Sigma M_B = 0; \qquad V_A = -\frac{M_{AB} + M_{BA}}{4}$$

$$\zeta + \Sigma M_C = 0; \qquad V_D = -\frac{M_{DC}}{4}$$

Thus, from Eq. 6,

$$10 + \frac{M_{AB} + M_{BA}}{4} + \frac{M_{DC}}{4} = 0 \qquad (7)$$

Substituting the slope-deflection equations into Eqs. (5) and (7) and simplifying yields

$$\theta_B = \frac{3}{4}\psi$$

$$10 + \frac{EI}{4}\left(\frac{3}{2}\theta_B - \frac{15}{4}\psi\right) = 0$$

Thus,

$$\theta_B = \frac{240}{21EI}$$

$$\psi = \frac{320}{21EI}$$

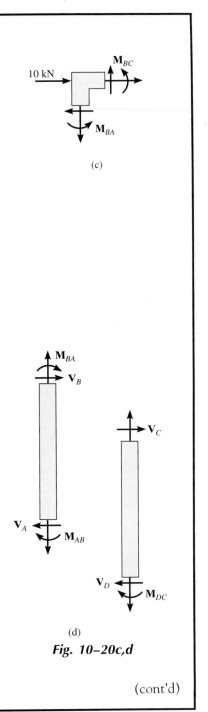

(c)

(d)

Fig. 10–20c,d

(cont'd)

Example 10–8 (continued)

Substituting these values into Eqs. (1)–(4), we have

$$M_{AB} = -17.1 \text{ kN} \cdot \text{m}$$
$$M_{BA} = -11.4 \text{ kN} \cdot \text{m}$$
$$M_{BC} = 11.4 \text{ kN} \cdot \text{m}$$
$$M_{DC} = -11.4 \text{ kN} \cdot \text{m}$$

Using these results, the end reactions on each member can be determined from the equations of equilibrium, Fig. 10–20e. The moment diagram for the frame is shown in Fig. 10–20f.

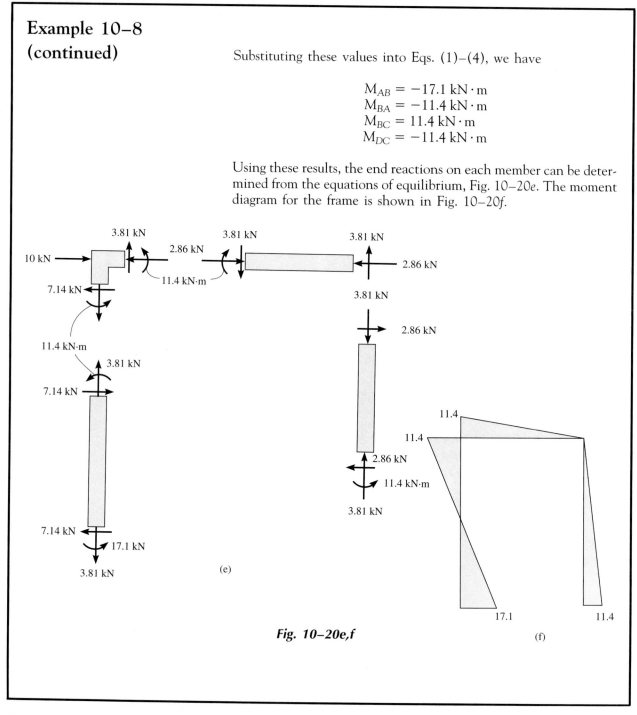

Fig. 10–20e,f

Example 10–9

Explain how the moments in each joint of the two-story frame shown in Fig. 10–21a are determined. EI is constant.

Solution

Slope-Deflection Equation. Since the supports at A and F are fixed, Eq. 10–8 applies for all six spans of the frame. No FEMs have to be calculated, since the applied loading acts at the joints. Here the loading displaces joints B and E an amount Δ_1, and C and D an amount $\Delta_1 + \Delta_2$. As a result, members AB and FE undergo rotations of $\psi_1 = \Delta_1/5$ and BC and ED undergo rotations of $\psi_2 = \Delta_2/5$.

Applying Eq. 10–8 to the frame yields

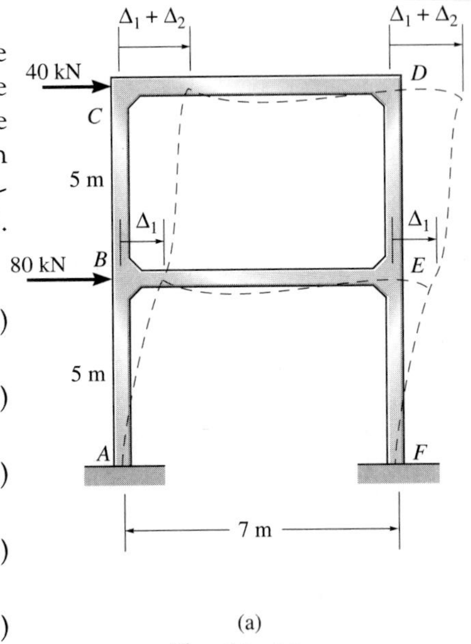

$$M_{AB} = 2E\left(\frac{I}{5}\right)[2(0) + \theta_B - 3\psi_1] + 0 \tag{1}$$

$$M_{BA} = 2E\left(\frac{I}{5}\right)[2\theta_B + 0 - 3\psi_1] + 0 \tag{2}$$

$$M_{BC} = 2E\left(\frac{I}{5}\right)[2\theta_B + \theta_C - 3\psi_2] + 0 \tag{3}$$

$$M_{CB} = 2E\left(\frac{I}{5}\right)[2\theta_C + \theta_B - 3\psi_2] + 0 \tag{4}$$

$$M_{CD} = 2E\left(\frac{I}{7}\right)[2\theta_C + \theta_D - 3(0)] + 0 \tag{5}$$

$$M_{DC} = 2E\left(\frac{I}{7}\right)[2\theta_D + \theta_C - 3(0)] + 0 \tag{6}$$

$$M_{BE} = 2E\left(\frac{I}{7}\right)[2\theta_B + \theta_E - 3(0)] + 0 \tag{7}$$

$$M_{EB} = 2E\left(\frac{I}{7}\right)[2\theta_E + \theta_B - 3(0)] + 0 \tag{8}$$

$$M_{ED} = 2E\left(\frac{I}{5}\right)[2\theta_E + \theta_D - 3\psi_2] + 0 \tag{9}$$

$$M_{DE} = 2E\left(\frac{I}{5}\right)[2\theta_D + \theta_E - 3\psi_2] + 0 \tag{10}$$

$$M_{FE} = 2E\left(\frac{I}{5}\right)[2(0) + \theta_E - 3\psi_1] + 0 \tag{11}$$

$$M_{EF} = 2E\left(\frac{I}{5}\right)[2\theta_E + 0 - 3\psi_1] + 0 \tag{12}$$

These 12 equations contain 18 unknowns.

(a)
Fig. 10–21a

(cont'd)

Example 10–9 (continued)

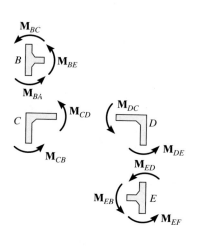

(b)

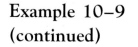

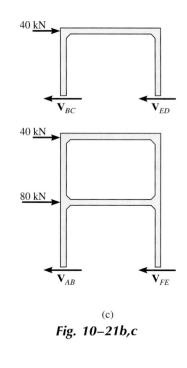

(c)

Fig. 10–21b,c

Equilibrium Equations. Moment equilibrium of joints B, C, D, and E, Fig. 10–21b, requires

$$M_{BA} + M_{BE} + M_{BC} = 0 \qquad (13)$$
$$M_{CB} + M_{CD} = 0 \qquad (14)$$
$$M_{DC} + M_{DE} = 0 \qquad (15)$$
$$M_{EF} + M_{EB} + M_{ED} = 0 \qquad (16)$$

As in the preceding examples, the shear at the base of all the columns for any story must balance the applied horizontal loads, Fig. 10–21c. This yields

$$\xrightarrow{+} \Sigma F_x = 0; \qquad 40 - V_{BC} - V_{ED} = 0$$
$$40 + \frac{M_{BC} + M_{CB}}{5} + \frac{M_{ED} + M_{DE}}{5} = 0 \qquad (17)$$
$$\xrightarrow{+} \Sigma F_x = 0; \qquad 40 + 80 - V_{AB} - V_{FE} = 0$$
$$120 + \frac{M_{AB} + M_{BA}}{5} + \frac{M_{EF} + M_{FE}}{5} = 0 \qquad (18)$$

Solution requires substituting Eqs. (1)–(12) into Eqs. (13)–(18), which yields six equations having six unknowns, ψ_1, ψ_2, θ_B, θ_C, θ_D, and θ_E. These equations can then be solved simultaneously. The results are resubstituted into Eqs. (1)–(12), which yield the moments at the joints.

Example 10–10

Determine the moments at each joint of the frame shown in Fig. 10–22a. *EI* is constant for each member.

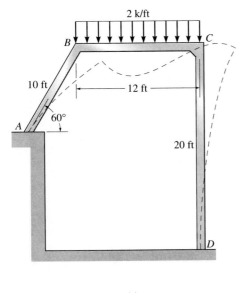

2 k/ft

(a)

Fig. 10–22a

Solution

Slope-Deflection Equations. Equation 10–8 applies to each of the three spans. The FEMs are

$$(\text{FEM})_{BC} = -\frac{wL^2}{12} = -\frac{2(12)^2}{12} = -24 \text{ k} \cdot \text{ft}$$

$$(\text{FEM})_{CB} = \frac{wL^2}{12} = \frac{2(12)^2}{12} = 24 \text{ k} \cdot \text{ft}$$

The sloping member *AB* causes the frame to sidesway to the right as shown in Fig. 10–22a. As a result, joints *B* and *C* are subjected to both rotation *and* displacement. The displacements are shown in

(cont'd)

Example 10–10
(continued)

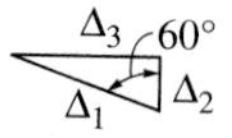

(c)

Fig. 10–22b, where B moves Δ_1 to B' and C moves Δ_3 to C'. These displacements cause the members to rotate as if they were "rigid bodies." The angles of rotation are ψ_1, ψ_3 (clockwise) and $-\psi_2$ (counterclockwise) as shown.* Hence,

$$\psi_1 = \frac{\Delta_1}{10} \qquad \psi_2 = -\frac{\Delta_2}{12} \qquad \psi_3 = \frac{\Delta_3}{20}$$

As shown in Fig. 10–22c, the three displacements can be related. For example, $\Delta_2 = 0.5\Delta_1$ and $\Delta_3 = 0.866\Delta_1$. Thus, from the above equations we have

$$\psi_2 = -0.417\psi_1 \qquad \psi_3 = 0.433\psi_1$$

Using these results, the slope-deflection equations for the frame are

$$M_{AB} = 2E\left(\frac{I}{10}\right)[2(0) + \theta_B - 3\psi_1] + 0 \tag{1}$$

$$M_{BA} = 2E\left(\frac{I}{10}\right)[2\theta_B + 0 - 3\psi_1] + 0 \tag{2}$$

$$M_{BC} = 2E\left(\frac{I}{12}\right)[2\theta_B + \theta_C - 3(-0.417\psi_1)] - 24 \tag{3}$$

$$M_{CB} = 2E\left(\frac{I}{12}\right)[2\theta_C + \theta_B - 3(-0.417\psi_1)] + 24 \tag{4}$$

$$M_{CD} = 2E\left(\frac{I}{20}\right)[2\theta_C + 0 - 3(0.433\psi_1)] + 0 \tag{5}$$

$$M_{DC} = 2E\left(\frac{I}{20}\right)[2(0) + \theta_C - 3(0.433\psi_1)] + 0 \tag{6}$$

These six equations contain nine unknowns.

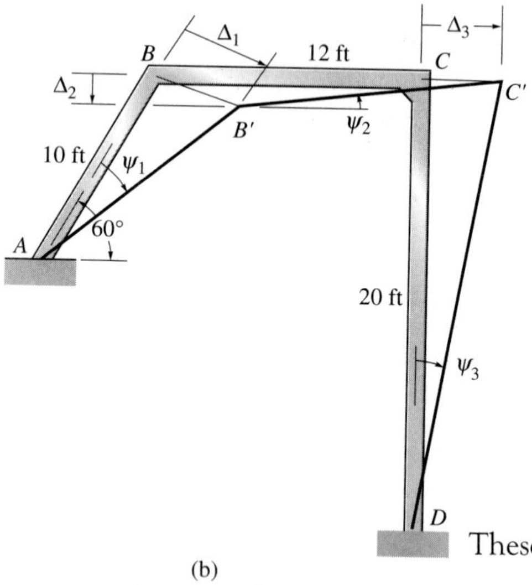

(b)

Fig. 10–22b,c

*Recall that distortions due to axial forces are neglected and the arc displacements BB' and CC' can be considered as straight lines, since ψ_1 and ψ_3 are actually very small.

Equations of Equilibrium. Moment equilibrium at joints B and C yields

$$M_{BA} + M_{BC} = 0 \qquad (7)$$
$$M_{CD} + M_{CB} = 0 \qquad (8)$$

The necessary third equilibrium equation can be obtained by summing moments about point O on the entire frame, Fig. 10–22d. This eliminates the normal forces $\mathbf{N}_A$ and $\mathbf{N}_D$, and therefore

$$\zeta + \Sigma M_O = 0;$$
$$M_{AB} + M_{DC} - \left(\frac{M_{AB} + M_{BA}}{10}\right)(34)$$
$$- \left(\frac{M_{DC} + M_{CD}}{20}\right)(40.78) - 24(6) = 0$$
$$-2.4M_{AB} - 3.4M_{BA} - 2.04M_{CD} - 1.04M_{DC} - 144 = 0 \quad (9)$$

Substituting Eqs. (2) and (3) into Eq. (7), Eqs. (4) and (5) into Eq. (8), and Eqs. (1), (2), (5), and (6) into Eq. (9) yields

$$0.733\theta_B + 0.167\theta_C - 0.392\psi_1 = \frac{24}{EI}$$
$$0.167\theta_B + 0.533\theta_C + 0.0786\psi_1 = -\frac{24}{EI}$$
$$-1.840\theta_B - 0.512\theta_C + 3.880\psi_1 = \frac{144}{EI}$$

Solving these equations simultaneously yields

$$EI\theta_B = 87.86 \qquad EI\theta_C = -82.57 \qquad EI\psi_1 = 67.88$$

Substituting these values into Eqs. (1)–(6), we have

$$M_{AB} = -23.2 \text{ k} \cdot \text{ft} \qquad M_{BC} = 5.63 \text{ k} \cdot \text{ft} \qquad M_{CD} = -25.3 \text{ k} \cdot \text{ft}$$
$$\text{Ans.}$$
$$M_{BA} = -5.63 \text{ k} \cdot \text{ft} \qquad M_{CB} = 25.3 \text{ k} \cdot \text{ft} \qquad M_{DC} = -17.1 \text{ k} \cdot \text{ft}$$
$$\text{Ans.}$$

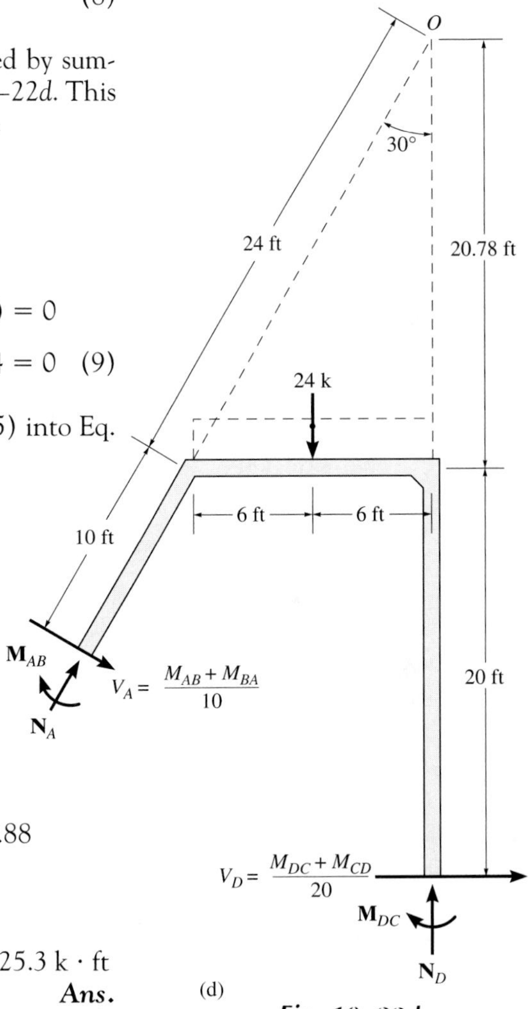

(d)

Fig. 10–22d

555

REFERENCES

Evans, L. T., "The Modified Slope-Deflection Equations," *Journal of the American Concrete Institute*, Oct. 1931.

Maney, G. A., "Studies in Engineering," University of Minnesota, Minneapolis, MN, Bulletin No. 1, 1915.

PROBLEMS

10–1. Determine the moments at A, B, and C. Assume the support at B is a roller and A and C are fixed. EI is constant.

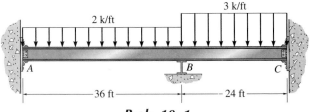

Prob. 10–1

10–2. Determine the moments at A, B, and C, then draw the moment diagram. EI is constant. Assume the support at B is a roller and A and C are fixed.

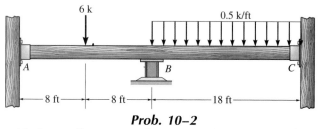

Prob. 10–2

10–3. Determine the moments at A, B, and C, then draw the moment diagram for the beam. The moment of inertia of each span is indicated in the figure. Assume the support at B is a roller and A and C are fixed. $E = 29(10^3)$ ksi.

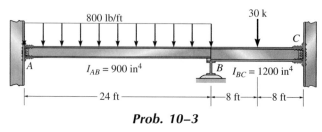

Prob. 10–3

*10–4. Determine the reactions at the supports, then draw the moment diagram. Assume A and D are pins and B and C are rollers. The support at B settles 0.03 ft. Take $E = 29(10^3)$ ksi and $I = 4500$ in⁴.

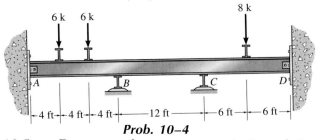

Prob. 10–4

10.5 Determine the moments at A, B, and C. The support at B settles 0.15 ft. $E = 29(10^3)$ ksi and $I = 8000$ in⁴. Assume the supports at B and C are rollers and A is fixed.

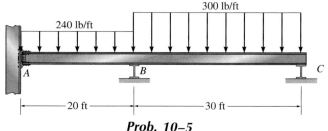

Prob. 10–5

10–6. Determine the internal moment in the beam at B, then draw the moment diagram. Assume A and B are rockers and C is a pin.

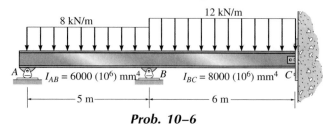

Prob. 10–6

10–7. Determine the reactions at A and D. Take EI to be the same for each member. Assume the supports at A and D are fixed.

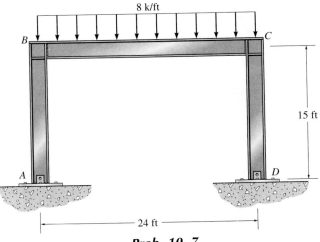

8 k/ft

B

C

15 ft

A

D

24 ft

Prob. 10–7

*10–8. Determine the moments at the ends of each member of the frame. Take $E = 29(10^3)$ ksi. The moment of inertia of each member is listed in the figure. Assume the joint at B is fixed, C is pinned, and A is fixed.

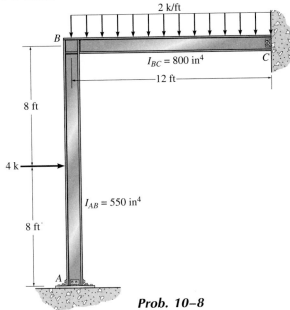

2 k/ft

B

$I_{BC} = 800 \text{ in}^4$

C

12 ft

8 ft

4 k

$I_{AB} = 550 \text{ in}^4$

8 ft

A

Prob. 10–8

10–9. Determine the moments acting at A and B. Assume A is fixed supported, B is a roller, and C is a pin. EI is constant.

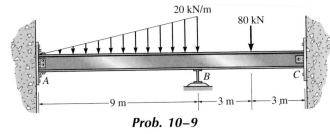

20 kN/m

80 kN

A

B

C

9 m

3 m

3 m

Prob. 10–9

10–10. Determine the moments at the supports, then draw the moment diagram. Assume A and D are fixed. EI is constant.

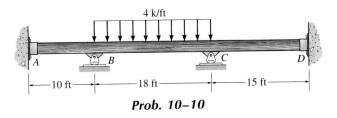

4 k/ft

A

B

C

D

10 ft

18 ft

15 ft

Prob. 10–10

10–11. Determine the moments at B and C of the overhanging beam, then draw the bending moment diagram. EI is constant. Assume the beam is supported by a pin at A and rollers at B and C.

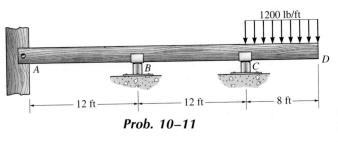

1200 lb/ft

A

B

C

D

12 ft

12 ft

8 ft

Prob. 10–11

*10–12.   Determine the moments at A, B, and C and then draw the moment diagram. EI is constant. Assume the support at B is a roller and A and C are fixed.

10–14.   Determine the moments acting at the ends of each member. Take $E = 29(10^3)$ ksi, $I_{ABC} = 700$ in^4, and $I_{BD} = 1100$ in^4. Assume A and D are pin-supported and C is fixed.

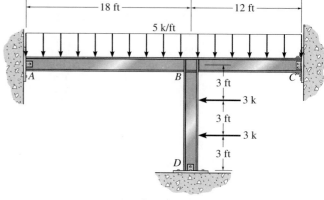

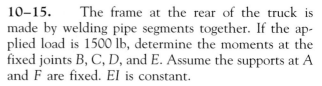

Prob. 10–14

10–13.   Determine the internal moments acting at each joint. Assume A, D, and E are pinned and B and C are fixed joints. Take $E = 29(10^3)$ ksi. The moment of inertia of each member is listed in the figure.

10–15.   The frame at the rear of the truck is made by welding pipe segments together. If the applied load is 1500 lb, determine the moments at the fixed joints B, C, D, and E. Assume the supports at A and F are fixed. EI is constant.

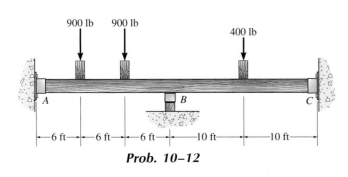

Prob. 10–12

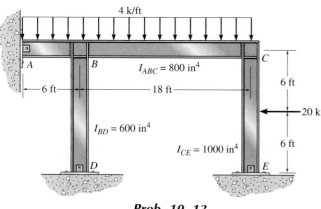

Prob. 10–13

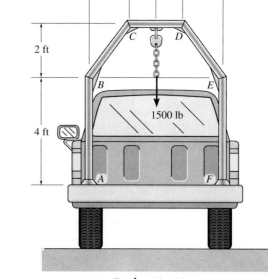

Prob. 10–15

*10–16. The wood frame is subjected to the load of 6 kN. Determine the moments at the fixed joints A, B, and D. The joint at C is pinned. EI is constant.

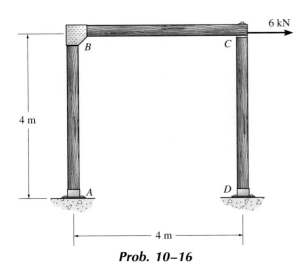

Prob. 10–16

10–18. When the 15 kN/m load is applied to the three-member frame the support at D settles 10 mm. Determine the moment acting at each of the fixed supports A, C, and D. The members are pin-connected at B. $E = 200$ GPa, and $I = 800(10^6)$ mm^4.

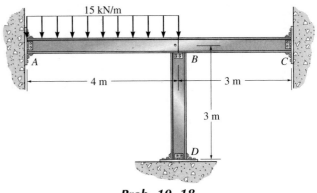

Prob. 10–18

10–17. Determine the moments at each joint of the frame, then draw the moment diagram for member BCE. Assume B, C, and E are fixed-connected and A and D are pins.

10–19. Determine the moments acting at the ends of each member. EI is the same for all members. Assume all joints are fixed.

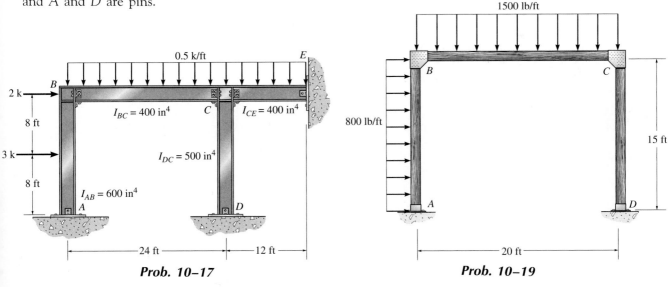

Prob. 10–17

Prob. 10–19

559

*10–20. Determine the moments at the ends of each member. Assume A and E are pins and B and D are fixed-connected joints. EI is the same for all members.

10–22. Determine the moments acting at the ends of each member of the frame. EI is the same for all members.

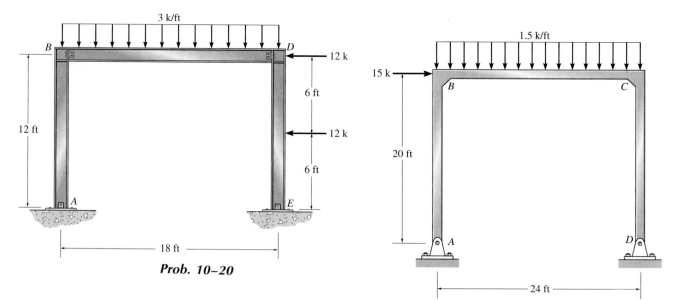

Prob. 10–20

10–21. Determine the moments acting at the ends of each member. Assume the supports at A and D are fixed. The moment of inertia of each member is indicated in the figure. $E = 29(10^3)$ ksi.

Prob. 10–22

10–23. Determine the moments acting at the supports A and D of the battered-column frame. Take $E = 29(10^3)$ ksi, $I = 600$ in^4.

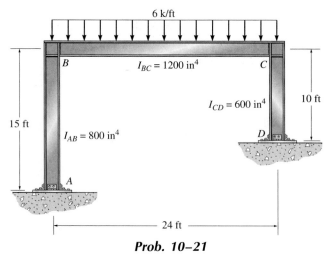

Prob. 10–21

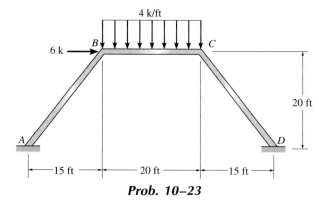

Prob. 10–23

*10–24. Wind loads are transmitted to the frame at joint *E*. If *A*, *B*, *E*, *D*, and *F* are all pin-connected and *C* is fixed-connected, determine the moments at joint *C* and draw the bending-moment diagrams for the girders *BCE*. *EI* is constant.

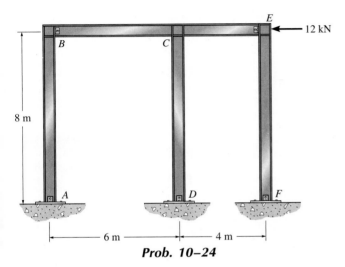

Prob. 10–24

The statically indeterminate loading in bridge girders that are continuous over their piers can be determined using the method of moment distribution.

11

Displacement Method of Analysis: Moment Distribution

The moment-distribution method is a displacement method of analysis that is easy to apply once certain elastic constants have been determined. In this chapter we will first state the important definitions and concepts for moment distribution and then apply the method to solve problems involving statically indeterminate beams and frames. Application to multistory frames is discussed in the last part of the chapter.

11.1 General Principles and Definitions

The method of analyzing beams and frames using moment distribution was developed by Hardy Cross, a professor of civil engineering at the University of Illinois. At the time this method was first published in 1932, it attracted immediate attention, and it has been recognized as one of the most notable advances in structural analysis during the twentieth century.

As will be explained in detail later, moment distribution is a method of successive approximations that may be carried out to any desired degree of accuracy. Essentially, the method begins by assuming each joint of a structure is fixed. Then, by unlocking and locking each joint in succession, the internal moments at the joints are "distributed" and balanced until the joints have rotated to their final or nearly final positions. It will be found that this process of calculation is both repetitive and easy to apply. Before explaining the techniques of moment distribution, however, certain definitions and concepts must be presented.

Sign Convention

We will establish the same sign convention as that established for the general displacement method and the slope-deflection equations: *Clockwise moments* that act *on the member* are considered *positive*, whereas *counterclockwise moments* are *negative*.

Fixed-End Moments (FEMs)

The moments at the "walls" or fixed joints of a loaded member are called *fixed-end moments*. These moments can be determined from the table given on the inside back cover, depending upon the type of loading on the member. For example, the beam loaded as shown in Fig. 11–1 has fixed-end moments of FEM = $PL/8 = 800(10)/8 = 1000$ N $\cdot$ m. Noting the action of these moments *on the beam* and applying our sign convention, it is seen that $M_{AB} = -1000$ N $\cdot$ m and $M_{BA} = +1000$ N $\cdot$ m.

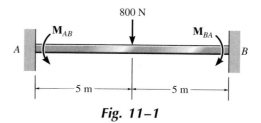

Fig. 11–1

Member Stiffness Factor

Consider the beam in Fig. 11–2, which is pinned at one end and fixed at the other. Application of the moment **M** causes the end A to rotate through an angle θ_A. In Chapter 10 we related M to θ_A using the conjugate-beam method. This resulted in Eq. 10–1, that is, M = $(4EI/L)\,\theta_A$. The term

$$K = \frac{4EI}{L}$$

Far End Fixed

$$(11\text{--}1)$$

is referred to as the *stiffness factor* at A and can be defined as the amount of moment M required to rotate the end A of the beam $\theta_A = 1$ rad.

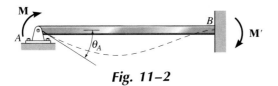

Fig. 11–2

Joint Stiffness Factor

If several members are fixed-connected to a joint and each of their far ends is fixed, then by the principle of superposition, *the total stiffness factor at the joint is the sum of the member stiffness factors at the joint,* that is, $K_T = \Sigma K$. For example, consider the frame joint A in Fig. 11–3. The numerical value of each member stiffness factor is determined from Eq. 11–1 and listed in the figure. Using these values, the total stiffness factor of joint A is $K_T = \Sigma K = 4000 + 5000 + 1000 = 10,000$. This value represents the amount of moment needed to rotate the joint through an angle of 1 rad.

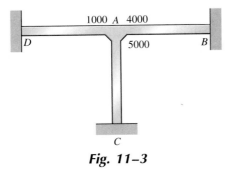

Fig. 11–3

Distribution Factor (DF)

If a moment **M** is applied to a fixed-connected joint, the connecting members will each supply a portion of the resisting moment necessary to satisfy moment equilibrium at the joint. That fraction of the total resisting moment supplied by the member is called the *distribution factor* (DF). To obtain its value, imagine the joint is fixed-connected to n members. If an applied moment **M** causes the joint to rotate an amount θ, then each member i rotates by this same amount. If the stiffness factor of the ith member is K_i, then the moment contributed by the member is $M_i = K_i\theta$. Since equilibrium requires

$$M = M_1 + M_2 + \cdots + M_i + \cdots + M_n$$
$$= K_1\theta + K_2\theta + \cdots + K_i\theta + \cdots + K_n\theta = \theta\Sigma K_i$$

then the distribution factor for the ith member is

$$DF_i = \frac{M_i}{M} = \frac{K_i\theta}{\theta\Sigma K_i}$$

Canceling the common term θ, it is seen that the distribution factor for a member is equal to the stiffness factor of the member divided by the total stiffness factor for the joint; that is, in general,

$$DF = \frac{K}{\Sigma K} \qquad (11-2)$$

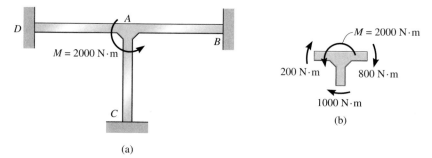

(a)

(b)

Fig. 11–4

For example, the distribution factors for members AB, AC, and AD at joint A in Fig. 11–3 are $DF_{AB} = 0.4$, $DF_{AC} = 0.5$, and $DF_{AD} = 0.1$. As a result, if $M = 2000$ N · m acts at joint A, Fig. 11–4a, the equilibrium moments exerted by the members on the joint are $M_{AB} = 0.4(2000) = 800$ N · m, $M_{AC} = 0.5(2000) = 1000$ N · m, and $M_{AD} = 0.1(2000) = 200$ N · m, as shown in Fig. 11–4b.

Member Relative Stiffness Factor

Quite often a beam or frame will be made from the same material so its modulus of elasticity E will be the *same* for all the members. If this is the case, the common factor $4E$ in Eq. 11–1 will *cancel* from the numerator and denominator of Eq. 11–2 when the distribution factor for a joint is determined. Hence, it is *easier* just to determine the member's *relative-stiffness factor*,

$$\boxed{\begin{array}{c} K_R = \dfrac{I}{L} \\ \text{Far End Fixed} \end{array}} \qquad (11\text{–}3)$$

and use this for the computations of the DF.

Carry-over Factor

Consider again the beam in Fig. 11–2. It was shown in Chapter 10 that $M_{AB} = (4EI/L)\theta_A$ (Eq. 10–1) and $M_{BA} = (2EI/L)\theta_A$ (Eq. 10–2). Solving for θ_A and equating then equations we get $M_{BA} = M_{AB}/2$. In other words, the moment **M** at the pin induces a moment of $\mathbf{M'} = \frac{1}{2}\mathbf{M}$ at the wall. The carry-over factor represents the fraction of **M** that is "carried over" from the pin to the wall. Hence, in the case of a beam with the far end fixed, the carry-over factor is $+\frac{1}{2}$. The plus sign indicates both moments act in the same direction.

The preceding definitions and results will now be used to explain the method of moment distribution.

11.2   Moment Distribution for Beams

Moment distribution is based on the principle of successively locking and unlocking the joints of a structure in order to allow the moments at the joints to be distributed and balanced. The best way to explain the method is by examples.

Consider the beam with a constant modulus of elasticity E and having the dimensions and loading shown in Fig. 11–5a. Before we begin, we must first determine the distribution factors at the two ends of each span. Using Eq. 11–1, $K = 4EI/L$, the *relative-stiffness factors* of these ends are

$$(K_R)_{AB} = \frac{4E(300)}{15} = 20 \text{ in}^4/\text{ft} \qquad (K_R)_{BC} = \frac{4E(600)}{20} = 30 \text{ in}^4/\text{ft}$$

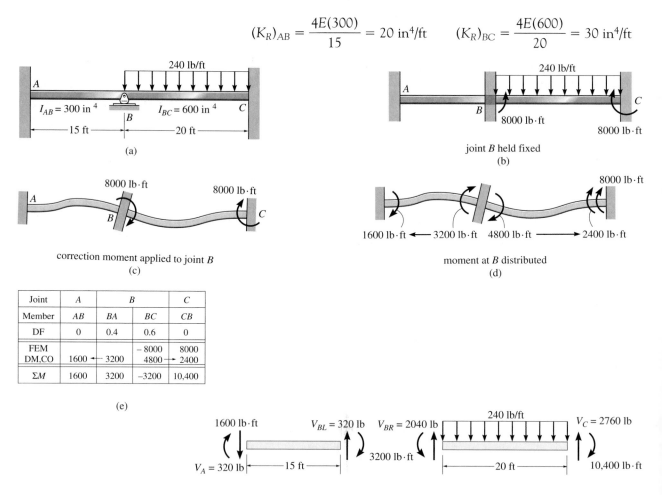

Joint	A	B		C
Member	AB	BA	BC	CB
DF	0	0.4	0.6	0
FEM			– 8000	8000
DM,CO	1600 ← 3200		4800 → 2400	
ΣM	1600	3200	–3200	10,400

(e)

Fig. 11–5   (f)

Thus, using Eq. 11–2, $DF = K/\Sigma K$, for the ends connected to joint B, we have

$$DF_{BA} = \frac{4E(20)}{4E(20) + 4E(30)} = 0.4$$

$$DF_{BC} = \frac{4E(30)}{4E(20) + 4E(30)} = 0.6$$

At the walls, joint A and joint C, the distribution factor depends on the member stiffness factor and the "stiffness factor" of the wall. Since in theory it would take an "infinite" size moment to rotate the wall one radian, the wall stiffness factor is infinite. Thus for joints A and C we have

$$DF_{AB} = \frac{4E(20)}{\infty + 4E(20)} = 0$$

$$DF_{CB} = \frac{4E(30)}{\infty + 4E(30)} = 0$$

Note that the above results could also have been obtained if the stiffness factor, $K_R = I/L$ (Eq. 11–3), had been used for the calculations. Furthermore, as long as a *consistent* set of units is used for the stiffness factor, the DF will always be dimensionless, and at a joint, except where it is located at a fixed wall, the sum of the DFs will always equal 1.

Having computed the DFs, we will now determine the FEMs. Only span BC is loaded, and using the table on the inside back cover for a uniform load, we have

$$(FEM)_{BC} = -\frac{wL^2}{12} = -\frac{240(20)^2}{12} = -8000 \text{ lb} \cdot \text{ft}$$

$$(FEM)_{CB} = \frac{wL^2}{12} = \frac{240(20)^2}{12} = 8000 \text{ lb} \cdot \text{ft}$$

We begin by assuming joint B is fixed or locked. The fixed-end moment at B then holds span BC in this fixed or locked position as shown in Fig. 11–5b. This, of course, does not represent the actual equilibrium situation at B, since the moments on *each side* of this joint must be equal but opposite. To correct this, we will apply an equal, but opposite moment of 8000 lb · ft to the joint and allow the joint to rotate freely, Fig. 11–5c. As a result, portions of this moment are distributed in spans BC and BA in accordance with the DFs (or stiffness) of these spans at the joint. Specifically, the moment in BC is $0.6(8000) = 4800 \text{ lb} \cdot \text{ft}$ and the moment in BA is $0.4(8000) = 3200 \text{ lb} \cdot \text{ft}$. Finally, due to the released rotation that takes place at B, these moments must be "carried over" since moments are developed at the far ends of the span. Using the carry-over factor of $+\frac{1}{2}$, the results are shown in Fig. 11–5d.

This example indicates the basic steps necessary when distributing moments at a joint: Determine the unbalanced moment acting at the initially "locked" joint, unlock the joint and apply an equal but opposite unbalanced moment to correct the equilibrium, distribute the moment among the connecting spans, and carry the moment in each span over to its other end. The steps are usually presented in tabular form as indicated in Fig. 11–5e. Here the notation, Dist, CO indicates a line where moments are distributed, then carried over. In this particular case only *one cycle* of moment distribution is necessary, since the wall supports at A and C "absorb" the moments and no further joints have to be balanced or unlocked to satisfy joint equilibrium. Once distributed in this manner, the moments at each joint are summed, yielding the final results shown on the bottom line of the table in Fig. 11–5e. Notice that joint B is now in equilibrium. Since M_{BC} is negative, this moment is applied to span BC in a counterclockwise sense as shown on free-body diagrams of the beam spans in Fig. 11–5f. With the end moments known, the end shears have been computed from the equations of equilibrium applied to each of these spans.

Consider now the same beam, except the support at C is a roller, Fig. 11–6a. In this case only *one member* is at joint C, so the distribution factor for member CB at joint C is

$$\text{DF}_{CB} = \frac{30}{30} = 1$$

The other distribution factors and the FEMs are the same as computed previously. They are listed on lines 1 and 2 of the table in Fig. 11–6b. Initially, we will assume joints B and C are locked. We begin by unlocking joint C and placing an equilibrating moment of $-8000 \text{ lb} \cdot \text{ft}$ at the joint. The entire moment is distributed in member CB since $(1)(-8000) \text{ lb} \cdot \text{ft} = -8000 \text{ lb} \cdot \text{ft}$. The arrow on line 3 indicates that $\frac{1}{2}(-8000) \text{ lb} \cdot \text{ft} = -4000 \text{ lb} \cdot \text{ft}$ is carried over to joint B since joint C has been allowed to rotate freely. Joint C is now *relocked*. Since the total moment at C is *balanced*, a line is placed under the -8000-lb $\cdot$ ft moment. We will now consider the unbalanced $-12,000$-lb $\cdot$ ft moment at joint B. Here for equilibrium, a $+12,000$-lb $\cdot$ ft moment is applied to B and this joint is unlocked such that portions of the

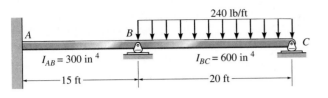

(a)

(b)

Joint	A	B		C	
Member	AB	BA	BC	CB	
DF	0	0.4	0.6	1	1
FEM			− 8000	8000	2
		− 4000	− 8000		3
	2400	4800	7200	3600	4
			− 1800	− 3600	5
	360	720	1080	540	6
			− 270	− 540	7
	54	108	162	81	8
			− 40.5	− 81	9
	8.1	16.2	24.3	12.1	10
			− 6.0	− 12.1	11
	1.2	2.4	3.6	1.8	12
			− 0.9	− 1.8	13
		0.4	0.5		14
ΣM	2823.3	5647.0	−5647.0	0	15

Joint	A	B		C	
Member	AB	BA	BC	CB	
DF	0	0.4	0.6	1	1
FEM			− 8000	8000	2
Dist.		3200	4800	− 8000	3
CO	1600		− 4000	2400	4
Dist.		1600	2400	− 2400	5
CO	800		− 1200	1200	6
Dist.		480	720	− 1200	7
CO	240		− 600	360	8
Dist.		240	360	− 360	9
CO	120		− 180	180	10
Dist.		72	108	− 180	11
CO	36		− 90	54	12
Dist.		36	54	− 54	13
CO	18		− 27	27	14
Dist.		10.9	16.2	− 27	15
CO	5.4		− 13.5	8.1	16
Dist.		5.4	8.1	− 8.1	17
CO	2.7		− 4.05	4.05	18
Dist.		1.62	2.43	− 4.05	19
CO	0.81		− 2.02	1.22	20
Dist.		0.80	1.22	− 1.22	21
CO	0.40		− 0.61	0.61	22
Dist.		0.24	0.37	− 0.61	23
ΣM	2823	5647	−5647	0	24

(c)

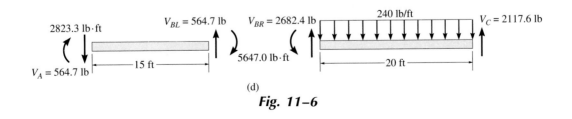

(d)

Fig. 11–6

moment are distributed into BA and BC, that is, $(0.4)(12,000) = 4800$ lb $\cdot$ ft and $(0.6)(12,000) = 7200$ lb $\cdot$ ft as shown on line 4. Also note that $+\frac{1}{2}$ of these moments must be carried over to the fixed wall A and roller C, since joint B has rotated. Joint B is now *relocked*. Again joint C is unlocked and the unbalanced moment at the roller is distributed as was done previously. The results are on line 5. Successively locking and unlocking joints B and C will essentially diminish the size of the moment to be balanced until it becomes negligible compared with the original moments, line 14. Each of the steps on lines 3 through 14 should be thoroughly understood. Summing the moments, the final results are shown on line 15, where it is seen that the final moments now satisfy joint equilibrium.

Rather than applying the moment distribution process successively to each joint, as illustrated here, it is also possible to apply it to all joints at the *same time*. This scheme is shown in the table in Fig. 11–6c. In this case, we start by fixing all the joints and then balancing and distributing the fixed-end moments at both joints B and C, line 3. Releasing joints B and C simultaneously (joint A is always fixed), the moments are then carried over to the end of each span, line 4. Again the joints are relocked, and the moments are balanced and distributed, line 5. Unlocking the joints allows the moments to be carried over, as shown in line 6. Continuing, we obtain the final results, as before, listed on line 24. By comparison, this method gives a slower convergence to the answer than does the previous method; however, in many cases this method will be more efficient to apply, and for this reason we will use it in the examples that follow. Finally, using the results in either Fig. 11–6b or 11–6c, the free-body diagrams of each beam span are drawn as shown in Fig. 11–6d.

Although several steps were involved in obtaining the final results in the example, the work required is rather methodical since it requires application of a series of arithmetical steps, rather than solving a set of equations as in the slope-deflection method. It should be noted, however, that the fundamental process of moment distribution follows the same procedure as the general displacement method. There the process is to establish load-displacement relations at each joint and then satisfy joint equilibrium requirements by determining the correct angular displacement for the joint (compatibility). Here, however, the equilibrium and compatibility of rotation at the joint is satisfied *directly*, using a "moment balance" process that incorporates the load-deflection relations. Further simplification for using moment distribution is possible, and this will be discussed in the next section.

Procedure for Analysis

The following procedure provides a general method for determining the end moments on beam spans using moment distribution.

Distribution Factors and Fixed-End Moments   The joints or nodes on the beam should be identified and the stiffness factors for each span at the joints should be calculated. Using these values the distribution factors can be determined from Eq. 11–2 (DF = $K_i/\Sigma K$). Remember that DF = 0 for a fixed end and DF = 1 for an *end* pin or roller support.

The fixed-end moments for each loaded span are computed using the table given on the inside back cover. Positive FEMs act clockwise on the span and negative FEMs act counterclockwise. For convenience, these values can be recorded in tabular form, similar to that shown in Fig. 11–6b.

Moment Distribution Process   Assume that all joints at which the moments in the connecting spans must be determined are *initially locked*. Then:

1. Determine the moment that is needed to put each joint in equilibrium.
2. Release or "unlock" the joints and distribute the moments into each connecting span.
3. Carry the moments in each span over to its other end by multiplying each moment by the carry-over factor $+\frac{1}{2}$.

By repeating this cycle of locking and unlocking the joints, it will be found that the moment corrections will diminish since the beam tends to achieve its final deflected shape. When a small enough value for the corrections is obtained, the process of cycling should be stopped with no "carry-over" of the last moments. Each column of FEMs, distributed moments, and carry-over moments should then be added. If this is done correctly, moment equilibrium at the joints will be achieved.

Example 11–1

Determine the internal moments at each support of the beam shown in Fig. 11–7a. EI is constant.

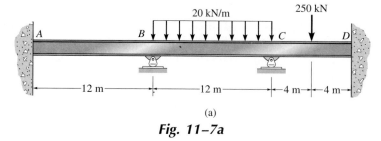

(a)

Fig. 11–7a

Solution

The distribution factors at each joint must be computed first.*

$$(K_R)_{AB} = \frac{4EI}{12} \qquad (K_R)_{BC} = \frac{4EI}{12} \qquad (K_R)_{CD} = \frac{4EI}{8}$$

Therefore,

$$DF_{AB} = DF_{DC} = 0 \qquad DF_{BA} = DF_{BC} = \frac{4EI/12}{4EI/12 + 4EI/12} = 0.5$$

$$DF_{CB} = \frac{4EI/12}{4EI/12 + 4EI/8} = 0.4 \qquad DF_{CD} = \frac{4EI/8}{4EI/12 + 4EI/8} = 0.6$$

The fixed-end moments are

$$(FEM)_{BC} = -\frac{wL^2}{12} = \frac{-20(12)^2}{12} = -240 \text{ kN} \cdot \text{m}$$

$$(FEM)_{CB} = \frac{wL^2}{12} = \frac{20(12)^2}{12} = 240 \text{ kN} \cdot \text{m}$$

$$(FEM)_{CD} = -\frac{PL}{8} = \frac{-250(8)}{8} = -250 \text{ kN} \cdot \text{m}$$

$$(FEM)_{DC} = \frac{PL}{8} = \frac{250(8)}{8} = 250 \text{ kN} \cdot \text{m}$$

*Here we have used the stiffness factor $4EI/L$; however, the relative stiffness factor I/L could also have been used.

Starting with the FEMs, line 4, Fig. 11–7*b*, the moments at joints *B* and *C* are distributed *simultaneously*, line 5. These moments are then carried over *simultaneously* to the respective ends of each span, line 6. The resulting moments are again simultaneously distributed and carried over, lines 7 and 8. The process is continued until the resulting moments are diminished an appropriate amount, line 13. The resulting moments are found by summation, line 14.

Placing the moments on each beam span and applying the equations of equilibrium yields the end shears shown in Fig. 11–7*c* and the bending-moment diagram for the entire beam, Fig. 11–7*d*.

Joint	*A*	*B*		*C*		*D*	1
Member	*AB*	*BA*	*BC*	*CB*	*CD*	*DC*	2
DF	0	0.5	0.5	0.4	0.6	0	3
FEM			−240	240	−250	250	4
Dist.		120	120	4	6		5
CO	60		2	60			6
Dist.		−1	−1	−24	−36	3	7
CO	−0.5		−12	0.5			8
Dist.		6	6	0.2	0.3	−18	9
CO	3		0.1	3			10
Dist.		−0.05	−0.05	−1.2	−1.8	0.2	11
CO	−0.02		−0.6	−0.02			12
Dist.		0.3	0.3	0.01	0.01	−0.9	13
ΣM	62.5	125.3	−125.3	281.5	−281.5	234.3	14

(b)

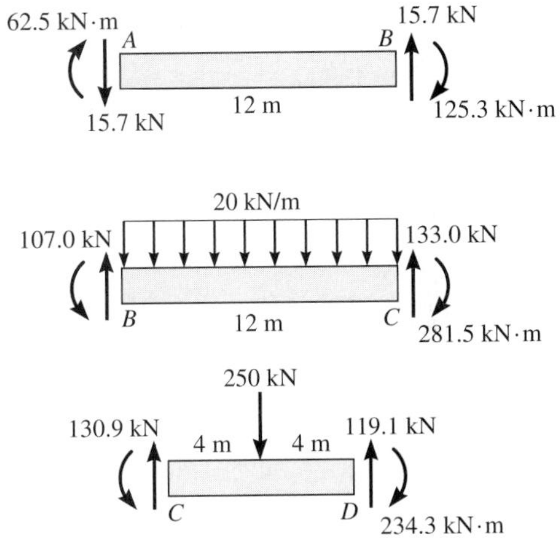

(c)

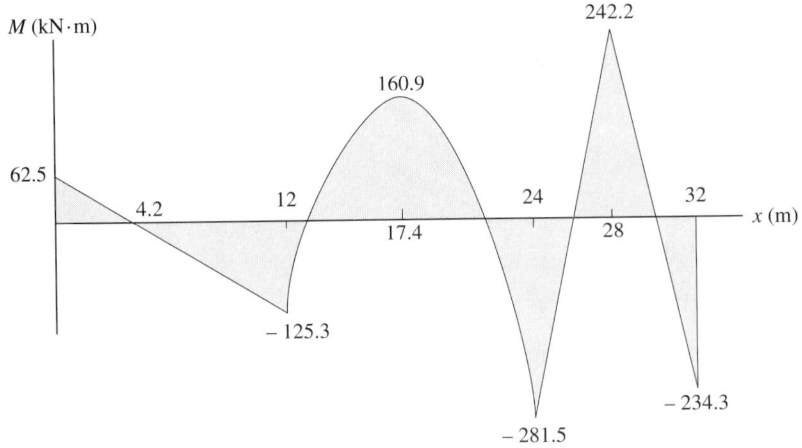

(d)

Fig. 11–7b,c,d

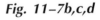

Example 11–2

Determine the internal moment at each support of the beam shown in Fig. 11–8a. The moment of inertia of each span is indicated.

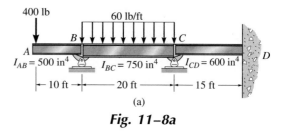

(a)

Fig. 11–8a

Solution

In this problem a moment does not get distributed in the overhanging span AB and so the distribution factor $(DF)_{BA} = 0$. The stiffness factors, distribution factors, and fixed-end moments are computed as follows:

$$(K_R)_{BC} = \frac{4E(750)}{20} = 150E$$

$$(K_R)_{CD} = \frac{4E(600)}{15} = 160E$$

$$(DF)_{BC} = 1 - (DF)_{BA} = 1 - 0 = 1$$

$$(DF)_{CB} = \frac{150E}{150E + 160E} = 0.484$$

$$(DF)_{CD} = \frac{160E}{150E + 160E} = 0.516$$

$$(DF)_{DC} = \frac{160E}{\infty + 160E} = 0$$

Due to the overhang,

$$(FEM)_{BA} = 400 \text{ lb}(10 \text{ ft}) = 4000 \text{ lb} \cdot \text{ft}$$

$$(FEM)_{BC} = -\frac{wL^2}{12} = -\frac{60(20)^2}{12} = -2000 \text{ lb} \cdot \text{ft}$$

$$(FEM)_{CB} = \frac{wL^2}{12} = \frac{60(20)^2}{12} = 2000 \text{ lb} \cdot \text{ft}$$

The overhanging span requires the internal moment to the left of B to be 4000 lb · ft. Balancing at joint B requires an internal moment of -4000 lb · ft to the right of B. As shown in the table -2000 lb · ft is added to BC in order to satisfy this condition. Following the alternative method of Example 11–1, the carry-over and distributing operations proceed in the usual manner as indicated.

Since the internal moments are known, the moment diagram for the beam can be constructed (Fig. 11–8c).

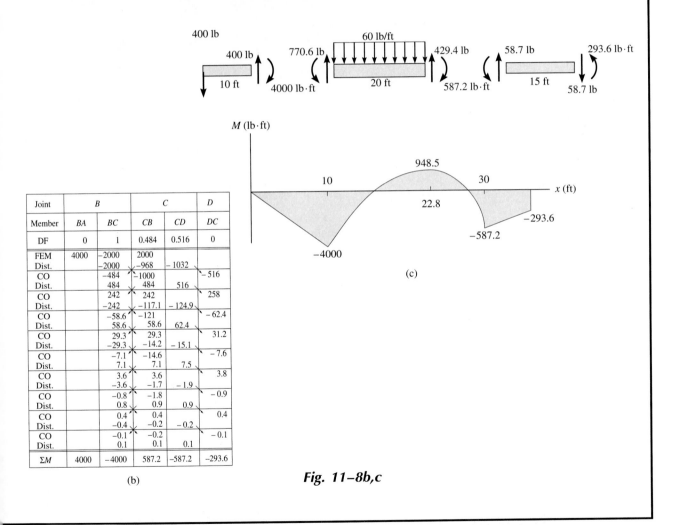

Joint	B		C		D
Member	BA	BC	CB	CD	DC
DF	0	1	0.484	0.516	0
FEM	4000	−2000	2000		
Dist.		−2000	−968	−1032	
CO		−484	−1000		−516
Dist.		484	484	516	
CO		242	242		258
Dist.		−242	−117.1	−124.9	
CO		−58.6	−121		−62.4
Dist.		58.6	58.6	62.4	
CO		29.3	29.3		31.2
Dist.		−29.3	−14.2	−15.1	
CO		−7.1	−14.6		−7.6
Dist.		7.1	7.1	7.5	
CO		3.6	3.6		3.8
Dist.		−3.6	−1.7	−1.9	
CO		−0.8	−1.8		−0.9
Dist.		0.8	0.9	0.9	
CO		0.4	0.4		0.4
Dist.		−0.4	−0.2	−0.2	
CO		−0.1	−0.2		−0.1
Dist.		0.1	0.1	0.1	
ΣM	4000	−4000	587.2	−587.2	−293.6

(b)

Fig. 11–8b,c

A multistory, reinforced-concrete building. Notice how the columns and floor slabs are formed and poured together, ensuring fixed-connected joints. This method is known as flat-plate construction.

11.3 Stiffness-Factor Modifications

In the previous examples of moment distribution we have considered each beam span to be constrained by a fixed support (locked joint) at its far end when distributing and carrying over the moments. For this reason we have computed the stiffness factors, distribution factors, and the carry-over factors based on the case shown in Fig. 11–9. Here, of course, the stiffness factor is $K = 4EI/L$ (Eq. 11–1), and the carry-over factor is $+\frac{1}{2}$.

In some cases it is possible to modify the stiffness factor of a particular beam span and thereby simplify the process of moment distribution. Three cases where this frequently occurs in practice will now be considered.

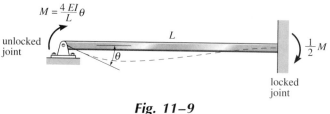

$$M = \frac{4\,EI}{L}\theta$$

unlocked joint

L

θ

$\frac{1}{2}M$

locked joint

Fig. 11–9

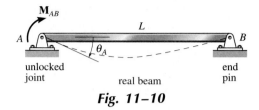

Fig. 11–10

Member Pin-Supported at Far End

Many indeterminate beams or members have their far-end span supported by an end pin (or roller) as in the case of joint B in Fig. 11–10. The stiffness factor of the span for this case was considered in Chapter 9, where it was shown by Eq. 9–4 that

$$M_{AB} = \frac{3EI}{L}\theta_A$$

Thus, the stiffness factor for this beam is

$$K = \frac{3EI}{L}$$

Far End Pinned
or Roller-Supported

$$(11\text{--}4)$$

Also, note that *the carry-over factor is zero*, since the pin at B does not support a moment. By comparison, then, *if the far end was fixed-supported, the stiffness factor $K = 4EI/L$ would have to be modified by $\frac{3}{4}$ to model the case of having the far end pin-supported*. If this modification is considered, the moment distribution process is simplified since the end pin does *not* have to be unlocked–locked successively when distributing the moments. Also, since the end span is pinned, the fixed-end moments for the span are computed using the values in the right column of the table on the inside back cover. The following example illustrates how to apply these simplifications.

579

Example 11–3

Determine the internal moments at the supports of the beam shown in Fig. 11–11a. The moment of inertia of the two spans is shown in the figure.

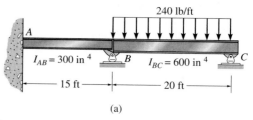

(a)

Fig. 11–11a

Solution I

Since the beam is roller-supported at its far end C, the stiffness of span BC will be computed on the basis of Eq. 11–4, $K = 3EI/L$. We have

$$K_{AB} = \frac{4EI}{L} = \frac{4E(300)}{15} = 80E$$

$$K_{BC} = \frac{3EI}{L} = \frac{3E(600)}{20} = 90E$$

Thus,

$$DF_{AB} = \frac{80E}{\infty + 80E} = 0$$

$$DF_{BA} = \frac{80E}{80E + 90E} = 0.4706$$

$$DF_{BC} = \frac{90E}{80E + 90E} = 0.5294$$

$$DF_{CB} = \frac{90E}{90E} = 1$$

The fixed-end moments for span BC are

$$(FEM)_{BC} = -\frac{wL^2}{12} = \frac{-240(20)^2}{12} = -8000 \text{ lb} \cdot \text{ft}$$

$$(FEM)_{CB} = \frac{wL^2}{12} = \frac{240(20)^2}{12} = 8000 \text{ lb} \cdot \text{ft}$$

The foregoing data are entered into the table in Fig. 11–11b. As noted, the moment at C is distributed first. Once this moment is carried over to joint B it becomes *unnecessary* to bring the distributed moment from joint B back to C, since the stiffness (or distribution factor) for BC has been *modified*. By comparison with Fig. 11–6b, this considerably simplifies the distribution.

Solution II

Further simplification of the distribution method for this problem is possible by realizing that a *single* fixed-end moment for the end span BC can be used. Using the right-hand column of the table on the inside back cover for a uniformly loaded span having one side fixed, the other pinned, we have

$$(\text{FEM})_{BC} = -\frac{wL^2}{8} = \frac{-240(20)^2}{8} = -12,000 \text{ lb} \cdot \text{ft}$$

Considering this value altogether eliminates the distribution of moment at joint C, as indicated by comparison of the tables in Fig. 11–11b and 11–11c.

Using the results, the beam's end shears and moment diagrams are shown in Fig. 11–11d.

Joint	A	B		C
Member	AB	BA	BC	CB
DF	0	0.4706	0.5294	1
FEM			–8000	8000
Dist.		3764.8	4235.2	–8000
CO	1882.4		–4000	
Dist.		1882.4	2117.6	
CO	941.2			
ΣM	2823.6	5647.2	–5647.2	0

(b)

Joint	A	B		C
Member	AB	BA	BC	CB
DF	0	0.4706	0.5294	1
FEM			–12,000	
Dist.		5647.2	6352.8	
CO	2823.6			
ΣM	2823.6	5647.2	–5647.2	0

(c)

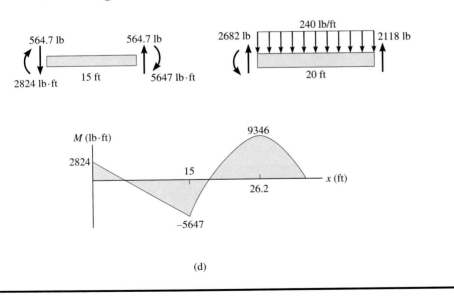

(d)

Fig. 11–11b,c,d

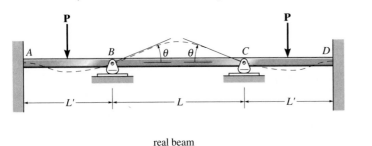

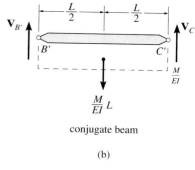

real beam

conjugate beam

(a)

(b)

Fig. 11–12

Symmetric Beam and Loading

If a beam is symmetric with respect to both its loading and geometry, the bending-moment diagram for the beam will also be symmetric. As a result, a modification of the stiffness factor for the center span can be made, so that moments in the beam only have to be distributed through joints lying on either half of the beam. To develop the appropriate stiffness-factor modification, consider the beam shown in Fig. 11–12a. Due to the symmetry, the internal moments at B and C are equal. Assuming this value to be **M**, the conjugate beam for span BC is shown in Fig. 11–12b. The slope θ at each end is therefore

$$\zeta + \Sigma M_C' = 0; \qquad -V_B'(L) + \frac{M}{EI}(L)\left(\frac{L}{2}\right) = 0$$

$$V_B' = \theta = \frac{ML}{2EI}$$

or

$$M = \frac{2EI}{L}\theta$$

The stiffness factor for the center span is therefore

$$K = \frac{2EI}{L} \tag{11–5}$$
Symmetric Beam and Loading

Thus, moments for only half the beam can be distributed provided the stiffness factor for the center span is computed using Eq. 11–5. By comparison, *the center span's stiffness factor will be one half that usually determined using* $K = 4EI/L$.

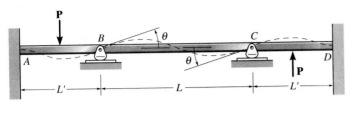

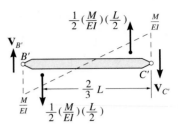

real beam

(a)

conjugate beam

(b)

Fig. 11–13

Symmetric Beam with Antisymmetric Loading

If a symmetric beam is subjected to antisymmetric loading, the result-ing moment diagram will be antisymmetric. As in the previous case, we can modify the stiffness factor of the center span so that only one half of the beam has to be considered for the moment-distribution analysis. Consider the beam in Fig. 11–13a. The conjugate beam for its center span BC is shown in Fig. 11–13b. Due to the antisymmetric loading, the internal moment at B is equal, but opposite to that at C. Assuming this value to be **M**, the slope θ at each end is determined as follows:

$$\zeta + \Sigma M'_C = 0; \quad -V'_B(L) + \frac{1}{2}\left(\frac{M}{EI}\right)\left(\frac{L}{2}\right)\left(\frac{2L}{3}\right) = 0$$

$$V'_B = \theta = \frac{ML}{6EI}$$

or

$$M = \frac{6EI}{L}\theta$$

The stiffness factor for the center span is, therefore,

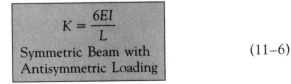

$$K = \frac{6EI}{L}$$

Symmetric Beam with
Antisymmetric Loading   (11–6)

Thus, when the stiffness factor for the beam's center span is computed using Eq. 11–6, the moments in only half the beam have to be distrib-uted. *Here the stiffness factor is one and a half times as large as that deter-mined using* $K = 4EI/L$.

Example 11–4

Determine the internal moments at the supports for the beam shown in Fig. 11–14a. EI is constant.

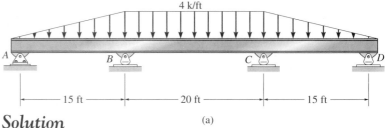

(a)

Solution

By inspection, the beam and loading are symmetrical. Thus, we will apply Eq. 11–5, $K = 2EI/L$, to compute the stiffness factor of the center span BC and therefore use only the left half of the beam for the analysis. The analysis can be shortened even further by using Eq. 11–4, $K = 3EI/L$, for computing the stiffness factor of segment AB since the end A is pinned. Furthermore, the distribution of moment at A can be skipped by using the FEM for a triangular loading on a span with one end fixed and the other pinned. Thus,

Fig. 11–14

$$K_{AB} = \frac{3EI}{15} \quad \text{(using Eq. 11–4)}$$

$$K_{BC} = \frac{2EI}{20} \quad \text{(using Eq. 11–5)}$$

$$(DF)_{AB} = \frac{3EI/15}{3EI/15} = 1$$

$$(DF)_{BA} = \frac{3EI/15}{3EI/15 + 2EI/20} = 0.667$$

$$(DF)_{BC} = \frac{2EI/20}{3EI/15 + 2EI/20} = 0.333$$

$$(FEM)_{BA} = \frac{wL^2}{15} = \frac{4(15)^2}{15} = 60 \text{ k} \cdot \text{ft}$$

$$(FEM)_{BC} = -\frac{wL^2}{12} = -\frac{4(20)^2}{12} = -133.3 \text{ k} \cdot \text{ft}$$

Joint	A	B	
Member	AB	BA	BC
DF	1	0.667	0.333
FEM Dist.		60 48.9	−133.3 24.4
ΣM	0	108.9	−108.9

These data are listed in the table in Fig. 11–14b. Computing the stiffness factors as shown above considerably reduces the analysis, since only joint B must be balanced and carry-overs to joints A and C are not necessary. Obviously, joint C is subjected to the same internal moment of 108.9 k · ft.

(b)

11.4 Moment Distribution for Frames: No Sidesway _____

Application of the moment-distribution method for frames having no sidesway follows the same procedure as that given for beams. To minimize the chance for errors, it is suggested that the analysis be arranged in a tabular form, as in the previous examples. Also, the distribution of moments can be shortened if the stiffness factor of a span can be modified as indicated in the previous section.

The framework for this steam-electric generating plant provides an example of a structure that can be modeled as having no sidesway. (*Courtesy of Bethlehem Steel Corporation*)

585

Example 11–5

Determine the internal moments at the joints of the frame shown in Fig. 11–15a. *EI* is constant.

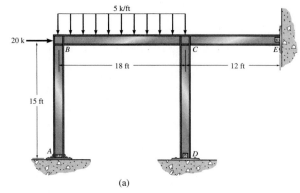

(a)

Fig. 11–15a ## Solution

By inspection, the pin at *E* will prevent the frame from sidesway. The stiffness factors of *CD* and *CE* can be computed using $K = 3EI/L$ since the far ends are pinned. Also, the 20-k load does not contribute a FEM since it is applied at joint *B*. Thus,

$$K_{AB} = \frac{4EI}{15} \qquad K_{BC} = \frac{4EI}{18} \qquad K_{CD} = \frac{3EI}{15} \qquad K_{CE} = \frac{3EI}{12}$$

$$(DF)_{AB} = 0$$

$$(DF)_{BA} = \frac{4EI/15}{4EI/15 + 4EI/18} = 0.545$$

$$(DF)_{BC} = 1 - 0.545 = 0.455$$

$$(DF)_{CB} = \frac{4EI/18}{4EI/18 + 3EI/15 + 3EI/12} = 0.330$$

$$(DF)_{CD} = \frac{3EI/15}{4EI/18 + 3EI/15 + 3EI/12} = 0.298$$

$$(DF)_{CE} = 1 - 0.330 - 0.298 = 0.372$$

$$(DF)_{DC} = 1 \quad (DF)_{EC} = 1$$

$$(FEM)_{BC} = \frac{-wL^2}{12} = \frac{-5(18)^2}{12} = -135 \text{ k} \cdot \text{ft}$$

$$(FEM)_{CB} = \frac{wL^2}{12} = \frac{5(18)^2}{12} = 135 \text{ k} \cdot \text{ft}$$

The data are shown in the table in Fig. 11–15b. Here the distribution of moments successively goes to joints B and C. The final moments are shown on the last line.

Using these data, the moment diagram for the frame is constructed in Fig. 11–15c.

Joint	A	B		C			D	E
Member	AB	BA	BC	CB	CD	CE	DC	EC
DF	0	0.545	0.455	0.330	0.298	0.372	1	1
FEM Dist.		73.6	−135 61.4	135 −44.6	−40.2	−50.2		
CO Dist.	36.8	12.2	−22.3 10.1	30.7 −10.1	−9.1	−11.5		
CO Dist.	6.1	2.8	−5.1 2.3	5.1 −1.7	−1.5	−1.9		
CO Dist.	1.4	0.4	−0.8 0.4	1.2 −0.4	−0.4	−0.4		
CO Dist.	0.2	0.1	−0.2 0.1	0.2 −0.1	0.0	−0.1		
ΣM	44.5	89.1	−89.1	115	−51.2	−64.1		

(b)

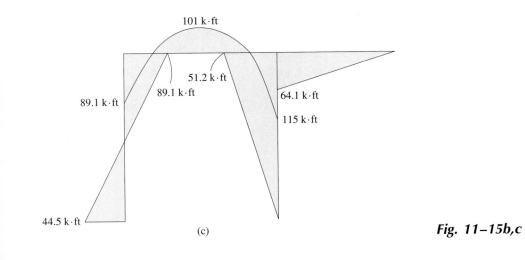

(c)

Fig. 11–15b,c

587

11.5 Moment Distribution for Frames: Sidesway

It has been shown in Sec. 10.4 that frames that are nonsymmetrical or subjected to nonsymmetrical loadings have a tendency to sidesway. An example of one such case is shown in Fig. 11–16a. Here the applied loading **P** will create unequal moments at joints B and C such that the frame will deflect an amount Δ to the right. To determine this deflection and the internal moments at the joints using moment distribution, we will use the principle of superposition. In this regard, the frame in Fig. 11–16b is first considered held from sidesway by applying an artificial joint support at C. Moment distribution is applied and then by statics the restraining force **R** is determined. The equal, but opposite, restraining force is then applied to the frame, Fig. 11–16c, and the moments in the frame are calculated. One method for doing this last step requires first *assuming* a numerical value for one of the internal moments, say M'_{BA}. Using moment distribution and statics, the deflection Δ' and external force **R**′ corresponding to the assumed value of M'_{BA} can then be determined. Since linear elastic deformations occur, the force **R**′ develops moments in the frame that are *proportional* to those developed by **R**. For example, if M'_{BA} and **R**′ are known, the moment at B developed by **R** will be $M_{BA} = R(M'_{BA}/R')$. Addition of the joint moments for both cases, Fig. 11–16b and c, will yield the actual moments in the frame, Fig. 11–16a.

Application of this technique is illustrated in the following examples.

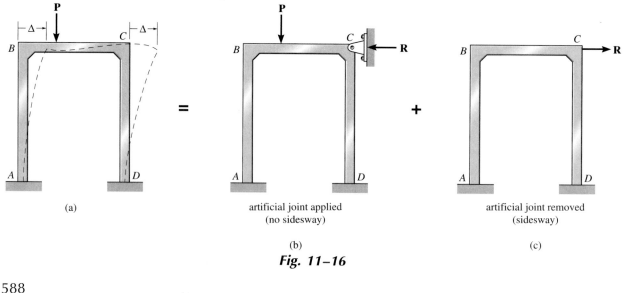

(a)	=	artificial joint applied (no sidesway)	+	artificial joint removed (sidesway)
		(b)		(c)

Fig. 11–16

Example 11–6

Determine the moments at each joint of the frame shown in Fig. 11–17a. EI is constant.

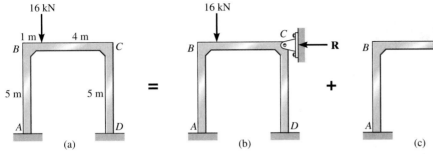

Fig. 11–17a,b,c,d,e

Joint	A	B		C		D
Member	AB	BA	BC	CB	CD	DC
DF	0	0.5	0.5	0.5	0.5	0
FEM			−10.24	2.56		
Dist.		5.12	5.12	−1.28	−1.28	
CO	2.56		−0.64	2.56		−0.64
Dist.		0.32	0.32	−1.28	−1.28	
CO	0.16		−0.64	0.16		−0.64
Dist.		0.32	0.32	−0.08	−0.08	
CO	0.16		−0.04	0.16		−0.04
Dist.		0.02	0.02	−0.08	−0.08	
ΣM	2.88	5.78	−5.78	2.72	−2.72	−1.32

(d)

Solution

First we consider the frame held from sidesway as shown in Fig. 11–17b. We have

$$(\text{FEM})_{BC} = -\frac{16(4)^2(1)}{(5)^2} = -10.24 \text{ kN} \cdot \text{m}$$

$$(\text{FEM})_{CB} = \frac{16(1)^2(4)}{(5)^2} = 2.56 \text{ kN} \cdot \text{m}$$

The stiffness factor of each span is computed on the basis of $4EI/L$ or by using the relative-stiffness factor I/L. The DFs and the moment distribution are shown in the table, Fig. 11–17d. Using these results, the equations of equilibrium are applied to the free-body diagram of the columns in order to determine $\mathbf{A}_x$ and $\mathbf{D}_x$, Fig. 11–17e. From the free-body diagram of the entire frame (not shown) the joint restraint $\mathbf{R}$ in Fig. 11–17b has a magnitude of

$$\Sigma F_x = 0; \qquad R = 1.73 - 0.81 = 0.92 \text{ kN}$$

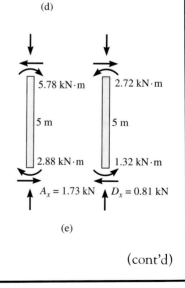

(e)

(cont'd)

Example 11–6
(continued)

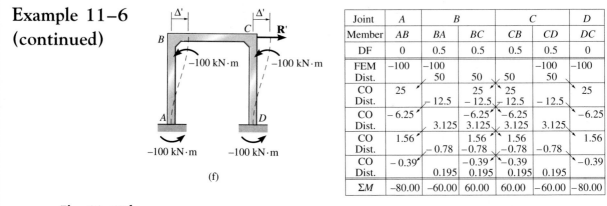

Fig. 11–17f,g

(f)

Joint	A	B		C		D
Member	AB	BA	BC	CB	CD	DC
DF	0	0.5	0.5	0.5	0.5	0
FEM	−100	−100			−100	−100
Dist.		50	50	50	50	
CO	25		25	25		25
Dist.		12.5	−12.5	12.5	−12.5	
CO	−6.25		−6.25	−6.25		−6.25
Dist.		3.125	3.125	3.125	3.125	
CO	1.56		1.56	1.56		1.56
Dist.		−0.78	−0.78	−0.78	−0.78	
CO	−0.39		−0.39	−0.39		−0.39
Dist.		0.195	0.195	0.195	0.195	
ΣM	−80.00	−60.00	60.00	60.00	−60.00	−80.00

(g)

An equal but opposite value of R = 0.92 kN must now be applied to the frame at C and the internal moments computed, Fig. 11–17c. To solve the problem of computing these moments, we will assume a force **R′** is applied at C, causing the frame to deflect **Δ′** as shown in Fig. 11–17f. Here the joints at B and C are *temporarily restrained from rotating*, and as a result the fixed-end moments at the ends of the columns are determined from the formula for deflection found on the inside back cover or by slope deflection, Eq. 9–5, that is,

$$M = \frac{6EI\Delta}{L^2}$$

Since *both* B and C happen to be displaced the same amount **Δ′**, and AB and DC have the *same* E, I, and L, the FEM in AB will be the *same* as that in DC. As shown in Fig. 11−17f, we will arbitrarily *assume* this fixed-end moment to be*

$$(FEM)_{AB} = (FEM)_{BA} = (FEM)_{CD} = (FEM)_{DC} = -100 \text{ kN} \cdot \text{m}$$

A *negative sign* is necessary since the moment must act *counterclockwise* on the column for deflection **Δ′** to the right. The value of **R′** associated with this −100 kN · m moment can now be determined. The moment distribution of the FEMs is shown in Fig. 11–17g.

---

*If E is given, we could also specify an arbitrary value for Δ′ and then calculate the numerical value for FEM = 6EIΔ′/L². The above method, however, where the FEM is assumed, is more efficient.

From equilibrium, the horizontal reactions at A and D are calculated, Fig. 11–17h. Thus, for the entire frame we require

$$\Sigma F_x = 0; \qquad R' = 28 + 28 = 56.0 \text{ kN}$$

Hence, $R' = 56.0$ kN creates the moments tabulated in Fig. 11–17g. Corresponding moments caused by $R = 0.92$ kN can be determined by proportion. Therefore, the resultant moment in the frame, Fig. 11–17a, is equal to the *sum* of those calculated for the frame in Fig. 11–17b plus the proportionate amount of those for the frame in Fig. 11–17c. We have

$$M_{AB} = 2.88 + \frac{0.92}{56.0}(-80) = 1.57 \text{ kN} \cdot \text{m} \qquad \textbf{Ans.}$$

$$M_{BA} = 5.78 + \frac{0.92}{56.0}(-60) = 4.79 \text{ kN} \cdot \text{m} \qquad \textbf{Ans.}$$

$$M_{BC} = -5.78 + \frac{0.92}{56.0}(60) = -4.79 \text{ kN} \cdot \text{m} \qquad \textbf{Ans.}$$

$$M_{CB} = 2.72 + \frac{0.92}{56.0}(60) = 3.71 \text{ kN} \cdot \text{m} \qquad \textbf{Ans.}$$

$$M_{CD} = -2.72 + \frac{0.92}{56.0}(-60) = -3.71 \text{ kN} \cdot \text{m} \qquad \textbf{Ans.}$$

$$M_{DC} = -1.32 + \frac{0.92}{56.0}(-80) = -2.63 \text{ kN} \cdot \text{m} \qquad \textbf{Ans.}$$

Using these results, the force reactions on each member of the frame and its moment diagram are shown in Fig. 11–17i and 11–17j, respectively.

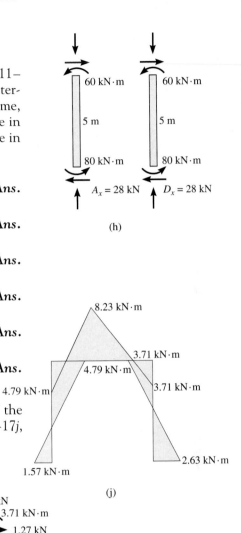

(h)

(j)

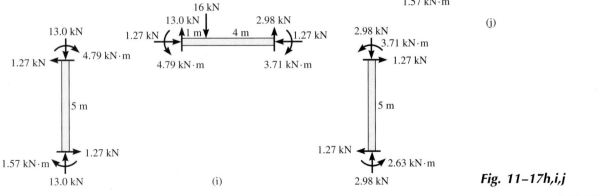

(i)

Fig. 11–17h,i,j

Example 11–7

Determine the moments at each joint of the frame shown in Fig. 11–18a. The moment of inertia of each member is indicated in the figure.

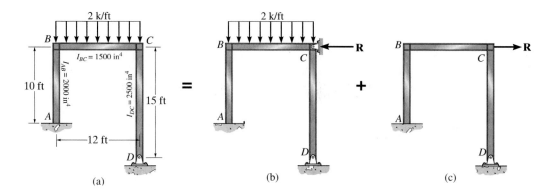

(a)   (b)   (c)

Solution

The frame is first held from sidesway as shown in Fig. 11–18b. The internal moments are computed at the joints as indicated in Fig. 11–18d. Here the stiffness factor of CD was computed using $3EI/L$ since there is a pin at D. Calculation of the horizontal reactions at A and D is shown in Fig. 11–18e. Thus, for the entire frame,

$$\Sigma F_x = 0; \qquad R = 2.89 - 1.00 = 1.89 \text{ k}$$

Joint	A	B		C		D
Member	AB	BA	BC	CB	CD	DC
DF	0	0.615	0.385	0.5	0.5	1
FEM			−24	24		
Dist.		14.76	9.24	−12	−12	
CO	7.38		−6	4.62		
Dist.		3.69	2.31	−2.31	−2.31	
CO	1.84		−1.16	1.16		
Dist.		0.713	0.447	−0.58	−0.58	
CO	0.357		−0.29	0.224		
Dist.		0.18	0.11	−0.11	−0.11	
ΣM	9.58	19.34	−19.34	15.00	−15.00	0

(d)

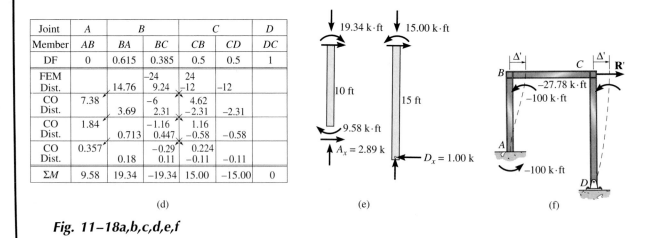

(e)   (f)

Fig. 11–18a,b,c,d,e,f

The opposite force is now applied to the frame as shown in Fig. 11–18c. As in the previous example, we will consider a force $\mathbf{R}'$ acting as shown in Fig. 11–18f. As a result, joints B and C are displaced by the same amount Δ'. The fixed-end moments for BA are computed from

$$(\text{FEM})_{AB} = (\text{FEM})_{BA} = -\frac{6EI\Delta}{L^2} = -\frac{6E(2000)\Delta'}{(10)^2}$$

However, from the table on the inside back cover, for CD we have

$$(\text{FEM})_{CD} = -\frac{3EI\Delta}{L^2} = -\frac{3E(2500)\Delta'}{(15)^2}$$

Assuming the FEM for AB is $-100\ \text{k} \cdot \text{ft}$ as shown in Fig. 11–18f, the *corresponding* FEM at C, causing the *same* Δ', is found by comparison, i.e.,

$$\Delta' = \frac{100(10)^2}{6E(2000)} = -\frac{(\text{FEM})_{CD}(15)^2}{3E(2500)}$$
$$(\text{FEM})_{CD} = -27.78\ \text{k} \cdot \text{ft}$$

Moment distribution for these FEMs is tabulated in Fig. 11–18g. Computation of the horizontal reactions at A and D is shown in Fig. 11–18h. Thus, for the entire frame,

$$\Sigma F_x = 0; \qquad R' = 11.0 + 1.55 = 12.55\ \text{k}$$

Fig. 11–18g,h

Joint	A	B		C		D
Member	AB	BA	BC	CB	CD	DC
DF	0	0.615	0.385	0.5	0.5	1
FEM	−100	−100			−27.78	
Dist.		61.5	38.5	13.89	13.89	
CO	30.75		6.94	19.25		
Dist.		−4.27	−2.67	−9.625	−9.625	
CO	−2.14		−4.81	−1.34		
Dist.		2.96	1.85	0.67	0.67	
CO	1.48		0.33	0.92		
Dist.		−0.20	−0.13	−0.46	−0.46	
ΣM	−69.91	−40.01	40.01	23.31	−23.31	0

(g)

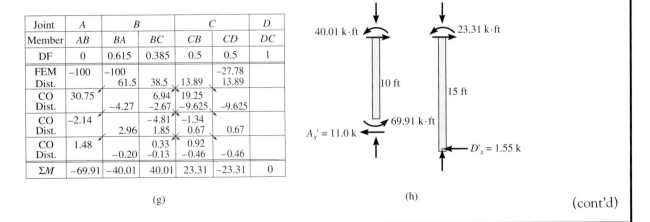

(h)

(cont'd)

Example 11–7
(continued)

The resultant moments in the frame are therefore

$$M_{AB} = 9.58 + \left(\frac{1.89}{12.55}\right)(-69.91) = -0.948 \text{ k} \cdot \text{ft} \qquad \textbf{Ans.}$$

$$M_{BA} = 19.34 + \left(\frac{1.89}{12.55}\right)(-40.01) = 13.3 \text{ k} \cdot \text{ft} \qquad \textbf{Ans.}$$

$$M_{BC} = -19.34 + \left(\frac{1.89}{12.55}\right)(40.01) = -13.3 \text{ k} \cdot \text{ft} \qquad \textbf{Ans.}$$

$$M_{CB} = 15.00 + \left(\frac{1.89}{12.55}\right)(23.31) = 18.5 \text{ k} \cdot \text{ft} \qquad \textbf{Ans.}$$

$$M_{CD} = -15.00 + \left(\frac{1.89}{12.55}\right)(-23.31) = -18.5 \text{ k} \cdot \text{ft} \quad \textbf{Ans.}$$

Example 11–8

Determine the moments at each joint of the frame shown in Fig. 11–19a. EI is constant.

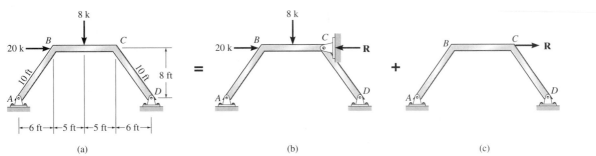

(a)   =   (b)   +   (c)

Solution

First sidesway is prevented by the restraining force **R,** Fig. 11–19b. The FEMs for member BC are

$$(FEM)_{BC} = -\frac{8(10)}{8} = -10 \text{ k} \cdot \text{ft} \quad (FEM)_{CB} = \frac{8(10)}{8} = 10 \text{ k} \cdot \text{ft}$$

Since spans AB and DC are pinned at their ends, the stiffness factor is computed using $3EI/L$. The moment distribution is shown in Fig. 11–19d. Using these results, the *horizontal reactions* at A and D must be determined. This is done using an equilibrium analysis of *each member*, Fig. 11–19e. Summing moments about points B and C on each leg, we have

Fig. 11–19a,b,c,d

$\zeta+\Sigma M_B = 0;\qquad 5.97 - A_x(8) + 4(6) = 0 \qquad A_x = 3.75 \text{ k}$

$\zeta+\Sigma M_C = 0;\qquad -5.97 + D_x(8) - 4(6) = 0 \qquad D_x = 3.75 \text{ k}$

Joint	A	B		C		D
Member	AB	BA	BC	CB	CD	DC
DF	1	0.429	0.571	0.571	0.429	1
FEM			−10	10		
Dist.		4.29	5.71	−5.71	−4.29	
CO			−2.86	2.86		
Dist.		1.23	1.63	−1.63	−1.23	
CO			−0.82	0.82		
Dist.		0.35	0.47	−0.47	−0.35	
CO			−0.24	0.24		
Dist.		0.10	0.13	−0.13	−0.10	
ΣM	0	5.97	−5.97	5.97	−5.97	0

(d)

(cont'd)

Example 11–8 (continued)

Thus, for the entire frame,

$$\Sigma F_x = 0; \qquad R = 3.75 - 3.75 + 20 = 20 \text{ k}$$

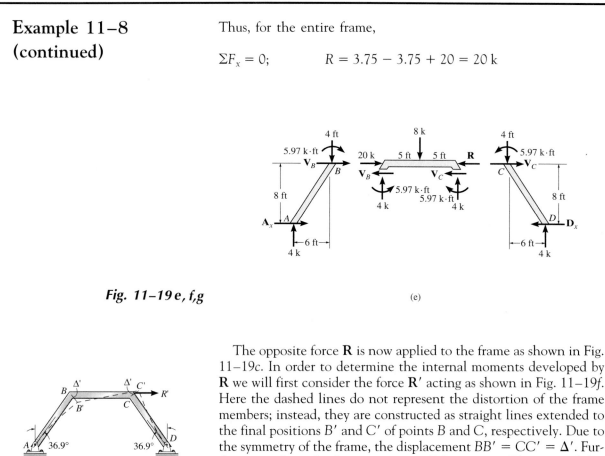

Fig. 11–19 e, f, g

(e)

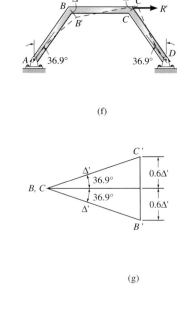

(f)

(g)

The opposite force **R** is now applied to the frame as shown in Fig. 11–19c. In order to determine the internal moments developed by **R** we will first consider the force **R'** acting as shown in Fig. 11–19f. Here the dashed lines do not represent the distortion of the frame members; instead, they are constructed as straight lines extended to the final positions B' and C' of points B and C, respectively. Due to the symmetry of the frame, the displacement $BB' = CC' = \Delta'$. Furthermore, these displacements cause BC to rotate. The vertical distance between B' and C' is $1.2\Delta'$, as shown on the displacement diagram, Fig. 11–19g. Since each span undergoes end-point displacements that cause the spans to rotate, fixed-end moments are induced in the spans. These are as follows:

$$(\text{FEM})_{BA} = (\text{FEM})_{CD} = -\frac{3EI\Delta'}{(10)^2}$$

$$(\text{FEM})_{BC} = (\text{FEM})_{CB} = \frac{6EI(1.2\Delta')}{(10)^2}$$

Notice that for BA and CD the moments are *negative* since clockwise rotation of the span causes a *counterclockwise* FEM. (See the FEM table on the inside back cover.)

Joint	A	B		C		D
Member	AB	BA	BC	CB	CD	DC
DF	1	0.429	0.571	0.571	0.429	1
FEM Dist.		−100 −60.06	240 −79.94	240 −79.94	−100 −60.06	
CO Dist.		17.15	−39.97 22.82	−39.97 22.82	17.15	
CO Dist.		−4.89	11.41 −6.52	11.41 −6.52	−4.89	
CO Dist.		1.40	−3.26 1.86	−3.26 1.86	1.40	
CO Dist.		−0.40	0.93 −0.53	0.93 −0.53	−0.40	
ΣM	0	−146.80	146.80	146.80	−146.80	0

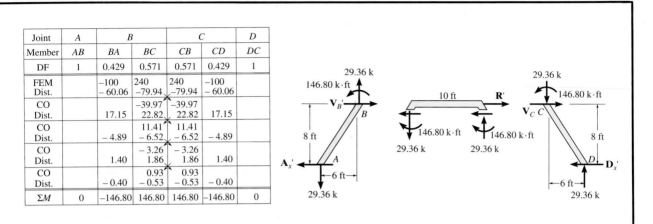

(h)   (i)

Fig. 11–19h,i

If we arbitrarily assign a value of $(FEM)_{BA} = (FEM)_{CD} = -100 \text{ k} \cdot \text{ft}$, then equating Δ' in the above formulas yields $(FEM)_{BC} = (FEM)_{CB} = 240 \text{ k} \cdot \text{ft}$. These moments are applied to the frame and distributed, Fig. 11–19h. Using these results, the equilibrium analysis is shown in Fig. 11–19i. For each leg, we have

$$\zeta + \Sigma M_B = 0; \quad A'_x(8) - 29.36(6) - 146.80 = 0 \qquad A'_x = 40.37 \text{ k}$$
$$\zeta + \Sigma M_C = 0; \quad D'_x(8) - 29.36(6) - 146.80 = 0 \qquad D'_x = 40.37 \text{ k}$$

Thus, for the entire frame,

$$\Sigma F_x = 0; \qquad R' = 40.37 + 40.37 = 80.74 \text{ k}$$

The resultant moments in the frame are therefore

$$M_{BA} = -5.97 + \left(\frac{20}{80.74}\right)(-146.80) = -30.4 \text{ k} \cdot \text{ft} \quad \textbf{Ans.}$$

$$M_{BC} = -5.97 + \left(\frac{20}{80.74}\right)(146.80) = 30.4 \text{ k} \cdot \text{ft} \qquad \textbf{Ans.}$$

$$M_{CB} = 5.97 + \left(\frac{20}{80.74}\right)(146.80) = 42.3 \text{ k} \cdot \text{ft} \qquad \textbf{Ans.}$$

$$M_{CD} = -5.97 + \left(\frac{20}{80.74}\right)(-146.80) = -42.3 \text{ k} \cdot \text{ft} \quad \textbf{Ans.}$$

11.6 Moment Distribution for Multistory Frames

The frames analyzed in the previous section all have a single joint displacement that can be related to the displacement of all the other joints. Quite often, multistory frameworks may have several *independent* joint displacements, and consequently the moment-distribution analysis will involve more computation. Consider, for example, the two-story frame shown in Fig. 11–20a. This structure can have two independent joint displacements, since the sidesway Δ_1 of the first story is independent of any displacement Δ_2 of the second story. Unfortunately, these displacements are not known initially, so the analysis must proceed on the basis of superposition, as discussed previously. In this case, two restraining forces $\mathbf{R}_1$ and $\mathbf{R}_2$ are applied, Fig. 11–20b, and the fixed-end moments are determined and distributed. Using the equations of equilibrium, the numerical values of $\mathbf{R}_1$ and $\mathbf{R}_2$ are then determined. Next, the restraint at the floor of the first story is removed and the floor is given a displacement Δ'. This displacement causes fixed-end moments (FEMs) in the frame, which can be assigned specific numerical values. By distributing these moments and using the equations of equilibrium, the associated numerical values of $\mathbf{R}_1'$ and $\mathbf{R}_2'$ can be determined. In a similar manner, the floor of the second story is then given a displacement Δ'', Fig. 11–20d. Assuming numerical values for the fixed-end moments, the moment distribution and equilibrium analysis will yield specific values of $\mathbf{R}_1''$ and $\mathbf{R}_2''$. Since the last two steps associated with Fig. 11–20c and d depend on *assumed* values of the FEMs, correction factors C' and C'' must be applied to the distributed moments. With reference to the restraining forces in Fig. 11–20, we require

$$0 = -R_2 - C'R_2' + C''R_2''$$
$$0 = -R_1 + C'R_1' - C''R_1''$$

Simultaneous solution of these equations yields the values of C' and C''. These correction factors are then multiplied by the internal joint moments found from the moment distribution in Fig. 11–20c and 11–20d. The resultant moments are then found by adding these corrected moments to those obtained for the frame in Fig. 11–20b.

Other types of frames having independent joint displacements can be analyzed using this same procedure; however, it must be admitted that the foregoing method does require quite a bit of numerical calculation. Although some techniques have been developed to shorten the calculations, it is best to solve these types of problems on a computer, preferably using a matrix analysis. The techniques for doing this will be discussed in Chapter 15.

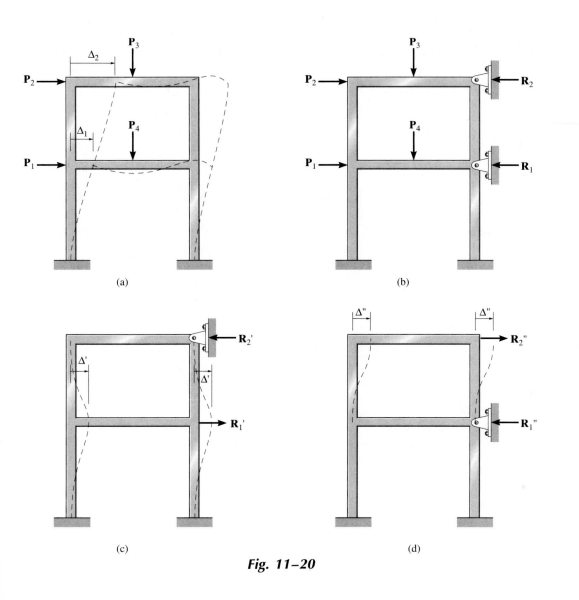

(a)

(b)

(c)

(d)

Fig. 11–20

REFERENCE

Cross, H., "Analysis of Continuous Frames by Distributing Fixed-End Moments," *Proceedings of the American Society of Civil Engineers*, 56, 1930, p. 919–28.

PROBLEMS

11–1. Determine the moments at the supports. EI is constant. Assume B is a roller and A and C are fixed.

*11–4. Determine the moments at each support and then draw the moment diagram. Assume A is fixed. EI is constant.

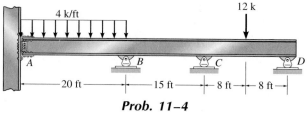

Prob. 11–4

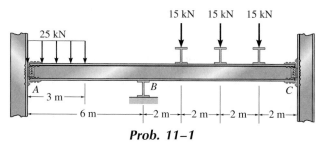

Prob. 11–1

11–2. Determine the moments at B and C. EI is constant. Assume B and C are rollers and A and D are pinned.

11–5. Determine the reactions at the supports and then draw the moment diagram. Assume A is fixed.

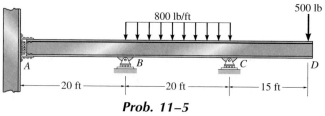

Prob. 11–5

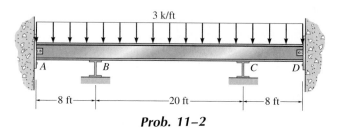

Prob. 11–2

11–3. Determine the reactions at the supports. Assume B and C are rollers and A is fixed. EI is constant.

11–6. Determine the moments at A, B, and C and then draw the moment diagram for the girder DE. EI is constant. Assume the support at B is a pin and A and C are rollers. The distributed load rests on simply supported floor boards that transmit the load to the floor beams.

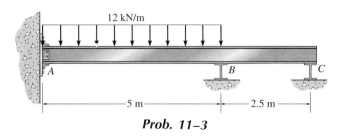

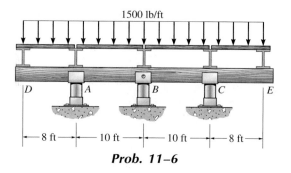

Prob. 11–3

Prob. 11–6

11–7. The beam is subjected to the loading shown. Determine the reactions at the supports, then draw the moment diagram. *EI* is constant.

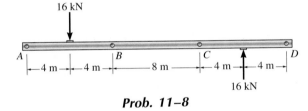

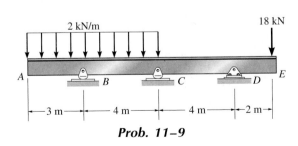

Prob. 11–7

11–10. Draw the moment diagram for each member. *EI* is constant. All joints are fixed-connected.

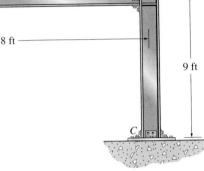

Prob. 11–10

*11–8. The bar is pin-connected at each indicated point. If the normal force in the bar can be neglected, determine the vertical reaction at each pin. *EI* is constant.

16 kN

A ——4 m——B——4 m——8 m——C——4 m——D——4 m——

16 kN

Prob. 11–8

11–9. Determine the moments at A, B, and C, then draw the moment diagram. *EI* is constant.

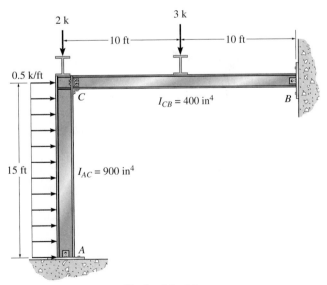

Prob. 11–9

11–11. Determine the horizontal and vertical components of reaction at the connections A and B. Assume A and B are pins and C is a fixed joint. Take $E = 29(10^3)$ ksi.

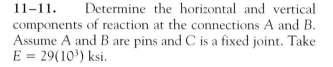

Prob. 11–11

601

*11–12. Determine the moments at the ends of each member. The members are fixed-connected at the supports and joint B. Take $E = 29(10^3)$ ksi. The moment of inertia of each member is given in the figure.

11–14. Determine the reactions at the supports. Assume A and C are pinned and B and D are fixed-connected. EI is the same for each member.

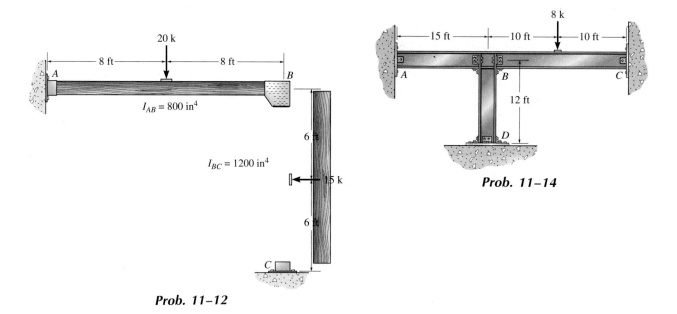

Prob. 11–12

Prob. 11–14

11–13. Determine the internal moments at the supports and draw the moment diagram for ABDE. Assume A is pinned, D is a roller and C is fixed. Take $I_{ABDE} = 1200$ in^4. and $I_{BC} = 800$ in^4. $E = 29(10^3)$ ksi.

11–15. The frame is made from pipe that is fixed-connected. If it supports the loading shown, determine the moments developed at each of the joints. EI is constant.

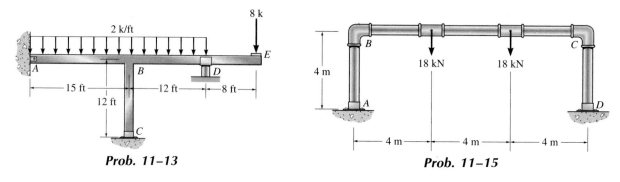

Prob. 11–13

Prob. 11–15

*11–16. Determine the moments at the ends of each member of the frame. The supports at A and C and joint B are fixed-connected. Take EI to be the same for each member.

11–18. Determine the moments at each joint of the gable frame. EI is constant. The roof load is transmitted to each of the purlins over simply supported sections of the roof decking. Assume the supports at A and E are pins.

11–19. Solve Prob. 11–18 assuming the supports at A and E are fixed.

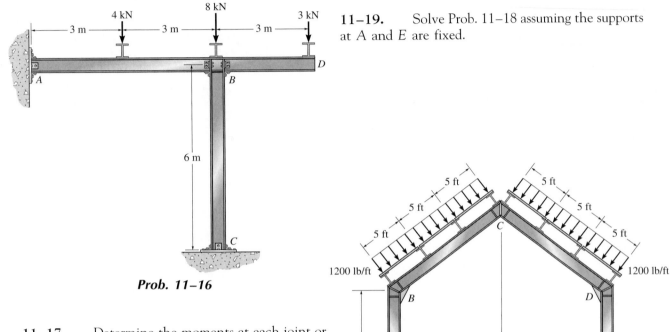

Prob. 11–16

11–17. Determine the moments at each joint or support of the battered column frame. EI is constant. The joints and supports are fixed-connected.

Probs. 11–18/11–19

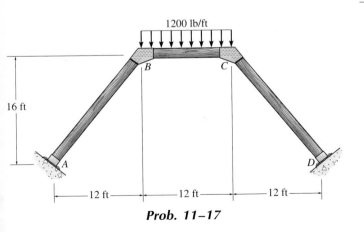

Prob. 11–17

*11–20. Determine the moments at the ends of each member of the frame. The members are fixed-connected at the supports and joints. *EI* is the same for each member.

11–22. Determine the moments at each joint and fixed support. *EI* is constant.

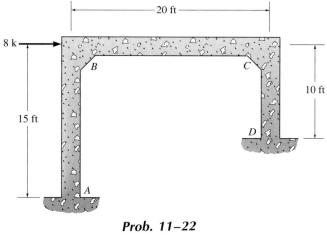

Prob. 11–22

Prob. 11–20

11–21. Determine the moments at each joint or support. There are fixed connections at *B* and *C* and fixed supports at *A* and *D*. *EI* is constant.

11–23. The side of the frame is subjected to the hydrostatic loading shown. Determine the moments at each joint and support. *EI* is constant.

Prob. 11–21

Prob. 11–23

*11–24. Determine the moment at each joint of the frame having battered columns. The supports at A and D are pins. EI is constant.

11–25. For the battered column frame in Prob. 11–24, determine the moment at each joint and the supports A and D. Assume the supports are fixed-connected and EI is constant.

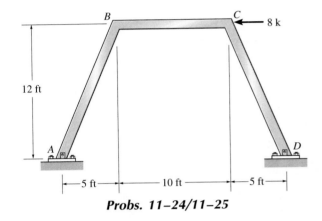

Probs. 11–24/11–25

The use of variable-moment-of-inertia girders has reduced considerably the deadweight loading of this bridge span. *(Photo courtesy Bethlehem Steel)*

12

Analysis of Beams and Frames Consisting of Nonprismatic Members

In previous chapters we have considered the indeterminate analysis of beams and frames composed of prismatic members, that is, those having a *constant* moment of inertia. Often, to save material, girders used for long spans on bridges and buildings are designed to have a variable moment of inertia. In this chapter we will apply the slope-deflection and moment-distribution methods to analyze beams and frames composed of such nonprismatic members. Actually, the general procedures for applying these methods are identical to those presented in the preceding two chapters. Here, though, we must develop appropriate carry-over factors, stiffness factors, and fixed-end moments, each of which depend on the geometry of the member's cross section. A general numerical technique for doing this will first be developed using the conjugate-beam method. This is followed by a discussion related to using tabular values often published in design literature. Finally, the analysis of statically indeterminate structures using the slope-deflection and moment-distribution methods will be discussed.

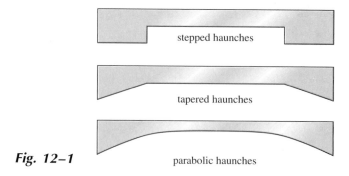

Fig. 12–1

stepped haunches

tapered haunches

parabolic haunches

12.1 Deflections of Nonprismatic Members

The most common forms of structural members that are nonprismatic have haunches that are either stepped, tapered, or parabolic, Fig. 12–1. We can use the principle of virtual work or Castigliano's theorem as discussed in Chapter 7 to compute their deflections. The equations are

$$\Delta = \int_0^l \frac{Mm}{EI}\,dx \qquad \text{or} \qquad \Delta = \int_0^l \frac{\partial M}{\partial P}\frac{M}{EI}\,dx$$

For a nonprismatic member the integration requires I to be expressed as a function of the length coordinate x. As a result, the member's geometry and loading may require evaluation of an integral that will be impossible to evaluate in closed form. Consequently, Simpson's rule or some other numerical technique will have to be used to carry out the integration.

It is also possible to use a geometrical technique such as the moment-area theorems or the conjugate-beam method to determine the deflection of a nonprismatic member. The following example illustrates the use of the conjugate-beam method.

A reinforced-concrete pier, made in the form of a haunched beam, is used to support the steel girders of the highway overpass.

Example 12–1

Use the conjugate-beam method to determine the deflection of the end A of the tapered beam shown in Fig. 12–2a. Assume that the beam has a thickness of 1 ft and $E = 4000$ ksi.

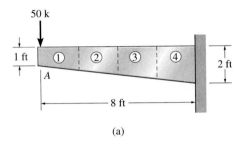

(a)

Solution

The beam will be segmented every 2 ft and we will identify the segments numerically as shown in Fig. 12–2a. The moment diagram is shown in Fig. 12–2b. Using $I = \frac{1}{12} bh^3$, the moment of inertia of the beam is computed at the 2-ft intervals and reported in Fig. 12–2c. From the data, the M/EI diagram is determined and shown on the conjugate beam in Fig. 12–2d. We are required to find $\Delta_A = M_{A'}$. To do this, the load is assumed (approximately) to consist of seven triangular areas indicated by the dashed lines. To simplify the calculation it is seen that three pairs of triangles have a common vertical "base" and the same "height." The centroid location of these pairs and the rightmost triangle is indicated by dots. Hence,

$$\zeta + \Sigma M_{A'} = 0;$$

$$\Delta_A = M_A' = \frac{1}{2}\left(\frac{613.50}{E}\right)(4)(2) + \frac{1}{2}\left(\frac{711.74}{E}\right)(4)(4)$$

$$+ \frac{1}{2}\left(\frac{671.14}{E}\right)(4)(6) + \frac{1}{2}\left(\frac{600}{E}\right)(2)(7.33)$$

$$= \frac{20,599.6}{E}$$

Substituting $E = 4000$ k/in^2 (144 in^2/ft^2), we have

$$\Delta_A = 0.0358 \text{ ft} = 0.429 \text{ in.} \qquad \textbf{Ans.}$$

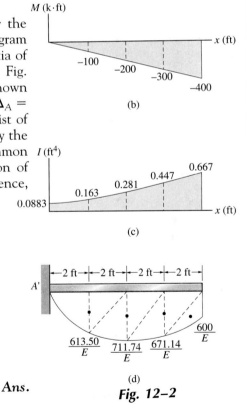

Fig. 12–2

609

12.2 Loading Properties of Nonprismatic Members Using the Conjugate-Beam Method

A conjugate-beam analysis is a practical numerical method that can be used to determine the fixed-end moments and stiffness and carry-over factors for a member that has a variable moment of inertia. Although the work may be a bit tedious, the method is illustrated here so that one can understand how to obtain numerical values for loading properties of a particular member that may not be available in the published literature.

Two numerical examples will serve to illustrate the method. In this regard, recall the following definitions.

> **Fixed-end moments (FEM):** The end moment reactions of a beam that is assumed fixed-supported, Fig. 12–3a.
>
> **Stiffness factor (K):** The magnitude of moment that must be applied to the end of a beam such that the end rotates through an angle of $\theta = 1$ rad. Here the moment is applied at the pin support of the beam, while the other end is assumed fixed, Fig. 12–3b.*

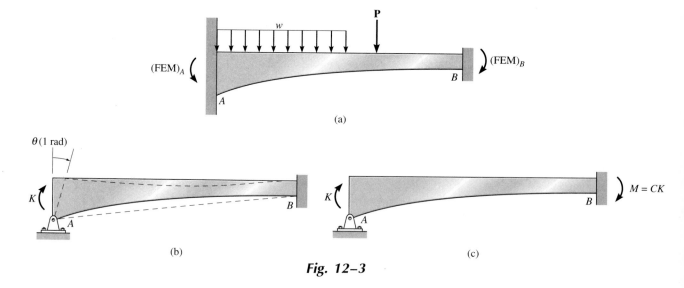

(a)

(b)

(c)

Fig. 12–3

*For illustrative purposes, the angle $\theta = 1$ rad (57.3°) is not drawn to scale on the beam.

Carry-over factor (COF): Represents the numerical fraction (C) of the moment that is "carried over" from the pin-supported end of a beam to the wall, Fig. 12–3c.

The computations for the stiffness and carry-over factors can be checked, in part, by noting an important relationship that exists between them. In this regard, consider the beam in Fig. 12–4 subjected to the loads and deflections shown. Application of the Maxwell-Betti reciprocal theorem† requires the work done by the loads in Fig. 12–4a acting through the displacements in Fig. 12–4b be equal to the work of the loads in Fig. 12–4b acting through the displacements in Fig. 12–4a, that is,

$$U_{AB} = U_{BA}$$
$$K_A(0) + C_{AB}K_A(1) = C_{BA}K_B(1) + K_B(0)$$

or

$$C_{AB}K_A = C_{BA}K_B \qquad (12-1)$$

Hence, once determined, the stiffness and carry-over factors must satisfy Eq. 12–1.

The haunched legs of this highway bridge provide a rigid framework for the side girders.

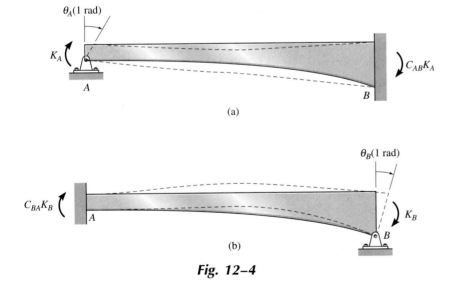

$\theta_A(1 \text{ rad})$

K_A

A

B

$C_{AB}K_A$

(a)

$C_{BA}K_B$

A

$\theta_B(1 \text{ rad})$

K_B

B

(b)

Fig. 12–4

† See Sec. 8.3.

Example 12–2

Determine the fixed-end moments for the beam shown in Fig. 12–5a. The cross-sectional area has a constant width of 1 ft. E is constant.

Solution

Since the real beam supports at A and B are fixed, the conjugate beam is free at its ends A' and B'. The moment diagrams will be plotted in parts, using the method of superposition discussed in Sec. 4.5. These parts shown on the right in Fig. 12–5b through 12–5d represent the moment diagrams for a simply-supported beam loaded with a concentrated force of 3 k and each of the end moments $\mathbf{M}_A$ and $\mathbf{M}_B$. For span AC, $I_{AC} = \frac{1}{12}(1)(2)^3 = 0.667$ ft^4, and for span CB, $I_{CB} = \frac{1}{12}(1)(1)^3 = 0.0833$ ft^4. Using these values, the M/EI diagrams (loadings) on the conjugate beam are shown on the left in Fig. 12–5b through 12–5d. The resultant forces caused by these "distributed" loads have been computed and are also shown in the figure. Since the slope and deflection at the end A (or B) of the real beam are equal to zero, we require

$$+\uparrow\Sigma F_y = 0; \quad \frac{45.56}{E} + \frac{121.5}{E} - \frac{5.06M_A}{E} - \frac{3.375M_A}{E}$$
$$- \frac{4.5M_A}{E} - \frac{5.06M_B}{E} - \frac{27M_B}{E} - \frac{4.5M_B}{E} = 0$$

or

$$167.06 - 12.94M_A - 36.56M_B = 0 \tag{1}$$

$$\curvearrowleft + \Sigma M_{A'} = 0; \quad -\frac{45.56}{E}(6) - \frac{121.5}{E}(10) + \frac{5.06M_A}{E}(3)$$
$$+ \frac{3.375M_A}{E}(4.5) + \frac{4.5M_A}{E}(10) + \frac{5.06M_B}{E}(6)$$
$$+ \frac{27M_B}{E}(10.5) + \frac{4.5M_B}{E}(11) = 0$$

or

$$-1488.36 + 75.37M_A + 363.36M_B = 0 \tag{2}$$

Solving Eqs. (1) and (2) simultaneously, we have

$$M_A = (FEM)_{AB} = 3.23 \text{ k} \cdot \text{ft} \qquad \textbf{Ans.}$$
$$M_B = (FEM)_{BA} = 3.43 \text{ k} \cdot \text{ft} \qquad \textbf{Ans.}$$

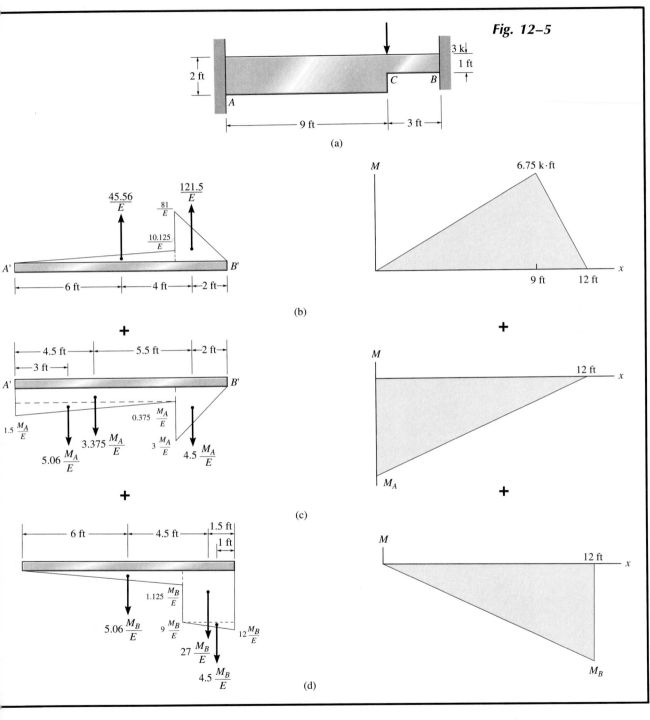

Fig. 12–5

613

Example 12–3

Determine the stiffness and carry-over factors for the end A of the beam shown in Fig. 12–6a. The cross-sectional area has a constant width of 1 ft. Take $E = 10(10^3)$ ksi.

Solution

The stiffness factor is determined by placing a pin (or roller) at the end A of the beam and calculating the moment $\mathbf{M}_A$ needed to rotate the beam $\theta_A = 1$ rad, Fig. 12–6b. The fraction C_{AB} of $\mathbf{M}_A$ necessary to hold the beam in equilibrium at B is the carry-over factor from A to B.

The moment diagrams and conjugate beam for the beam in Fig. 12–6b are shown in Fig. 12–6c. Data on the diagram are taken from Fig. 12–5c and 12–5d due to the similarity of loadings. Since a rotation $\theta_A = 1$ rad is required, the reaction at A' is $V_{A'} = 1$ as shown in Fig. 12–6c. Hence,

$$+\uparrow \Sigma F_y = 0; \quad -1 + 5.06 \frac{M_A}{E} + 3.375 \frac{M_A}{E} + 4.5 \frac{M_A}{E}$$

$$- 5.06 \frac{C_{AB}M_A}{E} - 27 \frac{C_{AB}M_A}{E} - 4.5 \frac{C_{AB}M_A}{E} = 0$$

or

$$-1 + 12.94 \frac{M_A}{E} - 36.56 \frac{C_{AB}M_A}{E} = 0 \qquad (1)$$

Since the deflection at A must be zero, we require

$$\uparrow + \Sigma M_{A'} = 0; \qquad -5.06 \frac{M_A}{E}(3) - 3.375 \frac{M_A}{E}(4.5)$$

$$- 4.5 \frac{M_A}{E}(10) + 5.06 \frac{C_{AB}M_A}{E}(6)$$

$$+ 27 \frac{C_{AB}M_A}{E}(10.5) + 4.5 \frac{C_{AB}M_A}{E}(11) = 0$$

Factoring out M_A/E and solving for the carry-over factor, we have

$$-75.37 + 363.36C_{AB} = 0 \qquad C_{AB} = 0.207 \qquad \textbf{Ans.}$$

Substituting into Eq. (1), setting $E = 10(10^3)$ ksi $= 1440(10^3)$ ksf, and solving for the stiffness factor yields

$$M_A = K_A = 268(10^3) \text{ k} \cdot \text{ft} \qquad \textbf{Ans.}$$

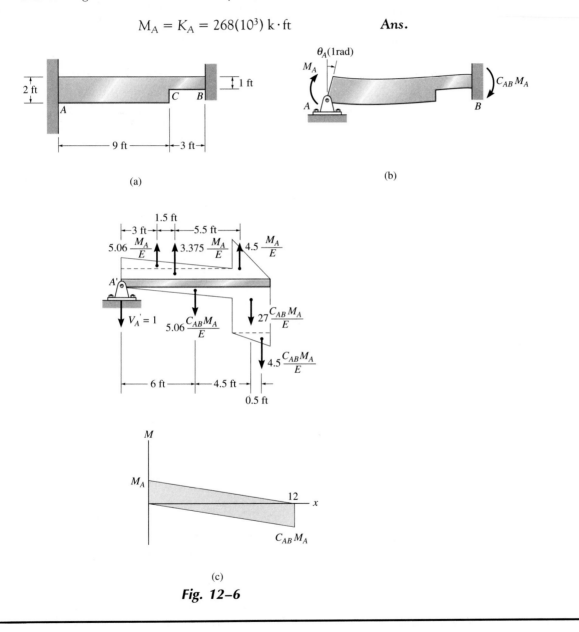

(a)

(b)

(c)

Fig. 12–6

615

Example 12–4

Determine the stiffness and carry-over factors for the end A of the tapered beam shown in Fig. 12–7a. The cross-sectional area has a constant width of 1 ft. E is constant.

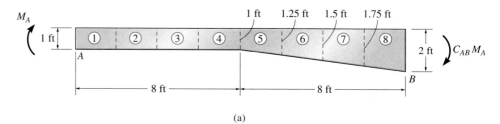

(a)

Solution

As an approximation, the beam will be segmented every 2 ft. The segments are identified numerically in Fig. 12–7a. The moment diagram is drawn in parts as that due to M_A and $C_{AB}M_A$, and the fractional values of M_A and $C_{AB}M_A$ are computed, Fig. 12–7b. The beam's moment of inertia is computed using $\frac{1}{12}bh^3$ and the values are shown in Fig. 12–7c. Using these values, the conjugate beam and computed loading (M/EI) are shown in Fig. 12–7d. Computations for the stiffness and carryover factors require a summation of vertical forces and moments about A'. For simplicity the "distributed" load-

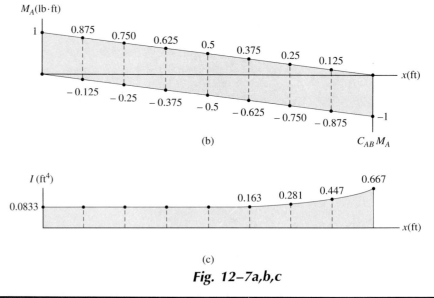

Fig. 12–7a,b,c

ing is segmented into triangles, as shown by the inclined dashed lines. From this it will be noted that pairs of triangles have a common vertical "base" and horizontal "height" (2 ft). The centroid of these pairs is indicated by a dot in the figure. The area (force) and moment computations are tabulated in Fig. 12–7e. Using these results and applying the equilibrium equations to the conjugate beam yields

$$+\uparrow \Sigma F_y = 0; \qquad -1 + 84.94\,\frac{M_A}{E} - 48.42C_{AB}\,\frac{M_A}{E} = 0$$

$$\zeta + \Sigma M_{A'} = 0; \qquad -383.20\,\frac{M_A}{E} + 398.56C_{AB}\,\frac{M_A}{E} = 0$$

Solving yields

$$C_{AB} = 0.961 \qquad\qquad \textbf{Ans.}$$

$$K_A = M_A = \frac{E}{38.39} \qquad\qquad \textbf{Ans.}$$

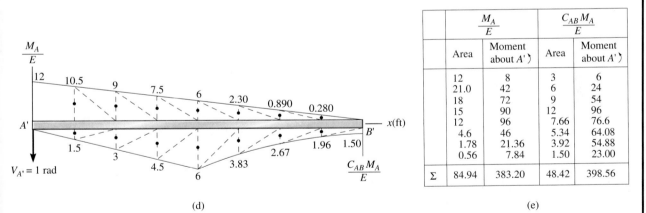

	$\dfrac{M_A}{E}$		$\dfrac{C_{AB}M_A}{E}$	
	Area	Moment about A')	Area	Moment about A')
	12	8	3	6
	21.0	42	6	24
	18	72	9	54
	15	90	12	96
	12	96	7.66	76.6
	4.6	46	5.34	64.08
	1.78	21.36	3.92	54.88
	0.56	7.84	1.50	23.00
Σ	84.94	383.20	48.42	398.56

(d)   (e)

Fig. 12–7d,e

617

12.3 Loading Properties of Nonprismatic Members Available from Publications

As noted in the previous section, considerable labor is often involved in determining the fixed-end moments, stiffness factors, and carry-over factors for a nonprismatic member. As a result, graphs and tables have been made available to determine this data for common shapes used in structural design. One such source is the *Handbook of Frame Constants*, published by the Portland Cement Association.* A portion of these tables, taken from this publication, is listed here as Tables 12–1 and 12–2. A more complete tabular form of the data is given in the PCA handbook along with the relevant derivations of formulas used.

The nomenclature in the tables is defined as follows:

a_A, a_B = ratio of the length of haunch at end A or B to the length of span

b = ratio of the distance from the concentrated load to end A to the length of span

C_{AB}, C_{BA} = carry-over factors of member AB at ends A and B, respectively

h_A, h_B = depth of member at ends A and B, respectively

h_C = depth of member at minimum section

I_C = moment of inertia of section at minimum depth

k_{AB}, k_{BA} = stiffness factor at ends A and B, respectively

L = length of member

M_{AB}, M_{BA} = fixed-end moment at ends A and B, respectively; specified in tables for uniform load w or concentrated force P

r_A, r_B = ratios for rectangular cross-sectional areas, where $r_A = (h_A - h_C)/h_C$, $r_B = (h_B - h_C)/h_C$

As noted, the fixed-end moments and carry-over factors are taken directly from the tables. The absolute stiffness factor can be determined using the tabulated stiffness factors and found from

$$K_A = \frac{k_{AB}EI_C}{L} \qquad K_B = \frac{k_{BA}EI_C}{L} \qquad (12-2)$$

Application of the use of the tables will be illustrated in Example 12–5.

---

*See the references at the end of the chapter.

Table 12–1   Straight Haunches—Constant Width

Note: All carry-over factors and fixed-end moment coefficients are negative and all stiffness factors are positive.

Right Haunch		Carry-over Factors		Stiffness Factors		Unif. Load FEM Coef. × wL²		Concentrated Load FEM—Coef. × PL, b = 0.1		b = 0.3		b = 0.5		b = 0.7		b = 0.9		Haunch Load at Left FEM Coef. × W_A L²		Haunch Load at Right FEM Coef. × W_B L²	
a_B	r_B	C_{AB}	C_{BA}	k_{AB}	k_{BA}	M_{AB}	M_{BA}	M_{AB}	M_{BA}	M_{AB}	M_{BA}	M_{AB}	M_{BA}	M_{AB}	M_{BA}	M_{AB}	M_{BA}	M_{AB}	M_{BA}	M_{AB}	M_{BA}
								$a_A = 0.3$ a_B = variable $r_A = 1.0$ r_B = variable													
0.1	0.4	0.499	0.784	8.79	5.60	0.1237	0.0728	0.0940	0.0030	0.2216	0.0337	0.2036	0.1025	0.0981	0.1466	0.0100	0.0867	0.0134	0.0007	0.0001	0.0016
	0.6	0.515	0.782	8.95	5.89	0.1213	0.0759	0.0938	0.0032	0.2203	0.0355	0.2000	0.1074	0.0931	0.1532	0.0079	0.0893	0.0134	0.0008	0.0001	0.0016
	1.0	0.535	0.779	9.18	6.31	0.1181	0.0803	0.0937	0.0033	0.2175	0.0378	0.1951	0.1143	0.0864	0.1622	0.0054	0.0928	0.0133	0.0008	0.0000	0.0016
	1.5	0.551	0.777	9.35	6.64	0.1157	0.0836	0.0935	0.0035	0.2173	0.0396	0.1910	0.1196	0.0813	0.1693	0.0035	0.0952	0.0133	0.0009	0.0000	0.0016
	2.0	0.561	0.775	9.46	6.85	0.1141	0.0857	0.0935	0.0037	0.2162	0.0411	0.1885	0.1230	0.0781	0.1737	0.0025	0.0966	0.0133	0.0009	0.0000	0.0017
0.2	0.4	0.543	0.766	9.19	6.52	0.1194	0.0791	0.0935	0.0034	0.2185	0.0384	0.1955	0.1147	0.0889	0.1601	0.0096	0.0870	0.0133	0.0008	0.0006	0.0058
	0.6	0.576	0.758	9.53	7.24	0.1152	0.0851	0.0934	0.0038	0.2158	0.0422	0.1883	0.1250	0.0798	0.1729	0.0075	0.0898	0.0133	0.0009	0.0005	0.0060
	1.0	0.622	0.748	10.06	8.37	0.1089	0.0942	0.0931	0.0042	0.2118	0.0480	0.1771	0.1411	0.0668	0.1919	0.0047	0.0935	0.0132	0.0011	0.0004	0.0062
	1.5	0.660	0.740	10.52	9.38	0.1037	0.1018	0.0927	0.0047	0.2085	0.0530	0.1678	0.1550	0.0559	0.2078	0.0028	0.0961	0.0130	0.0012	0.0002	0.0064
	2.0	0.684	0.734	10.83	10.09	0.1002	0.1069	0.0924	0.0050	0.2062	0.0565	0.1614	0.1645	0.0487	0.2185	0.0019	0.0974	0.0129	0.0013	0.0001	0.0065
0.3	0.4	0.579	0.741	9.47	7.40	0.1175	0.0822	0.0934	0.0037	0.2164	0.0419	0.1909	0.1225	0.0856	0.1649	0.0100	0.0861	0.0133	0.0009	0.0022	0.0118
	0.6	0.629	0.726	9.98	8.64	0.1120	0.0902	0.0931	0.0042	0.2126	0.0477	0.1808	0.1379	0.0747	0.1807	0.0080	0.0888	0.0132	0.0010	0.0018	0.0124
	1.0	0.705	0.705	10.85	10.85	0.1034	0.1034	0.0924	0.0052	0.2063	0.0577	0.1640	0.1640	0.0577	0.2063	0.0052	0.0924	0.0131	0.0013	0.0013	0.0131
	1.5	0.771	0.689	11.70	13.10	0.0956	0.1157	0.0917	0.0062	0.2002	0.0675	0.1483	0.1892	0.0428	0.2294	0.0033	0.0953	0.0129	0.0015	0.0008	0.0137
	2.0	0.817	0.678	12.33	14.85	0.0901	0.1246	0.0913	0.0069	0.1957	0.0750	0.1368	0.2080	0.0326	0.2455	0.0022	0.0968	0.0128	0.0017	0.0006	0.0141
0.4	0.4	0.604	0.713	9.64	8.16	0.1167	0.0826	0.0933	0.0040	0.2153	0.0440	0.1883	0.1260	0.0851	0.1626	0.0104	0.0849	0.0133	0.0009	0.0052	0.0186
	0.6	0.669	0.689	10.28	9.97	0.1109	0.0915	0.0930	0.0047	0.2107	0.0515	0.1766	0.1444	0.0743	0.1779	0.0084	0.0874	0.0132	0.0011	0.0044	0.0196
	1.0	0.777	0.657	11.46	13.55	0.1010	0.1071	0.0922	0.0058	0.2022	0.0655	0.1560	0.1784	0.0572	0.2033	0.0059	0.0910	0.0130	0.0015	0.0033	0.0212
	1.5	0.878	0.631	12.74	17.72	0.0914	0.1232	0.0910	0.0075	0.1933	0.0811	0.1349	0.2148	0.0420	0.2272	0.0039	0.0938	0.0128	0.0019	0.0024	0.0227
	2.0	0.952	0.614	13.80	21.39	0.0841	0.1360	0.0904	0.0089	0.1861	0.0948	0.1182	0.2448	0.0315	0.2444	0.0026	0.0955	0.0126	0.0022	0.0017	0.0237
0.5	0.4	0.616	0.686	9.75	8.76	0.1161	0.0818	0.0932	0.0041	0.2144	0.0449	0.1864	0.1255	0.0846	0.1587	0.0107	0.0840	0.0133	0.0010	0.0097	0.0254
	0.6	0.692	0.653	10.46	11.09	0.1101	0.0905	0.0931	0.0049	0.2094	0.0532	0.1740	0.1444	0.0743	0.1723	0.0087	0.0863	0.0131	0.0012	0.0085	0.0270
	1.0	0.827	0.607	11.85	16.13	0.0999	0.1066	0.0920	0.0064	0.1997	0.0699	0.1521	0.1803	0.0579	0.1952	0.0063	0.0894	0.0129	0.0016	0.0067	0.0295
	1.5	0.966	0.572	13.49	22.79	0.0896	0.1246	0.0905	0.0086	0.1889	0.0907	0.1289	0.2219	0.0439	0.2173	0.0044	0.0923	0.0126	0.0022	0.0051	0.0320
	2.0	1.076	0.548	14.99	29.43	0.0814	0.1401	0.0898	0.0106	0.1790	0.1116	0.1095	0.2587	0.0339	0.2342	0.0031	0.0940	0.0124	0.0026	0.0039	0.0339
								$a_A = 0.2$ a_B = variable $r_A = 1.5$ r_B = variable													
0.1	0.4	0.523	0.730	7.63	5.46	0.1205	0.0736	0.0967	0.0016	0.2212	0.0335	0.1924	0.1060	0.0906	0.1491	0.0090	0.0870	0.0064	0.0001	0.0001	0.0016
	0.6	0.539	0.728	7.77	5.75	0.1184	0.0767	0.0967	0.0017	0.2201	0.0351	0.1889	0.1109	0.0861	0.1556	0.0072	0.0896	0.0064	0.0001	0.0001	0.0016
	1.0	0.560	0.726	7.96	6.14	0.1154	0.0810	0.0966	0.0018	0.2186	0.0373	0.1842	0.1178	0.0799	0.1646	0.0049	0.0929	0.0064	0.0001	0.0000	0.0016
	1.5	0.577	0.724	8.10	6.46	0.1132	0.0843	0.0965	0.0019	0.2174	0.0390	0.1806	0.1232	0.0751	0.1716	0.0032	0.0953	0.0064	0.0001	0.0000	0.0016
	2.0	0.587	0.722	8.19	6.66	0.1118	0.0863	0.0965	0.0019	0.2166	0.0402	0.1782	0.1265	0.0721	0.1760	0.0023	0.0966	0.0064	0.0001	0.0000	0.0017
0.2	0.4	0.569	0.714	7.97	6.35	0.1166	0.0799	0.0966	0.0019	0.2186	0.0377	0.1847	0.1183	0.0821	0.1626	0.0088	0.0873	0.0064	0.0001	0.0006	0.0058
	0.6	0.603	0.707	8.26	7.04	0.1127	0.0858	0.0965	0.0021	0.2163	0.0413	0.1778	0.1288	0.0736	0.1752	0.0068	0.0901	0.0064	0.0001	0.0005	0.0060
	1.0	0.652	0.698	8.70	8.12	0.1069	0.0947	0.0963	0.0023	0.2127	0.0468	0.1675	0.1449	0.0616	0.1940	0.0043	0.0937	0.0064	0.0002	0.0004	0.0062
	1.5	0.691	0.691	9.08	9.08	0.1021	0.1021	0.0962	0.0025	0.2097	0.0515	0.1587	0.1587	0.0515	0.2097	0.0025	0.0962	0.0064	0.0002	0.0002	0.0064
	2.0	0.716	0.686	9.34	9.75	0.0990	0.1071	0.0960	0.0028	0.2077	0.0547	0.1528	0.1681	0.0449	0.2202	0.0017	0.0975	0.0064	0.0002	0.0001	0.0065
0.3	0.4	0.607	0.692	8.21	7.21	0.1148	0.0829	0.0965	0.0021	0.2168	0.0409	0.1801	0.1263	0.0789	0.1674	0.0091	0.0866	0.0064	0.0002	0.0020	0.0118
	0.6	0.659	0.678	8.65	8.40	0.1098	0.0907	0.0964	0.0024	0.2135	0.0464	0.1706	0.1418	0.0688	0.1831	0.0072	0.0892	0.0064	0.0002	0.0017	0.0123
	1.0	0.740	0.660	9.38	10.52	0.1018	0.1037	0.0961	0.0028	0.2078	0.0559	0.1550	0.1678	0.0530	0.2085	0.0047	0.0927	0.0064	0.0002	0.0012	0.0130
	1.5	0.809	0.645	10.09	13.56	0.0947	0.1156	0.0958	0.0033	0.2024	0.0651	0.1403	0.1928	0.0393	0.2311	0.0029	0.0950	0.0063	0.0003	0.0008	0.0137
	2.0	0.857	0.636	10.62	14.32	0.0897	0.1242	0.0955	0.0038	0.1985	0.0720	0.1296	0.2119	0.0299	0.2469	0.0020	0.0968	0.0063	0.0003	0.0005	0.0141
0.4	0.4	0.634	0.667	8.37	7.96	0.1141	0.0833	0.0964	0.0022	0.2158	0.0429	0.1779	0.1297	0.0787	0.1650	0.0095	0.0853	0.0064	0.0002	0.0048	0.0188
	0.6	0.703	0.645	8.91	9.71	0.1087	0.0920	0.0962	0.0026	0.2117	0.0500	0.1669	0.1482	0.0686	0.1802	0.0077	0.0877	0.0064	0.0002	0.0041	0.0197
	1.0	0.817	0.616	9.92	13.17	0.0996	0.1073	0.0958	0.0033	0.2043	0.0633	0.1475	0.1821	0.0528	0.2054	0.0053	0.0912	0.0064	0.0002	0.0030	0.0213
	1.5	0.924	0.593	11.02	17.18	0.0908	0.1229	0.0954	0.0042	0.1964	0.0780	0.1278	0.2182	0.0387	0.2289	0.0036	0.0936	0.0063	0.0003	0.0022	0.0228
	2.0	1.003	0.578	11.93	20.70	0.0842	0.1353	0.0950	0.0049	0.1900	0.0904	0.1121	0.2479	0.0290	0.2459	0.0025	0.0956	0.0063	0.0004	0.0016	0.0238
0.5	0.4	0.649	0.641	8.46	8.57	0.1137	0.0823	0.0964	0.0022	0.2151	0.0437	0.1763	0.1290	0.0785	0.1610	0.0097	0.0843	0.0064	0.0002	0.0090	0.0257
	0.6	0.731	0.612	9.08	10.84	0.1081	0.0908	0.0961	0.0027	0.2106	0.0516	0.1647	0.1478	0.0688	0.1744	0.0080	0.0865	0.0064	0.0002	0.0079	0.0272
	1.0	0.875	0.571	10.29	15.76	0.0989	0.1065	0.0956	0.0037	0.2021	0.0677	0.1442	0.1836	0.0540	0.1969	0.0058	0.0897	0.0063	0.0003	0.0062	0.0297
	1.5	1.023	0.539	11.73	22.25	0.0895	0.1237	0.0951	0.0048	0.1924	0.0874	0.1225	0.2247	0.0408	0.2189	0.0041	0.0923	0.0062	0.0004	0.0047	0.0322
	2.0	1.141	0.519	13.06	28.73	0.0820	0.1385	0.0945	0.0060	0.1836	0.1060	0.1044	0.2609	0.0316	0.2355	0.0030	0.0941	0.0062	0.0005	0.0037	0.0341

619

Table 12–2 Parabolic Haunches—Constant Width

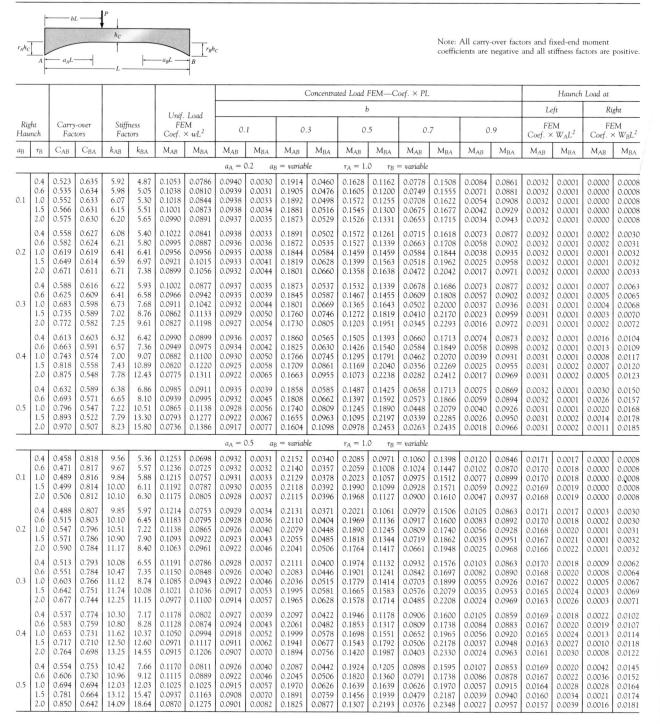

Note: All carry-over factors and fixed-end moment coefficients are negative and all stiffness factors are positive.

Right Haunch a_B	r_B	Carry-over Factors C_{AB}	C_{BA}	Stiffness Factors k_{AB}	k_{BA}	Unif. Load FEM Coef. × wL^2 M_{AB}	M_{BA}	b=0.1 M_{AB}	M_{BA}	b=0.3 M_{AB}	M_{BA}	b=0.5 M_{AB}	M_{BA}	b=0.7 M_{AB}	M_{BA}	b=0.9 M_{AB}	M_{BA}	Haunch Load at Left FEM Coef. × $W_A L^2$ M_{AB}	M_{BA}	Haunch Load at Right FEM Coef. × $W_B L^2$ M_{AB}	M_{BA}
										$a_A = 0.2$	$a_B =$ variable	$r_A = 1.0$	$r_B =$ variable								
0.1	0.4	0.523	0.635	5.92	4.87	0.1053	0.0786	0.0940	0.0030	0.1914	0.0460	0.1628	0.1162	0.0778	0.1508	0.0084	0.0861	0.0032	0.0001	0.0000	0.0008
	0.6	0.535	0.634	5.98	5.05	0.1038	0.0810	0.0939	0.0031	0.1905	0.0476	0.1605	0.1200	0.0749	0.1555	0.0071	0.0881	0.0032	0.0001	0.0000	0.0008
	1.0	0.552	0.633	6.07	5.30	0.1018	0.0844	0.0938	0.0033	0.1892	0.0498	0.1572	0.1255	0.0708	0.1622	0.0054	0.0908	0.0032	0.0001	0.0000	0.0008
	1.5	0.566	0.631	6.15	5.51	0.1001	0.0873	0.0938	0.0034	0.1881	0.0516	0.1545	0.1300	0.0675	0.1677	0.0042	0.0929	0.0032	0.0001	0.0000	0.0008
	2.0	0.575	0.630	6.20	5.65	0.0990	0.0891	0.0937	0.0035	0.1873	0.0529	0.1526	0.1331	0.0653	0.1715	0.0034	0.0943	0.0032	0.0001	0.0000	0.0008
0.2	0.4	0.558	0.627	6.08	5.40	0.1022	0.0841	0.0938	0.0033	0.1891	0.0502	0.1572	0.1261	0.0715	0.1618	0.0073	0.0877	0.0032	0.0001	0.0002	0.0030
	0.6	0.582	0.624	6.21	5.80	0.0995	0.0887	0.0936	0.0036	0.1872	0.0535	0.1527	0.1339	0.0663	0.1708	0.0058	0.0902	0.0032	0.0001	0.0002	0.0031
	1.0	0.619	0.619	6.41	6.41	0.0956	0.0956	0.0935	0.0038	0.1844	0.0584	0.1459	0.1459	0.0584	0.1844	0.0038	0.0935	0.0032	0.0001	0.0001	0.0032
	1.5	0.649	0.614	6.59	6.97	0.0921	0.1015	0.0933	0.0041	0.1819	0.0628	0.1399	0.1563	0.0518	0.1962	0.0025	0.0958	0.0032	0.0001	0.0001	0.0032
	2.0	0.671	0.611	6.71	7.38	0.0899	0.1056	0.0932	0.0044	0.1801	0.0660	0.1358	0.1638	0.0472	0.2042	0.0017	0.0971	0.0032	0.0001	0.0000	0.0033
0.3	0.4	0.588	0.616	6.22	5.93	0.1002	0.0877	0.0937	0.0035	0.1873	0.0537	0.1532	0.1339	0.0678	0.1686	0.0073	0.0877	0.0032	0.0001	0.0007	0.0063
	0.6	0.625	0.609	6.41	6.58	0.0966	0.0942	0.0935	0.0039	0.1845	0.0587	0.1467	0.1455	0.0609	0.1808	0.0057	0.0902	0.0032	0.0001	0.0005	0.0065
	1.0	0.683	0.598	6.73	7.68	0.0911	0.1042	0.0932	0.0044	0.1801	0.0669	0.1365	0.1643	0.0502	0.2000	0.0037	0.0936	0.0031	0.0001	0.0004	0.0068
	1.5	0.735	0.589	7.02	8.76	0.0862	0.1133	0.0929	0.0050	0.1760	0.0746	0.1272	0.1819	0.0410	0.2170	0.0023	0.0959	0.0031	0.0001	0.0003	0.0070
	2.0	0.772	0.582	7.25	9.61	0.0827	0.1198	0.0927	0.0054	0.1730	0.0805	0.1203	0.1951	0.0345	0.2293	0.0016	0.0972	0.0031	0.0001	0.0002	0.0072
0.4	0.4	0.613	0.603	6.32	6.42	0.0990	0.0911	0.0936	0.0037	0.1860	0.0565	0.1505	0.1393	0.0660	0.1713	0.0074	0.0873	0.0032	0.0001	0.0016	0.0104
	0.6	0.663	0.591	6.57	7.36	0.0949	0.0975	0.0934	0.0042	0.1825	0.0630	0.1426	0.1540	0.0584	0.1849	0.0058	0.0898	0.0032	0.0001	0.0013	0.0109
	1.0	0.743	0.574	7.00	9.07	0.0882	0.1100	0.0930	0.0050	0.1766	0.0745	0.1295	0.1791	0.0462	0.2070	0.0039	0.0931	0.0031	0.0001	0.0008	0.0117
	1.5	0.818	0.558	7.43	10.89	0.0820	0.1220	0.0925	0.0058	0.1709	0.0861	0.1169	0.2040	0.0356	0.2269	0.0025	0.0955	0.0031	0.0002	0.0007	0.0120
	2.0	0.875	0.548	7.78	12.43	0.0775	0.1311	0.0922	0.0065	0.1663	0.0955	0.1073	0.2238	0.0282	0.2412	0.0017	0.0969	0.0031	0.0002	0.0005	0.0123
0.5	0.4	0.632	0.589	6.38	6.86	0.0985	0.0911	0.0935	0.0039	0.1858	0.0585	0.1487	0.1425	0.0658	0.1713	0.0075	0.0869	0.0032	0.0001	0.0030	0.0150
	0.6	0.693	0.571	6.65	8.10	0.0939	0.0995	0.0932	0.0045	0.1808	0.0662	0.1397	0.1592	0.0573	0.1866	0.0059	0.0894	0.0032	0.0001	0.0026	0.0157
	1.0	0.796	0.547	7.22	10.51	0.0865	0.1138	0.0928	0.0056	0.1740	0.0809	0.1245	0.1890	0.0448	0.2079	0.0040	0.0926	0.0031	0.0001	0.0020	0.0168
	1.5	0.893	0.522	7.79	13.30	0.0793	0.1277	0.0922	0.0067	0.1655	0.0963	0.1095	0.2197	0.0339	0.2285	0.0026	0.0950	0.0031	0.0002	0.0014	0.0178
	2.0	0.970	0.507	8.23	15.80	0.0736	0.1386	0.0917	0.0077	0.1604	0.1098	0.0978	0.2453	0.0263	0.2435	0.0018	0.0966	0.0031	0.0002	0.0011	0.0185
										$a_A = 0.5$	$a_B =$ variable	$r_A = 1.0$	$r_B =$ variable								
0.1	0.4	0.458	0.818	9.56	5.36	0.1253	0.0698	0.0932	0.0031	0.2152	0.0340	0.2085	0.0971	0.1060	0.1398	0.0120	0.0846	0.0171	0.0017	0.0000	0.0008
	0.6	0.471	0.817	9.67	5.57	0.1236	0.0725	0.0932	0.0032	0.2140	0.0357	0.2059	0.1008	0.1024	0.1447	0.0102	0.0870	0.0170	0.0018	0.0000	0.0008
	1.0	0.489	0.816	9.84	5.88	0.1215	0.0757	0.0931	0.0033	0.2129	0.0378	0.2023	0.1057	0.0975	0.1512	0.0077	0.0899	0.0170	0.0018	0.0000	0.0008
	1.5	0.499	0.814	10.00	6.11	0.1192	0.0787	0.0930	0.0035	0.2118	0.0392	0.1990	0.1099	0.0928	0.1571	0.0059	0.0922	0.0169	0.0019	0.0000	0.0008
	2.0	0.506	0.812	10.10	6.30	0.1175	0.0805	0.0928	0.0037	0.2115	0.0396	0.1968	0.1127	0.0900	0.1610	0.0047	0.0937	0.0168	0.0019	0.0000	0.0008
0.2	0.4	0.488	0.807	9.85	5.97	0.1214	0.0753	0.0929	0.0034	0.2131	0.0371	0.2021	0.1061	0.0979	0.1506	0.0105	0.0863	0.0171	0.0017	0.0003	0.0030
	0.6	0.515	0.803	10.10	6.45	0.1183	0.0795	0.0928	0.0036	0.2110	0.0404	0.1969	0.1136	0.0917	0.1600	0.0083	0.0892	0.0170	0.0018	0.0002	0.0030
	1.0	0.547	0.796	10.51	7.22	0.1138	0.0865	0.0926	0.0040	0.2079	0.0448	0.1890	0.1245	0.0809	0.1740	0.0056	0.0928	0.0168	0.0020	0.0001	0.0031
	1.5	0.571	0.786	10.90	7.90	0.1093	0.0922	0.0923	0.0043	0.2055	0.0485	0.1818	0.1344	0.0719	0.1862	0.0035	0.0951	0.0167	0.0021	0.0001	0.0032
	2.0	0.590	0.784	11.17	8.40	0.1063	0.0961	0.0922	0.0046	0.2041	0.0506	0.1764	0.1417	0.0661	0.1948	0.0025	0.0968	0.0166	0.0022	0.0001	0.0032
0.3	0.4	0.513	0.793	10.08	6.55	0.1191	0.0786	0.0928	0.0037	0.2111	0.0400	0.1974	0.1132	0.0932	0.1576	0.0103	0.0863	0.0170	0.0018	0.0009	0.0062
	0.6	0.551	0.784	10.47	7.35	0.1150	0.0848	0.0926	0.0040	0.2083	0.0446	0.1901	0.1241	0.0842	0.1697	0.0082	0.0890	0.0168	0.0020	0.0008	0.0064
	1.0	0.603	0.766	11.12	8.74	0.1085	0.0943	0.0922	0.0046	0.2036	0.0515	0.1779	0.1414	0.0703	0.1899	0.0055	0.0926	0.0167	0.0022	0.0005	0.0067
	1.5	0.642	0.751	11.74	10.08	0.1021	0.1036	0.0917	0.0053	0.1995	0.0581	0.1665	0.1583	0.0576	0.2079	0.0035	0.0953	0.0165	0.0024	0.0003	0.0069
	2.0	0.677	0.744	12.25	11.15	0.0977	0.1100	0.0914	0.0057	0.1965	0.0628	0.1578	0.1714	0.0485	0.2208	0.0024	0.0969	0.0163	0.0026	0.0003	0.0071
0.4	0.4	0.537	0.774	10.30	7.17	0.1178	0.0802	0.0927	0.0039	0.2097	0.0422	0.1946	0.1178	0.0906	0.1600	0.0105	0.0859	0.0169	0.0018	0.0022	0.0102
	0.6	0.583	0.759	10.80	8.28	0.1128	0.0874	0.0924	0.0043	0.2061	0.0482	0.1853	0.1317	0.0809	0.1738	0.0084	0.0883	0.0167	0.0020	0.0019	0.0107
	1.0	0.653	0.731	11.62	10.37	0.1050	0.0994	0.0918	0.0052	0.1999	0.0578	0.1698	0.1551	0.0652	0.1965	0.0056	0.0920	0.0165	0.0024	0.0013	0.0114
	1.5	0.717	0.710	12.50	12.60	0.0971	0.1117	0.0911	0.0062	0.1941	0.0677	0.1543	0.1792	0.0506	0.2178	0.0037	0.0948	0.0163	0.0027	0.0010	0.0118
	2.0	0.764	0.698	13.25	14.55	0.0915	0.1206	0.0907	0.0070	0.1894	0.0756	0.1420	0.1987	0.0403	0.2330	0.0024	0.0963	0.0161	0.0030	0.0008	0.0122
0.5	0.4	0.554	0.753	10.42	7.66	0.1170	0.0811	0.0926	0.0040	0.2087	0.0442	0.1924	0.1205	0.0898	0.1595	0.0107	0.0853	0.0169	0.0020	0.0042	0.0145
	0.6	0.606	0.730	10.96	9.12	0.1115	0.0889	0.0922	0.0046	0.2045	0.0506	0.1820	0.1360	0.0791	0.1738	0.0086	0.0878	0.0167	0.0022	0.0036	0.0152
	1.0	0.694	0.694	12.03	12.03	0.1025	0.1025	0.0915	0.0057	0.1970	0.0626	0.1639	0.1639	0.0626	0.1970	0.0057	0.0915	0.0164	0.0028	0.0028	0.0164
	1.5	0.781	0.664	13.12	15.47	0.0937	0.1163	0.0908	0.0070	0.1891	0.0759	0.1456	0.1939	0.0479	0.2187	0.0039	0.0940	0.0160	0.0034	0.0021	0.0174
	2.0	0.850	0.642	14.09	18.64	0.0870	0.1275	0.0901	0.0082	0.1825	0.0877	0.1307	0.2193	0.0376	0.2348	0.0027	0.0957	0.0157	0.0039	0.0016	0.0181

12.4   Moment Distribution for Structures Having Nonprismatic Members

Once the fixed-end moments and stiffness and carryover factors for the nonprismatic members of a structure have been determined, application of the moment-distribution method follows the same procedure as outlined in Chapter 11. In this regard, recall that the distribution of moments may be shortened if a member stiffness factor is modified to account for conditions of end-span pin support and structure symmetry or antisymmetry. Similar modifications can also be made to nonprismatic members.

Beam Pin-Supported at Far End

Consider the beam in Fig. 12–8a, which is pinned at its far end B. The absolute stiffness factor K'_A is the moment applied at A such that it rotates the beam at A through 1 rad. It can be determined as follows. First assume that B is temporarily fixed and a moment K_A is applied at A, Fig. 12–8b. The moment induced at B is $C_{AB}K_A$, where C_{AB} is the carry-over factor from A to B. Second, since B is not to be fixed, application of the opposite moment $C_{AB}K_A$ to the beam, Fig. 12–8c, will induce a moment $C_{BA}C_{AB}K_A$ at end A. By superposition, the result of these two applications of moment yields the beam loaded as shown in Fig. 12–8a. Hence it can be seen that the absolute stiffness factor of the beam at A is

$$\boxed{K'_A = K_A(1 - C_{AB}C_{BA})} \qquad (12\text{–}3)$$

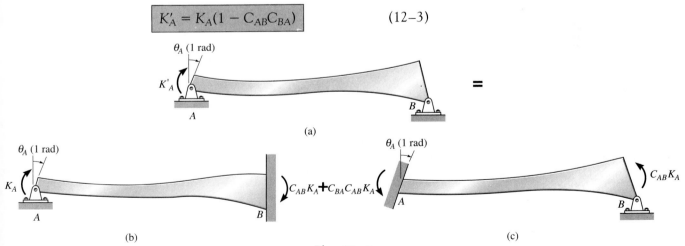

Fig. 12–8

Here K_A is the absolute stiffness factor of the beam, assuming it to be fixed at the far end B. For example, in the case of a prismatic beam, $K_A = 4EI/L$ and $C_{AB} = C_{BA} = \frac{1}{2}$. Substituting into Eq. 12–3 yields $K'_A = 3EI/L$, the same as Eq. 11–4.

Symmetric Beam and Loading

Here we must determine the moment K'_A needed to rotate end A $\theta_A = +1$ rad while $\theta_B = -1$ rad, Fig. 12–9a. In this case we first assume that end B is fixed and apply the moment K_A at A, Fig. 12–9b. Next we apply a negative moment K_B to end B assuming that end A is fixed. This results in a moment of $C_{BA}K_B$ at end A as shown in Fig. 12–9c. Superposition of these two applications of moment at A yields the results of Fig. 12–9a. We require

$$K'_A = K_A - C_{BA}K_B$$

Using Eq. 12–1 ($C_{BA}K_B = C_{AB}K_A$), we can also write

$$\boxed{K'_A = K_A(1 - C_{AB})} \tag{12–4}$$

In the case of a prismatic beam, $K_A = 4EI/L$ and $C_{AB} = \frac{1}{2}$, so that $K'_A = 2EI/L$, which is the same as Eq. 11–5.

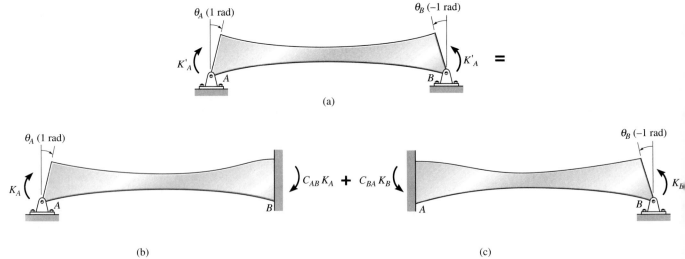

Fig. 12–9

Symmetric Beam with Antisymmetric Loading

In the case of a symmetric beam with antisymmetric loading, we must determine K'_A such that equal rotations occur at the ends of the beam, Fig. 12–10a. To do this, we first fix end B and apply the moment K_A at A, Fig. 12–10b. Likewise, application of K_B at end B while end A is fixed is shown in Fig. 12–10c. Superposition of both cases yields the results of Fig. 12–10a. Hence,

$$K'_A = K_A + C_{BA}K_B$$

or, using Eq. 12–1 ($C_{BA}K_B = C_{AB}K_A$), we have for the absolute stiffness

$$K'_A = K_A(1 + C_{AB}) \qquad (12–5)$$

Substituting the data for a prismatic member, $K_A = 4EI/L$ and $C_{AB} = \frac{1}{2}$, yields $K'_A = 6EI/L$, which is the same as Eq. 11–6.

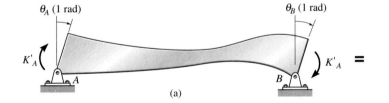

(a)

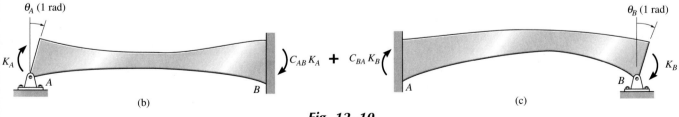

(b)

(c)

Fig. 12–10

623

Relative Joint Translation of Beam

Fixed-end moments are developed in a nonprismatic member if it has a relative joint translation Δ between its ends A and B, Fig. 12–11a. In order to determine these moments, we proceed as follows. First consider the ends A and B to be pin-connected and allow the beam to be displaced a distance Δ such that the end rotations are $\theta_A = \theta_B = \Delta/L$, Fig. 12–11$b$. Second, assume that B is fixed and apply a moment of $M'_A = -K_A(\Delta/L)$ to end A such that it rotates the end $\theta_A = -\Delta/L$, Fig. 12–11c. Third, assume that A is fixed and apply a moment $M'_B = -K_B(\Delta/L)$ to end B such that it rotates the end $\theta_B = -\Delta/L$, Fig.

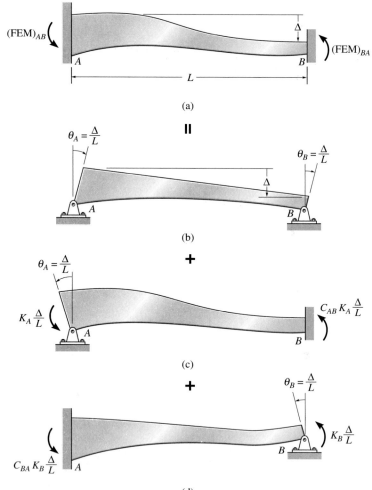

(a)

$\mathbf{\|}$

(b)

$\mathbf{+}$

(c)

$\mathbf{+}$

(d)

Fig. 12–11

12–11*d*. Since the total sum of these three operations yields the condition shown in Fig. 12–11*a*, we have at *A*

$$(\text{FEM})_{AB} = -K_A \frac{\Delta}{L} - C_{BA} K_B \frac{\Delta}{L}$$

Applying Eq. 12–1 ($C_{BA}K_B = C_{AB}K_A$), yields

$$(\text{FEM})_{AB} = -K_A \frac{\Delta}{L}(1 + C_{AB}) \qquad (12\text{–}6)$$

A similar expression can be written for end *B*. Recall that for a prismatic member $K_A = 4EI/L$ and $C_{AB} = \frac{1}{2}$. Thus $(\text{FEM})_{AB} = -6EI\Delta/L^2$, which is the same as Eq. 9–5.

If end *B* is pinned rather than fixed, Fig. 12–12, the fixed-end moment at *A* can be determined in a manner similar to that described above. The result is

$$(\text{FEM})'_{AB} = -K_A \frac{\Delta}{L}(1 - C_{AB}C_{BA}) \qquad (12\text{–}7)$$

Here it is seen that for a prismatic member this equation gives $(\text{FEM})_{AB} = -3EI\Delta/L^2$, which is the same as that listed on the inside back cover.

The following example illustrates application of the moment-distribution method to structures having nonprismatic members. Once the fixed-end moments and stiffness and carry-over factors have been determined, and the stiffness factor modified according to the equations given above, the procedure for analysis is the same as that discussed in Chapter 11.

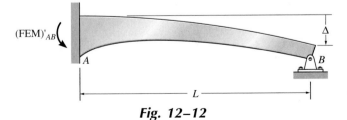

Fig. 12–12

625

Example 12–5

Determine the internal moments at the supports of the beam shown in Fig. 12–13a. The beam has a thickness of 1 ft and E is constant.

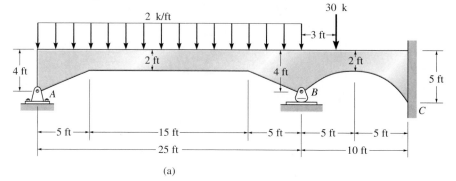

Fig. 12–13a

(a)

Solution

Since the haunches are parabolic, we will use Table 12–2 to obtain the moment-distribution properties of the beam.

Span AB

$$a_A = a_B = \frac{5}{25} = 0.2 \qquad r_A = r_B = \frac{4 - 2}{2} = 1.0$$

Entering Table 12–2 with these ratios, we find

$$C_{AB} = C_{BA} = 0.619$$
$$k_{AB} = k_{BA} = 6.41$$

Using Eqs. 12–2,

$$K_{AB} = K_{BA} = \frac{kEI_C}{L} = \frac{6.41E\,(\frac{1}{12})(1)(2)^3}{25} = 0.171E$$

Since the far end of span BA is pinned, we will modify the stiffness factor of BA using Eq. 12–3. We have

$$K'_{BA} = K_{BA}(1 - C_{AB}C_{BA}) = 0.171E[1 - 0.619(0.619)] = 0.105E$$

Uniform load, Table 12–2,

$$(FEM)_{AB} = -(0.0956)(2)(25)^2 = -119.50 \text{ k} \cdot \text{ft}$$
$$(FEM)_{BA} = 119.50 \text{ k} \cdot \text{ft}$$

Span BC

$$a_B = a_C = \frac{5}{10} = 0.5 \qquad r_B = \frac{4-2}{2} = 1.0$$

$$r_C = \frac{5-2}{2} = 1.5$$

From Table 12–2 we find

$$C_{BC} = 0.781 \qquad C_{CB} = 0.664$$
$$k_{BC} = 13.12 \qquad k_{CB} = 15.47$$

Thus, from Eqs. 12–2,

$$K_{BC} = \frac{kEI_C}{L} = \frac{13.12E\,(\frac{1}{12})(1)(2)^3}{10} = 0.875E$$

$$K_{CB} = \frac{kEI_C}{L} = \frac{15.47E\,(\frac{1}{12})(1)(2)^3}{10} = 1.031E$$

Concentrated load:

$$b = \frac{3}{10} = 0.3$$
$$(\text{FEM})_{BC} = -0.1891(30)(10) = -56.73 \text{ k}\cdot\text{ft}$$
$$(\text{FEM})_{CB} = 0.0759(30)(10) = 22.77 \text{ k}\cdot\text{ft}$$

Using the foregoing values for the stiffness factors, the distribution factors are computed and entered in the table, Fig. 12–13b. The moment distribution follows the same procedure outlined in Chapter 11. The results in k·ft are shown on the last line of the table.

Joint	A	B		C
Member	AB	BA	BC	CB
K	0.171E	0.105E	0.875E	1.031E
DF	1	0.107	0.893	0
COF	0.619	0.619	0.781	0.664
FEM	−119.50	119.50	−56.73	22.77
Dist.	+119.50	−6.72	−56.05	
CO		73.97		−43.78
Dist.		−7.91	−66.06	
CO				−51.59
ΣM	0	178.84	−178.84	−72.60

(b)

Fig. 12–13b

12.5 Slope-Deflection Equations for Nonprismatic Members

The slope-deflection equations for prismatic members were developed in Chapter 10. In this section we will generalize the form of these equations so that they apply as well to nonprismatic members. To do this, we will use the results of the previous section and proceed to formulate the equations in the same manner discussed in Chapter 10, that is, considering the effects caused by the loads, relative joint displacement, and each joint rotation separately, and then superimposing the results.

Loads. Loads are specified by the fixed-end moments $(FEM)_{AB}$ and $(FEM)_{BA}$ acting at the ends A and B of the span. Positive moments act clockwise.

Relative Joint Translation. When a relative displacement Δ at the joints occurs, the induced moments are determined from Eq. 12–6. At end A this moment is $-K_A \Delta / L (1 + C_{AB})$ and at end B it is $-K_B \Delta / L (1 + C_{BA})$.

Rotation at A. If end A rotates θ_A, the required moment in the span at A is $K_A \theta_A$. Also, this induces a moment of $C_{AB} K_A \theta_A = C_{BA} K_B \theta_A$ at end B.

Rotation at B. If end B rotates θ_B, a moment of $K_B \theta_B$ must act at end B, and the moment induced at end A is $C_{BA} K_B \theta_B = C_{AB} K_A \theta_B$.

The total end moments caused by these effects yield the generalized slope-deflection equations, which can therefore be written as

$$M_{AB} = K_A \left[\theta_A + C_{AB} \theta_B - \frac{\Delta}{L}(1 + C_{AB}) \right] + (FEM)_{AB}$$

$$M_{BA} = K_B \left[\theta_B + C_{BA} \theta_A - \frac{\Delta}{L}(1 + C_{BA}) \right] + (FEM)_{BA}$$

Since these two equations are similar, we can express them as a single equation. Referring to one end of the span as the near end (N) and the other end as the far end (F), and representing the member rotation as $\psi = \Delta / L$, we have

$$\boxed{M_N = K_N(\theta_N + C_N \theta_F - \psi(1 + C_N)) + (FEM)_N} \quad (12\text{–}8)$$

Here

M_N = internal moment at the near end of the span; this moment is positive clockwise when acting on the span

K_N = absolute stiffness of the near end determined from tables or by calculation

θ_N, θ_F = near and far end slopes of the span at the supports; the angles are measured in *radians* and are *positive clockwise*

ψ = span cord rotation due to a linear displacement, $\psi = \Delta/L$; this angle is measured in *radians* and is *positive clockwise*

$(FEM)_N$ = fixed-end moment at the near-end support; the moment is *positive* clockwise when acting on the span and is obtained from tables or by calculations

Application of the above equation follows the same procedure outlined in Chapter 10 and therefore will not be discussed here. In particular, note that Eq. 12–8 reduces to Eq. 10–2 when applied to members that are prismatic.

A continuous, reinforced-concrete highway bridge.

REFERENCES

Coughey, R. A. and R. S. Cebula, "Constants for Design of Continuous Girders with Abrupt Changes in Moments of Inertia," *Bulletin 176*, Iowa Engineering Experiment Station, Ames, Iowa.

Handbook of Frame Constants, Portland Cement Association, Chicago, 1958.

Laurson, P., "Calculation of Continuous Beams of Variable Section," *Engineering News-Record*, April 15, 1926, p. 604.

PROBLEMS

12–1. Determine approximately the stiffness and carry-over factors for the steel beam having a moment of inertia of 600 in⁴ and a depth of 10 in. The beam is partially reinforced by 1 in. × 10 in. flange cover plates at each end. Take $E = 29(10^3)$ ksi.

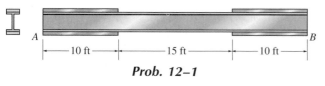

Prob. 12–1

12–2. Determine approximately the stiffness and carry-over factors for the steel beam having a moment of inertia of $94.9(10^6)$ mm⁴ and a depth of 222 mm. The beam is partially reinforced by 15 mm × 200 mm flange cover plates at each end. Take $E = 200$ GPa.

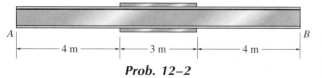

Prob. 12–2

12–3. Determine the fixed-end moments at A and C for the composite beam. $E = 29(10^3)$ ksi.

*12–4. Determine the stiffness K_A and carry-over factor C_{AC} for the beam. A and C are considered fixed-supported.

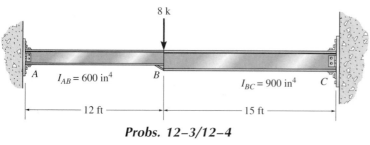

Probs. 12–3/12–4

12–5. The tapered girder is made of steel having a $\frac{1}{2}$-in. web plate and welded to flange plates which are 8 in. × 1 in. Determine the approximate carry-over factor and stiffness at the ends A and B. Segment the girder every 4 ft for the calculations. Take $E = 29(10^3)$ ksi.

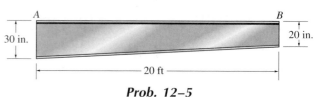

Prob. 12–5

12–6. Determine approximately the stiffness and carry-over factors for the laminated wood beam. Take $E = 11$ GPa. The beam has a thickness of 300 mm. Segment the beam every 1 m for the calculation.

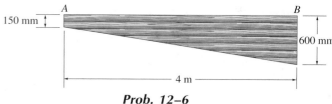

Prob. 12–6

12–7. Determine the slope at the end B of the cantilever beam. The cross section is rectangular having a constant width of 1 ft. Take $E = 29(10^3)$ ksi. Segment the beam every 5 ft for the calculation.

*12–8. Determine approximately the stiffness and carry-over factors for the end B of the beam.

12–10. The column AB serves to support a beam rail C for a light industrial building. Determine the stiffness and carry-over factors at ends A and B. $E = 29(10^3)$ ksi.

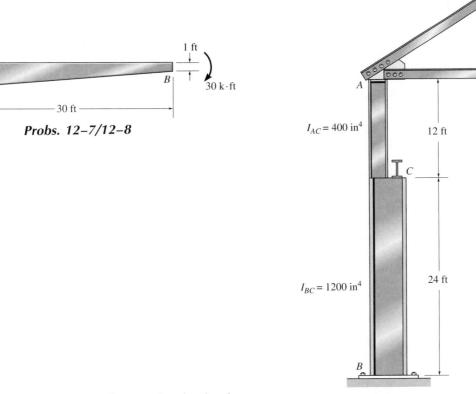

Probs. 12–7/12–8

$I_{AC} = 400$ in^4

$I_{BC} = 1200$ in^4

Prob. 12–10

12–9. Draw the moment diagram for the fixed-end straight-haunched beam. $E = 1.9(10^3)$ ksi.

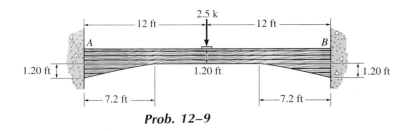

Prob. 12–9

12–11. Determine the moments at A, B, and C by the moment-distribution method. Use Table 12–1 to determine the necessary beam properties. E is constant. Assume the supports at A and C are fixed and the roller support at B is on a rigid base. The girder has a thickness of 1 ft.

*12–12. Solve Prob. 12–11 using the slope-deflection equations.

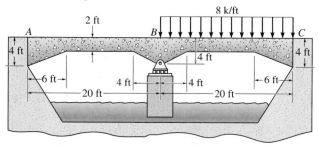

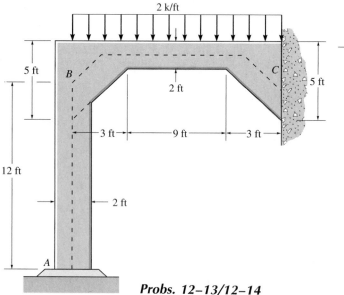

Probs. 12–11/12–12

12–13. Use the moment-distribution method to determine the moments each joint of the frame. E is constant and the members have a thickness of 1 ft. The supports at A and C are fixed. Use Table 12–1.

12–14. Solve Prob. 12–13 using the slope-deflection equations.

Probs. 12–13/12–14

12–15. Use the moment-distribution method to determine the moments at each joint of the symmetric parabolic-haunched frame. E is constant. Use Table 12–1 to compute the necessary frame constants. The members are each 1 ft thick.

*12–16. Solve Prob. 12–15 using the slope-deflection equations.

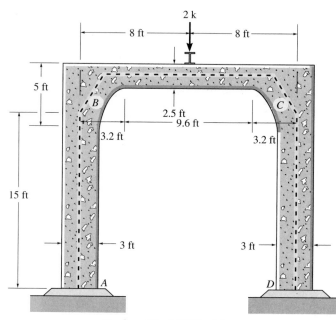

Probs. 12–15/12–16

12–17.   Use the moment-distribution method to determine the moments at each joint of the symmetric bridge frame. *E* is constant. Use Table 12–1 to compute the necessary frame constants. The members are each 1 ft thick.

12–18.   Solve Prob. 12–17 using the slope-deflection equations.

12–19.   Use the moment-distribution method to determine the moments at each joint of the frame. Assume that *E* is constant and the members have a thickness of 1 ft. The supports at *A* and *C* are pinned and the joints at *B* and *D* are fixed-connected. Use Table 12–1.

*12–20.   Solve Prob. 12–19 using the slope-deflection equations.

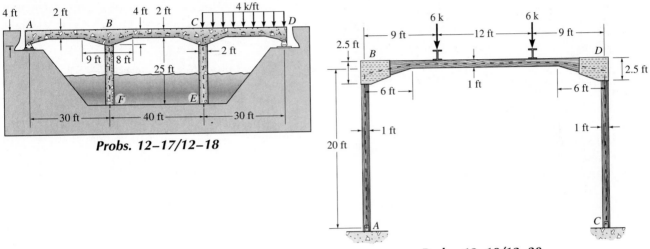

Probs. 12–17/12–18

Probs. 12–19/12–20

Matrix algebra plays an important part in the computer analysis of structures. Some programs provide graphic displays of deflections, illustrated here on the screen of this microcomputer.

Matrix Algebra for Structural Analysis

With the recent accessibility of microcomputers, application of matrix algebra for the analysis of structures has become widespread. Matrix algebra provides an appropriate tool for the analysis, since it is relatively easy to formulate the solution in a concise form and then perform the actual matrix manipulations using a computer. For this reason it is important that the structural engineer be familiar with the fundamental operations of this type of mathematics.

13.1 Basic Definitions and Types of Matrices

Matrix

A *matrix* is a rectangular arrangement of numbers having m rows and n columns. The numbers, which are called *elements*, are assembled within brackets. For example, the **A** matrix is written as

$$\mathbf{A} = \begin{bmatrix} a_{11} & a_{12} & \cdots & a_{1n} \\ a_{21} & a_{22} & \cdots & a_{2n} \\ & & \vdots & \\ a_{m1} & a_{m2} & \cdots & a_{mn} \end{bmatrix}$$

Such a matrix is said to have an *order* of $m \times n$ (m by n). Notice that the first subscript for an element denotes its row position and the second subscript denotes its column position. In general, then, a_{ij} is the element located in the ith row and jth column.

Row Matrix

If the matrix consists only of elements in a single row, it is called a *row matrix*. For example, a $1 \times n$ row matrix is written as

$$\mathbf{A} = [a_1 \quad a_2 \quad \cdots \quad a_n]$$

Here only a single subscript is used to denote an element, since the row subscript is always understood to be equal to 1, that is, $a_1 = a_{11}$, $a_2 = a_{12}$, and so on.

Column Matrix

A matrix with elements stacked in a single column is called a *column matrix*. The $m \times 1$ column matrix is

$$\mathbf{A} = \begin{bmatrix} a_1 \\ a_2 \\ \vdots \\ a_m \end{bmatrix}$$

Here the subscript notation symbolizes $a_1 = a_{11}$, $a_2 = a_{21}$, and so on.

Square Matrix

When the number of rows in a matrix equals the number of columns, the matrix is referred to as a *square matrix*. An $n \times n$ square matrix would be

$$\mathbf{A} = \begin{bmatrix} a_{11} & a_{12} & \cdots & a_{1n} \\ a_{21} & a_{22} & \cdots & a_{2n} \\ & & \vdots & \\ a_{n1} & a_{n2} & \cdots & a_{nn} \end{bmatrix}$$

Diagonal Matrix

When all the elements of a square matrix are zero except along the main diagonal, running down from left to right, the matrix is called a *diagonal matrix*. For example,

$$\mathbf{A} = \begin{bmatrix} a_{11} & 0 & 0 \\ 0 & a_{22} & 0 \\ 0 & 0 & a_{33} \end{bmatrix}$$

Unit or Identity Matrix

The *unit* or *identity matrix* is a diagonal matrix with all the diagonal elements equal to unity. For example,

$$\mathbf{I} = \begin{bmatrix} 1 & 0 & 0 \\ 0 & 1 & 0 \\ 0 & 0 & 1 \end{bmatrix}$$

Null Matrix

If all the elements of the matrix are zero, $a_{ij} = 0$, the matrix is referred to as a *null matrix*. For example,

$$\mathbf{O} = \begin{bmatrix} 0 & 0 & 0 \\ 0 & 0 & 0 \\ 0 & 0 & 0 \end{bmatrix}$$

Symmetric Matrix

A *square matrix* is symmetric provided $a_{ij} = a_{ji}$. For example,

$$\mathbf{A} = \begin{bmatrix} 3 & 5 & 2 \\ 5 & -1 & 4 \\ 2 & 4 & 8 \end{bmatrix}$$

13.2 Matrix Operations _____

Equality of Matrices

Matrices **A** and **B** are said to be equal if they are of the same order and each of their corresponding elements are equal, that is, $a_{ij} = b_{ij}$. For example, if

$$\mathbf{A} = \begin{bmatrix} 2 & 6 \\ 4 & -3 \end{bmatrix} \qquad \mathbf{B} = \begin{bmatrix} 2 & 6 \\ 4 & -3 \end{bmatrix}$$

then **A = B.**

Addition and Subtraction of Matrices

Two matrices can be added together or subtracted from one another if they are of the same order. The result is obtained by adding or subtracting the corresponding elements. For example, if

$$\mathbf{A} = \begin{bmatrix} 6 & 7 \\ 2 & -1 \end{bmatrix} \qquad \mathbf{B} = \begin{bmatrix} -5 & 8 \\ 1 & 4 \end{bmatrix}$$

then

$$\mathbf{A} + \mathbf{B} = \begin{bmatrix} 1 & 15 \\ 3 & 3 \end{bmatrix} \qquad \mathbf{A} - \mathbf{B} = \begin{bmatrix} 11 & -1 \\ 1 & -5 \end{bmatrix}$$

Multiplication by a Scalar

When a matrix is multiplied by a scalar, each element of the matrix is multiplied by the scalar. For example, if

$$\mathbf{A} = \begin{bmatrix} 4 & 1 \\ 6 & -2 \end{bmatrix} \qquad k = -6$$

then

$$k\mathbf{A} = \begin{bmatrix} -24 & -6 \\ -36 & 12 \end{bmatrix}$$

Matrix Multiplication

Two matrices $\mathbf{A}$ and $\mathbf{B}$ can be multiplied together only if they are *conformable*. This condition is satisfied if the number of *columns* in $\mathbf{A}$ *equals* the number of *rows* in $\mathbf{B}$. For example, if

$$\mathbf{A} = \begin{bmatrix} a_{11} & a_{12} \\ a_{21} & a_{22} \end{bmatrix} \qquad \mathbf{B} = \begin{bmatrix} b_{11} & b_{12} & b_{13} \\ b_{21} & b_{22} & b_{23} \end{bmatrix}$$

then $\mathbf{AB}$ can be determined since $\mathbf{A}$ has two columns and $\mathbf{B}$ has two rows. Notice, however, that $\mathbf{BA}$ is not possible. Why? It will be shown below that multiplication of matrix $\mathbf{A}$ having an order of $(m \times n)$ by matrix $\mathbf{B}$ having an order of $(n \times q)$ will yield a matrix $\mathbf{C}$ having an order of $(m \times q)$, that is,

$$\begin{array}{ccc} \mathbf{A} & \mathbf{B} & = \quad \mathbf{C} \\ (m \times n)(n \times q) & (m \times q) \end{array} \qquad (13\text{--}1)$$

The elements of matrix $\mathbf{C}$ are found using the elements a_{ij} of $\mathbf{A}$ and b_{ij} of $\mathbf{B}$ as follows:

$$c_{ij} = \sum_{k=1}^{n} a_{ik} b_{kj} \qquad (13\text{--}2)$$

The methodology of this formula can be explained by a few simple examples. Consider

$$\mathbf{A} = \begin{bmatrix} 2 & 4 & 3 \\ -1 & 6 & 1 \end{bmatrix} \qquad \mathbf{B} = \begin{bmatrix} 2 \\ 6 \\ 7 \end{bmatrix}$$

By inspection, the product $\mathbf{C} = \mathbf{AB}$ is possible since the matrices are conformable, that is, $\mathbf{A}$ has three columns and $\mathbf{B}$ has three rows. By Eq. 13–1, the multiplication will yield matrix $\mathbf{C}$ having two rows and one column. The results are obtained as follows:

c_{11}: Multiply the elements in the first row of $\mathbf{A}$ by corresponding elements in the column of $\mathbf{B}$ and add the results; that is,

$$c_{11} = c_1 = 2(2) + 4(6) + 3(7) = 49$$

c_{21}: Multiply the elements in the second row of **A** by corresponding elements in the column of **B** and add the results; that is,

$$c_{21} = c_2 = -1(2) + 6(6) + 1(7) = 41$$

Thus

$$\mathbf{C} = \begin{bmatrix} 49 \\ 41 \end{bmatrix}$$

As a second example, consider

$$\mathbf{A} = \begin{bmatrix} 5 & 3 \\ 4 & 1 \\ -2 & 8 \end{bmatrix} \qquad \mathbf{B} = \begin{bmatrix} 2 & 7 \\ -3 & 4 \end{bmatrix}$$

Here again the product $\mathbf{C} = \mathbf{AB}$ can be found since **A** has two columns and **B** has two rows. The resulting matrix **C** will have three rows and two columns. The elements are obtained as follows:

$c_{11} = 5(2) + 3(-3) = 1$ (first row of **A** times first column of **B**)

$c_{12} = 5(7) + 3(4) = 47$ (first row of **A** times second column of **B**)

$c_{21} = 4(2) + 1(-3) = 5$ (second row of **A** times first column of **B**)

$c_{22} = 4(7) + 1(4) = 32$ (second row of **A** times second column of **B**)

$c_{31} = -2(2) + 8(-3) = -28$ (third row of **A** times first column of **B**)

$c_{32} = -2(7) + 8(4) = 18$ (third row of **A** times second column of **B**)

This scheme for multiplication follows application of Eq. 13–2. Thus,

$$\mathbf{C} = \begin{bmatrix} 1 & 47 \\ 5 & 32 \\ -28 & 18 \end{bmatrix}$$

Notice also that **BA** does not exist, since written in this manner the matrices are nonconformable.

The following rules therefore apply to matrix multiplication.

1. Unless the matrices are diagonal, in general the product of two matrices is not commutative:

$$\mathbf{AB} \neq \mathbf{BA} \qquad (13\text{--}3)$$

2. The distributive law is valid:

$$\mathbf{A}(\mathbf{B} + \mathbf{C}) = \mathbf{AB} + \mathbf{AC} \qquad (13\text{--}4)$$

3. The associative law is valid:

$$\mathbf{A}(\mathbf{BC}) = (\mathbf{AB})\mathbf{C} \qquad (13\text{--}5)$$

Transposed Matrix

A matrix may be transposed by interchanging its rows and columns. For example, if

$$\mathbf{A} = \begin{bmatrix} a_{11} & a_{12} & a_{13} \\ a_{21} & a_{22} & a_{23} \\ a_{31} & a_{32} & a_{33} \end{bmatrix} \qquad \mathbf{B} = \begin{bmatrix} b_1 & b_2 & b_3 \end{bmatrix}$$

Then

$$\mathbf{A}^T = \begin{bmatrix} a_{11} & a_{21} & a_{31} \\ a_{12} & a_{22} & a_{32} \\ a_{13} & a_{23} & a_{33} \end{bmatrix} \qquad \mathbf{B}^T = \begin{bmatrix} b_1 \\ b_2 \\ b_3 \end{bmatrix}$$

Notice that $\mathbf{AB}$ is nonconformable and so the product does not exist. ($\mathbf{A}$ has three columns and $\mathbf{B}$ has one row.) Alternatively, multiplication $\mathbf{AB}^T$ is possible since here the matrices are conformable ($\mathbf{A}$ has three columns and $\mathbf{B}^T$ has three rows). The following properties for transposed matrices hold:

$$(\mathbf{A} + \mathbf{B})^T = \mathbf{A}^T + \mathbf{B}^T \qquad (13\text{--}6)$$
$$(k\mathbf{A})^T = k\mathbf{A}^T \qquad (13\text{--}7)$$
$$(\mathbf{AB})^T = \mathbf{B}^T\mathbf{A}^T \qquad (13\text{--}8)$$

This last identity will be illustrated by example. If

$$\mathbf{A} = \begin{bmatrix} 6 & 2 \\ 1 & -3 \end{bmatrix} \qquad \mathbf{B} = \begin{bmatrix} 4 & 3 \\ 2 & 5 \end{bmatrix}$$

Then, by Eq. 13–8,

$$\left(\begin{bmatrix} 6 & 2 \\ 1 & -3 \end{bmatrix} \begin{bmatrix} 4 & 3 \\ 2 & 5 \end{bmatrix} \right)^T = \begin{bmatrix} 4 & 2 \\ 3 & 5 \end{bmatrix} \begin{bmatrix} 6 & 1 \\ 2 & -3 \end{bmatrix}$$

$$\left(\begin{bmatrix} 28 & 28 \\ -2 & -12 \end{bmatrix} \right)^T = \begin{bmatrix} 28 & -2 \\ 28 & -12 \end{bmatrix}$$

$$\begin{bmatrix} 28 & -2 \\ 28 & -12 \end{bmatrix} = \begin{bmatrix} 28 & -2 \\ 28 & -12 \end{bmatrix}$$

Matrix Partitioning

A matrix can be subdivided into submatrices by partitioning. For example,

$$\mathbf{A} = \begin{bmatrix} a_{11} & a_{12} & a_{13} & a_{14} \\ a_{21} & a_{22} & a_{23} & a_{24} \\ a_{31} & a_{32} & a_{33} & a_{34} \end{bmatrix} = \begin{bmatrix} \mathbf{A}_{11} & \mathbf{A}_{12} \\ \mathbf{A}_{21} & \mathbf{A}_{22} \end{bmatrix}$$

Here the submatrices are

$$\mathbf{A}_{11} = [a_{11}] \qquad \mathbf{A}_{12} = [a_{12} \quad a_{13} \quad a_{14}]$$

$$\mathbf{A}_{21} = \begin{bmatrix} a_{21} \\ a_{31} \end{bmatrix} \qquad \mathbf{A}_{22} = \begin{bmatrix} a_{22} & a_{23} & a_{24} \\ a_{32} & a_{33} & a_{34} \end{bmatrix}$$

The rules of matrix algebra apply to partitioned matrices provided the partitioning is conformable. For example, corresponding submatrices of **A** and **B** can be added or subtracted, provided they have an equal number of rows and columns. Likewise, matrix multiplication is possible provided the respective number of columns and rows of both **A** and **B** and their submatrices are equal. For instance,

$$\mathbf{A} = \begin{bmatrix} 4 & 1 & -1 \\ -2 & 0 & -5 \\ 6 & 3 & 8 \end{bmatrix} \qquad \mathbf{B} = \begin{bmatrix} 2 & -1 \\ 0 & 8 \\ 7 & 4 \end{bmatrix}$$

Here the product **AB** exists, since the number of columns of **A** equals the number of rows of **B** (three). Likewise, the partitioned matrices are conformable for multiplication since **A** is partitioned into two columns and **B** is partitioned into two rows, that is,

$$\mathbf{AB} = \begin{bmatrix} \mathbf{A}_{11} & \mathbf{A}_{12} \\ \mathbf{A}_{21} & \mathbf{A}_{22} \end{bmatrix} \begin{bmatrix} \mathbf{B}_{11} \\ \mathbf{B}_{21} \end{bmatrix} = \begin{bmatrix} \mathbf{A}_{11}\mathbf{B}_{11} + \mathbf{A}_{12}\mathbf{B}_{21} \\ \mathbf{A}_{21}\mathbf{B}_{11} + \mathbf{A}_{22}\mathbf{B}_{21} \end{bmatrix}$$

Multiplication of the submatrices yields

$$\mathbf{A}_{11}\mathbf{B}_{11} = \begin{bmatrix} 4 & 1 \\ -2 & 0 \end{bmatrix} \begin{bmatrix} 2 & -1 \\ 0 & 8 \end{bmatrix} = \begin{bmatrix} 8 & 4 \\ -4 & 2 \end{bmatrix}$$

$$\mathbf{A}_{12}\mathbf{B}_{21} = \begin{bmatrix} -1 \\ -5 \end{bmatrix} [7 \quad 4] = \begin{bmatrix} -7 & -4 \\ -35 & -20 \end{bmatrix}$$

$$\mathbf{A}_{21}\mathbf{B}_{11} = [6 \quad 3] \begin{bmatrix} 2 & -1 \\ 0 & 8 \end{bmatrix} = [12 \quad 18]$$

$$\mathbf{A}_{22}\mathbf{B}_{21} = [8][7 \quad 4] = [56 \quad 32]$$

Thus,

$$\mathbf{AB} = \begin{bmatrix} \begin{bmatrix} 8 & 4 \\ -4 & 2 \end{bmatrix} + \begin{bmatrix} -7 & -4 \\ -35 & -20 \end{bmatrix} \\ [12 \quad 18] + [56 \quad 32] \end{bmatrix} = \begin{bmatrix} 1 & 0 \\ -39 & -18 \\ 68 & 50 \end{bmatrix}$$

13.3   Determinants

In the next section we will discuss how to invert a matrix, since this operation requires an evaluation of the determinant of the matrix, we will now discuss some of the basic properties of determinants.

A determinant is a square array of numbers enclosed within vertical bars. For example, an nth-order determinant, having n rows and n columns, is

$$|A| = \begin{vmatrix} a_{11} & a_{12} & \cdots & a_{1n} \\ a_{21} & a_{22} & \cdots & a_{2n} \\ & & \vdots & \\ a_{n1} & a_{n2} & \cdots & a_{nn} \end{vmatrix} \tag{13–9}$$

Evaluation of this determinant leads to a single numerical value which can be determined using *Laplace's expansion*. This method makes use of the determinant's minors and cofactors. Specifically, each element a_{ij} of a determinant of nth order has a *minor* M_{ij} which is a determinant of order $n - 1$. This determinant (minor) remains when the *i*th row and *j*th column in which the a_{ij} element is contained is canceled out. If the minor is multiplied by $(-1)^{i+j}$ it is called the cofactor of a_{ij} and is denoted as

$$|C_{ij}| = (-1)^{i+j}|M_{ij}| \qquad (13\text{--}10)$$

For example, consider the third-order determinant

$$\begin{vmatrix} a_{11} & a_{12} & a_{13} \\ a_{21} & a_{22} & a_{23} \\ a_{31} & a_{32} & a_{33} \end{vmatrix}$$

The cofactors for the elements in the first row are

$$|C_{11}| = (-1)^{1+1}\begin{vmatrix} a_{22} & a_{23} \\ a_{32} & a_{33} \end{vmatrix} = \begin{vmatrix} a_{22} & a_{23} \\ a_{32} & a_{33} \end{vmatrix}$$

$$|C_{12}| = (-1)^{1+2}\begin{vmatrix} a_{21} & a_{23} \\ a_{31} & a_{33} \end{vmatrix} = \begin{vmatrix} a_{21} & a_{23} \\ a_{31} & a_{33} \end{vmatrix}$$

$$|C_{13}| = (-1)^{1+3}\begin{vmatrix} a_{21} & a_{22} \\ a_{31} & a_{32} \end{vmatrix} = \begin{vmatrix} a_{21} & a_{22} \\ a_{31} & a_{32} \end{vmatrix}$$

Laplace's expansion for a determinant of order n, Eq. 13–9, states that the numerical value represented by the determinant is equal to the sum of the products of the elements of any row or column and their respective cofactors, i.e.,

$$D = a_{1j}C_{1j} + a_{2j}C_{2j} + \cdots + a_{nj}C_{nj} \qquad (j = 1, 2, \ldots, \text{or } n)$$

or

$$(13\text{--}11)$$

$$D = a_{1j}C_{1j} + a_{2j}C_{2j} + \cdots + a_{nj}C_{nj} \qquad (j = 1, 2, \ldots, \text{or } n)$$

For application, it is seen that due to the cofactors the number D is defined in terms of n determinants of order $n - 1$ each. These determinants can each be reevaluated using the same formula, whereby one must evaluate $(n - 1)$ determinants of order $(n - 2)$, and so on. The process of evaluation continues until the remaining determinants to be evaluated reduce to the second order, whereby the cofactors of the elements are single elements of $|D|$. Consider, for example, the following second-order determinant

$$|D| = \begin{vmatrix} 3 & 5 \\ -1 & 2 \end{vmatrix}$$

We can evaluate $|D|$ along the top row of elements, which yields

$$D = 3(-1)^{1+1}(2) + 5(-1)^{1+2}(-1) = 11$$

Or, for example, using the second column of elements, we have

$$D = 5(-1)^{1+2}(-1) + 2(-1)^{2+2}(3) = 11$$

Rather than using Eqs. 13–11, it is perhaps easier to realize that the evaluation of a second-order determinant can be performed by multiplying the elements of the diagonal, from top left down to right, and subtract from this the product of the elements from the top right down to left, i.e., follow the arrow,

$$D = \begin{vmatrix} 3 & 5 \\ -1 & 2 \end{vmatrix} = 3(2) - 5(-1) = 11$$

Consider next the third-order determinant

$$|D| = \begin{vmatrix} 1 & 3 & -1 \\ 4 & 2 & 6 \\ -1 & 0 & 2 \end{vmatrix}$$

Using Eq. 13–11, we can evaluate $|D|$ using the elements along the top row, which yields

$$D = (1)(-1)^{1+1}\begin{vmatrix} 2 & 6 \\ 0 & 2 \end{vmatrix} + (3)(-1)^{1+2}\begin{vmatrix} 4 & 6 \\ -1 & 2 \end{vmatrix} + (-1)(-1)^{1+3}\begin{vmatrix} 4 & 2 \\ -1 & 0 \end{vmatrix}$$

$$= 1(4 - 0) - 3(8 + 6) - 1(0 + 2) = -40$$

It is also possible to evaluate $|D|$ using the elements along the first column, i.e.,

$$D = 1(-1)^{1+1}\begin{vmatrix} 2 & 6 \\ 0 & 2 \end{vmatrix} + 4(-1)^{2+1}\begin{vmatrix} 3 & -1 \\ 0 & 2 \end{vmatrix} + (-1)(-1)^{3+1}\begin{vmatrix} 3 & -1 \\ 2 & 6 \end{vmatrix}$$

$$= 1(4 - 0) - 4(6 - 0) - 1(18 + 2) = -40$$

As an exercise try to evaluate $|D|$ using the elements along the second row.

Rules for Evaluation

The following two rules may be used to simplify the evaluation of a determinant:

1. A multiple common to a row or column may be factored out of the determinant and later multiplied by $|D|$.
2. If corresponding elements in two rows or columns of a determinant are proportional, then $D = 0$.

13.4   Inverse of a Matrix

Consider the following set of three linear equations:

$$a_{11}x_1 + a_{12}x_2 + a_{13}x_3 = c_1$$
$$a_{21}x_1 + a_{22}x_2 + a_{23}x_3 = c_2$$
$$a_{31}x_1 + a_{32}x_2 + a_{33}x_3 = c_3$$

which can be written in matrix form as

$$\begin{bmatrix} a_{11} & a_{12} & a_{13} \\ a_{21} & a_{22} & a_{23} \\ a_{31} & a_{32} & a_{33} \end{bmatrix}\begin{bmatrix} x_1 \\ x_2 \\ x_3 \end{bmatrix} = \begin{bmatrix} c_1 \\ c_2 \\ c_3 \end{bmatrix} \qquad (13\text{--}12)$$

or

$$\mathbf{Ax} = \mathbf{C} \qquad (13\text{--}13)$$

One would think that a solution for x could be determined by dividing $\mathbf{C}$ by $\mathbf{A}$; however, division is not possible in matrix algebra. Instead, one multiplies by the inverse of the matrix. The *inverse* of the matrix $\mathbf{A}$ is another matrix of the same order and symbolically written as $\mathbf{A}^{-1}$. It has the following property,

$$\mathbf{A}\mathbf{A}^{-1} = \mathbf{A}^{-1}\mathbf{A} = \mathbf{I}$$

where $\mathbf{I}$ is an identity matrix. Multiplying both sides of Eq. 13–13 by $\mathbf{A}^{-1}$, we obtain

$$\mathbf{A}^{-1}\mathbf{A}\mathbf{x} = \mathbf{A}^{-1}\mathbf{C}$$

Since $\mathbf{A}^{-1}\mathbf{A}\mathbf{x} = \mathbf{I}\mathbf{x} = \mathbf{x}$, we have

$$\mathbf{x} = \mathbf{A}^{-1}\mathbf{C} \tag{13–14}$$

Provided $\mathbf{A}^{-1}$ can be obtained, a solution for $\mathbf{x}$ is possible.

For hand calculation the method used to formulate $\mathbf{A}^{-1}$ can be developed using Cramer's rule. The development will not be given here; instead, only the results are given.* In this regard, the elements in the matrices of Eq. 13–14 can be written as

$$\mathbf{x} = \mathbf{A}^{-1}\mathbf{C}$$

$$\begin{bmatrix} x_1 \\ x_2 \\ x_3 \end{bmatrix} = \frac{1}{|A|} \begin{bmatrix} C_{11} & C_{21} & C_{31} \\ C_{12} & C_{22} & C_{32} \\ C_{13} & C_{23} & C_{33} \end{bmatrix} \begin{bmatrix} c_1 \\ c_2 \\ c_3 \end{bmatrix} \tag{13–15}$$

The square matrix containing the cofactors C_{ij} is called the *adjoint matrix*. By comparison it can be seen that the inverse matrix $\mathbf{A}^{-1}$ is obtained from $\mathbf{A}$ by first replacing each element a_{ij} by its cofactor $|C_{ij}|$, then transposing the resulting matrix, yielding the adjoint matrix, and finally multiplying the adjoint matrix by $1/|A|$.

To illustrate how to obtain $\mathbf{A}^{-1}$ numerically, we will consider the solution of the following set of linear equations:

$$\begin{aligned} x_1 - x_2 + x_3 &= -1 \\ -x_1 + x_2 + x_3 &= -1 \\ x_1 + 2x_2 - 2x_3 &= 5 \end{aligned} \tag{13–16}$$

*See Kreyszig, in References.

Here

$$
A = \begin{bmatrix} 1 & -1 & 1 \\ -1 & 1 & 1 \\ 1 & 2 & -2 \end{bmatrix}
$$

The cofactor matrix for A is

$$
C = \begin{bmatrix}
\begin{vmatrix} 1 & 1 \\ 2 & -2 \end{vmatrix} & -\begin{vmatrix} -1 & 1 \\ 1 & -2 \end{vmatrix} & \begin{vmatrix} -1 & 1 \\ 1 & 2 \end{vmatrix} \\[6pt]
-\begin{vmatrix} -1 & 1 \\ 2 & -2 \end{vmatrix} & \begin{vmatrix} 1 & 1 \\ 1 & -2 \end{vmatrix} & -\begin{vmatrix} 1 & -1 \\ 1 & 2 \end{vmatrix} \\[6pt]
\begin{vmatrix} -1 & 1 \\ 1 & 1 \end{vmatrix} & -\begin{vmatrix} 1 & 1 \\ -1 & 1 \end{vmatrix} & \begin{vmatrix} 1 & -1 \\ -1 & 1 \end{vmatrix}
\end{bmatrix}
$$

The adjoint matrix is

$$
C^T = \begin{bmatrix} -4 & 0 & -2 \\ -1 & -3 & -2 \\ -3 & -3 & 0 \end{bmatrix}
$$

Since

$$
A = \begin{vmatrix} 1 & -1 & 1 \\ -1 & 1 & 1 \\ 1 & 2 & -2 \end{vmatrix} = -6
$$

The inverse of A is, therefore,

$$
A^{-1} = -\frac{1}{6} \begin{bmatrix} -4 & 0 & -2 \\ -1 & -3 & -2 \\ -3 & -3 & 0 \end{bmatrix}
$$

Solution of Eqs. 13–16 yields

$$
\begin{bmatrix} x_1 \\ x_2 \\ x_3 \end{bmatrix} = -\frac{1}{6} \begin{bmatrix} -4 & 0 & -2 \\ -1 & -3 & -2 \\ -3 & -3 & 0 \end{bmatrix} \begin{bmatrix} -1 \\ -1 \\ 5 \end{bmatrix}
$$

$$x_1 = -\tfrac{1}{6}[(-4)(-1) + 0(-1) + (-2)(5)] = 1$$
$$x_2 = -\tfrac{1}{6}[(-1)(-1) + (-3)(-1) + (-2)(5)] = 1$$
$$x_3 = -\tfrac{1}{6}[(-3)(-1) + (-3)(-1) + (0)(5)] = -1$$

Obviously, the numerical calculations are quite expanded for larger sets of equations. For this reason, computers are used in structural analysis to determine the inverse of matrices.

13.5   The Gauss Method for Solving Simultaneous Equations

When many simultaneous linear equations have to be solved, the Gauss elimination method may be used because of its numerical efficiency. Application of this method requires solving one of a set of n equations for an unknown, say x_1, in terms of all the other unknowns, $x_2, x_3, \ldots, x_n$. Substituting this so-called *pivotal equation* into the remaining equations leaves one with a set of $n - 1$ equations with $n - 1$ unknowns. Repeating the process by solving one of these equations for x_2 in terms of the $n - 2$ remaining unknowns $x_3, x_4, \ldots, x_n$ forms the second pivotal equation. This equation is then substituted into the other equations, leaving one with a set of $n - 3$ equations with $n - 3$ unknowns. The process is repeated until one is left with a pivotal equation having one unknown, which is then solved. The other unknowns are then determined by successive back substitution into the other pivotal equations. To improve the accuracy of solution, when developing each pivotal equation one should always select the equation of the set having the *largest* numerical coefficient for the unknown one is trying to eliminate. The process will now be illustrated by an example.

Solve the following set of equations using Gauss elimination:

$$-2x_1 + 8x_2 + 2x_3 = 2 \qquad (13\text{--}17)$$

$$2x_1 - x_2 + x_3 = 2 \qquad (13\text{--}18)$$

$$4x_1 - 5x_2 + 3x_3 = 4 \qquad (13\text{--}19)$$

We will begin by eliminating x_1. The largest coefficient of x_1 is in Eq. 13–19; hence, we will take it to be the pivotal equation. Solving for x_1, we have

$$x_1 = 1 + 1.25x_2 - 0.75x_3 \qquad (13\text{--}20)$$

Substituting into Eqs. 13–17 and 13–18 and simplifying yields

$$2.75x_2 + 1.75x_3 = 2 \qquad (13\text{–}21)$$

$$1.5\ x_2 - 0.5\ x_3 = 0 \qquad (13\text{–}22)$$

Next we eliminate x_2. Choosing Eq. 13–21 for the pivotal equation since the coefficient of x_2 is largest here, we have

$$x_2 = 0.727 - 0.636x_3 \qquad (13\text{–}23)$$

Substituting this equation into Eq. 13–22 and simplifying yields the final pivotal equation, which can be solved for x_3. This yields

$$x_3 = 0.75 \qquad \qquad \textbf{Ans.}$$

Substituting this value into the pivotal Eq. 13–23 gives

$$x_2 = 0.25 \qquad \qquad \textbf{Ans.}$$

Finally, from pivotal Eq. 13–20 we get

$$x_1 = 0.75 \qquad \qquad \textbf{Ans.}$$

Although the Gauss method, as illustrated here, is the accepted and standard method for computer use, for structural analysis, further reduction in the computational work is possible using improvements to this technique developed by Banachiewicz and Cholesky as stated in Bodewig.

REFERENCES

Bodewig, E., *Matrix Calculus*, 2nd ed. North Holland Publishing Company, Amsterdam, 1959.

Kreyszig, E., *Advanced Engineering Mathematics*, John Wiley & Sons, Inc., New York, 1962.

PROBLEMS

13–1.   If $\mathbf{A} = \begin{bmatrix} 4 & 2 & -3 \\ 6 & 1 & 5 \end{bmatrix}$ and

$\mathbf{B} = \begin{bmatrix} 4 & -1 & 0 \\ 2 & 0 & 8 \end{bmatrix}$, determine $\mathbf{A} + \mathbf{B}$ and $\mathbf{A} - 2\mathbf{B}$.

13–2.   If $\mathbf{A} = [6 \quad 1 \quad 3]$ and $\mathbf{B} = [1 \quad 6 \quad 3]$, show that $(\mathbf{A} + \mathbf{B})^T = \mathbf{A}^T + \mathbf{B}^T$.

13–3.   If $\mathbf{A} = \begin{bmatrix} 3 & 5 \\ -2 & 7 \end{bmatrix}$, determine $\mathbf{A} + \mathbf{A}^T$.

*13–4. If $A = \begin{bmatrix} 1 \\ 0 \\ 5 \end{bmatrix}$ and $B = [2 \quad -1 \quad 3]$, determine **AB**.

13–5. If $A = \begin{bmatrix} 6 & 2 & 2 \\ -5 & 1 & 1 \\ 0 & 3 & 1 \end{bmatrix}$ and

$B = \begin{bmatrix} -1 & 3 & 1 \\ 2 & -5 & 1 \\ 0 & 7 & 5 \end{bmatrix}$, determine **AB**.

13–6. Determine **BA** for the matrices of Prob. 13–5.

13–7. If $A = \begin{bmatrix} 5 & 7 \\ -2 & 1 \end{bmatrix}$ and $B = \begin{bmatrix} 6 \\ 7 \end{bmatrix}$, determine **AB**.

*13–8. If $A = \begin{bmatrix} 1 & 8 & 4 \\ 1 & 2 & 3 \end{bmatrix}$ and $B = \begin{bmatrix} 3 \\ 2 \\ -6 \end{bmatrix}$, determine **AB**.

13–9. If $A = \begin{bmatrix} 2 & 7 & 3 \\ -2 & 1 & 0 \end{bmatrix}$ and $B = \begin{bmatrix} 6 \\ 9 \\ -1 \end{bmatrix}$, determine **AB**.

13–10. If $A = \begin{bmatrix} 6 & 4 & 2 \\ 2 & 1 & 1 \\ 0 & -3 & 1 \end{bmatrix}$ and

$B = \begin{bmatrix} -1 & 3 & -2 \\ 2 & 4 & 1 \\ 0 & 7 & 5 \end{bmatrix}$, determine **AB**.

13–11. If $A = \begin{bmatrix} 2 & 5 \\ 1 & 3 \end{bmatrix}$, determine AA^T.

*13–12. Show that the distributive law is valid, i.e., $A(B + C) = AB + AC$, if $A = \begin{bmatrix} 2 & 1 & 6 \\ 4 & 5 & 3 \end{bmatrix}$,

$B = \begin{bmatrix} 3 \\ 1 \\ -6 \end{bmatrix}$, $C = \begin{bmatrix} 5 \\ -1 \\ 2 \end{bmatrix}$.

13–13. Show that the associative law is valid, i.e., $A(BC) = (AB)C$, if

$A = \begin{bmatrix} 2 & 1 & 6 \\ 4 & 5 & 3 \end{bmatrix}$, $B = \begin{bmatrix} 3 \\ 1 \\ -6 \end{bmatrix}$, $C = [5 \quad -1 \quad 2]$.

13–14. Evaluate the determinants $\begin{vmatrix} 2 & 5 \\ 7 & 1 \end{vmatrix}$ and $\begin{vmatrix} 1 & 3 & 5 \\ 2 & 7 & 1 \\ 3 & 8 & 6 \end{vmatrix}$.

13–15. If $A = \begin{bmatrix} 5 & 1 \\ 3 & -2 \end{bmatrix}$, determine A^{-1}.

*13–16. If $A = \begin{bmatrix} 0 & 1 & 3 \\ 2 & 5 & 0 \\ 1 & -1 & 2 \end{bmatrix}$, determine A^{-1}.

13–17. Solve the equations $-x_1 + 4x_2 + x_3 = 1$, $2x_1 - x_2 + x_3 = 2$, and $4x_1 - 5x_2 + 3x_3 = 4$ using the matrix equation $X = A^{-1}C$.

13–18. Solve the equations in Prob. 13–17 using the Gauss elimination method.

13–19. Solve the equations $x_1 - x_2 + x_3 = -1$, $-x_1 + x_2 + x_3 = -1$, and $x_1 + 2x_2 - 2x_3 = 5$ using the matrix equation $X = A^{-1}B$.

*13–20. Solve the equations in Prob. 13–19 using the Gauss elimination method.

The space-truss analysis of transmission towers can be performed using the stiffness method of matrix analysis.

Truss Analysis Using the Stiffness Method

In this chapter we will explain the basic fundamentals of using the stiffness method for analyzing structures. It will be shown that this method, although tedious to do by hand, is quite suited for use on a computer. Examples of specific applications to planar trusses will then be given. Beams and framed structures will be discussed in the next chapter.

14.1 Fundamentals of the Stiffness Method

Each of the classical methods of indeterminate analysis that were presented in previous chapters can be classified as being either a force or a displacement method. As noted in Chapter 8, a *force method* is based on first specifying the internal or external redundant forces and then solving for these forces using *conditions of compatibility* for displacement, along with load-displacement relationships. Once the forces on the structure are determined, displacements can then be computed using one of the methods in Chapter 7. A *displacement method*, on the other hand, first requires an identification of the number of unknown degrees of freedom for the structure. These displacements are then determined from *equations of equilibrium*. After calculating the displacements, the unknown external and internal forces for the structure are determined from the compatibility and load-displacement relationships.

The stiffness method of matrix analysis, to be used in this and the next chapter, is a displacement method of analysis. A force method of matrix analysis, as outlined in Sec. 8.9, can also be used to analyze structures; however, this method will not be presented here. There are several reasons for this. Most important, the displacement or stiffness method can be used to analyze both statically determinate and indeterminate structures, whereas the force method requires a different procedure for each of these two cases. Also, the displacement method yields the displacements and forces directly, whereas with the force method the displacements are not obtained directly. Furthermore, it is generally much easier to formulate the necessary matrices for the computer operations using the displacement method; and once this is done, the computer calculations can be performed efficiently.

Application of the stiffness method of analysis requires subdividing the structure into a series of discrete *finite elements* and identifying their end points as *nodes*. For truss analysis, the finite elements are represented by each of the members that compose the truss, and the nodes represent the joints. The force-displacement properties of each element are determined and then related to one another using the force equilibrium equations written at the nodes. These relationships, for the entire structure, are then grouped together into what is called the *structure stiffness matrix* **K**. Once it is established, the unknown displacements of the nodes can then be determined for any given loading on the structure. When these displacements are known, the external and internal forces in the structure can be calculated using the force-displacement relations for each member.

Before developing a formal procedure for applying the stiffness method of analysis, it is first necessary to establish some preliminary definitions and concepts.

Member and Node Identification

One of the first steps when applying the stiffness method is to identify the elements or members of the structure and their nodes. We will specify each member by a number enclosed within a square, and use a number enclosed within a circle to identify the nodes. Also, the "near" and "far" ends of the member must be identified. This will be done using an arrow written along the member, with the head of the arrow directed toward the far end. Examples of member, node, and "direction" identification for a truss are shown in Fig. 14–1a. These assignments have all been done *arbitrarily.**

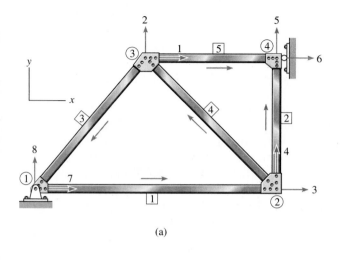

(a)

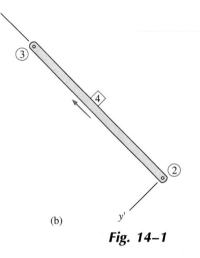

(b)

Fig. 14–1

Global and Member Coordinates

Since loads and displacements are vector quantities, it is necessary to establish a coordinate system in order to specify their correct sense of direction. Here we will use two different types of coordinate systems. A single *global* or *structure coordinate system,* using x, y axes, will be used to specify the sense of each of the external force and displacement components at the nodes, Fig. 14–1a. A *local* or *member coordinate system* will be used for each member to specify the sense of direction of its displacements and internal loadings. This system will be identified using x', y' axes with the origin at the "near" node and the x' axis extending toward the "far" node. An example for truss member 4 is shown in Fig. 14–1b.

Degrees of Freedom

The unconstrained degrees of freedom for the structure represent the primary unknowns in the stiffness method, and therefore these must be identified. As a general rule there are two degrees of freedom, or two possible displacements, for each truss joint (node). For application,

*For large trusses, matrix manipulations using **K** are actually more efficient using selective numbering of the members in a wave pattern, that is, starting from top to bottom, then bottom to top, etc.

655

each degree of freedom should be specified on the structure using a code number, shown at the joint or node, and referenced to its positive global coordinate direction using an associated arrow. For example, the truss in Fig. 14–1a has eight degrees of freedom, which have been identified by the "code numbers" 1 through 8 as shown. Of these eight possible displacements, 1 through 5 represent unknown or *unconstrained degrees of freedom*, and 6 through 8 represent *constrained degrees of freedom*. Due to the constraints, the displacements here are zero. For later application, *the lowest code numbers will always be used to identify the unknown displacements (unconstrained degrees of freedom) and the highest code numbers will be used to identify the known displacements (constrained degrees of freedom)*. The reason for choosing this method of identification has to do with the convenience of later partitioning the structure stiffness matrix, so that the unknown displacements can be found in the most direct manner.

Once the truss is labeled and the code numbers are specified as indicated above, the structure stiffness matrix **K** can then be determined. To do this we must first establish a *member stiffness matrix* **k′** for each member of the truss. This matrix is used to express the member's load-displacement relations in terms of the *local coordinates*. Since all the members of the truss are not in the same direction, we must develop a means for transforming these quantities from each member's local x', y' coordinate system to the structure's global x, y coordinate system. This can be done using *force and displacement transformation matrices*. Once established, the elements of the member stiffness matrix can be transformed from local to global coordinates and then used to create the structure stiffness matrix. There are actually two ways to make this transformation. We can first construct a *composite element stiffness matrix* **k**$_c$ that represents the load-deflection relations for *all* the members of the truss in terms of the local coordinates. Then we can develop a *structure transformation matrix* specifically for the truss under consideration, and transform **k**$_c$ from its local coordinates to the global coordinates. This will yield **K** directly. Unfortunately, this method does not work well on a computer due to the uniqueness of **K** for each truss. Instead, we will use the alternative method, referred to as the *direct stiffness approach*; that is, each member stiffness matrix will be transformed *separately* from local to global coordinates. When the global stiffness matrices for all the truss members have been determined, the structure stiffness matrix **K** will be formulated by assembling together each of the member stiffness matrices. Using **K**, as stated above, we can then determine the node displacements first, followed by the support reactions and the member forces. We will now elaborate on the development of this method.

14.2   Truss-Member Stiffness Matrix _____

In this section we will establish the stiffness matrix for a single truss member using local x', y' coordinates, oriented as shown in Fig. 14–2. This matrix will represent the load-displacement relations between the ends of the member when the member is subjected to its various displacements and loadings.

A truss member can only be displaced along its axis (x' axis) since the loads are applied along this axis. Two independent displacements are therefore possible. In Fig. 14–2a, the positive displacement d_N is imposed on the near end of the member while the far end is held pinned. The forces developed at the ends of the members are

$$q'_N = \frac{AE}{L}d_N \qquad q'_F = -\frac{AE}{L}d_N$$

Note that q'_F is negative since for equilibrium it acts in the negative x' direction. Likewise, a positive displacement of d_F at the far end, keeping the near end pinned, Fig. 14–2b, results in member forces of

$$q''_N = -\frac{AE}{L}d_F \qquad q''_F = \frac{AE}{L}d_F$$

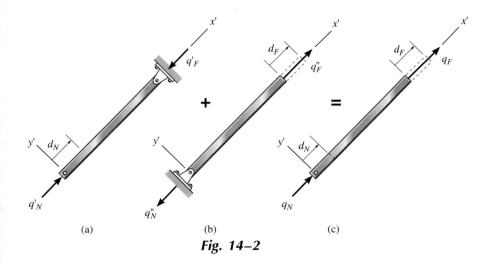

(a)   +   (b)   =   (c)

Fig. 14–2

By superposition, Fig. 14–2c, the resultant effect of both d_N and d_F is

$$q_N = \frac{AE}{L}d_N - \frac{AE}{L}d_F \qquad (14-1)$$

$$q_F = \frac{AE}{L}d_F - \frac{AE}{L}d_N \qquad (14-2)$$

These load-displacement equations may be written in matrix form as

$$\begin{bmatrix} q_N \\ q_F \end{bmatrix} = \frac{AE}{L}\begin{bmatrix} 1 & -1 \\ -1 & 1 \end{bmatrix}\begin{bmatrix} d_N \\ d_F \end{bmatrix}$$

or

$$\boxed{\mathbf{q} = \mathbf{k'd}} \qquad (14-3)$$

where

$$\mathbf{k'} = \frac{AE}{L}\begin{bmatrix} 1 & -1 \\ -1 & 1 \end{bmatrix} \qquad (14-4)$$

This matrix, $\mathbf{k'}$, is called the *member stiffness matrix*, and it is of the same form for each member of the truss. The four elements that comprise it are called *member stiffness influence coefficients*, k'_{ij}. Physically, k'_{ij} represents the force at joint i when a *unit displacement* is imposed only at joint j. For example, if the far joint is held fixed, the force at the near joint—when this joint undergoes a displacement of $d_N = 1$—is determined from $i = j = 1$ and from Eq. 14–1,

$$q_N = k'_{11} = \frac{AE}{L}$$

Likewise, the force at the far joint is determined from $i = 2, j = 1$, so that

$$q_F = k'_{21} = -\frac{AE}{L}$$

Note that these two terms represent the first column of the member stiffness matrix. In the same manner, the second column of this matrix represents the forces in the member only when the far end of the member undergoes a unit displacement. From the development of Eq. 14–4, it should be noted that both equilibrium and compatibility of deformation of the member are satisfied.

14.3 Displacement and Force Transformation Matrices

Since a truss is composed of many members (elements), we will now develop a method for transforming the member forces **q** and displacements **d** defined in local coordinates to a global or structure x, y coordinate system for the entire truss. For the sake of convention, we will consider positive x to the right and positive y upward. The smallest angles between the *positive x, y* global axes and the *positive x'* local axis will be defined as θ_x and θ_y as shown in Fig. 14–3. The cosines of these angles will be used in the matrix analysis that follows. These will be identified as $\lambda_x = \cos\theta_x$, $\lambda_y = \cos\theta_y$. Numerical values for λ_x and λ_y can easily be generated by a computer once the coordinates of the near end N and far end F of the member have been specified. For example, consider member NF of the truss shown in Fig. 14–4. Here the coordinates of N and F are specified from the origin of the global coordinate system.* Used as input data,

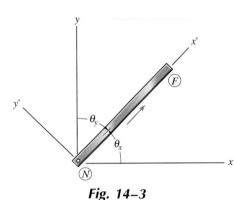

Fig. 14–3

$$\lambda_x = \cos\theta_x = \frac{x_F - x_N}{L} = \frac{x_F - x_N}{\sqrt{(x_F - x_N)^2 + (y_F - y_N)^2}} \qquad (14\text{–}5)$$

$$\lambda_y = \cos\theta_y = \frac{y_F - y_N}{L} = \frac{y_F - y_N}{\sqrt{(x_F - x_N)^2 + (y_F - y_N)^2}} \qquad (14\text{–}6)$$

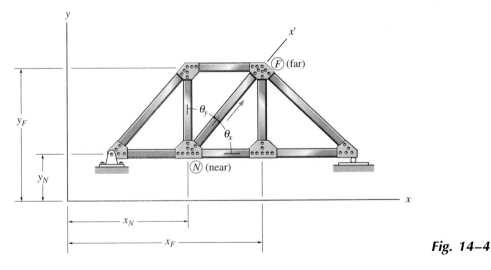

Fig. 14–4

*The origin can be located at any convenient point. Usually, however, it is located where the x, y coordinates of all the nodes will be *positive*, as shown in Fig. 14–4.

The algebraic signs in these "generalized" equations will automatically account for members that are oriented in other quadrants of the x-y plane.

Displacement Transformation Matrix

In global coordinates each end of the member can have two degrees of freedom or independent displacements; namely, joint N has D_{Nx} and D_{Ny}, Fig. 14–5a and 14–5b, and joint F has D_{Fx} and D_{Fy}, Fig. 14–5c and 14–5d. We will now consider each of these global displacements separately, in order to determine each component displacement along the member. When the far end is held pinned and the near end is given a global displacement D_{Nx}, Fig. 14–5a, the corresponding dis-

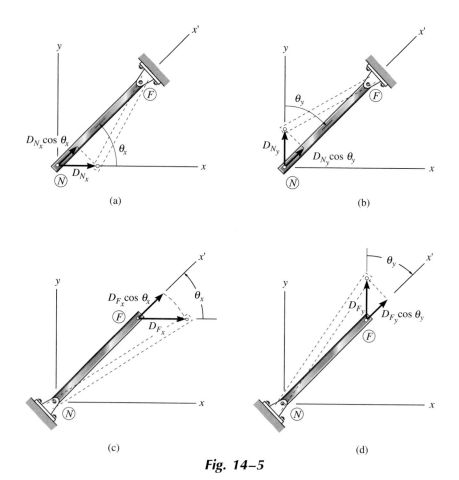

(a)

(b)

(c)

(d)

Fig. 14–5

placement (deformation) along the bar is $D_{Nx} \cos \theta_x$.* Likewise, a global displacement D_{Ny} will cause the bar to be displaced $D_{Ny} \cos \theta_y$ along the x' axis, Fig. 14–5b. The effect of *both* global displacements is therefore

$$d_N = D_{Nx} \cos \theta_x + D_{Ny} \cos \theta_y$$

In a similar manner, positive displacements D_{Fx} and D_{Fy} successively applied at the far end F, while the near end is held pinned, Fig. 14–5c and 14–5d, will cause the member to be displaced

$$d_F = D_{Fx} \cos \theta_x + D_{Fy} \cos \theta_y$$

Letting $\lambda_x = \cos \theta_x$ and $\lambda_y = \cos \theta_y$ represent the *direction cosines* for the member, we have

$$d_N = D_{Nx}\lambda_x + D_{Ny}\lambda_y$$
$$d_F = D_{Fx}\lambda_x + D_{Fy}\lambda_y$$

which can be written in matrix form as

$$\begin{bmatrix} d_N \\ d_F \end{bmatrix} = \begin{bmatrix} \lambda_x & \lambda_y & 0 & 0 \\ 0 & 0 & \lambda_x & \lambda_y \end{bmatrix} \begin{bmatrix} D_{Nx} \\ D_{Ny} \\ D_{Fx} \\ D_{Fy} \end{bmatrix} \tag{14–7}$$

or

$$\boxed{\mathbf{d} = \mathbf{TD}} \tag{14–8}$$

where

$$\mathbf{T} = \begin{bmatrix} \lambda_x & \lambda_y & 0 & 0 \\ 0 & 0 & \lambda_x & \lambda_y \end{bmatrix} \tag{14–9}$$

From the above derivation, $\mathbf{T}$ transforms the four global x, y displacements $\mathbf{D}$ into the two local x' displacements $\mathbf{d}$. Hence, $\mathbf{T}$ is referred to as the *displacement transformation matrix*.

---

*The change in θ_x or θ_y will be neglected, since it is very small.

Force Transformation Matrix

Consider now application of the force q_N to the near end of the member, the far end held pinned, Fig. 14–6a. Here the global force components of q_N at N are

$$Q_{Nx} = q_N \cos \theta_x \qquad Q_{Ny} = q_N \cos \theta_y$$

Likewise, if q_F is applied to the bar, Fig. 14–6b, the global force components at F are

$$Q_{Fx} = q_F \cos \theta_x \qquad Q_{Fy} = q_F \cos \theta_y$$

Using the direction cosines $\lambda_x = \cos \theta_x$, $\lambda_y = \cos \theta_y$, these equations become

$$
\begin{aligned}
Q_{Nx} &= q_N \lambda_x & Q_{Ny} &= q_N \lambda_y \\
Q_{Fx} &= q_F \lambda_x & Q_{Fy} &= q_F \lambda_y
\end{aligned}
$$

which can be written in matrix form as

$$
\begin{bmatrix} Q_{Nx} \\ Q_{Ny} \\ Q_{Fx} \\ Q_{Fy} \end{bmatrix} =
\begin{bmatrix} \lambda_x & 0 \\ \lambda_y & 0 \\ 0 & \lambda_x \\ 0 & \lambda_y \end{bmatrix}
\begin{bmatrix} q_N \\ q_F \end{bmatrix}
\tag{14–10}
$$

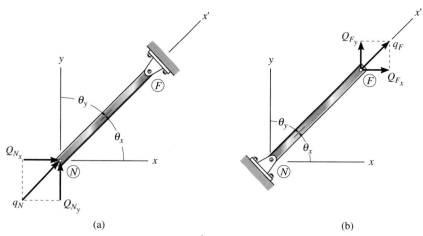

(a)   (b)

Fig. 14–6

or

$$Q = T^T q \qquad (14\text{–}11)$$

where

$$T^T = \begin{bmatrix} \lambda_x & 0 \\ \lambda_y & 0 \\ 0 & \lambda_x \\ 0 & \lambda_y \end{bmatrix} \qquad (14\text{–}12)$$

In this case T^T transforms the two local (x') forces q acting at the ends of the member into the four global (x, y) force components Q. By comparison, this *force transformation matrix* is the transpose of the displacement transformation matrix, Eq. 14–9.

14.4   Member Global Stiffness Matrix

We will now combine the results of the preceding sections and determine the stiffness matrix for a member which relates the member's global force components Q to its global displacements D. If we substitute Eq. 14–8 ($d = TD$) into Eq. 14–3 ($q = k'd$), we can determine the member's forces q in terms of the global displacements D at its end points, namely,

$$q = k'TD \qquad (14\text{–}13)$$

Substituting this equation into Eq. 14–11 $Q = T^T q$ yields the final result,

$$Q = T^T k' T D$$

or

$$Q = kD \qquad (14\text{–}14)$$

where

$$\mathbf{k} = \mathbf{T}^T \mathbf{k}' \mathbf{T} \qquad (14\text{--}15)$$

The matrix $\mathbf{k}$ is the *member's stiffness matrix* in global coordinates. Since $\mathbf{T}^T$, $\mathbf{T}$, and $\mathbf{k}'$ are known, we have

$$\mathbf{k} = \begin{bmatrix} \lambda_x & 0 \\ \lambda_y & 0 \\ 0 & \lambda_x \\ 0 & \lambda_y \end{bmatrix} \frac{AE}{L} \begin{bmatrix} 1 & -1 \\ -1 & 1 \end{bmatrix} \begin{bmatrix} \lambda_x & \lambda_y & 0 & 0 \\ 0 & 0 & \lambda_x & \lambda_y \end{bmatrix}$$

Performing the matrix operations yields

$$\mathbf{k} = \frac{AE}{L} \begin{array}{cccc} & N_x & N_y & F_x & F_y \\ \begin{bmatrix} \lambda_x^2 & \lambda_x\lambda_y & -\lambda_x^2 & -\lambda_x\lambda_y \\ \lambda_x\lambda_y & \lambda_y^2 & -\lambda_x\lambda_y & -\lambda_y^2 \\ -\lambda_x^2 & -\lambda_x\lambda_y & \lambda_x^2 & \lambda_x\lambda_y \\ -\lambda_x\lambda_y & -\lambda_y^2 & \lambda_x\lambda_y & \lambda_y^2 \end{bmatrix} & \begin{array}{c} N_x \\ N_y \\ F_x \\ F_y \end{array} \end{array} \qquad (14\text{--}16)$$

The *location* of each element in this 4×4 symmetric matrix is referenced with each global degree of freedom associated with the near end N, followed by the far end F. This is indicated by the code number notation along the rows and columns, that is, N_x, N_y, F_x, F_y. Like $\mathbf{k}'$, here $\mathbf{k}$ represents the force-displacement relations for the member when the components of force and displacement at the ends of the member are in the global or x, y directions. Each of the terms in the matrix is therefore a *stiffness influence coefficient* $\mathbf{k}_{ij}$, which denotes the x or y force component at i needed to cause an associated *unit* x or y displacement component at j. As a result, each identified column of the matrix represents the four force components developed at the ends of the member when the identified end undergoes a unit displacement related to its matrix column. For example, a unit displacement $D_{Nx} = 1$ will create the four force components on the member shown in the first column of the matrix.

14.5   Structure Stiffness Matrix

Once all the member stiffness matrices are formed in global coordinates, it becomes necessary to assemble them in the proper order so that the structure stiffness matrix $\mathbf{K}$ for the entire truss can be found. This process of combining the member matrices depends on careful identification of the elements in each matrix. As discussed in the previous section, this is done by designating the rows and columns of the matrix by the *four* code numbers N_x, N_y, F_x, F_y used to identify the two global degrees of freedom that can occur at each end of the member (see Eq. 14–16). The structure stiffness matrix will then have an order that will be equal to the highest code number assigned to the structure, since this represents the total number of degrees of freedom for the structure. When the $\mathbf{k}$ matrices are assembled, each element in $\mathbf{k}$ will then be placed in its *same* row and column designation in the structure stiffness matrix $\mathbf{K}$. In particular, when two or more members are connected to the *same* joint or node, then some of the elements of each of the $\mathbf{k}$ matrices will be assigned to the same position in the $\mathbf{K}$ matrix. When this occurs, the elements assigned to the common location must be added together algebraically. The reason for this becomes clear if one realizes that each element of the $\mathbf{k}$ matrix represents the resistance of the member to an applied force at its end. In this way, adding these resistances in the x or y direction when forming the $\mathbf{K}$ matrix is symbolic of determining the *total resistance* of each joint to a unit displacement in the x or y direction.

This method of assembling the member matrices to form the structure stiffness matrix will now be demonstrated by two numerical examples. Although this process is somewhat tedious when done by hand, it is rather easy to program on a computer.

The structural frame work of this aircraft hangar is constructed entirely of trusses in order to significantly reduce the weight of the structure. *(Courtesy of Bethlehem Steel Corporation)*

665

Example 14–1

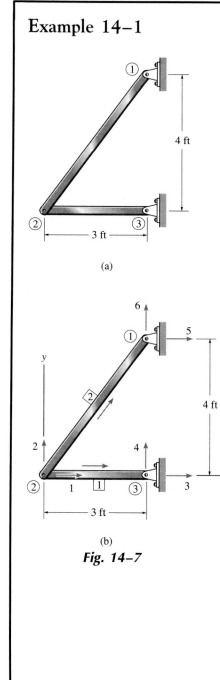

(a)

6

5

y

2   4

②   1   ①   ③   3

3 ft

(b)

Fig. 14–7

4 ft

3 ft

4 ft

Determine the structure stiffness matrix for the two-member truss shown in Fig. 14–7a. AE is constant.

Solution

By inspection, joint ② will have two unknown displacement components, whereas joints ① and ③ are constrained from displacement. Consequently, the displacement components at joint ② are code-numbered first, followed by those at joints ③ and ①, Fig. 14–7b. The origin of the global coordinate system can be located at any point. For convenience, we will choose joint ② as shown. The members are identified arbitrarily and arrows are written along the two members to identify the near and far ends of each member. The direction cosines and the stiffness matrix for each member can now be determined.

Member 1. Since ② is the near end and ③ is the far end, then by Eqs. 14–5 and 14–6, we have

$$\lambda_x = \frac{3 - 0}{3} = 1 \qquad \lambda_y = \frac{0 - 0}{3} = 0$$

Using Eq. 14–16, dividing each element by $L = 3$ ft, we have

$$\mathbf{k}_1 = AE \begin{bmatrix} \overset{1}{0.333} & \overset{2}{0} & \overset{3}{-0.333} & \overset{4}{0} \\ 0 & 0 & 0 & 0 \\ -0.333 & 0 & 0.333 & 0 \\ 0 & 0 & 0 & 0 \end{bmatrix} \begin{matrix} 1 \\ 2 \\ 3 \\ 4 \end{matrix}$$

The calculations can be checked in part by noting that $\mathbf{k}_1$ is *symmetric*. Note that the rows and columns in $\mathbf{k}_1$ are identified by the x, y degrees of freedom at the near end, followed by the far end, that is, 1, 2, 3, 4, respectively, for member 1, Fig. 14–7b. This is done in order to identify the elements for assembly into the **K** matrix.

Member 2. Since ② is the near end and ① is the far end, we have

$$\lambda_x = \frac{3 - 0}{5} = 0.6 \qquad \lambda_y = \frac{4 - 0}{5} = 0.8$$

Thus Eq. 14–16 with $L = 5$ ft becomes

$$
\mathbf{k}_2 = AE
\begin{array}{cccc}
1 & 2 & 5 & 6 \\
\end{array}
\begin{bmatrix}
0.072 & 0.096 & -0.072 & -0.096 \\
0.096 & 0.128 & -0.096 & -0.128 \\
-0.072 & -0.096 & 0.072 & 0.096 \\
-0.096 & -0.128 & 0.096 & 0.128
\end{bmatrix}
\begin{array}{c}
1 \\ 2 \\ 5 \\ 6
\end{array}
$$

Here the rows and columns are identified as 1, 2, 5, 6, since these numbers represent, respectively, the x, y degrees of freedom at the near and far ends of member 2.

Structure Stiffness Matrix. This matrix has an order of 6×6 since there are six designated degrees of freedom for the truss, Fig. 14–7b. Corresponding elements of the above two matrices are added algebraically to form the structure stiffness matrix. Perhaps the assembly process is easier to see if the missing numerical columns and rows in $\mathbf{k}_1$ and $\mathbf{k}_2$ are expanded with zeros to form two 6×6 matrices. Since

$$\mathbf{K} = \mathbf{k}_1 + \mathbf{k}_2$$

$$
\mathbf{K} = AE
\begin{bmatrix}
0.333 & 0 & -0.333 & 0 & 0 & 0 \\
0 & 0 & 0 & 0 & 0 & 0 \\
-0.333 & 0 & 0.333 & 0 & 0 & 0 \\
0 & 0 & 0 & 0 & 0 & 0 \\
0 & 0 & 0 & 0 & 0 & 0 \\
0 & 0 & 0 & 0 & 0 & 0
\end{bmatrix}
+ AE
\begin{bmatrix}
0.072 & 0.096 & 0 & 0 & -0.072 & -0.096 \\
0.096 & 0.128 & 0 & 0 & -0.096 & -0.128 \\
0 & 0 & 0 & 0 & 0 & 0 \\
0 & 0 & 0 & 0 & 0 & 0 \\
-0.072 & -0.096 & 0 & 0 & 0.072 & 0.096 \\
-0.096 & -0.128 & 0 & 0 & 0.096 & 0.128
\end{bmatrix}
$$

$$
\mathbf{K} = AE
\begin{bmatrix}
0.405 & 0.096 & -0.333 & 0 & -0.072 & -0.096 \\
0.096 & 0.128 & 0 & 0 & -0.096 & -0.128 \\
-0.333 & 0 & 0.333 & 0 & 0 & 0 \\
0 & 0 & 0 & 0 & 0 & 0 \\
-0.072 & -0.096 & 0 & 0 & 0.072 & 0.096 \\
-0.096 & -0.128 & 0 & 0 & 0.096 & 0.128
\end{bmatrix}
$$

If a computer is used for this operation, generally one starts with $\mathbf{K}$ having all zero elements; then as the member global stiffness matrices are generated, they are placed directly into their respective element positions in the $\mathbf{K}$ matrix, rather than developing the member stiffness matrices, storing them, then assembling them.

Example 14–2

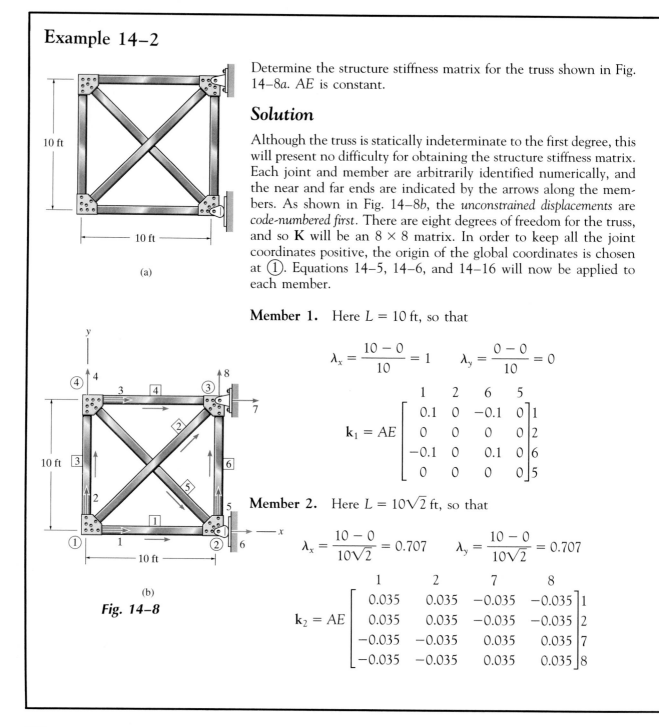

Determine the structure stiffness matrix for the truss shown in Fig. 14–8a. AE is constant.

Solution

Although the truss is statically indeterminate to the first degree, this will present no difficulty for obtaining the structure stiffness matrix. Each joint and member are arbitrarily identified numerically, and the near and far ends are indicated by the arrows along the members. As shown in Fig. 14–8b, the *unconstrained displacements* are *code-numbered first*. There are eight degrees of freedom for the truss, and so **K** will be an 8×8 matrix. In order to keep all the joint coordinates positive, the origin of the global coordinates is chosen at ①. Equations 14–5, 14–6, and 14–16 will now be applied to each member.

Member 1. Here $L = 10$ ft, so that

$$\lambda_x = \frac{10 - 0}{10} = 1 \qquad \lambda_y = \frac{0 - 0}{10} = 0$$

$$\mathbf{k}_1 = AE \begin{array}{c} \quad \\ \begin{bmatrix} 0.1 & 0 & -0.1 & 0 \\ 0 & 0 & 0 & 0 \\ -0.1 & 0 & 0.1 & 0 \\ 0 & 0 & 0 & 0 \end{bmatrix} \begin{array}{c} 1 \\ 2 \\ 6 \\ 5 \end{array}$$

with columns labeled $1 \quad 2 \quad 6 \quad 5$

Member 2. Here $L = 10\sqrt{2}$ ft, so that

$$\lambda_x = \frac{10 - 0}{10\sqrt{2}} = 0.707 \qquad \lambda_y = \frac{10 - 0}{10\sqrt{2}} = 0.707$$

$$\mathbf{k}_2 = AE \begin{bmatrix} 0.035 & 0.035 & -0.035 & -0.035 \\ 0.035 & 0.035 & -0.035 & -0.035 \\ -0.035 & -0.035 & 0.035 & 0.035 \\ -0.035 & -0.035 & 0.035 & 0.035 \end{bmatrix} \begin{array}{c} 1 \\ 2 \\ 7 \\ 8 \end{array}$$

with columns labeled $1 \quad 2 \quad 7 \quad 8$

Fig. 14–8

Member 3.   Here $L = 10$ ft, so that

$$\lambda_x = \frac{0 - 0}{10} = 0 \qquad \lambda_y = \frac{10 - 0}{10} = 1$$

$$\mathbf{k}_3 = AE \begin{array}{cccc} 1 & 2 & 3 & 4 \\ \left[\begin{array}{cccc} 0 & 0 & 0 & 0 \\ 0 & 0.1 & 0 & -0.1 \\ 0 & 0 & 0 & 0 \\ 0 & -0.1 & 0 & 0.1 \end{array}\right] & \begin{array}{c} 1 \\ 2 \\ 3 \\ 4 \end{array} \end{array}$$

Member 4.   Here $L = 10$ ft, so that

$$\lambda_x = \frac{10 - 0}{10} = 1 \qquad \lambda_y = \frac{10 - 10}{10} = 0$$

$$\mathbf{k}_4 = AE \begin{array}{cccc} 3 & 4 & 7 & 8 \\ \left[\begin{array}{cccc} 0.1 & 0 & -0.1 & 0 \\ 0 & 0 & 0 & 0 \\ -0.1 & 0 & 0.1 & 0 \\ 0 & 0 & 0 & 0 \end{array}\right] & \begin{array}{c} 3 \\ 4 \\ 7 \\ 8 \end{array} \end{array}$$

Member 5.   Here $L = 10\sqrt{2}$ ft, so that

$$\lambda_x = \frac{10 - 0}{10\sqrt{2}} = 0.707 \qquad \lambda_y = \frac{0 - 10}{10\sqrt{2}} = -0.707$$

$$\mathbf{k}_5 = AE \begin{array}{cccc} 3 & 4 & 6 & 5 \\ \left[\begin{array}{cccc} 0.035 & -0.035 & -0.035 & 0.035 \\ -0.035 & 0.035 & 0.035 & -0.035 \\ -0.035 & 0.035 & 0.035 & -0.035 \\ 0.035 & -0.035 & -0.035 & 0.035 \end{array}\right] & \begin{array}{c} 3 \\ 4 \\ 6 \\ 5 \end{array} \end{array}$$

(cont'd)

Example 14–2
(continued)

Member 6. Here $L = 10$ ft, so that

$$\lambda_x = \frac{10 - 10}{10} = 0 \qquad \lambda_y = \frac{10 - 0}{10} = 1$$

$$\mathbf{k}_6 = AE \begin{array}{cccc} 6 & 5 & 7 & 8 \\ \begin{bmatrix} 0 & 0 & 0 & 0 \\ 0 & 0.1 & 0 & -0.1 \\ 0 & 0 & 0 & 0 \\ 0 & -0.1 & 0 & 0.1 \end{bmatrix} & \begin{array}{c} 6 \\ 5 \\ 7 \\ 8 \end{array} \end{array}$$

Structure Stiffness Matrix. The foregoing six matrices can now be assembled into the 8×8 **K** matrix by algebraically adding their corresponding elements. For example, $(k_{11})_1 = AE(0.1)$, $(k_{11})_2 = AE(0.035)$, $(k_{11})_3 = (k_{11})_4 = (k_{11})_5 = (k_{11})_6 = 0$. Thus, $K_{11} = AE(0.1 + 0.035) = AE(0.135)$, and so on. The final result is thus,

$$\mathbf{K} = AE \begin{array}{cccccccc} 1 & 2 & 3 & 4 & 5 & 6 & 7 & 8 \\ \begin{bmatrix} 0.135 & 0.035 & 0 & 0 & 0 & -0.1 & -0.035 & -0.035 \\ 0.035 & 0.135 & 0 & -0.1 & 0 & 0 & -0.035 & -0.035 \\ 0 & 0 & 0.135 & -0.035 & 0.035 & -0.035 & -0.1 & 0 \\ 0 & -0.1 & -0.035 & 0.135 & -0.035 & 0.035 & 0 & 0 \\ 0 & 0 & 0.035 & -0.035 & 0.135 & -0.035 & 0 & -0.1 \\ -0.1 & 0 & -0.035 & 0.035 & -0.035 & 0.135 & 0 & 0 \\ -0.035 & -0.035 & -0.1 & 0 & 0 & 0 & 0.135 & 0.035 \\ -0.035 & -0.035 & 0 & 0 & -0.1 & 0 & 0.035 & 0.135 \end{bmatrix} & \begin{array}{c} 1 \\ 2 \\ 3 \\ 4 \\ 5 \\ 6 \\ 7 \\ 8 \end{array} \end{array} \quad \mathbf{Ans.}$$

14.6   Application of the Stiffness Method for Truss Analysis

Once the structure stiffness matrix is formed using the methods of the preceding section, we can then use it to determine the joint displacements, external force reactions, and the internal member forces. Since we have always assigned the lowest code numbers to identify the unconstrained degrees of freedom, this will allow us now to partition Eq. 14–14, $Q = KD$ in the following form*:

$$\left[\frac{Q_k}{Q_u} \right] = \left[\begin{array}{c|c} K_{11} & K_{12} \\ \hline K_{21} & K_{22} \end{array} \right] \left[\frac{D_u}{D_k} \right] \qquad (14\text{–}17)$$

Here

Q_k, D_k = *known* external loads and displacements; the loads here exist on the truss as part of the problem, and the displacements are generally specified as zero due to support constraints such as pins or rollers

Q_u, D_u = *unknown* loads and displacements; the loads here represent the unknown support reactions, and the displacements are at joints where motion is unconstrained in a particular direction

$\quad K$ = structure stiffness matrix, which is partitioned to be compatible with the partitioning of Q and D

Expanding Eq. 14–17 yields

$$Q_k = K_{11}D_u + K_{12}D_k \qquad (14\text{–}18)$$
$$Q_u = K_{21}D_u + K_{22}D_k \qquad (14\text{–}19)$$

Most often $D_k = 0$ since the supports are not displaced. Provided this is the case, Eq. 14–18 becomes

$$Q_k = K_{11}D_u$$

---

*This partitioning scheme will become obvious in the numerical examples that follow.

671

Since the elements in the partitioned matrix $\mathbf{K}_{11}$ represent the *total resistance* at a truss joint to a unit displacement in either the x or y direction, then the above equation symbolizes the collection of all the *force equilibrium equations* applied to the joints where the external loads are zero or have a known value ($\mathbf{Q}_k$). Solving for $\mathbf{D}_u$, we have

$$\mathbf{D}_u = [\mathbf{K}_{11}]^{-1}\mathbf{Q}_k \qquad (14\text{--}20)$$

From this equation we can obtain a direct solution for all the unknown joint displacements. Substituting this result into Eq. 14–19 with $\mathbf{D}_k = \mathbf{0}$ yields

$$\mathbf{Q}_u = \mathbf{K}_{21}\mathbf{D}_u \qquad (14\text{--}21)$$

from which the unknown support reactions can be determined. The member forces can be determined using Eq. 14–13, namely

$$\mathbf{q} = \mathbf{k}'\mathbf{TD}$$

Expanding this equation yields

$$\begin{bmatrix} q_N \\ q_F \end{bmatrix} = \frac{AE}{L} \begin{bmatrix} 1 & -1 \\ -1 & 1 \end{bmatrix} \begin{bmatrix} \lambda_x & \lambda_y & 0 & 0 \\ 0 & 0 & \lambda_x & \lambda_y \end{bmatrix} \begin{bmatrix} D_{Nx} \\ D_{Ny} \\ D_{Fx} \\ D_{Fy} \end{bmatrix}$$

Since $q_N = -q_F$ for equilibrium, only one of the forces has to be found. Here we will determine q_F, the one that exerts tension in the member, Fig. 14–6*b*. Thus,

$$q_F = \frac{AE}{L}[-\lambda_x \quad -\lambda_y \quad \lambda_x \quad \lambda_y] \begin{bmatrix} D_{Nx} \\ D_{Ny} \\ D_{Fx} \\ D_{Fy} \end{bmatrix} \qquad (14\text{--}22)$$

In particular, if the computed result using this equation is negative, the member is then in compression.

Procedure for Analysis

The following method provides a means for determining the unknown displacements and support reactions for a truss using the stiffness method of matrix analysis. This procedure is suitable for *any type* of truss, be it simple, compound, or complex. Also, it is applicable to either statically determinate or statically indeterminate trusses.

Notation   Establish the x, y global coordinate system. The origin is usually located at one of the joints such that the coordinates for all the other joints are positive. Identify each joint and member numerically, and arbitrarily specify the near and far ends of each member symbolically by directing an arrow along the member with the head directed toward the far end. Also, specify the two code numbers at each joint, using the *lowest numbers* to identify *unconstrained degrees of freedom*, followed by the *highest numbers* to identify the *constrained degrees of freedom*. From the problem, establish $\mathbf{D}_k$ and $\mathbf{Q}_k$.

Structure Stiffness Matrix   For each member determine λ_x and λ_y and the member stiffness matrix using Eq. 14–16. Assemble these matrices to form the stiffness matrix for the entire truss as explained in Sec. 14.5. As a partial check of the calculations, the member and structure stiffness matrices should be *symmetric*.

Displacements and Loads   Partition the structure stiffness matrix as indicated by Eq. 14–17. Then determine the unknown joint displacements $\mathbf{D}_u$ using Eq. 14–20, the support reactions $\mathbf{Q}_u$ using Eq. 14–21, and each member force $\mathbf{q}_F$ using Eq. 14–22.

The following three examples numerically illustrate this procedure.

Example 14–3

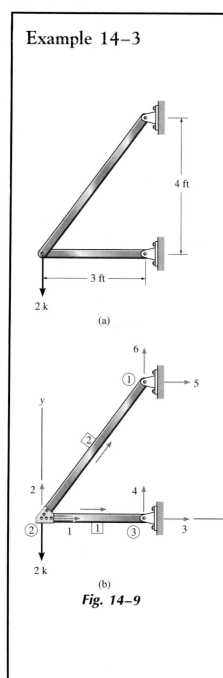

4 ft

3 ft

2 k

(a)

6

①

5

y

②

2

1

② 1 1 ③ 3

2 k

(b)

Fig. 14–9

Determine the force in each member of the two-member truss shown in Fig. 14–9a. AE is constant.

Solution

Notation. The origin of x, y and the numbering of the joints and members are shown in Fig. 14–9b. Also, the near and far ends of each member are identified by arrows, and code numbers are used at each joint. By inspection it is seen that the known external displacements are $D_3 = D_4 = D_5 = D_6 = 0$. Also, the known external loads are $Q_1 = 0$, $Q_2 = -2$ k. Hence,

$$\mathbf{D}_k = \begin{bmatrix} 0 \\ 0 \\ 0 \\ 0 \end{bmatrix} \begin{matrix} 3 \\ 4 \\ 5 \\ 6 \end{matrix} \qquad \mathbf{Q}_k = \begin{bmatrix} 0 \\ -2 \end{bmatrix} \begin{matrix} 1 \\ 2 \end{matrix}$$

Structure Stiffness Matrix. Using the same notation as used here, this matrix has been developed in Example 14–1.

Displacements and Loads. Writing Eq. 14–17, $\mathbf{Q} = \mathbf{KD}$, for the truss we have

$$\begin{bmatrix} 0 \\ -2 \\ \hline Q_3 \\ Q_4 \\ Q_5 \\ Q_6 \end{bmatrix} = AE \begin{bmatrix} 0.405 & 0.096 & -0.333 & 0 & -0.072 & -0.096 \\ 0.096 & 0.128 & 0 & 0 & -0.096 & -0.128 \\ \hline -0.333 & 0 & 0.333 & 0 & 0 & 0 \\ 0 & 0 & 0 & 0 & 0 & 0 \\ -0.072 & -0.096 & 0 & 0 & 0.072 & 0.096 \\ -0.096 & -0.128 & 0 & 0 & 0.096 & 0.128 \end{bmatrix} \begin{bmatrix} D_1 \\ D_2 \\ \hline 0 \\ 0 \\ 0 \\ 0 \end{bmatrix}$$

$$(1)$$

From this equation we can now identify $\mathbf{K}_{11}$ and thereby determine $\mathbf{D}_u$. It is seen that the matrix multiplication, like Eq. 14–18, yields

$$\begin{bmatrix} 0 \\ -2 \end{bmatrix} = AE \begin{bmatrix} 0.405 & 0.096 \\ 0.096 & 0.128 \end{bmatrix} \begin{bmatrix} D_1 \\ D_2 \end{bmatrix} + \begin{bmatrix} 0 \\ 0 \end{bmatrix}$$

Here it is easy to solve by a direct expansion,

$$0 = AE(0.405D_1 + 0.096D_2)$$
$$-2 = AE(0.096D_1 + 0.128D_2)$$

Physically these equations represent $\Sigma F_x = 0$ and $\Sigma F_y = 0$ applied to joint ②. Solving, we get

$$D_1 = \frac{4.505}{AE} \qquad D_2 = \frac{-19.003}{AE}$$

By inspection of Fig. 14–9b, one would indeed expect a rightward and downward displacement to occur at joint ② as indicated by the positive and negative signs of these answers.

Using these results, the support reactions are now obtained from Eq. (1), written in the form of Eq. 14–19 (or Eq. 14–21) as

$$\begin{bmatrix} Q_3 \\ Q_4 \\ Q_5 \\ Q_6 \end{bmatrix} = AE \begin{bmatrix} -0.333 & 0 \\ 0 & 0 \\ -0.072 & -0.096 \\ -0.096 & -0.128 \end{bmatrix} \frac{1}{AE} \begin{bmatrix} 4.505 \\ -19.003 \end{bmatrix} + \begin{bmatrix} 0 \\ 0 \\ 0 \\ 0 \end{bmatrix}$$

or

$$Q_3 = -0.333(4.505) = -1.5 \text{ k}$$
$$Q_4 = 0$$
$$Q_5 = -0.072(4.505) - 0.096(-19.003) = 1.5 \text{ k}$$
$$Q_6 = -0.096(4.505) - 0.128(-19.003) = 2.0 \text{ k}$$

The force in each member is found from Eq. 14–22. Using the data for λ_x and λ_y in Example 14–1, we have

Member 1: $\lambda_x = 1, \ \lambda_y = 0, \ L = 3 \text{ ft}$

$$q_1 = \frac{AE}{3}[-1 \quad 0 \quad 1 \quad 0]\frac{1}{AE}\begin{bmatrix} 4.505 \\ -19.003 \\ 0 \\ 0 \end{bmatrix}\begin{matrix} 1 \\ 2 \\ 3 \\ 4 \end{matrix}$$

$$= \frac{1}{3}[-4.505] = -1.5 \text{ k} \qquad \textbf{Ans.}$$

Member 2: $\lambda_x = 0.6, \ \lambda_y = 0.8, \ L = 5 \text{ ft}$

$$q_2 = \frac{AE}{5}[-0.6 \quad -0.8 \quad 0.6 \quad 0.8]\frac{1}{AE}\begin{bmatrix} 4.505 \\ -19.003 \\ 0 \\ 0 \end{bmatrix}\begin{matrix} 1 \\ 2 \\ 5 \\ 6 \end{matrix}$$

$$= \frac{1}{5}[-0.6(4.505) - 0.8(-19.003)] = 2.5 \text{ k} \qquad \textbf{Ans.}$$

These answers can of course be verified by equilibrium, applied at joint ②.

Example 14–4

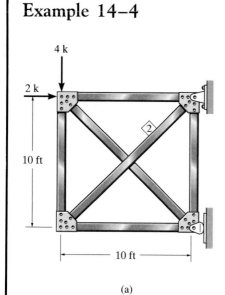

(a)

Determine the support reactions and the force in member 2 of the truss shown in Fig. 14–10a. AE is constant.

Solution

Notation. The joints and members are numbered and the origin of the x, y axes is established at ①, Fig.14–10b. Also, arrows are used to reference the near and far ends of each member. Using the code numbers, where the lowest numbers denote the unconstrained degrees of freedom, Fig. 14–10b, we have

$$\mathbf{D}_k = \begin{bmatrix} 0 \\ 0 \\ 0 \\ 0 \end{bmatrix} \begin{matrix} 6 \\ 7 \\ 8 \end{matrix} \qquad \mathbf{Q}_k = \begin{bmatrix} 0 \\ 0 \\ 2 \\ -4 \\ 0 \end{bmatrix} \begin{matrix} 1 \\ 2 \\ 3 \\ 4 \\ 5 \end{matrix}$$

Structure Stiffness Matrix. This matrix has been determined in Example 14–2 using the same notation as in Fig. 14–10b.

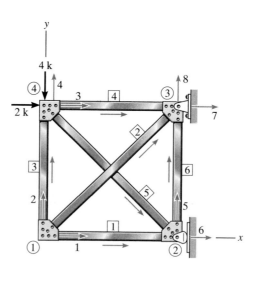

(b)

Fig. 14–10

Displacements and Loads. For this problem $\mathbf{Q} = \mathbf{KD}$ is

$$
\begin{bmatrix}
0 \\
0 \\
2 \\
-4 \\
0 \\
\hline
Q_6 \\
Q_7 \\
Q_8
\end{bmatrix}
= AE
\begin{bmatrix}
0.135 & 0.035 & 0 & 0 & 0 & -0.1 & -0.035 & -0.035 \\
0.035 & 0.135 & 0 & -0.1 & 0 & 0 & -0.035 & -0.035 \\
0 & 0 & 0.135 & -0.035 & 0.035 & -0.035 & -0.1 & 0 \\
0 & -0.1 & -0.035 & 0.135 & -0.035 & 0.035 & 0 & 0 \\
0 & 0 & 0.035 & -0.035 & 0.135 & -0.035 & 0 & -0.1 \\
\hline
-0.1 & 0 & -0.035 & 0.035 & -0.035 & 0.135 & 0 & 0 \\
-0.035 & -0.035 & -0.1 & 0 & 0 & 0 & 0.135 & 0.035 \\
-0.035 & -0.035 & 0 & 0 & -0.1 & 0 & 0.035 & 0.135
\end{bmatrix}
\begin{bmatrix}
D_1 \\
D_2 \\
D_3 \\
D_4 \\
D_5 \\
\hline
0 \\
0 \\
0
\end{bmatrix}
$$
$$(1)$$

Multiplying so as to formulate the unknown displacement equation 14–18, we get

$$
\begin{bmatrix}
0 \\
0 \\
2 \\
-4 \\
0
\end{bmatrix}
= AE
\begin{bmatrix}
0.135 & 0.035 & 0 & 0 & 0 \\
0.035 & 0.135 & 0 & -0.1 & 0 \\
0 & 0 & 0.135 & -0.035 & 0.035 \\
0 & -0.1 & -0.035 & 0.135 & -0.035 \\
0 & 0 & 0.035 & -0.035 & 0.135
\end{bmatrix}
\begin{bmatrix}
D_1 \\
D_2 \\
D_3 \\
D_4 \\
D_5
\end{bmatrix}
+
\begin{bmatrix}
0 \\
0 \\
0 \\
0 \\
0
\end{bmatrix}
$$

Expanding and solving the equations for the displacements yields

$$
\begin{bmatrix}
D_1 \\
D_2 \\
D_3 \\
D_4 \\
D_5
\end{bmatrix}
= \frac{1}{AE}
\begin{bmatrix}
17.94 \\
-69.20 \\
-2.06 \\
-87.14 \\
-22.06
\end{bmatrix}
$$

(cont'd)

Example 14–4
(continued)

Developing Eq. 14–19 from Eq. (1) using the calculated results, we have

$$
\begin{bmatrix} Q_6 \\ Q_7 \\ Q_8 \end{bmatrix} = AE
\begin{bmatrix}
-0.1 & 0 & -0.035 & 0.035 & -0.035 \\
-0.035 & -0.035 & -0.1 & 0 & 0 \\
-0.035 & -0.035 & 0 & 0 & -0.1
\end{bmatrix}
\frac{1}{AE}
\begin{bmatrix}
17.94 \\
-69.20 \\
-2.06 \\
-87.14 \\
-22.06
\end{bmatrix}
+
\begin{bmatrix}
0 \\
0 \\
0
\end{bmatrix}
$$

Expanding and computing the support reactions yields

$$
\begin{aligned}
Q_6 &= -4.0 \text{ k} & \textit{Ans.} \\
Q_7 &= 2.0 \text{ k} & \textit{Ans.} \\
Q_8 &= 4.0 \text{ k} & \textit{Ans.}
\end{aligned}
$$

The negative sign for Q_6 indicates that the roller support reaction acts in the negative x direction. The force in member 2 is found from Eq. 14–22, where from Example 14–2, $\lambda_x = 0.707$, $\lambda_y = 0.707$, $L = 10\sqrt{2}$ ft. Thus,

$$
q_2 = \frac{AE}{10\sqrt{2}}[-0.707 \quad -0.707 \quad 0.707 \quad 0.707]\frac{1}{AE}
\begin{bmatrix}
17.94 \\
-69.20 \\
0 \\
0
\end{bmatrix}
$$

$$
= 2.56 \text{ k} \qquad\qquad \textit{Ans.}
$$

Example 14–5

Determine the force in each member of the truss in Fig. 14–11a if the support at joint ① settles *downward* 25 mm. Take $AE = 8(10^3)$ kN.

Solution

Notation. For convenience the origin of the global coordinates in Fig. 14–11b is established at joint ② and as usual, the lowest coding numbers are used to reference the unconstrained degrees of freedom. Thus,

$$\mathbf{D}_k = \begin{bmatrix} 0 \\ -0.025 \\ 0 \\ 0 \\ 0 \\ 0 \end{bmatrix} \begin{matrix} 3 \\ 4 \\ 5 \\ 6 \\ 7 \\ 8 \end{matrix} \qquad \mathbf{Q}_k = \begin{bmatrix} 0 \\ 0 \end{bmatrix} \begin{matrix} 1 \\ 2 \end{matrix}$$

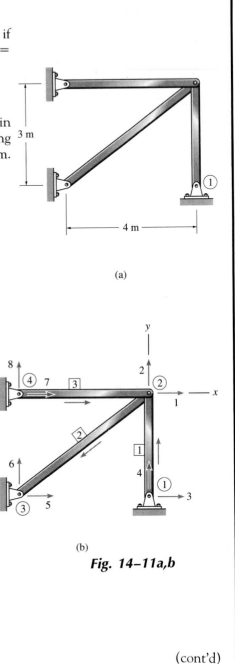

(a)

Structure Stiffness Matrix. Using Eq. 14–16, we have

Member 1: $\lambda_x = 0, \lambda_y = -1, L = 3$ m, so that

$$\mathbf{k}_1 = AE \begin{bmatrix} 0 & 0 & 0 & 0 \\ 0 & 0.333 & 0 & -0.333 \\ 0 & 0 & 0 & 0 \\ 0 & -0.333 & 0 & 0.333 \end{bmatrix} \begin{matrix} 3 \\ 4 \\ 1 \\ 2 \end{matrix}$$

with column headings $\begin{matrix} 3 & 4 & 1 & 2 \end{matrix}$

Member 2: $\lambda_x = -0.8, \lambda_y = -0.6, L = 5$ m, so that

$$\mathbf{k}_2 = AE \begin{bmatrix} 0.128 & 0.096 & -0.128 & -0.096 \\ 0.096 & 0.072 & -0.096 & -0.072 \\ -0.128 & -0.096 & 0.128 & 0.096 \\ -0.096 & -0.072 & 0.096 & 0.072 \end{bmatrix} \begin{matrix} 1 \\ 2 \\ 5 \\ 6 \end{matrix}$$

with column headings $\begin{matrix} 1 & 2 & 5 & 6 \end{matrix}$

Member 3: $\lambda_x = -1, \lambda_y = 0, L = 4$ m, so that

$$\mathbf{k}_3 = AE \begin{bmatrix} 0.25 & 0 & -0.25 & 0 \\ 0 & 0 & 0 & 0 \\ -0.25 & 0 & 0.25 & 0 \\ 0 & 0 & 0 & 0 \end{bmatrix} \begin{matrix} 7 \\ 8 \\ 1 \\ 2 \end{matrix}$$

with column headings $\begin{matrix} 7 & 8 & 1 & 2 \end{matrix}$

(b)

Fig. 14–11a,b

(cont'd)

Example 14–5
(continued)

By assembling these matrices, the structure stiffness matrix becomes

$$
K = AE
\begin{array}{c}
\begin{array}{cccccccc} 1 & 2 & 3 & 4 & 5 & 6 & 7 & 8 \end{array} \\
\begin{bmatrix}
0.378 & 0.096 & 0 & 0 & -0.128 & -0.096 & -0.25 & 0 \\
0.096 & 0.405 & 0 & -0.333 & -0.096 & -0.072 & 0 & 0 \\
0 & 0 & 0 & 0 & 0 & 0 & 0 & 0 \\
0 & -0.333 & 0 & 0.333 & 0 & 0 & 0 & 0 \\
-0.128 & -0.096 & 0 & 0 & 0.128 & 0.096 & 0 & 0 \\
-0.096 & -0.072 & 0 & 0 & 0.096 & 0.072 & 0 & 0 \\
-0.25 & 0 & 0 & 0 & 0 & 0 & 0.25 & 0 \\
0 & 0 & 0 & 0 & 0 & 0 & 0 & 0
\end{bmatrix}
\begin{array}{c} 1 \\ 2 \\ 3 \\ 4 \\ 5 \\ 6 \\ 7 \\ 8 \end{array}
\end{array}
$$

Displacement and Loads.   Here $Q = KD$ yields

$$
\begin{bmatrix} 0 \\ 0 \\ \overline{Q_3} \\ Q_4 \\ Q_5 \\ Q_6 \\ Q_7 \\ Q_8 \end{bmatrix}
= AE
\begin{bmatrix}
0.378 & 0.096 & 0 & 0 & -0.128 & -0.096 & -0.25 & 0 \\
0.096 & 0.405 & 0 & -0.333 & -0.096 & -0.072 & 0 & 0 \\
0 & 0 & 0 & 0 & 0 & 0 & 0 & 0 \\
0 & -0.333 & 0 & 0.333 & 0 & 0 & 0 & 0 \\
-0.128 & -0.096 & 0 & 0 & 0.128 & 0.096 & 0 & 0 \\
-0.096 & -0.072 & 0 & 0 & 0.096 & 0.072 & 0 & 0 \\
-0.25 & 0 & 0 & 0 & 0 & 0 & 0.25 & 0 \\
0 & 0 & 0 & 0 & 0 & 0 & 0 & 0
\end{bmatrix}
\begin{bmatrix} D_1 \\ D_2 \\ \overline{0} \\ -0.025 \\ 0 \\ 0 \\ 0 \\ 0 \end{bmatrix}
$$

Developing the solution for the displacements, Eq. 14–18, we have

$$
\begin{bmatrix} 0 \\ 0 \end{bmatrix} = AE \begin{bmatrix} 0.378 & 0.096 \\ 0.096 & 0.405 \end{bmatrix} \begin{bmatrix} D_1 \\ D_2 \end{bmatrix} + AE \begin{bmatrix} 0 & 0 & -0.128 & -0.096 & -0.25 & 0 \\ 0 & -0.333 & -0.096 & -0.072 & 0 & 0 \end{bmatrix} \begin{bmatrix} 0 \\ -0.025 \\ 0 \\ 0 \\ 0 \\ 0 \end{bmatrix}
$$

which yields

$$0 = AE[(0.378D_1 + 0.096D_2) + 0]$$
$$0 = AE[(0.096D_1 + 0.4053D_2) + 0.00833]$$

Solving these equations simultaneously gives

$$D_1 = 0.00556 \text{ m}$$
$$D_2 = -0.021875 \text{ m}$$

Although the support reactions do not have to be calculated, if needed they can be found from the expansion as defined by Eq. 14–19. Using Eq. 14–22 to determine the member forces yields

Member 1: $\lambda_x = 0$, $\lambda_y = -1$, $L = 3$, $AE = 8(10^3)$ kN, so that

$$q_1 = \frac{8(10^3)}{3}[0 \quad 1 \quad 0 \quad -1]\begin{bmatrix} 0.00556 \\ -0.021875 \\ 0 \\ -0.025 \end{bmatrix}$$

$$= \frac{8(10^3)}{3}(-0.021875 + 0.025)$$

$$= 8.34 \text{ kN} \qquad\qquad\qquad \textbf{Ans.}$$

Member 2: $\lambda_x = -0.8$, $\lambda_y = -0.6$, $L = 5$, $AE = 8(10^3)$ kN, so that

$$q_2 = \frac{8(10^3)}{5}[0.8 \quad 0.6 \quad -0.8 \quad -0.6]\begin{bmatrix} 0.00556 \\ -0.0219 \\ 0 \\ 0 \end{bmatrix}$$

$$= \frac{8(10^3)}{5}(0.00444 - 0.0131) = -13.9 \text{ kN} \qquad \textbf{Ans.}$$

Member 3: $\lambda_x = -1$, $\lambda_y = 0$, $L = 4$, $AE = 8(10^3)$ kN, so that

$$q_3 = \frac{8(10^3)}{4}[1 \quad 0 \quad -1 \quad 0]\begin{bmatrix} 0.00556 \\ -0.0219 \\ 0 \\ 0 \end{bmatrix} = \frac{8(10^3)}{4}(0.00556)$$

$$= 11.1 \text{ kN} \qquad\qquad\qquad\qquad\qquad\qquad \textbf{Ans.}$$

These results are shown on the free-body diagram of joint ②, Fig. 14–11, which can be checked to be in equilibrium.

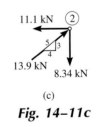

(c)

Fig. 14–11c

14.7 Space Truss Analysis

The analysis of statically determinate space trusses was discussed in Sec. 3.9. We can, of course, analyze these types of trusses and those that are statically indeterminate by using the same procedure discussed in the preceding section. To account for the three-dimensional aspects of the problem, however, additional elements must be included in the transformation matrix **T.** In this regard, consider the truss member shown in Fig. 14–12. The stiffness matrix for the member defined in terms of the local coordinate x' is given by Eq. 14–4. Furthermore, by inspection of Fig. 14–12, the direction cosines between the global and local coordinates can be found using equations analogous to Eqs. 14–5 and 14–6, that is,

$$\lambda_x = \cos \theta_x = \frac{x_F - x_N}{L}$$

$$= \frac{x_F - x_N}{\sqrt{(x_F - x_N)^2 + (y_F - y_N)^2 + (z_F - z_N)^2}} \quad (14\text{–}23)$$

$$\lambda_y = \cos \theta_y = \frac{y_F - y_N}{L}$$

$$= \frac{y_F - y_N}{\sqrt{(x_F - x_N)^2 + (y_F - y_N)^2 + (z_F - z_N)^2}} \quad (14\text{–}24)$$

$$\lambda_z = \cos \theta_z = \frac{z_F - z_N}{L}$$

$$= \frac{z_F - z_N}{\sqrt{(x_F - x_N)^2 + (y_F - y_N)^2 + (z_F - z_N)^2}} \quad (14\text{–}25)$$

As a result of the third dimension, the transformation matrix, Eq. 14–9, becomes

$$\mathbf{T} = \begin{bmatrix} \lambda_x & \lambda_y & \lambda_z & 0 & 0 & 0 \\ 0 & 0 & 0 & \lambda_x & \lambda_y & \lambda_z \end{bmatrix}$$

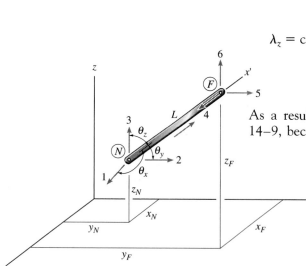

Fig. 14–12

Substituting this and Eq. 14–4 into Eq. 14–15, $\mathbf{k} = \mathbf{T}^T\mathbf{k}'\mathbf{T}$, yields

$$
\mathbf{k} = \begin{bmatrix} \lambda_x & 0 \\ \lambda_y & 0 \\ \lambda_z & 0 \\ 0 & \lambda_x \\ 0 & \lambda_y \\ 0 & \lambda_z \end{bmatrix} \frac{AE}{L} \begin{bmatrix} 1 & -1 \\ -1 & 1 \end{bmatrix} \begin{bmatrix} \lambda_x & \lambda_y & \lambda_z & 0 & 0 & 0 \\ 0 & 0 & 0 & \lambda_x & \lambda_y & \lambda_z \end{bmatrix}
$$

Carrying out the matrix multiplication yields the *symmetric* matrix

$$
\mathbf{k} = \frac{AE}{L} \begin{array}{cccccc} N_x & N_y & N_z & F_x & F_y & F_z \end{array}
$$

$$
\mathbf{k} = \frac{AE}{L} \begin{bmatrix}
\lambda_x^2 & \lambda_x\lambda_y & \lambda_x\lambda_z & -\lambda_x^2 & -\lambda_x\lambda_y & -\lambda_x\lambda_z \\
\lambda_y\lambda_x & \lambda_y^2 & \lambda_y\lambda_z & -\lambda_y\lambda_x & -\lambda_y^2 & -\lambda_y\lambda_z \\
\lambda_z\lambda_x & \lambda_z\lambda_y & \lambda_z^2 & -\lambda_z\lambda_x & -\lambda_z\lambda_y & -\lambda_z^2 \\
-\lambda_x^2 & -\lambda_x\lambda_y & -\lambda_x\lambda_z & \lambda_x^2 & \lambda_x\lambda_y & \lambda_x\lambda_z \\
-\lambda_y\lambda_x & -\lambda_y^2 & -\lambda_y\lambda_z & \lambda_y\lambda_x & \lambda_y^2 & \lambda_y\lambda_z \\
-\lambda_z\lambda_x & -\lambda_z\lambda_y & -\lambda_z^2 & \lambda_z\lambda_x & \lambda_z\lambda_y & \lambda_z^2
\end{bmatrix} \begin{array}{c} N_x \\ N_y \\ N_z \\ F_x \\ F_y \\ F_z \end{array}
$$

$$(14-26)$$

This equation represents the *member stiffness matrix* expressed in *global coordinates*. The code numbers along the rows and columns reference the x, y, z directions at the near end, N_x, N_y, N_z, followed by those at the far end, F_x, F_y, F_z, for the member.

For computer programming using these matrices, it is generally more efficient to use Eq. 14–26 than to carry out the matrix multiplication $\mathbf{T}^T\mathbf{k}'\mathbf{T}$ for each member. As stated previously, computer storage space is saved if the "structure" stiffness matrix $\mathbf{K}$ is first initiated with all zero elements; then as the elements of each member stiffness matrix are generated, they are placed directly into their respective positions in $\mathbf{K}$. After the structure stiffness matrix has been developed, the procedure outlined in Sec. 14.6 can be followed to determine the joint displacements, support reactions, and internal member forces.

REFERENCES

Tezcan, S. "Computer Analysis of Plane and Space Structures," *Structural Division* ASCE, 143–173, April, 1966.

Wilson, E. "The Use of Minicomputers in Structural Analysis," *Computers and Structures*, 2, 695–698, 1980.

PROBLEMS

14–1. Determine the stiffness matrix **K** for the truss. Take $A = 0.5$ in^2 and $E = 29(10^3)$ ksi for each member.

14–2. Determine the vertical displacement at joint ③ of the truss in Prob. 14–1.

14–3. Determine the force in each member of the truss in Prob. 14–1.

14–6. Determine the stiffness matrix **K** for the truss. Take $A = 0.0015$ m^2 and $E = 200$ GPa for each member.

14–7. Determine the force in member ⑥ of the truss on Prob. 14–6.

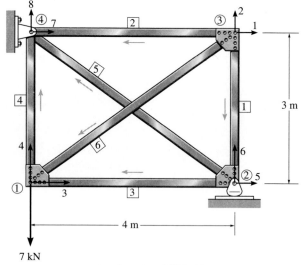

Probs. 14–6/14–7

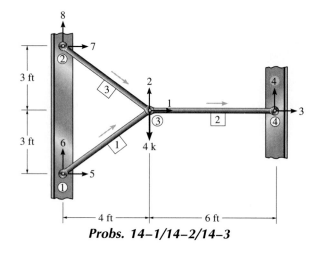

Probs. 14–1/14–2/14–3

*14–4. Determine the stiffness matrix **K** for the truss. AE is constant. Assume all joints are pinned.

14–5. Determine the vertical deflection of joint ③ and the force in member ① of the truss in Prob. 14–4.

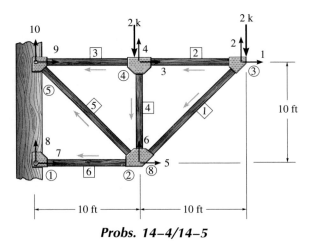

Probs. 14–4/14–5

*14–8. Determine the stiffness matrix **K** for the truss. *AE* is constant.

14–9. Determine the force in members ⬜1 and ⬜5 of the truss in Prob. 14–8.

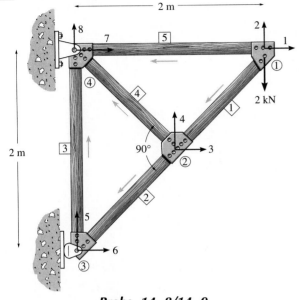
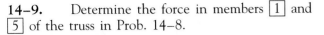

Probs. 14–8/14–9

14–10. Determine the stiffness matrix **K** for the truss. *AE* is constant.

14–11. Determine the force in each member of the truss in Prob. 14–10.

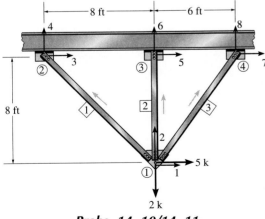

Probs. 14–10/14–11

*14–12. Determine the stiffness matrix **K** for the truss. Take $A = 0.75$ in^2, $E = 29(10^3)$ ksi. Assume all joints are pin-connected.

*14–13. Determine the vertical deflection of joint ① and the force in member ⬜2 of the truss in Prob. 14–12.

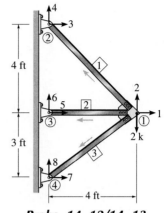

Probs. 14–12/14–13

14–14. Determine the stiffness matrix **K** for the truss. Take $A = 0.0015$ m^2 and $E = 200$ GPa for each member.

14–15. Determine the vertical deflection at joint ② and the force in member ⬜4 of the truss in Prob. 14–14.

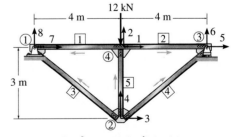

Probs. 14–14/14–15

See Probs. 15–22 and 15–23 (p. 716) for application of the STRAN program.

685

Building frames that are highly indeterminate are best solved using a computer program that is formulated from the stiffness method of matrix analysis. *(Peter Vandermark/Stock, Boston)*

Beam and Plane Frame Analysis Using the Stiffness Method

The concepts presented in the previous chapter will be extended here and applied to the analysis of beams and plane frames. It will be shown that once the member stiffness matrix and the transformation matrix have been developed, the procedure for application is exactly the same as that for trusses. A structural analysis of three-dimensional frameworks requires a simple extension of the concepts presented here and is discussed in books devoted to matrix analysis.

15.1 Preliminary Remarks

Before we show how the stiffness method applies to beams and frames, we will first discuss some preliminary concepts and definitions related to these structures.

Member and Node Identification

In order to apply the stiffness method to beams and frames, we must first determine how to subdivide the structure into its component finite elements. In general, the nodes of each element are located at a support, at a corner or joint, where an external force is applied, or where the linear or rotational displacement at a point (or node) is to be determined. For example, consider the frame in Fig. 15–1a. Using the same scheme as that for trusses, the five nodes are specified numerically within a circle, and the four elements are identified numerically within a square. Also, notice the "near" and "far" ends of each member are identified by the arrows written alongside each member. In particular, the nodal point 4 is chosen, in order to determine the three components of displacement at this point.

Global and Member Coordinates

The global or structure coordinate system will be identified using x, y, z axes that generally have their origin at a node, and are positioned so that the nodes at other points on the structure all have positive coordinates, Fig. 15–1a. The local or member x', y', z' coordinates have their origin at the "near" end of each member and the positive x' axis is directed towards the "far" end. Figure 15–1b shows these coordinates for element 3. In both cases we have used a right-handed coordinate system, so that if the fingers of the right hand are curled from the $x(x')$ axis towards the $y(y')$ axis, the thumb points in the positive direction of the $z(z')$ axis, which is directed out of the page.

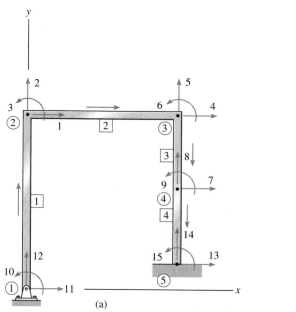

(a)

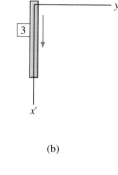

(b)

Fig. 15–1

Degrees of Freedom

Once the members and nodes have been identified, and the global coordinate system has been established, the degrees of freedom for the structure can be determined.

Frames.   When deriving the classical methods of analysis, we neglected the deformation of frame members caused by axial force and shear, and considered only the effect of bending. This was justified since axial forces or shear, in general, do not contribute significantly to the deflection of frame members. In the development that follows, however, we can easily provide a more exact analysis of the frame by incorporating both bending and axial force displacements in the stiffness method.* As a result, each node of a frame member will have *three degrees of freedom*, each of which is identified with a code number. Like the case for trusses, the *lowest code numbers* are used to identify the *unknown displacements* (unconstrained degrees of freedom), and the *highest numbers* are used to identify the *known displacements* (constrained degrees of freedom). An example of code-number labeling for a frame is also shown in Fig. 15–1a. Here the frame has 15 degrees of freedom, for which code numbers 1 through 10 represent unknown displacements and 11 through 15 represent known displacements, which in this case are zero.

Beams.   If we neglect the effects of axial force and shear and consider only deflections of beams caused by bending, as in the classical analysis, the size of the structure stiffness matrix will be rather small. Furthermore, if the beam does not have an overhang, or the supports do not have a transverse displacement, such as a settlement, then each node, *if it is located at a support,* has only one degree of freedom, represented as an angular displacement. Thus the continuous beam shown in Fig. 15–2 is labeled with three nodes and two members, and has three degrees of freedom. Code numbers 1 and 2 indicate the unknown angular displacements, and code number 3 indicates the known angular displacement (zero).

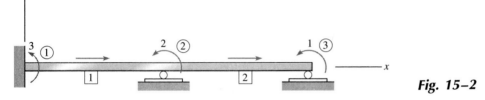

Fig. 15–2

*The lesser effect of shear can also be included in the analysis. This is discussed in books devoted to matrix structural analysis.

Intermediate Member Loading

If an element of a frame or beam supports a lateral load between its nodes, it will be convenient for a matrix analysis if the effects of this load are converted to an equivalent loading at the nodes. This is because the stiffness method, like all displacement methods, is based on writing equilibrium equations for the nodes, and therefore, if the load conversion is made, the equilibrium equations can be written in a simple form.

In order to show how to handle a case of lateral loading, we will consider the element of the beam or frame to be subjected to a constant distributed loading as shown in Fig. 15–3a. By the principle of superposition, this load can be represented by (1) the element loaded with the fixed-end moments and shears at the nodes of the element, Fig. 15–3b, and (2) the element, assumed to be fixed-supported and subjected to the actual load and its fixed-supported reactions, Fig. 15–3c. The matrix analysis is performed only for the loading shown in Fig. 15–3b, since the loadings in the fixed-supported case can be determined directly. In other words, once the matrix analysis of the loading in Fig. 15–3b is complete, the actual internal loadings and displacements at points along the element can be obtained by superposition of the effects caused by the nodal forces, Fig. 15–3b, and the distributed load and its fixed-end reactions, Fig. 15–3c. The fixed-end reactions for other cases of loading are given on the inside back cover. Application of this technique is illustrated numerically in Examples 15–1 and 15–3.

Development of the stiffness method for beams and frames follows the same procedure as that for trusses. First we must establish the member stiffness matrix, then the transformation matrices for displacements and loads. By combining these matrices, we can then form the structure stiffness matrix, from which we can determine the unknown displacements and loadings.

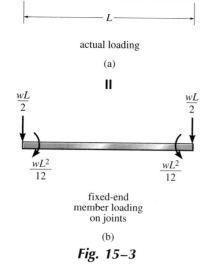

actual loading

(a)

$\parallel$

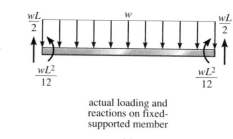

(b)

fixed-end member loading on joints

(c)

actual loading and reactions on fixed-supported member

Fig. 15–3

15.2   Frame-Member Stiffness Matrix

In this section we will develop the stiffness matrix for a frame member referenced from the local x', y', z' coordinate system, Fig. 15–4. The origin of coordinates is placed at the "near" end N, and the positive x' axis extends toward the "far" end F. There are three reactions at each end of the element consisting of axial forces $q_{Nx'}$ and $q_{Fx'}$, shear forces $q_{Ny'}$ and $q_{Fy'}$, and bending moments $q_{Nz'}$ and $q_{Fz'}$. These loadings all act in the positive coordinate directions. In particular, $q_{Nz'}$ and $q_{Fz'}$ are positive *counterclockwise*, since by the right-hand rule the moment vectors are then directed along the positive z' axis, which is out of the page.

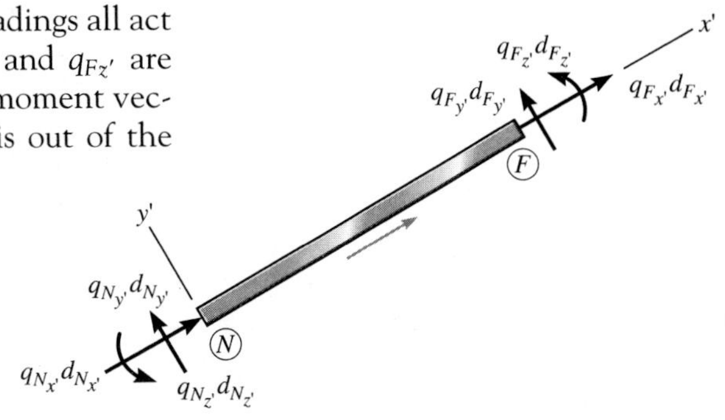

positive sign convention

Fig. 15–4

Linear and angular displacements associated with these loadings also follow this same positive sign convention. We will now impose each of these displacements separately and then determine the loadings acting on the member caused by each displacement.

x' Displacements

If the member undergoes a displacement $d_{Nx'}$, or a displacement $d_{Fx'}$, axial forces shown in Figs. 15–5a and 15–5b are developed at the ends of the member.

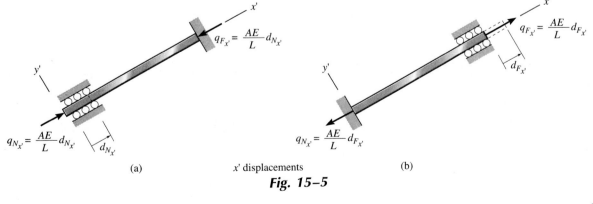

(a) x' displacements   (b)

Fig. 15–5

y' Displacements

When a positive displacement $d_{Ny'}$ is imposed while other possible displacements are prevented, the resulting shear forces and bending moments that are created are shown in Fig. 15–6a. In particular, the moment has been developed in Sec. 9.1 as Eq. 9–5. Likewise, when $d_{Fy'}$ is imposed, the required shears and bending moments are given in Fig. 15–6b.

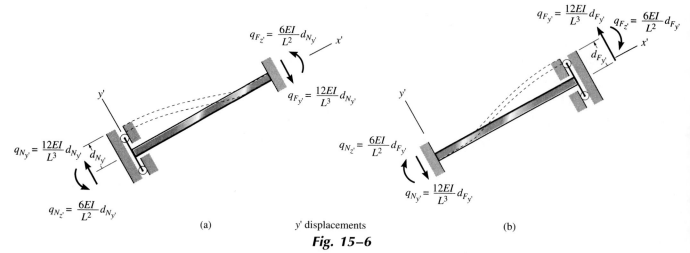

(a) y' displacements   (b)

Fig. 15–6

z' Rotations

If a positive rotation $d_{Nz'}$ is imposed while all other possible displacements are prevented, the required shears and moments necessary for the deformation are shown in Fig. 15–7a. In particular, the moment results have been developed in Sec. 9.1 as Eqs. 9–1 and 9–3. Likewise, when $d_{Fz'}$ is imposed, the resultant loadings are shown in Fig. 15–7b.

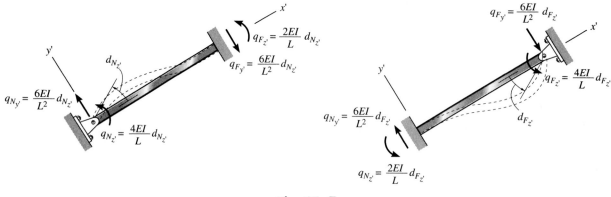

(a)   **Fig. 15–7**   (b)

By superposition, if the above results in Figs. 15–5 through 15–7 are added, the resulting six load-displacement relations for the member can be expressed in matrix form as

$$
\begin{bmatrix} q_{Nx'} \\ q_{Ny'} \\ q_{Nz'} \\ q_{Fx'} \\ q_{Fy'} \\ q_{Fz'} \end{bmatrix} =
\begin{bmatrix}
\dfrac{AE}{L} & 0 & 0 & -\dfrac{AE}{L} & 0 & 0 \\[2mm]
0 & \dfrac{12EI}{L^3} & \dfrac{6EI}{L^2} & 0 & -\dfrac{12EI}{L^3} & \dfrac{6EI}{L^2} \\[2mm]
0 & \dfrac{6EI}{L^2} & \dfrac{4EI}{L} & 0 & -\dfrac{6EI}{L^2} & \dfrac{2EI}{L} \\[2mm]
-\dfrac{AE}{L} & 0 & 0 & \dfrac{AE}{L} & 0 & 0 \\[2mm]
0 & -\dfrac{12EI}{L^3} & -\dfrac{6EI}{L^2} & 0 & \dfrac{12EI}{L^3} & -\dfrac{6EI}{L^2} \\[2mm]
0 & \dfrac{6EI}{L^2} & \dfrac{2EI}{L} & 0 & -\dfrac{6EI}{L^2} & \dfrac{4EI}{L}
\end{bmatrix}
\begin{bmatrix} d_{Nx'} \\ d_{Ny'} \\ d_{Nz'} \\ d_{Fx'} \\ d_{Fy'} \\ d_{Fz'} \end{bmatrix}
$$

where the column headings are Nx', Ny', Nz', Fx', Fy', Fz'.

$$(15\text{–}1)$$

These equations can also be written in abbreviated form as

$$\mathbf{q} = \mathbf{k'd} \qquad (15\text{–}2)$$

The symmetric matrix $\mathbf{k'}$ in Eq. 15–1 is referred to as the *member stiffness matrix*. The 36 influence coefficients $\mathbf{k}_{ij}$ that comprise it account for the axial and shear force and bending-moment displacements of the member. Physically these coefficients represent the load on the member when the member undergoes a specified unit displacement. For example, if $d_{Nx'} = 1$, Fig. 15–5a, *while all other displacements are zero*, the member will be subjected only to forces $q_{Nx'} = AE/L$ and $q_{Fx'} = -AE/L$, as indicated by the first column of the $\mathbf{k'}$ matrix. In a similar manner, the other columns of the $\mathbf{k'}$ matrix are the member loadings for unit displacements identified by the degree-of-freedom coding listed above the columns. From the development, both equilibrium and compatibility of displacements have been satisfied.

15.3   Displacement and Force Transformation Matrices

As in the case for trusses, we must be able to transform the internal member loads $\mathbf{q}$ and deformations $\mathbf{d}$ from local x', y', z' coordinates to global x, y, z coordinates. For this reason transformation matrices are needed.

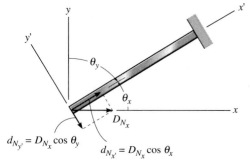

$d_{Ny'} = D_{N_x} \cos \theta_y$

$d_{N_{x'}} = D_{N_x} \cos \theta_x$

(a)

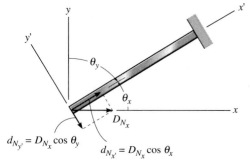

$d_{Ny'} = D_{N_y} \cos \theta_x$

$d_{N_{x'}} = D_{N_y} \cos \theta_y$

(b)

Fig. 15–8

Displacement Transformation Matrix

Consider the frame member shown in Fig. 15–8a. Here it is seen that a global coordinate displacement D_{Nx} creates local coordinate displacements

$$d_{Nx'} = D_{Nx} \cos \theta_x \qquad d_{Ny'} = -D_{Nx} \cos \theta_y$$

Likewise, a global coordinate displacement D_{Ny}, Fig. 15–8b, creates local coordinate displacements of

$$d_{Nx'} = D_{Ny} \cos \theta_y \qquad d_{Ny'} = D_{Ny} \cos \theta_x$$

Finally, since the z' and z axes are coincident, that is, directed out of the page, a rotation of D_{Nz} about z causes a corresponding rotation $d_{Nz'}$ about z'. Thus,

$$d_{Nz'} = D_{Nz}$$

In a similar manner, if global displacements D_{Fx} in the x direction, D_{Fy} in the y direction, and a rotation D_{Fz} are imposed on the far end of the member, the resulting transformation equations are, respectively,

$$
\begin{aligned}
d_{Fx'} &= D_{Fx} \cos \theta_x & d_{Fy'} &= -D_{Fx} \cos \theta_y \\
d_{Fx'} &= D_{Fy} \cos \theta_y & d_{Fy'} &= D_{Fy} \cos \theta_x \\
d_{Fz'} &= D_{Fz} &&
\end{aligned}
$$

Letting $\lambda_x = \cos \theta_x$, $\lambda_y = \cos \theta_y$ represent the direction cosines of the member, we can write the above equations in matrix form as

$$
\begin{bmatrix}
d_{Nx'} \\
d_{Ny'} \\
d_{Nz'} \\
d_{Fx'} \\
d_{Fy'} \\
d_{Fz'}
\end{bmatrix}
=
\begin{bmatrix}
\lambda_x & \lambda_y & 0 & 0 & 0 & 0 \\
-\lambda_y & \lambda_x & 0 & 0 & 0 & 0 \\
0 & 0 & 1 & 0 & 0 & 0 \\
0 & 0 & 0 & \lambda_x & \lambda_y & 0 \\
0 & 0 & 0 & -\lambda_y & \lambda_x & 0 \\
0 & 0 & 0 & 0 & 0 & 1
\end{bmatrix}
\begin{bmatrix}
D_{Nx} \\
D_{Ny} \\
D_{Nz} \\
D_{Fx} \\
D_{Fy} \\
D_{Fz}
\end{bmatrix}
\qquad (15\text{–}3)
$$

or

$$\mathbf{d} = \mathbf{TD} \qquad (15\text{–}4)$$

By inspection, **T** transforms the six global x, y, z displacements **D** into the six local x', y', z' displacements **d**. Hence **T** is referred to as the *displacement transformation matrix*.

Force Transformation Matrix

If we now apply each component of load to the near end of the member, we can determine how to transform the load components from local to global coordinates. Applying $q_{Nx'}$, Fig. 15–9a, it can be seen that

$$Q_{Nx} = q_{Nx'} \cos \theta_x \qquad Q_{Ny} = q_{Nx'} \cos \theta_y$$

If $q_{Ny'}$ is applied, Fig. 15–9b, then its components are

$$Q_{Nx} = -q_{Ny'} \cos \theta_y \qquad Q_{Ny} = q_{Ny'} \cos \theta_x$$

Finally, since $q_{Nz'}$ is collinear with Q_{Nz}, we have

$$Q_{Nz} = q_{Nz'}$$

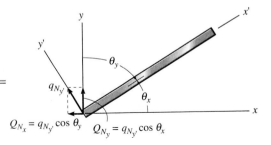

(a)

In a similar manner, end loads of $q_{Fx'}$, $q_{Fy'}$, $q_{Fz'}$ will yield the following respective components:

$$\begin{aligned} Q_{Fx} &= q_{Fx'} \cos \theta_x & Q_{Fy} &= q_{Fx'} \cos \theta_y \\ Q_{Fx} &= -q_{Fy'} \cos \theta_y & Q_{Fy} &= q_{Fy'} \cos \theta_x \\ Q_{Fz} &= q_{Fz'} \end{aligned}$$

These equations, assembled in matrix form with $\lambda_x = \cos \theta_x$, $\lambda_y = \cos \theta_y$, yield

$$\begin{bmatrix} Q_{Nx} \\ Q_{Ny} \\ Q_{Nz} \\ Q_{Fx} \\ Q_{Fy} \\ Q_{Fz} \end{bmatrix} = \begin{bmatrix} \lambda_x & -\lambda_y & 0 & 0 & 0 & 0 \\ \lambda_y & \lambda_x & 0 & 0 & 0 & 0 \\ 0 & 0 & 1 & 0 & 0 & 0 \\ 0 & 0 & 0 & \lambda_x & -\lambda_y & 0 \\ 0 & 0 & 0 & \lambda_y & \lambda_x & 0 \\ 0 & 0 & 0 & 0 & 0 & 1 \end{bmatrix} \begin{bmatrix} q_{Nx'} \\ q_{Ny'} \\ q_{Nz'} \\ q_{Fx'} \\ q_{Fy'} \\ q_{Fz'} \end{bmatrix} \qquad (15\text{–}5)$$

(b)

Fig. 15–9

or

$$\mathbf{Q} = \mathbf{T}^T \mathbf{q} \qquad (15\text{–}6)$$

Here, as stated, $\mathbf{T}^T$ transforms the six member loads expressed in local coordinates into the six loadings expressed in global coordinates.

695

15.4 Frame-Member Global Stiffness Matrix

The results of the previous section will now be combined in order to determine the stiffness matrix for a member that relates the global loadings $\mathbf{Q}$ to the global displacements $\mathbf{D}$. To do this, substitute Eq. 15–4 ($\mathbf{d} = \mathbf{TD}$), into Eq. 15–2 ($\mathbf{q} = \mathbf{k'd}$). We have

$$\boxed{\mathbf{q} = \mathbf{k'TD}} \tag{15–7}$$

Here the member forces $\mathbf{q}$ are related to the global displacements $\mathbf{D}$. Substituting this result into Eq. 15–6 ($\mathbf{Q} = \mathbf{T}^T\mathbf{q}$), yields the final result,

$$\mathbf{Q} = \mathbf{T}^T\mathbf{k'TD} \tag{15–8}$$

or

$$\mathbf{Q} = \mathbf{kD}$$

where

$$\boxed{\mathbf{k} = \mathbf{T}^T\mathbf{k'T}} \tag{15–9}$$

Here $\mathbf{k}$ represents the global stiffness matrix for the member. We can obtain its value in generalized form using Eqs. 15–5, 15–1, and 15–3, and performing the matrix operations. This yields the final result,

$$
\mathbf{k} =
\begin{bmatrix}
\left(\dfrac{AE}{L}\lambda_x^2 + \dfrac{12EI}{L^3}\lambda_y^2\right) & \left(\dfrac{AE}{L} - \dfrac{12EI}{L^3}\right)\lambda_x\lambda_y & -\dfrac{6EI}{L^2}\lambda_y & -\left(\dfrac{AE}{L}\lambda_x^2 + \dfrac{12EI}{L^3}\lambda_y^2\right) & -\left(\dfrac{AE}{L} - \dfrac{12EI}{L^3}\right)\lambda_x\lambda_y & -\dfrac{6EI}{L^2}\lambda_y \\[2ex]
\left(\dfrac{AE}{L} - \dfrac{12EI}{L^3}\right)\lambda_x\lambda_y & \left(\dfrac{AE}{L}\lambda_y^2 + \dfrac{12EI}{L^3}\lambda_x^2\right) & \dfrac{6EI}{L^2}\lambda_x & -\left(\dfrac{AE}{L} - \dfrac{12EI}{L^3}\right)\lambda_x\lambda_y & -\left(\dfrac{AE}{L}\lambda_y^2 + \dfrac{12EI}{L^3}\lambda_x^2\right) & \dfrac{6EI}{L^2}\lambda_x \\[2ex]
-\dfrac{6EI}{L^2}\lambda_y & \dfrac{6EI}{L^2}\lambda_x & \dfrac{4EI}{L} & \dfrac{6EI}{L^2}\lambda_y & -\dfrac{6EI}{L^2}\lambda_x & \dfrac{2EI}{L} \\[2ex]
-\left(\dfrac{AE}{L}\lambda_x^2 + \dfrac{12EI}{L^3}\lambda_y^2\right) & -\left(\dfrac{AE}{L} - \dfrac{12EI}{L^3}\right)\lambda_x\lambda_y & \dfrac{6EI}{L^2}\lambda_y & \left(\dfrac{AE}{L}\lambda_x^2 + \dfrac{12EI}{L^3}\lambda_y^2\right) & \left(\dfrac{AE}{L} - \dfrac{12EI}{L^3}\right)\lambda_x\lambda_y & \dfrac{6EI}{L^2}\lambda_y \\[2ex]
-\left(\dfrac{AE}{L} - \dfrac{12EI}{L^3}\right)\lambda_x\lambda_y & -\left(\dfrac{AE}{L}\lambda_y^2 + \dfrac{12EI}{L^3}\lambda_x^2\right) & -\dfrac{6EI}{L^2}\lambda_x & \left(\dfrac{AE}{L} - \dfrac{12EI}{L^3}\right)\lambda_x\lambda_y & \left(\dfrac{AE}{L}\lambda_y^2 + \dfrac{12EI}{L^3}\lambda_x^2\right) & -\dfrac{6EI}{L^2}\lambda_x \\[2ex]
-\dfrac{6EI}{L^2}\lambda_y & \dfrac{6EI}{L^2}\lambda_x & \dfrac{2EI}{L} & \dfrac{6EI}{L^2}\lambda_y & -\dfrac{6EI}{L^2}\lambda_x & \dfrac{4EI}{L}
\end{bmatrix}
\begin{matrix} N_x \\[2ex] N_y \\[2ex] N_z \\[2ex] F_x \\[2ex] F_y \\[2ex] F_z \end{matrix}
$$

Columns: $N_x \quad N_y \quad N_z \quad F_x \quad F_y \quad F_z$

$$\tag{15–10}$$

Note that this 6 × 6 matrix is *symmetric*. Furthermore, the location of each element is associated with the coding at the near end, N_x, N_y, N_z, followed by that of the far end, F_x, F_y, F_z, which is listed at the top of the columns and along the rows. Like the **k′** matrix, each column of the **k** matrix represents the global coordinate loads on the member at the nodes that are necessary to resist a unit displacement in the direction defined by the coding of the column. For example, the first column of **k** represents the global coordinate loadings at the near and far ends caused by a *unit displacement* at the near end in the x direction, that is, N_x.

15.5   Beam-Member Global Stiffness Matrix

It has been stated previously that provided the supports do not undergo transverse displacement, e.g., settlement, or the beam does not have an overhang, then in general each of the *supports* of the beam will have only one degree of freedom, namely, an angular displacement. This being the case, the member stiffness matrix of a beam element can be determined by canceling the rows and columns of the frame matrix (Eq. 15–10) that apply to displacements along N_x, N_y and F_x, F_y, since the beam supports do not have any degrees of freedom in these directions. Also, AE/L does not apply since axial displacements and loads are not considered. As a result, the beam stiffness matrix is represented by four elements, that is,

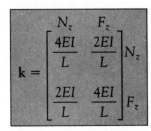

$$\mathbf{k} = \begin{array}{c} \\ \end{array} \begin{array}{cc} N_z & F_z \end{array} \\ \left[\begin{array}{cc} \dfrac{4EI}{L} & \dfrac{2EI}{L} \\[2ex] \dfrac{2EI}{L} & \dfrac{4EI}{L} \end{array} \right] \begin{array}{c} N_z \\[2ex] F_z \end{array}$$

$$(15–11)$$

It should be noted that this matrix is *equivalent* to the member stiffness matrix in local coordinates, **k′**, since the rotations are not transformed, that is, **k** = **k′T**.

15.6   Application of the Stiffness Method for Beam and Frame Analysis

Now that **k** has been developed, we can formulate a procedure for applying the stiffness method to solve problems involving beams and frames.

Structure Stiffness Matrix

Once all the member stiffness matrices have been found, we must assemble them into the structure stiffness matrix **K**. This process depends on first knowing the *location* of each element in the member stiffness matrix. In this regard, recall that the rows and columns of each **k** matrix (Eq. 15–10) are identified by the three code numbers at the near end of the member (N_x, N_y, N_z) followed by those at the far end (F_x, F_y, F_z). Therefore, when assembling the matrices, each element must be placed in the same location of the **K** matrix. In this way, **K** will have an order that will be equal to the highest code number assigned to the structure, since this represents the total number of degrees of freedom for the structure. When several members are connected to a node, they will have the same position in the **K** matrix, and therefore these member stiffness influence coefficients must be algebraically added together to determine the nodal stiffness influence coefficient for the structure. This is necessary since each coefficient represents the nodal resistance of the structure in a particular direction (x, y, or z) when a unit displacement (x, y, or z) occurs either at the same or at another node. For example, K_{26} represents the load in the direction and at the location of code number "2" when a unit displacement occurs in the direction and at the location of code number "6."

Procedure for Analysis

The following method provides a means of determining the displacements, support reactions, and internal loadings for the members or finite elements of a statically determinate or statically indeterminate beam or frame.

Notation Divide the structure into finite elements and arbitrarily identify each element and its nodes. Use a number written in a circle for a node and a number written in a square for a member. Usually an element extends between points of support, points of concentrated loads, corners or joints, or to points where internal loadings or displacements are to be determined. Specify the near and far ends of each element symbolically by directing an arrow along the element, with the head directed toward the far end.

Establish the x, y, z global coordinate system, usually for convenience with the origin at a nodal point on one of the elements and the axis located such that all the nodes have positive coordinates. At each nodal point of a frame, specify numerically the three x, y, z coding components. If a continuous beam having no overhang or transverse support displacement is considered, and the nodes are at the supports, use a code number only to identify the angular displacement at each support. In all cases use the *lowest numbers* to identify all the unconstrained degrees of freedom, followed by the remaining or highest numbers to identify the degrees of freedom that are constrained. From the problem, establish the known displacements D_k and known external loads Q_k.

Structure Stiffness Matrix Apply Eq. 15–10 to determine the stiffness matrix for each element expressed in global coordinates. In particular, the direction cosines λ_x and λ_y are determined from the x, y coordinates of the ends of the element, Eqs. 14–5 and 14–6.

After each member stiffness matrix is determined, and the rows and columns are identified with the appropriate coding numbers as explained above, assemble the matrices to determine the structure stiffness matrix **K.** As a check, the element *and* structure stiffness matrices should all be *symmetric*.

Displacements and Loads Partition the stiffness matrix as indicated by Eq. 14–17. Expansion then leads to

$$Q_k = K_{11}D_u + K_{12}D_k$$
$$Q_u = K_{21}D_u + K_{22}D_k$$

These equations express the force and moment equilibrium of each node.

The unknown displacements $\mathbf{D}_u$ are determined from the first of these equations. Using these values, the support reactions $\mathbf{Q}_u$ are computed from the second equation. Finally, the internal loadings $\mathbf{q}$ at the ends of the members can be computed by combining Eqs. 15–2 and 15–4, which yields

$$\mathbf{q} = \mathbf{k'TD}$$

The following examples numerically illustrate this procedure.

Example 15–1

Determine the moment developed at support A of the beam shown in Fig. 15–10a. EI is constant.

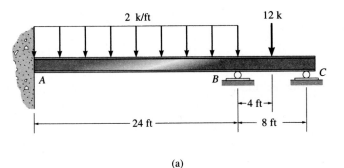

2 k/ft   12 k

B C

A

4 ft

24 ft   8 ft

(a)

Fig. 15–10a,b

Solution

Notation. The beam has two elements and three nodes, which are identified in Fig. 15–10b. The coding numbers 1 through 3 are indicated such that the *lowest numbers 1 and 2 identify the unconstrained degrees of freedom*.

The matrix analysis requires that the external loading be applied at the nodes and therefore the distributed and concentrated loads are replaced by their equivalent fixed-end moments, which are determined from the table on the inside back cover. (See Example 10–2.) Using superposition, the results of the matrix analysis for the loading in Fig. 15–10b will be modified by the loads shown in Fig. 15–10c. From Fig. 15–10b, the known displacement and load matrices are

$$\mathbf{D}_k = [0]3 \qquad \mathbf{Q}_k = \begin{bmatrix} 12 \\ 84 \end{bmatrix} \begin{matrix} 1 \\ 2 \end{matrix}$$

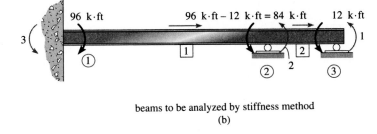

96 k·ft   96 k·ft − 12 k·ft = 84 k·ft   12 k·ft

3

①   1   ②   2   2   ③   1

beams to be analyzed by stiffness method

(b)

(cont'd)

**Example 15–1
(continued)**

Structure Stiffness Matrix. Each of the two member stiffness matrices is determined from Eq. 15–11. We have

$$
\mathbf{k}_1 = \begin{array}{cc} & \\ \end{array}
\begin{bmatrix} \dfrac{4EI}{24} & \dfrac{2EI}{24} \\[2ex] \dfrac{2EI}{24} & \dfrac{4EI}{24} \end{bmatrix} \begin{array}{c} 3 \\[2ex] 2 \end{array}
\qquad
\mathbf{k}_2 = \begin{bmatrix} \dfrac{4EI}{8} & \dfrac{2EI}{8} \\[2ex] \dfrac{2EI}{8} & \dfrac{4EI}{8} \end{bmatrix} \begin{array}{c} 2 \\[2ex] 1 \end{array}
$$

with column labels $3\ \ 2$ for $\mathbf{k}_1$ and $2\ \ 1$ for $\mathbf{k}_2$.

We can now assemble these matrices into the structure stiffness matrix. Notice that $K_{22} = 4EI/24 + 4EI/8 = 16EI/24$, so that

$$
\mathbf{K} = \begin{array}{ccc} 1 & 2 & 3 \end{array}
\begin{bmatrix} \dfrac{4EI}{8} & \dfrac{2EI}{8} & 0 \\[2ex] \dfrac{2EI}{8} & \dfrac{16EI}{24} & \dfrac{2EI}{24} \\[2ex] 0 & \dfrac{2EI}{24} & \dfrac{4EI}{24} \end{bmatrix} \begin{array}{c} 1 \\[2ex] 2 \\[2ex] 3 \end{array}
$$

Displacements and Loads. We require

$$\mathbf{Q} = \mathbf{KD}$$

$$
\begin{bmatrix} 12 \\ 84 \\ \hline Q_3 \end{bmatrix} =
\left[\begin{array}{cc|c}
\dfrac{4EI}{8} & \dfrac{2EI}{8} & 0 \\[2ex]
\dfrac{2EI}{8} & \dfrac{16EI}{24} & \dfrac{2EI}{24} \\[1ex]
\hline
0 & \dfrac{2EI}{24} & \dfrac{4EI}{24}
\end{array} \right]
\begin{bmatrix} D_1 \\ D_2 \\ \hline 0 \end{bmatrix}
$$

Expanding this equation yields

$$12 = \frac{4EI}{8}D_1 + \frac{2EI}{8}D_2$$

$$84 = \frac{2EI}{8}D_1 + \frac{16EI}{24}D_2$$

$$Q_3 = \frac{2EI}{24}D_2$$

Solving,

$$D_1 = -\frac{48}{EI}$$

$$D_2 = \frac{144}{EI}$$

$$Q_3 = 12 \text{ k} \cdot \text{ft}$$

The positive answers represent a counterclockwise rotation, which is in accordance with our sign convention. The actual moment at A must include the fixed-supported *reaction* of $+96 \text{ lb} \cdot \text{ft}$ shown in Fig. 15–10c, along with the calculated result for Q_3. Thus,

$$M_{AB} = 12 \text{ k} \cdot \text{ft} + 96 \text{ k} \cdot \text{ft} = 108 \text{ k} \cdot \text{ft} \curvearrowleft \qquad \textbf{Ans.}$$

This result compares with that determined in Example 10–2.

Although not required here, we can determine the internal moment at B by considering, for example, member 1, node 2, Fig. 15–10b. Since $\mathbf{k}_1 = \mathbf{k'T}$, Eq. 15–7 becomes

$$\mathbf{q} = \mathbf{k}_1 \mathbf{D}$$

$$\begin{bmatrix} q_3 \\ q_2 \end{bmatrix} = \begin{bmatrix} \dfrac{4EI}{24} & \dfrac{2EI}{24} \\[2mm] \dfrac{2EI}{24} & \dfrac{4EI}{24} \end{bmatrix} \begin{bmatrix} 0 \\[2mm] \dfrac{144}{EI} \end{bmatrix} \begin{matrix} 3 \\[4mm] 2 \end{matrix}$$

$$q_3 = 12 \text{ k} \cdot \text{ft}$$

$$q_2 = 24 \text{ k} \cdot \text{ft}$$

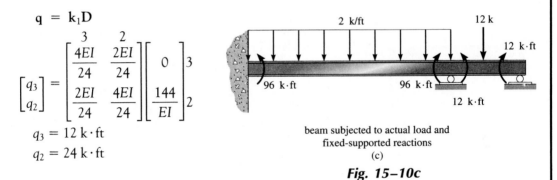

beam subjected to actual load and
fixed-supported reactions

(c)

Fig. 15–10c

so that

$$M_{BA} = 24 \text{ k} \cdot \text{ft} - 96 \text{ k} \cdot \text{ft} = -72 \text{ k} \cdot \text{ft} = 72 \text{ k} \cdot \text{ft} \curvearrowright$$

Example 15–2

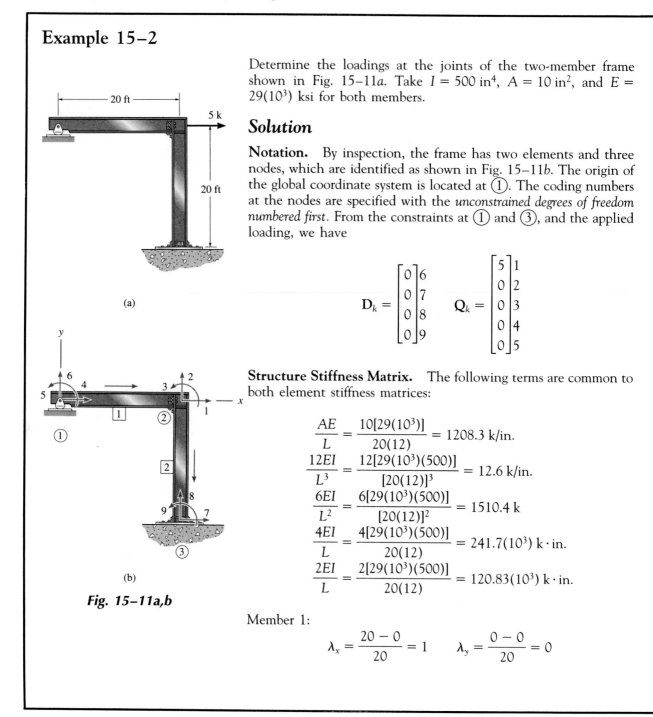

(a)

(b)

Fig. 15–11a,b

Determine the loadings at the joints of the two-member frame shown in Fig. 15–11a. Take $I = 500$ in⁴, $A = 10$ in², and $E = 29(10^3)$ ksi for both members.

Solution

Notation.   By inspection, the frame has two elements and three nodes, which are identified as shown in Fig. 15–11b. The origin of the global coordinate system is located at ①. The coding numbers at the nodes are specified with the *unconstrained degrees of freedom numbered first.* From the constraints at ① and ③, and the applied loading, we have

$$\mathbf{D}_k = \begin{bmatrix} 0 \\ 0 \\ 0 \\ 0 \\ 0 \end{bmatrix} \begin{matrix} 6 \\ 7 \\ 8 \\ 9 \end{matrix} \qquad \mathbf{Q}_k = \begin{bmatrix} 5 \\ 0 \\ 0 \\ 0 \\ 0 \end{bmatrix} \begin{matrix} 1 \\ 2 \\ 3 \\ 4 \\ 5 \end{matrix}$$

Structure Stiffness Matrix.   The following terms are common to both element stiffness matrices:

$$\frac{AE}{L} = \frac{10[29(10^3)]}{20(12)} = 1208.3 \text{ k/in.}$$

$$\frac{12EI}{L^3} = \frac{12[29(10^3)(500)]}{[20(12)]^3} = 12.6 \text{ k/in.}$$

$$\frac{6EI}{L^2} = \frac{6[29(10^3)(500)]}{[20(12)]^2} = 1510.4 \text{ k}$$

$$\frac{4EI}{L} = \frac{4[29(10^3)(500)]}{20(12)} = 241.7(10^3) \text{ k} \cdot \text{in.}$$

$$\frac{2EI}{L} = \frac{2[29(10^3)(500)]}{20(12)} = 120.83(10^3) \text{ k} \cdot \text{in.}$$

Member 1:

$$\lambda_x = \frac{20 - 0}{20} = 1 \qquad \lambda_y = \frac{0 - 0}{20} = 0$$

Substituting the data into Eq. 15–10, we have

$$
\mathbf{k}_1 = \begin{array}{cccccc}
4 & 6 & 5 & 1 & 2 & 3 \\
\end{array}
\begin{bmatrix}
1208.3 & 0 & 0 & -1208.3 & 0 & 0 \\
0 & 12.6 & 1510.4 & 0 & -12.6 & 1510.4 \\
0 & 1510.4 & 241.7(10^3) & 0 & -1510.4 & 120.83(10^3) \\
-1208.3 & 0 & 0 & 1208.3 & 0 & 0 \\
0 & -12.6 & -1510.4 & 0 & 12.6 & -1510.4 \\
0 & 1510.4 & 120.83(10^3) & 0 & -1510.4 & 241.7(10^3)
\end{bmatrix}
\begin{array}{c}
4 \\ 6 \\ 5 \\ 1 \\ 2 \\ 3
\end{array}
$$

The rows and columns of this 6×6 matrix are identified by the three x, y, z coding numbers, first at the near end and followed by the far end, that is, 4, 6, 5, 1, 2, 3, respectively, Fig. 15–11b. This is done for later assembly of the elements.

Member 2:
$$
\lambda_x = \frac{20 - 20}{20} = 0 \qquad \lambda_y = \frac{-20 - 0}{20} = -1
$$

Substituting the data into Eq. 15–10 yields

$$
\mathbf{k}_2 = \begin{array}{cccccc}
1 & 2 & 3 & 7 & 8 & 9 \\
\end{array}
\begin{bmatrix}
12.6 & 0 & 1510.4 & -12.6 & 0 & 1510.4 \\
0 & 1208.3 & 0 & 0 & -1208.3 & 0 \\
1510.4 & 0 & 241.7(10^3) & -1510.4 & 0 & 120.83(10^3) \\
-12.6 & 0 & -1510.4 & 12.6 & 0 & -1510.4 \\
0 & -1208.3 & 0 & 0 & 1208.3 & 0 \\
1510.4 & 0 & 120.83(10^3) & -1510.4 & 0 & 241.7(10^3)
\end{bmatrix}
\begin{array}{c}
1 \\ 2 \\ 3 \\ 7 \\ 8 \\ 9
\end{array}
$$

As usual, column and row identification is referenced by the three code numbers in x, y, z sequence for the near and far ends, respectively, that is, 1, 2, 3, then 7, 8, 9, Fig. 15–11b.

(cont'd)

705

Example 15–2
(continued)

The structure stiffness matrix is determined by assembling $\mathbf{k}_1$ and $\mathbf{k}_2$. The result, shown partitioned, since $\mathbf{Q} = \mathbf{KD}$, is

$$
\begin{bmatrix} 5 \\ 0 \\ 0 \\ 0 \\ 0 \\ \hline Q_6 \\ Q_7 \\ Q_8 \\ Q_9 \end{bmatrix}
=
\left[\begin{array}{ccccc|cccc}
 & 1 & 2 & 3 & 4 & 5 & 6 & 7 & 8 & 9 \\
1220.9 & 0 & 1510.4 & -1208.3 & 0 & 0 & -12.6 & 0 & 1510.4 \\
0 & 1220.9 & -1510.4 & 0 & -1510.4 & -12.6 & 0 & -1208.3 & 0 \\
1510.4 & -1510.4 & 483.4(10^3) & 0 & 120.83(10^3) & 1510.4 & -1510.4 & 0 & 120.83(10^3) \\
-1208.3 & 0 & 0 & 1208.3 & 0 & 0 & 0 & 0 & 0 \\
0 & -1510.4 & 120.83(10^3) & 0 & 241.7(10^3) & 1510.4 & 0 & 0 & 0 \\
\hline
0 & -12.6 & 1510.4 & 0 & 1510.4 & 12.6 & 0 & 0 & 0 \\
-12.6 & 0 & -1510.4 & 0 & 0 & 0 & 12.6 & 0 & -1510.4 \\
0 & -1208.3 & 0 & 0 & 0 & 0 & 0 & 1208.3 & 0 \\
1510.4 & 0 & 120.83(10^3) & 0 & 0 & 0 & -1510.4 & 0 & 241.7(10^3)
\end{array}\right]
\begin{bmatrix} D_1 \\ D_2 \\ D_3 \\ D_4 \\ D_5 \\ \hline 0 \\ 0 \\ 0 \\ 0 \end{bmatrix}
\quad (1)
$$

Displacements and Loads. Expanding to determine the displacements yields

$$
\begin{bmatrix} 5 \\ 0 \\ 0 \\ 0 \\ 0 \end{bmatrix}
=
\begin{bmatrix}
1220.9 & 0 & 1510.4 & -1208.3 & 0 \\
0 & 1220.9 & -1510.4 & 0 & -1510.4 \\
1510.4 & -1510.4 & 483.4(10^3) & 0 & 120.8(10^3) \\
-1208.3 & 0 & 0 & 1208.3 & 0 \\
0 & -1510.4 & 120.83(10^3) & 0 & 241.7(10^3)
\end{bmatrix}
\begin{bmatrix} D_1 \\ D_2 \\ D_3 \\ D_4 \\ D_5 \end{bmatrix}
+
\begin{bmatrix} 0 \\ 0 \\ 0 \\ 0 \\ 0 \end{bmatrix}
$$

Solving, we obtain

$$
\begin{bmatrix} D_1 \\ D_2 \\ D_3 \\ D_4 \\ D_5 \end{bmatrix}
=
\begin{bmatrix}
0.696 \text{ in.} \\
-1.55(10^{-3}) \text{ in.} \\
-2.488(10^{-3}) \text{ rad} \\
0.696 \text{ in.} \\
1.234(10^{-3}) \text{ rad}
\end{bmatrix}
$$

Using these results, the support reactions are determined from Eq. (1) as follows:

$$
\begin{bmatrix} Q_6 \\ Q_7 \\ Q_8 \\ Q_9 \end{bmatrix}
=
\begin{bmatrix}
0 & -12.6 & 1510.4 & 0 & 1510.4 \\
-12.6 & 0 & -1510.4 & 0 & 0 \\
0 & -1208.3 & 0 & 0 & 0 \\
1510.4 & 0 & 120.83(10^3) & 0 & 0
\end{bmatrix}
\begin{bmatrix}
0.696 \\
-1.55(10^{-3}) \\
-2.488(10^{-3}) \\
0.696 \\
1.234(10^{-3})
\end{bmatrix}
+
\begin{bmatrix} 0 \\ 0 \\ 0 \\ 0 \end{bmatrix}
=
\begin{bmatrix}
-1.87\, k \\
-5.00\, k \\
1.87\, k \\
750\, k \cdot in.
\end{bmatrix}
$$

Ans.

The internal loadings at node ② can be determined by applying Eq. 15–7 to member 1. Here $\mathbf{k}_1'$ is defined by Eq. 15–1 and $\mathbf{T}_1$ by Eq. 15–3. Thus,

$$
\mathbf{q}_1 = \mathbf{k}_1'\mathbf{T}_1\mathbf{D} =
\begin{array}{c}
\begin{array}{cccccc} 4 & \quad 6 & \quad 5 & \quad\quad 1 & \quad 2 & \quad\quad 3 \end{array} \\
\left[
\begin{array}{cccccc}
1208.3 & 0 & 0 & -1208.3 & 0 & 0 \\
0 & 12.6 & 1510.4 & 0 & -12.6 & 1510.4 \\
0 & 1510.4 & 241.7(10^3) & 0 & -1510.4 & 120.83(10^3) \\
-1208.3 & 0 & 0 & 1208.3 & 0 & 0 \\
0 & -12.6 & -1510.4 & 0 & 12.6 & -1510.4 \\
0 & 1510.4 & 120.83(10^3) & 0 & -1510.4 & 241.7(10^3)
\end{array}
\right]
\end{array}
\begin{bmatrix}
1 & 0 & 0 & 0 & 0 & 0 \\
0 & 1 & 0 & 0 & 0 & 0 \\
0 & 0 & 1 & 0 & 0 & 0 \\
0 & 0 & 0 & 1 & 0 & 0 \\
0 & 0 & 0 & 0 & 1 & 0 \\
0 & 0 & 0 & 0 & 0 & 1
\end{bmatrix}
\begin{bmatrix}
0.696 \\
0 \\
1.234(10^{-3}) \\
0.696 \\
-1.55(10^{-3}) \\
-2.488(10^{-3})
\end{bmatrix}
\begin{matrix}
4 \\ 6 \\ 5 \\ 1 \\ 2 \\ 3
\end{matrix}
$$

Note the appropriate arrangement of the elements in the matrices as indicated by the coding numbers alongside the columns and rows. Solving yields

$$
\begin{bmatrix}
q_4 \\ q_6 \\ q_5 \\ q_1 \\ q_2 \\ q_3
\end{bmatrix}
=
\begin{bmatrix}
0 \\ -1.87\ \text{k} \\ 0 \\ 0 \\ 1.87\ \text{k} \\ -450\ \text{k·in.}
\end{bmatrix}
$$

Ans.

The above results are shown in Fig. 15–11c. The directions of these vectors are in accordance with the positive directions defined in Fig. 15–3. Furthermore, the origin of the local x', y', z' axes is at the near end of the member. In a similar manner, the free-body diagram of member 2 is shown in Fig. 15–11d.

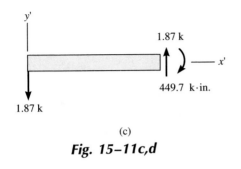

(c)

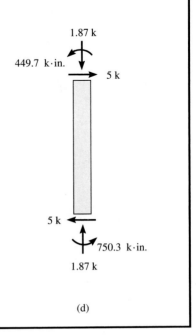

(d)

Fig. 15–11c,d

707

Example 15–3

Determine the loadings at the ends of each member of the frame shown in Fig. 15–12a. Take $I = 600$ in^4, $A = 12$ in^2, and $E = 29(10^3)$ ksi for each member.

Solution

Notation. To perform a matrix analysis, the distributed loading acting on the horizontal member will be replaced by equivalent end moments and shears computed from statics and the table listed on the inside back cover. Then using superposition, the results obtained for the frame in Fig. 15–12b will be modified for this member by the loads shown in Fig. 15–12c.

As shown in Fig. 15–12b, the nodes and members are numbered and the origin of the global coordinate system is placed at node ①. As usual, the code numbers are specified with numbers assigned first to the unconstrained degrees of freedom. Thus,

$$D_k = \begin{bmatrix} 0 & 4 \\ 0 & 5 \\ 0 & 6 \\ 0 & 7 \\ 0 & 8 \\ 0 & 9 \end{bmatrix} \qquad Q_k = \begin{bmatrix} 0 & 1 \\ -30 & 2 \\ -1200 & 3 \end{bmatrix}$$

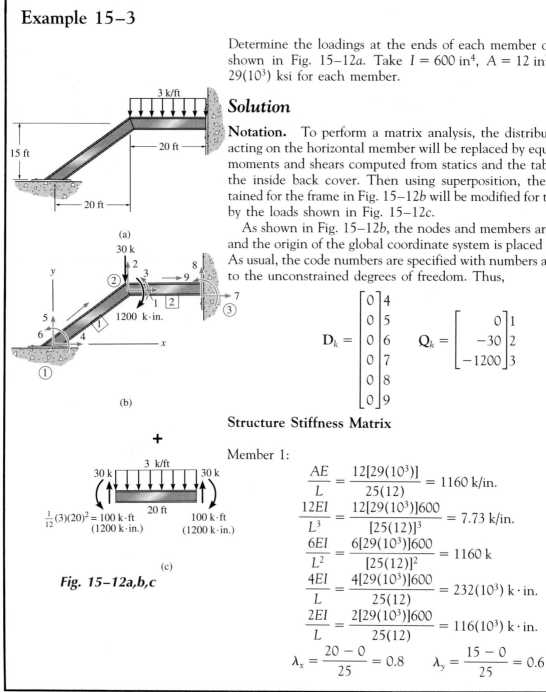

Structure Stiffness Matrix

Member 1:

$$\frac{AE}{L} = \frac{12[29(10^3)]}{25(12)} = 1160 \text{ k/in.}$$

$$\frac{12EI}{L^3} = \frac{12[29(10^3)]600}{[25(12)]^3} = 7.73 \text{ k/in.}$$

$$\frac{6EI}{L^2} = \frac{6[29(10^3)]600}{[25(12)]^2} = 1160 \text{ k}$$

$$\frac{4EI}{L} = \frac{4[29(10^3)]600}{25(12)} = 232(10^3) \text{ k} \cdot \text{in.}$$

$$\frac{2EI}{L} = \frac{2[29(10^3)]600}{25(12)} = 116(10^3) \text{ k} \cdot \text{in.}$$

$$\lambda_x = \frac{20 - 0}{25} = 0.8 \qquad \lambda_y = \frac{15 - 0}{25} = 0.6$$

708

Applying Eq. 15–10, we have

$$\mathbf{k}_1 = \begin{array}{c} \\ \\ \\ \\ \\ \\ \end{array} \begin{bmatrix} \overset{4}{745.18} & \overset{5}{553.09} & \overset{6}{-696} & \overset{1}{-745.18} & \overset{2}{-553.09} & \overset{3}{-696} \\ 553.09 & 422.55 & 928 & -553.09 & -422.55 & 928 \\ -696 & 928 & 232(10^3) & 696 & -928 & 116(10^3) \\ -745.18 & -553.09 & 696 & 745.18 & 553.09 & 696 \\ -553.09 & -422.55 & -928 & 553.09 & 422.55 & -928 \\ -696 & 928 & 116(10^3) & 696 & -928 & 232(10^3) \end{bmatrix} \begin{array}{c} 4 \\ 5 \\ 6 \\ 1 \\ 2 \\ 3 \end{array}$$

Member 2:

$$\frac{AE}{L} = \frac{12[29(10^3)]}{20(12)} = 1450 \text{ k/in.}$$

$$\frac{12EI}{L^3} = \frac{12[29(10^3)]600}{[20(12)]^3} = 15.10 \text{ k/in.}$$

$$\frac{6EI}{L^2} = \frac{6[29(10^3)]600}{[20(12)]^2} = 1812.50 \text{ k}$$

$$\frac{4EI}{L} = \frac{4[29(10^3)]600}{20(12)} = 2.90(10^5) \text{ k} \cdot \text{in.}$$

$$\frac{2EI}{L} = \frac{2[29(10^3)]600}{20(12)} = 1.45(10^5) \text{ k} \cdot \text{in.}$$

$$\lambda_x = \frac{40 - 20}{20} = 1 \qquad \lambda_y = \frac{15 - 15}{20} = 0$$

Thus, Eq. 15–10 becomes

$$\mathbf{k}_2 = \begin{bmatrix} \overset{1}{1450} & \overset{2}{0} & \overset{3}{0} & \overset{7}{-1450} & \overset{8}{0} & \overset{9}{0} \\ 0 & 15.10 & 1812.50 & 0 & -15.10 & 1812.50 \\ 0 & 1812.50 & 290(10^3) & 0 & -1812.50 & 145(10^3) \\ -1450 & 0 & 0 & 1450 & 0 & 0 \\ 0 & -15.10 & -1812.50 & 0 & 15.10 & -1812.50 \\ 0 & 1812.50 & 145(10^3) & 0 & -1812.50 & 290(10^3) \end{bmatrix} \begin{array}{c} 1 \\ 2 \\ 3 \\ 7 \\ 8 \\ 9 \end{array}$$

(cont'd)

Example 15–3 (continued)

The structure stiffness matrix, written in terms of $\mathbf{Q} = \mathbf{KD}$, is

$$
\begin{bmatrix} 0 \\ -30 \\ -1200 \\ \hline Q_4 \\ Q_5 \\ Q_6 \\ Q_7 \\ Q_8 \\ Q_9 \end{bmatrix}
=
\begin{bmatrix}
2195.18 & 553.09 & 696 & -745.18 & -553.09 & 696 & -1450 & 0 & 0 \\
553.09 & 437.65 & 884.5 & -553.09 & -422.55 & -928 & 0 & -15.10 & 1812.50 \\
696 & 884.5 & 522(10^3) & -696 & 928 & 116(10^3) & 0 & -1812.50 & 145(10^3) \\
-745.18 & -553.09 & -696 & 745.18 & 553.09 & -696 & 0 & 0 & 0 \\
-553.09 & -422.55 & 928 & 553.09 & 422.55 & 928 & 0 & 0 & 0 \\
696 & -928 & 116(10^3) & -696 & 928 & 232(10^3) & 0 & 0 & 0 \\
-1450 & 0 & 0 & 0 & 0 & 0 & 1450 & 0 & 0 \\
0 & -15.10 & -1812.50 & 0 & 0 & 0 & 0 & 15.10 & -1812.50 \\
0 & 1812.50 & 145(10^3) & 0 & 0 & 0 & 0 & -1812.50 & 290(10^3)
\end{bmatrix}
\begin{bmatrix} D_1 \\ D_2 \\ D_3 \\ \hline 0 \\ 0 \\ 0 \\ 0 \\ 0 \\ 0 \end{bmatrix}
\quad (1)
$$

Displacements and Loads. Expanding to determine the displacements yields

$$
\begin{bmatrix} 0 \\ -30 \\ -1200 \end{bmatrix}
=
\begin{bmatrix}
2195.18 & 553.09 & 696 \\
553.09 & 437.65 & 884.5 \\
696 & 884.5 & 522(10^3)
\end{bmatrix}
\begin{bmatrix} D_1 \\ D_2 \\ D_3 \end{bmatrix}
+
\begin{bmatrix} 0 \\ 0 \\ 0 \end{bmatrix}
$$

Solving, we obtain

$$
\begin{bmatrix} D_1 \\ D_2 \\ D_3 \end{bmatrix}
=
\begin{bmatrix} 0.0247 \text{ in.} \\ -0.0954 \text{ in.} \\ -0.00217 \text{ rad} \end{bmatrix}
$$

Using these results, the support reactions are determined from Eq. (1) as follows:

$$
\begin{bmatrix} Q_4 \\ Q_5 \\ Q_6 \\ Q_7 \\ Q_8 \\ Q_9 \end{bmatrix}
=
\begin{bmatrix}
-745.18 & -553.09 & -696 \\
-553.09 & -422.55 & 928 \\
696 & -928 & 116(10^3) \\
-1450 & 0 & 0 \\
0 & -15.10 & -1812.50 \\
0 & 1812.50 & 145(10^3)
\end{bmatrix}
\begin{bmatrix} 0.0247 \\ -0.0954 \\ -0.00217 \end{bmatrix}
+
\begin{bmatrix} 0 \\ 0 \\ 0 \\ 0 \\ 0 \\ 0 \end{bmatrix}
\begin{bmatrix} Q_4 \\ Q_5 \\ Q_6 \\ Q_7 \\ Q_8 \\ Q_9 \end{bmatrix}
=
\begin{bmatrix} 35.87 \text{ k} \\ 24.64 \text{ k} \\ -146.00 \text{ k} \cdot \text{in.} \\ -35.85 \text{ k} \\ 5.37 \text{ k} \\ -487.60 \text{ k} \cdot \text{in.} \end{bmatrix}
$$

The internal loadings can be determined from Eq. 15–7 applied to members 1 and 2. For example, for member 1, $\mathbf{q} = \mathbf{k}_1' \mathbf{T} \mathbf{D}$ or

$$
\begin{bmatrix} q_4 \\ q_5 \\ q_6 \\ q_1 \\ q_2 \\ q_3 \end{bmatrix} =
\begin{array}{c}
\begin{array}{cccccc} \quad 4 & \quad 5 & \quad 6 & \quad 1 & \quad 2 & \quad 3 \end{array} \\
\begin{bmatrix}
1160 & 0 & 0 & -1160 & 0 & 0 \\
0 & 7.73 & 1160 & 0 & -7.73 & 1160 \\
0 & 1160 & 232(10^3) & 0 & -1160 & 116(10^3) \\
-1160 & 0 & 0 & 1160 & 0 & 0 \\
0 & -7.73 & -1160 & 0 & 7.73 & -1160 \\
0 & 1160 & 116(10^3) & 0 & -1160 & 232(10^3)
\end{bmatrix}
\end{array}
\begin{bmatrix}
0.8 & 0.6 & 0 & 0 & 0 & 0 \\
-0.6 & 0.8 & 0 & 0 & 0 & 0 \\
0 & 0 & 1 & 0 & 0 & 0 \\
0 & 0 & 0 & 0.8 & 0.6 & 0 \\
0 & 0 & 0 & -0.6 & 0.8 & 0 \\
0 & 0 & 0 & 0 & 0 & 1
\end{bmatrix}
\begin{bmatrix} 0 \\ 0 \\ 0 \\ 0.0247 \\ -0.0954 \\ -0.00217 \end{bmatrix}
\begin{matrix} 4 \\ 5 \\ 6 \\ 1 \\ 2 \\ 3 \end{matrix}
$$

Note that $\mathbf{k}_1'$ is determined from Eq. 15–1, and $\mathbf{T}_1$ from Eq. 15–3. The coding numbers indicate the rows and columns for the near and far ends of the member, respectively, that is, 4, 5, 6, then 1, 2, 3, Fig. 15–12b. Thus,

$$
\begin{bmatrix} q_4 \\ q_5 \\ q_6 \\ q_1 \\ q_2 \\ q_3 \end{bmatrix} =
\begin{bmatrix}
43.5 \text{ k} \\
-1.81 \text{ k} \\
-146 \text{ k} \cdot \text{in.} \\
-43.5 \text{ k} \\
1.81 \text{ k} \\
-398 \text{ k} \cdot \text{in.}
\end{bmatrix}
$$

Ans.

These results are shown in Fig. 15–12d. Note that the sign convention to be followed is defined in Fig. 15–4, where the origin of the local coordinates is placed at the near end of the member, that is, node ①.

A similar analysis is performed for member 2. The results are shown at the top of Fig. 15–12e. To these we must superimpose the loadings of Fig. 15–12c, so that the final results for member 2 are as shown at the bottom of Fig. 15–12e.

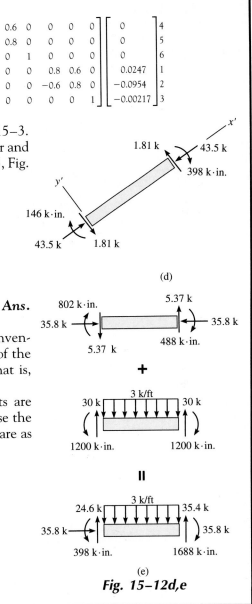

(d)

(e)

Fig. 15–12d,e

REFERENCES

Tezcan, S., "Computer Analysis of Plane and Space Structures," *Structural Division* ASCE, 143–173, April, 1966.

Wilson, E., "The Use of Minicomputers in Structural Analysis," *Computers and Structures*, 2, 695–698, 1980.

PROBLEMS

15–1. Determine the moments at the supports. *EI* is constant. Assume joint ② is a roller.

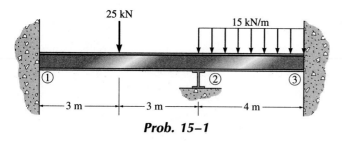

Prob. 15–1

15–2. Determine the internal moment in the beam at ① and ②. *EI* is constant. Assume ② and ③ are rollers.

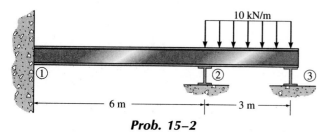

Prob. 15–2

15–3. Determine the moments at ② and ③. *EI* is constant. Assume ② and ③ are rollers and ① and ④ are pinned.

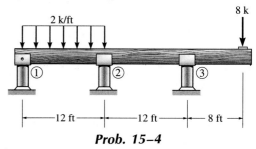

Prob. 15–3

*15–4. Determine the reactions at the supports ①, ②, and ③. Assume ① is pinned and ② and ③ are rollers. *EI* is constant.

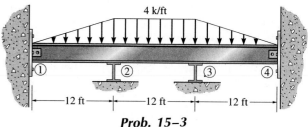

Prob. 15–4

15–5. Determine the reactions at the supports ①, ②, and ③. Assume ① is pinned, and ② and ③ are rollers. *EI* is constant.

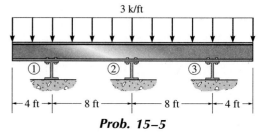

Prob. 15–5

15–6. Determine the stiffness matrix **K** for the frame. Take $E = 29(10^3)$ ksi. $I = 650$ in^4, $A = 20$ in^2 for each member. The joints at ② and ③ are fixed-connected.

15–7. Determine the components of displacement at joint ① in Prob. 15–6.

15–10. Determine the stiffness matrix **k** for each member of the frame. Take $E = 29(10^3)$ ksi, $I = 700$ in^4, $A = 30$ in^2 for each member. Joint ① is fixed-connected.

15–11. Determine the support reactions at ① and ③ in Prob. 15–10.

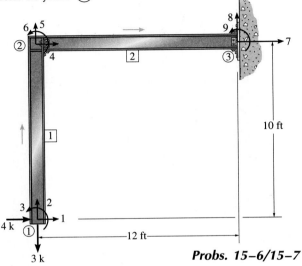

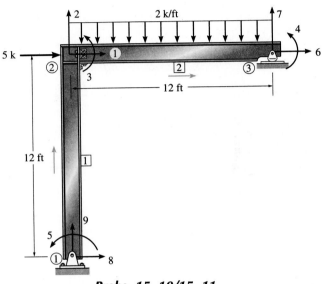

Probs. 15–6/15–7

*15–8. Determine the stiffness matrix **K** for the frame. Take $E = 29(10^3)$ ksi, $I = 600$ in^4, $A = 10$ in^2 for each member. Assume joints ① and ③ are pinned; joint ② is fixed.

15–9 Determine the rotation at ① and ③ and the support reactions in Prob. 15–8.

Probs. 15–10/15–11

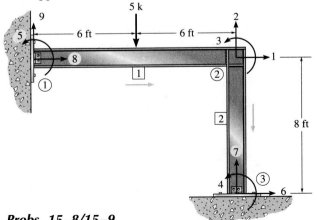

Probs. 15–8/15–9

*15–12.   Determine the stiffness matrix **k** for each member of the frame. Take $E = 10.5$ GPa, $I = 15(10^6)$ mm^4, $A = 7.5(10^3)$ mm^2 for each member. Joints ① and ② are fixed-connected, and ③ is pinned.

15–13.   Determine the support reactions at ① and ③ in Prob. 15–12.

15–14.   Determine the stiffness matrix **k** for each member of the frame. Take $E = 29(10^3)$ ksi, $I = 700$ in^4, $A = 15$ in^2 for each member. Joints ② and ③ are fixed-connected.

15–15.   Determine the reactions at the supports ① and ④ in Prob. 15–14.

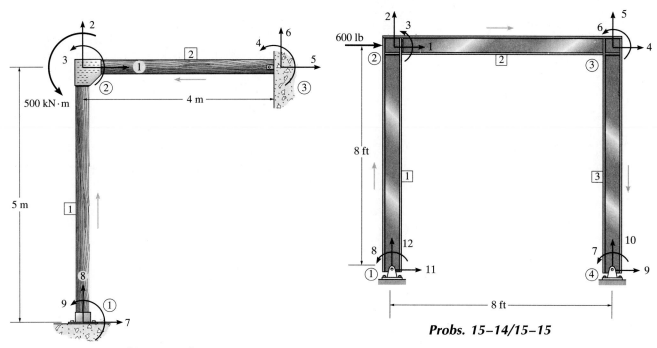

Probs. 15–12/15–13

Probs. 15–14/15–15

*15–16. Determine the stiffness matrix **k** for each member of the frame. Take $E = 29(10^3)$ ksi, $I = 450$ in⁴, $A = 8$ in² for each member. All joints are fixed-connected.

15–17. Determine the horizontal displacement of joint ② in Prob. 15–16. Also, compute the support reactions.

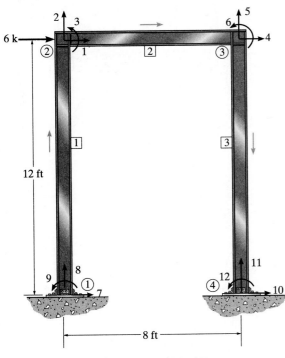

Probs. 15–16/15–17

15–18. Use the STRAN program to determine the internal moments at each nodal point. Establish the shear and moment diagrams. EI is constant.

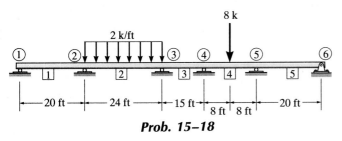

Prob. 15–18

15–19. Use the STRAN program to determine the horizontal and vertical forces, and the moment reactions at joint ①. EI is constant.

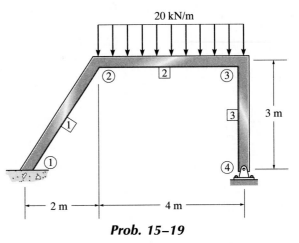

Prob. 15–19

*15–20. Use the STRAN program to determine the horizontal and vertical forces, and the moment reactions at joint ①. EI is constant.

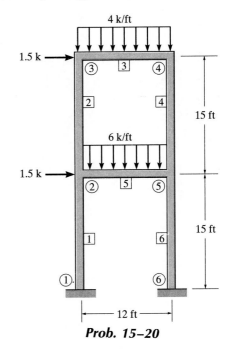

Prob. 15–20

715

15–21. Use the STRAN program to determine the horizontal and vertical forces, and the moment reactions at joint ①. *EI* is constant.

15–23. Use the STRAN program to determine the force in member $\boxed{9}$ of the truss. *AE* is constant.

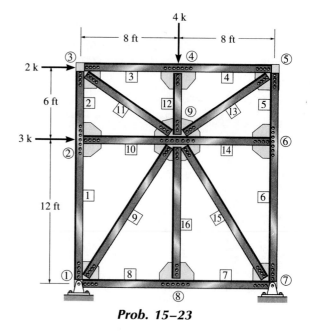

Prob. 15–23

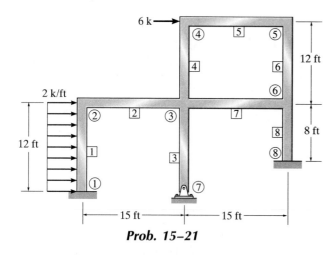

Prob. 15–21

*15–22. Use the STRAN program to determine the force in member $\boxed{11}$ of the truss. *AE* is constant.

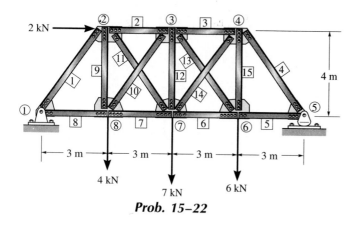

Prob. 15–22

Answers*

Chapter 1

1–1. 760 lb/ft
1–2. $p_x = 4.12$ psf, $p_y = 7.14$ psf
1–3. 87 lb/ft^2
1–5. D.L. = 431 lb/ft, L.L. = 750 lb/ft
1–6. D.L. = 8.10 k, L.L. = 37.5 k
1–7. 108.5 lb/ft^2
1–9. 1.54 k, $x = 4$ ft, $y = 9$ ft
1–10. 986 lb, $x = 6$ ft, $y = 3.5$ ft
1–11. 19.7 k
1–13. **(a)** 11.25 k, **(b)** 4.50 k
1–14. **(a)** 32.0 k, **(b)** 22.0 k
1–15. 94.5 k

Chapter 2

2–1. 475 lb/ft, At A and H, 4750 lb; At B, C, D, E, F, G, 9500 lb
2–2. 67.5 lb/ft, ends 101.2 lb, rest 202.5 lb
2–3. BG 200 lb/ft, ABCD 1-k loads in ctr., 0.5-k loads at ends.
2–5. BG ctr. 7 ft, 320 lb/ft, ABCD at B 1760 lb, $w_{max} = 160$ lb/ft.
2–6. BE ctr. 4 ft, 3.22 k/ft, FED at ctr. 11.3 k, $w_{max} = 1.61$ k/ft.
2–7. BE 2.14 k/ft, FED 12.9 k in ctr.
2–9. **(a)** statically indeterminate to 2°; **(b)** statically indeterminate to 1°; **(c)** statically determinate; **(d)** unstable
2–10. **(a)** statically determinate; **(b)** unstable; **(c)** statically indeterminate to 3°

*Note: Answers to every fourth problem are omitted.

2–11. **(a)** statically indeterminate to 3°; **(b)** statically indeterminate to 1°; **(c)** statically indeterminate to 1°; **(d)** unstable
2–13. **(a)** unstable; **(b)** statically determinate; **(c)** statically indeterminate to 1°; **(d)** statically indeterminate to 2°
2–14. **(a)** statically determinate; **(b)** unstable; **(c)** indeterminate to 4°
2–15. **(a)** statically indeterminate to 8°; **(b)** statically determinate; **(c)** statically indeterminate to 6°; **(d)** statically indeterminate to 6°; **(e)** statically indeterminate to 6°; **(f)** statically indeterminate to 8°
2–17. $B_y = 44.0$ kN, $A_x = 24$ kN, $A_y = 36$ kN
2–18. $B_y = 3.70$ k, $A_x = 6.00$ k, $A_y = 5.80$ k
2–19. $A_x = 0$, $B_y = 14$ k, $A_y = 22$ k
2–21. $F_B = 15.2$ k, $A_x = 4.17$ k, $A_y = 10.0$ k
2–22. $B_y = 48.3$ k, $A_y = 40.2$ k, $A_x = 0$
2–23. $A_y = 1.11$ k, $B_x = 0$, $B_y = 590$ lb
2–25. $A_y = 11.1$ k, $C_y = 4.88$ k, $C_x = 5$ k
2–26. $F_A = 11.2$ k, $B_x = 9.70$ k, $B_y = 4.10$ k
2–27. $A_y = 4$ kN, $C_y = 6.67$ kN, $E_x = 0$, $E_y = 5.33$ kN, $M_E = 5.33$ kN·m
2–29. $A_y = 7.36$ k, $B_y = 16.6$ k, $B_x = 0.5$ k
2–30. $A_y = 2$ kN, $B_y = 12$ kN, $B_x = 0$, $M_B = 32$ kN·m
2–31. $B_x = 8.25$ k, $B_y = 5$ k, $M_A = 555$ k·ft, $A_x = 8.25$ k, $A_y = 26$ k
2–33. $A_x = P/2$, $A_y = 2P/3$, $C_x = P/2$, $C_y = P/3$
2–34. $A_x = 12$ kN, $A_y = 10$ kN, $B_y = 16$ kN, $B_x = 0$, $C_x = 12$ kN, $C_y = 6$ kN, $M_C = 36$ kN·m
2–35. $A_x = 8.62$ kN, $A_y = 26.5$ kN, $C_x = 0.625$ kN, $C_y = 17.5$ kN
2–37. $D_x = 0$, $D_y = 0.75wL$, $A_x = wL$, $A_y = 0.75wL$, $M_A = wL^2/2$
2–38. $B_y = 16.6$ k, $A_x = 29.0$ k, $A_y = 17.0$ k
2–39. $A_x = 522$ lb, $A_y = 1473$ lb, $C_x = 678$ lb, $C_y = 1973$ lb

717

Chapter 3

3–1. (a) statically determinate; (b) externally unstable; (c) statically indeterminate to 5°; (d) statically indeterminate to 1°; (e) externally unstable

3–2. (a) statically indeterminate to 4°; (b) statically determinate, (c) statically indeterminate to 1°; (d) internally and externally unstable

3–3. (a) statically indeterminate to 3°; (b) unstable; (c) statically indeterminate to 2°; (d) unstable

3–5. $F_{AJ} = 7.26$ k (C), $F_{AB} = 4.54$ k (T), $F_{JI} = 7.26$ k (C), $F_{JB} = 2$ k (C), $F_{BI} = 2.15$ k (T), $F_{BC} = 3.74$ k (T), $F_{CI} = 0$, $F_{CD} = 3.74$ k (T), $F_{FG} = 10.1$ k (C), $F_{FE} = 8.53$ k (T), $F_{GH} = 10.1$ k (C), $F_{GE} = 3$ k (C), $F_{EH} = 3.84$ k (T), $F_{ED} = 6.13$ k (T), $F_{DI} = 5.10$ k (T), $F_{DH} = 4.50$ k (C), $F_{HI} = 7.23$ k (C)

3–6. $F_{GF} = 1.89$ k (C), $F_{EF} = 1.60$ k (C), $F_{HG} = 1.89$ k (C), $F_{GE} = 0$, $F_{HE} = 0$, $F_{ED} = 1.60$ k (T), $F_{IH} = 1.89$ k (C), $F_{HD} = 0$, $F_{IC} = 0$, $F_{CB} = 1.60$ k (T), $F_{ID} = 0$, $F_{CD} = 1.60$ k (T), $F_{AJ} = 2.57$ k (C), $F_{AB} = 1.60$ k (T), $F_{IJ} = 2.57$ k (C), $F_{BJ} = 0$, $F_{BI} = 0$

3–7. $F_{AI} = 4.04$ k (C), $F_{AB} = 3.75$ k (T), $F_{BC} = 3.75$ k (T), $F_{FE} = 12.1$ k (C), $F_{DE} = 7.75$ k (T), $F_{FD} = 0$, $F_{CD} = 7.75$ k (T), $F_{HG} = 2.15$ k (C), $F_{HI} = 0.798$ k (T), $F_{CF} = 4.04$ k (C), $F_{GF} = 8.08$ k (C), $F_{CI} = 0.269$ k (T), $F_{CG} = 1.40$ k (T), $F_{IG} = 5.92$ k (C), $F_{BI} = 0$

3–9. $F_{ED} = 8.33$ kN (T), $F_{CD} = 6.67$ kN (C), $F_{BC} = 6.67$ kN (C), $F_{CE} = 5$ kN (T), $F_{GF} = 20$ kN (T), $F_{GA} = 15$ kN (T), $F_{AF} = 18.0$ kN (C), $F_{AB} = 10.0$ kN (C), $F_{BE} = 4.17$ kN (C), $F_{FB} = 7.50$ kN (T), $F_{FE} = 12.5$ kN (T)

3–10. $F_{AL} = 7.42$ kN (C), $F_{AB} = 5.25$ kN (T), $F_{BL} = 0$, $F_{BC} = 5.25$ kN (T), $F_{LC} = 1.41$ kN (C), $F_{LK} = 6.01$ kN (C), $F_{CK} = 1.00$ kN (T), $F_{CD} = 4.25$ kN (T), $F_{KD} = 2.24$ kN (C), $F_{KJ} = 4.59$ kN (C), $F_{JD} = 4.00$ kN (T), $F_{HG} = 7.42$ kN (C), $F_{FG} = 5.25$ kN (T), $F_{HF} = 0$, $F_{EF} = 5.25$ kN (T), $F_{HE} = 1.41$ kN (C), $F_{HI} = 6.01$ kN (C), $F_{EI} = 1.00$ kN (T), $F_{DE} = 4.25$ kN (T), $F_{ID} = 2.24$ kN (C), $F_{JI} = 4.59$ kN (C)

3–11. $F_{EA} = 2.31$ kN (C), $F_{EC} = 2.31$ kN (T), $F_{AD} = 2.24$ kN (T), $F_{AB} = 3.16$ kN (C), $F_{BC} = 3.16$ kN (C), $F_{BD} = 3.16$ kN (C), $F_{DC} = 2.24$ kN (T)

3–13. $F_{CB} = 8.10$ kN (T), $F_{IJ} = 11.4$ kN (C), $F_{IB} = 3.82$ kN (T)

3–14. $F_{DE} = 2000$ lb (T), $F_{DI} = 667$ lb (C), $F_{JI} = 1566$ lb (C)

3–15. $F_{FC} = 1.67$ k (T), $F_{BC} = 1.39$ k (T), $F_{EF} = 2.44$ k (C)

3–17. $F_{CD} = 2.92$ k (T), $F_{LK} = 3.02$ k (C), $F_{LD} = 0.100$ k (T), $F_{DK} = 0.133$ k (C)

3–18. $F_{DC} = 11.7$ k (T), $F_{FG} = 16.6$ k (C), $F_{FC} = 4.86$ k (T)

3–19. $F_{GB} = 2.50$ k (C), $F_{GF} = 5.48$ k (C), $F_{BC} = 3.25$ k (T)

3–21. $F_{JH} = 0$, $F_{IH} = 3.02$ k (C), $F_{CD} = 2.63$ k (T)

3–22. $F_{JH} = 0.722$ k (T), $F_{IB} = 1.73$ k (T), $F_{BJ} = 1.73$ k (C)

3–23. $F_{CD} = 24.0$ kN (T), $F_{CF} = 13.4$ kN (C), $F_{ED} = 13.4$ kN (C), $F_{AE} = 9.49$ kN (C), $F_{AC} = 4.24$ kN (T), $F_{EC} = 12.7$ kN (T)

3–25. $F_{BE} = 10$ kN (C), $F_{FC} = 10$ kN (C), $F_{FE} = 16$ kN (T), $F_{AB} = 8$ kN (C), $F_{FB} = 2$ kN (C), $F_{AF} = 10$ kN (T), $F_{ED} = 10$ kN (T), $F_{EC} = 2$ kN (C), $F_{CD} = 8$ kN (C)

3–26. $F_{AC} = 5.41$ k (T), $F_{AD} = 6.36$ k (C), $F_{DB} = 5.41$ k (T), $F_{DC} = 9.00$ k (T), $F_{CB} = 6.36$ k (C)

3–27. $F_{AB} = 3$ kN (T), $F_{FD} = 4.24$ kN (C), $F_{ED} = 0$, $F_{AF} = 4.24$ kN (C), $F_{EA} = 4$ kN (C), $F_{ED} = F_{DC} = 0$, $F_{EF} = F_{CG} = 0$, $F_{DF} = F_{DG} = 4.24$ kN (C), $F_{AF} = F_{GB} = 4.24$ kN (C), $F_{AB} = 3$ kN (T), $F_{EA} = F_{CB} = 4$ kN (C)

3–29. $F_{AF} = 646$ lb (C), $F_{AB} = 580$ lb (C), $F_{EB} = 820$ lb (T), $F_{BC} = 580$ lb (C), $F_{EF} = 473$ lb (C), $F_{CF} = 580$ lb (T), $F_{CD} = 1593$ lb (C), $F_{ED} = 1166$ lb (C), $F_{DA} = 1428$ lb (T)

3–30. $F_{AB} = 0$, $F_{AG} = 1.50$ k (C), $F_{GB} = 0.707$ k (T), $F_{GL} = 0.500$ k (T), $F_{GI} = 0.707$ k (C), $F_{LI} = 0.707$ k (T), $F_{LK} = 0.500$ k (T), $F_{IK} = 0.707$ k (C), $F_{IF} = 0.707$ k (T), $F_{BF} = 2.12$ k (T), $F_{BC} = 1.00$ k (C), $F_{FC} = 0.707$ k (T), $F_{FH} = 2.12$ k (T), $F_{KH} = 0.707$ k (T), $F_{KJ} = 1.50$ k (C),

$F_{JH} = 2.12$ k (T), $F_{CD} = 0$, $F_{DE} = 0.500$ k (C),
$F_{CE} = 0.707$ k (C), $F_{HE} = 0.707$ k (T),
$F_{JE} = 1.50$ k (C)

3–31. $F_{AB} = 3.905$ kN (C), $F_{BC} = 2.50$ kN (C),
$F_{CD} = 2.50$ kN (C),
$F_{BG} = F_{GF} = F_{GH} = F_{CH} = F_{HE} = 0$

3–33. $F_{AC} = F_{AE} = 0$, $F_{AB} = 4$ kN (T),
$F_{BE} = 5.66$ kN (T), $F_{BD} = 2$ kN (C), $F_{DE} = 0$,
$F_{DC} = 0$, $F_{CE} = 0$

3–34. $F_{CE} = 707$ lb (T), $F_{CD} = 500$ lb (C), $F_{CB} = 0$,
$F_{BE} = 0$, $F_{BF} = 1.13$ k (T), $F_{BA} = 800$ lb (C),
$F_{ED} = 500$ lb (C), $F_{EF} = 0$, $F_{FD} = 0$,
$F_{FA} = 800$ lb (C), $F_{DA} = 0$

3–35. $F_{AD} = 2.47$ k (T), $F_{AB} = F_{AC} = 1.22$ k (C)

Chapter 4

4–1. $V_C = -1$ k, $N_C = 0$, $M_C = 56$ k·ft, $V_D = -1$ k,
$N_D = 0$, $M_D = 48$ k·ft

4–2. $V_{max} = 7$, $M_{max} = 56$

4–3. $N_B = 0$, $V_B = 28.8$ k, $M_B = -115$ k·ft

4–5. $N_C = -4$ k, $V_C = -1$ k, $M_C = -8$ k·ft,
$N_D = -4$ k, $V_D = 1.15$ k, $M_D = -2.6$ k·ft

4–6. $V_{max} = 2.65$, $M_{max} = -14$

4–7. $N_C = -3.60$ kN, $V_C = -0.0933$ kN,
$M_C = -0.0933$ kN·m, $N_D = 0$, $V_D = -4.89$ kN,
$M_D = -10.2$ kN·m

4–9. $N_C = 20$ kN, $V_C = 70.6$ kN, $M_C = -292$ kN·m

4–10. $V_{max} = 94.6$, $M_{max} = -540$

4–11. $N_D = 0$, $V_D = 0$, $M_D = 9$ k·ft, $N_E = 0$,
$V_E = -7$ k, $M_E = -12$ k·ft

4–13. $N_C = 0$, $V_C = 0.5$ k, $M_C = 3.60$ k·ft

4–14. $V_{max} = 10.1$, $M_{max} = -60$

4–15. $N_C = 0$, $V_C = 2014$ lb, $M_C = -15000$ lb·ft,
$N_D = 0$, $V_D = 1114$ lb, $M_D = 3771$ lb·ft

4–17. $V = \frac{4}{9}x^2 - 8x + 36$, $M = \frac{8}{54}x^3 - 4x^2 + 36x - 108$

4–18. $V_{max} = 36$, $M_{max} = -108$

4–19. $V = -3.33x^2 - 800$, $M = -1.11x^3 - 800x - 1200$

4–21. $V = 2960 - 400x$, $M = -200x^2 + 2960x + 600$

4–22. $V_{max} = -3040$, $M_{max} = 11552$

4–23. $V = 500 - 2x^3$, $M = 500x - 0.5x^4$

4–25. $V_{max} = \pm1200$, $M_{max} = 6400$

4–26. 3.00 k/ft, 18 ft

4–27. $V = -3.25$, $M = -3.25x + 14$

4–29. $V = 120 - 4x$, $M = -2x^2 + 120x - 1650$

4–30. $V_{max} = \pm60$, $M_{max} = -300$

4–31. $V_{max} = 1050$, $M_{max} = 3912$

4–33. $V_{max} = 35.8$, $M_{max} = -146$

4–34. $V_{max} = \pm36$, $M_{max} = -108$

4–35. $V_{max} = 48$, $M_{max} = -576$

4–37. $V_{max} = \pm25$, $M_{max} = -45$

4–38. ABC, $V_{max} = 13.2$, $M_{max} = 17.5$; BCD,
$V_{max} = -11.8$, $M_{max} = 13.6$

4–39. $V_{max} = 0$, $M_{max} = 0$; BC, $V_{max} = -9.20$,
$M_{max} = 52.9$; $V_{max} = 3$, $M_{max} = 24$

4–41. BC, $V_{max} = 5$, $M_{max} = 40$; BA, $V_{max} = 12$,
$M_{max} = -90$; DC, $V_{max} = 0$, $M_{max} = 0$

4–42. AB, $V_{max} = 2.22$, $M_{max} = 15.5$; BC, $V_{max} = 0.583$,
$M_{max} = 10.6$

4–43. AB, $V_{max} = 13.3$, $M_{max} = 26.7$; BC, $V_{max} = -8$,
$M_{max} = 9.24$

Chapter 5

5–1. $A_x = 2.72$ k, $A_y = 3.78$ k, $C_x = 0.276$ k,
$C_y = 0.216$ k

5–2. $F_B = 6.77$ k, $F_A = 10.8$ k, $F_C = 11.7$ k

5–3. $B_x = 46.67$ k, $B_y = 5$ k, $A_x = 46.7$ k, $A_y = 95$ k,
$C_x = 46.7$ k, $C_y = 85$ k

5–5. $C_y = 6.29$ k, $F_{AC} = 0.723$ k, $A_x = 1.69$ k,
$A_y = 6.17$ k

5–6. $A_x = 1.58$ k, $A_y = 4$ k, $C_x = 1.58$ k, $C_y = 4$ k

5–7. $T_{BC} = 4.67$ k, $T_{AB} = 8.30$ k, $T_{CD} = 8.81$ k, 20.2 ft

5–9. 3.53 m, $P = 0.8$ kN, $T_{DE} = 8.17$ kN

5–10. 51.9 lb/ft

5–11. 1.36 kN/m

5–13. $V_{max} = -1.91$, $M_{max} = 3.56$

5–14. $F_{min} = 100$ k, $F_{max} = 117$ k, $F = 10$ k

5–15. $V_{max} = \pm5$, $M_{max} = 6.25$

Chapter 6

6–1,2. **(a)** $(0, 1)$, $(35, 0)$ **(b)** $(0, 0)$, $(10^-, -0.286)$,
$(10^+, 0.714)$, $(35, 0)$

6–5,6. **(a)** $(0, 1)$, $(10, 1)$ **(b)** $(0, 0)$, $(5, 0)$, $(10, -5)$

6–9,10. $(0, 0)$, $(15, 0)$, $(20, -5)$

6–13,14. **(a)** $(0, 0)$, $(10^-, -0.5)$, $(10^+, 0.5)$, $(20, 0)$,
$(30, 0)$ **(b)** $(0, 0)$, $(10, 5)$, $(20, 0)$, $(30, 0)$
(c) $(0, 0)$, $(10, 0)$, $(20, 0)$, $(30, 1)$

6–17,18. $(0, 0)$, $(4, 2)$, $(8, 0)$, $(12, -2)$

6–21. **(a)** -390 k·ft; **(b)** 12 k; **(c)** -110 k·ft

6–22. **(a)** 30 kN·m; **(b)** 4750 N

6–23. **(a)** 32.8 k; **(b)** 12.5 k; **(c)** 125 k·ft

6–25. (a) $-13.1\,\text{k}\cdot\text{ft}$; (b) $188\,\text{lb}$

6–26. $-4000\,\text{lb}\cdot\text{ft}$

6–27. (a) $-86.4\,\text{k}\cdot\text{ft}$; (b) $5.40\,\text{k}$

6–29. (a) $(0,0)$, $(20,-6)$, $(40,0.2)$, $(50,0)$ (b) $(0,0)$, $(30,12)$, $(50,0)$

6–30. $-27\,\text{k}\cdot\text{ft}$

6–31. $-6.75\,\text{k}$

6–33. (a) $(0,0.5)$, $(5,0)$, $(10,0.5)$, $(15,0)$, $(20,-0.5)$
(b) $(0,-2.5)$, $(5,0)$, $(10,2.5)$, $(15,0)$, $(20,-2.5)$

6–34. $2.73\,\text{k}$, $61.25\,\text{k}\cdot\text{ft}$

6–35. (a) $(0,-1.33)$, $(4,0)$, $(12,2.67)$, $(16,0)$
(b) $(0,0.333)$, $(4,0)$, $(8,0.667)$, $(16,0)$

6–37. $10.7\,\text{kN}\cdot\text{m}$

6–38. (a) $16.7\,\text{k}$; (b) $6.67\,\text{k}$; (c) $46.7\,\text{k}\cdot\text{ft}$

6–39. $-6\,\text{k}$

6–41. $(0,0)$, $(20,0.577)$, $(40,0.577)$, $(60,0)$

6–42. $(0,0)$, $(20,0.385)$, $(40,-0.385)$, $(60,0)$

6–43. $(0,0)$, $(40,-0.770)$, $(60,0)$

6–45. $(0,0)$, $(12,2)$, $(24,0)$

6–46. $(0,0)$, $(8,-1.78)$, $(24,0)$

6–47. $15.1\,\text{k}$ (T)

6–49. $-32\,\text{k}$ (C)

6–50. $-27\,\text{kN}$ (C)

6–51. $-10.1\,\text{kN}$ (C)

6–53. $24\,\text{ft}$, $16.8\,\text{k}$ (C)

6–54. $36\,\text{ft}$, $16.8\,\text{k}$ (C)

6–55. $24\,\text{ft}$, $15\,\text{k}$ (T)

6–57. $3.50\,\text{k}$, $70\,\text{k}\cdot\text{ft}$

6–58. $-8.75\,\text{k}\cdot\text{ft}$

6–59. $1.46\,\text{k}$

6–61. $2.37\,\text{k}$ (T)

6–62. $1.92\,\text{k}$ (T)

6–63. $3.00\,\text{k}$ (C)

6–65. $33.4\,\text{k}$ (C)

6–66. $33.4\,\text{k}$ (T)

6–67. $23.2\,\text{k}$ (T)

6–69. $64.5\,\text{kN}\cdot\text{m}$

6–70. $10\,\text{kN}$, $-39\,\text{kN}\cdot\text{m}$

6–71. $164\,\text{kN}\cdot\text{m}$

6–73. $764\,\text{k}\cdot\text{ft}$

6–74. $10.5\,\text{k}\cdot\text{ft}$

Chapter 7

7–1. $F_{FB} = 250\,\text{lb}$ (T), $F_{AE} = 250\,\text{lb}$ (C),
$F_{FE} = 200\,\text{lb}$ (C), $F_{AB} = 200\,\text{lb}$ (T), $F_{AF} = 350\,\text{lb}$ (C), $F_{BD} = 250\,\text{lb}$ (T), $F_{CE} = 250\,\text{lb}$ (C), $F_{DE} = 200\,\text{lb}$ (C), $F_{BC} = 200\,\text{lb}$ (T),

$F_{CD} = 450\,\text{lb}$ (C), $F_{BE} = 300\,\text{lb}$ (C)

7–2. $F_{AE} = 0$, $F_{FB} = 500\,\text{lb}$ (T), $F_{AB} = 0$,
$F_{FE} = 400\,\text{lb}$ (C), $F_{AF} = 500\,\text{lb}$ (C),
$F_{BD} = 500\,\text{lb}$ (T), $F_{CE} = 0$, $F_{BC} = 0$,
$F_{CD} = 600\,\text{lb}$ (C), $F_{BE} = 600\,\text{lb}$ (C),
$F_{ED} = 400\,\text{lb}$ (C)

7–3. $F_{HB} = 5.89\,\text{k}$ (T), $F_{AG} = 5.89\,\text{k}$ (C),
$F_{AB} = 9.17\,\text{k}$ (T), $F_{AH} = 14.2\,\text{k}$ (C),
$F_{HG} = 4.17\,\text{k}$ (C), $F_{GC} = 1.18\,\text{k}$ (C),
$F_{BF} = 1.18\,\text{k}$ (T), $F_{GF} = 7.5\,\text{k}$ (C), $F_{GB} = 5\,\text{k}$ (C),
$F_{BC} = 12.5\,\text{k}$ (T), $F_{EC} = 8.25\,\text{k}$ (T),
$F_{DF} = 8.25\,\text{k}$ (T), $F_{CD} = 5.83\,\text{k}$ (T),
$F_{ED} = 15.8\,\text{k}$ (T), $F_{FE} = 0.833\,\text{k}$ (C), $F_{FC} = 5\,\text{k}$ (C)

7–5. $F_{EC} = 6.67\,\text{kN}$ (C), $F_{DF} = 6.67\,\text{kN}$ (T),
$F_{DC} = 5.33\,\text{kN}$ (C), $F_{EF} = 5.33\,\text{kN}$ (T),
$F_{DE} = 4\,\text{kN}$ (C), $F_{AC} = 15\,\text{kN}$ (T),
$F_{BF} = 15\,\text{kN}$ (C), $F_{AB} = 9\,\text{kN}$ (T),
$F_{CB} = 22.7\,\text{kN}$ (C), $F_{AF} = 22.7\,\text{kN}$ (T),
$F_{CF} = 5\,\text{kN}$ (C)

7–6. $F_{EC} = 0$, $F_{DF} = 13.3\,\text{kN}$ (T), $F_{ED} = 8\,\text{kN}$ (C),
$F_{EF} = 0$, $F_{DC} = 10.7\,\text{kN}$ (C), $F_{BF} = 0$,
$F_{AC} = 30\,\text{kN}$ (T), $F_{AB} = 0$, $F_{CB} = 34.7\,\text{kN}$ (C),
$F_{AF} = 10.7\,\text{kN}$ (T), $F_{CF} = 18\,\text{kN}$ (C)

7–7. 0, $55.7\,\text{k}\cdot\text{ft}$, $41.25\,\text{k}$

7–9. $M_A = 16.2\,\text{k}\cdot\text{ft}$, $A_x = 0$, $A_y = 12\,\text{k}$, $M_B = 9\,\text{k}\cdot\text{ft}$,
$B_x = 0$, $B_y = 16\,\text{k}$, $M_C = 7.20\,\text{k}\cdot\text{ft}$, $C_x = 0$,
$C_y = 4\,\text{k}$

7–10. $M_A = M_D = Ph/6$, $M_B = M_C = Ph/3$

7–11. (a) $M_A = M_D = 0$, $M_B = M_C = 36\,\text{k}\cdot\text{ft}$;
(b) $M_B = M_C = 18\,\text{k}\cdot\text{ft}$, $M_A = M_D = 18\,\text{k}\cdot\text{ft}$;
(c) $M_A = M_D = 12\,\text{k}\cdot\text{ft}$, $M_B = M_C = 24\,\text{k}\cdot\text{ft}$

7–13. $M_D = 2\,\text{kN}\cdot\text{m}$, $M_C = 2\,\text{kN}\cdot\text{m}$

7–14. $F_{CF} = 1.77\,\text{k}$ (T)

7–15. $F_{CE} = 1.06\,\text{k}$ (T)

7–17. $F_{CG} = 40.6\,\text{kN}$ (T), $F_{EG} = 35.0\,\text{kN}$ (C),
$F_{CD} = 5\,\text{kN}$ (C)

7–18. $F_{CG} = 21.9\,\text{kN}$ (T), $F_{EG} = 20\,\text{kN}$ (C),
$F_{CD} = 5\,\text{kN}$ (C)

7–19. $F_{FG} = 70\,\text{kN}$ (C), $F_{CG} = 54.2\,\text{kN}$ (T),
$F_{CD} = 22.5\,\text{kN}$ (T), $F_{GD} = 54.2\,\text{kN}$ (C),
$F_{GH} = 5\,\text{kN}$ (C), $F_{DH} = 54.2\,\text{kN}$ (T),
$F_{DE} = 42.5\,\text{kN}$ (C), $F_{HE} = 54.2\,\text{kN}$ (C),
$F_{HI} = 60\,\text{kN}$ (T), $A_x = B_x = 15\,\text{kN}$,
$A_y = B_y = 43.3\,\text{kN}$

7–21. $F_{CE} = 6.41\,\text{k}$ (T), $F_{DF} = 1.875\,\text{k}$ (C),
$F_{DE} = 6.50\,\text{k}$ (C), $F_{EF} = 2.25\,\text{k}$ (T),
$F_{EH} = 0.5\,\text{k}$ (C), $F_{FH} = 1.875\,\text{k}$ (C), $F_{FG} = 0$,
$F_{HJ} = 1.875\,\text{k}$ (T), $F_{HI} = 3.50\,\text{k}$ (C),

$F_{JI} = 2.25$ k (C), $F_{JK} = 1.875$ k (T),
$F_{IL} = 6.41$ k (C), $F_{IK} = 2.50$ k (T)

7–22. $F_{EC} = 17.0$ k (T), $F_{GH} = 19.5$ k (C), $F_{CD} = 0$,
$F_{EH} = 17.0$ k (T), $F_{HI} = 4.50$ k (T), $F_{FI} = 0$,
$F_{HF} = 17.0$ k (C), $F_{FD} = 17.0$ k (C), $V_A = 7.5$ k,
$A_y = 12$ k, $M_A = 45$ k·ft, $B_y = 12$ k, $M_B = 45$ k·ft,
$V_B = 7.5$ k

7–23. $A_x = B_x = 5$ k, $A_y = B_y = 6.25$ k, $F_{DE} = 8.84$ k (C),
$F_{CE} = 10$ k (T), $F_{DF} = 8.75$ k (C),
$F_{EF} = 8.84$ k (T), $F_{EG} = 2.50$ k (C),
$F_{FG} = 8.84$ k (C), $F_{FH} = 3.75$ k (T),
$F_{GH} = 8.84$ k (T), $F_{GI} = 15.0$ k (C)

7–25. $M_{max} = \pm18$

7–26. $M_{max} = \pm13.3$

7–27. $A_x = 3.25$ k, $A_y = 5.50$ k, $M_A = 29.25$ k·ft

7–29. $M_{max} = \pm42$ k·ft

7–30. $M_{max} = \pm30.6$ k·ft

7–31. $M_{max} = \pm3.33$ k·ft

7–33. $PQRST$, $M_{max} = \pm3.75$ k·ft, $BGLQ$,
$M_{max} = \pm30$ k·ft

7–34. $PQRST$, $M_{max} = \pm5.26$ k·ft, $BGLQ$,
$M_{max} = \pm24.6$ k·ft

Chapter 8

8–1,2,3,5. $\theta_B = \dfrac{1000}{EI}$ kN·m², $\Delta_B = \dfrac{5833}{EI}$ kN·m³

8–6. $\dfrac{PL^3}{48EI}$, $\dfrac{PL^2}{16EI}$

8–7. $\dfrac{PL^3}{48EI}$, $\dfrac{PL^2}{16EI}$

8–9. $\theta_A = -\dfrac{wL^3}{24\,EI}$, $\Delta_B = -\dfrac{5wL^4}{384\,EI}$

8–10,11. $\theta_A = \dfrac{wL^3}{24\,EI}$, $\Delta_B = \dfrac{5wL^4}{384\,EI}$

8–13. $\dfrac{5\,wL^4}{384\,EI}$

8–14. $\dfrac{ML^2}{8\,EI}$

8–15,17,18. 0.0385 rad, 263 mm

8–19. 0.263 m

8–21. 0.263 m

8–22,23,25,26. 0.000466 rad, 0.587 in.

8–27,29,30,31. $\dfrac{1500}{EI}$ kN·m³

8–33,34,35. $\theta_C = -\dfrac{333}{EI}$ lb·ft², $\Delta_C = -\dfrac{3333}{EI}$ lb·ft³

8–37,38,39. 16.9 mm

8–41,42,43. 0.0112 rad, 0.782 in.

8–45,46,47. $\dfrac{50,625}{EI}$ k·ft³, $\dfrac{3937.5}{EI}$ k·ft²

8–49,50. $\dfrac{833}{EI}$ k·ft³, $\dfrac{250}{EI}$ k·ft²

8–51. $5.37(10^{-3})$ rad

8–53. $\Delta_D = \dfrac{146}{EI}$ kN·m³, $\theta_{BL} = \dfrac{12.5}{EI}$ kN·m²,

$\theta_{BR} = \dfrac{4.17}{EI}$ kN·m²

8–54. $\dfrac{167}{EI}$ k·ft³, $\dfrac{33.3}{EI}$ k·ft²

8–55. $\dfrac{33.3}{EI}$ k·ft², $\dfrac{167}{EI}$ k·ft³

8–57. $\dfrac{167}{EI}$ k·ft³

8–58,59. 23.0 mm

8–61. 0.856 in.

8–62,63. 0.0401 in.

8–65. 0.0096 in.

8–66. 1.06 in.

8–67. $\dfrac{1.89(10^3)}{AE}$ kN·m

8–69. $\dfrac{1.51(10^6)}{AE}$ lb·in.

8–70. 0.0768 in.

8–71. 1.12 in.

8–73. $\dfrac{2100}{AE}$ lb·ft

8–74,75. 0.808 in.

8–77. $\dfrac{63.3(10^3)}{AE}$ lb·ft

8–78,79. $\dfrac{9352}{AE}$ k·in.

8–81,82. $(\Delta_C)_h = 0.179$ in., $(\Delta_C)_v = 0.448$ in.

8–83. 18.7 mm

8–85. $(\Delta_C)_h = 0.656$ in., $(\Delta_C)_v = 0.425$ in.

8–86. $(\Delta_C)_h = 0.657$ in., $(\Delta_C)_v = 0.425$ in.

8–87. $(\Delta_C)_h = 0.656$ in., $(\Delta_C)_v = 0.425$ in.

8–89,90. 2.81 mm

8–91. $0.414(10^{-3})$ rad

8–93,94. $\dfrac{440}{EI}$ k·ft³

8–95. $\dfrac{79.1 \text{ k·ft}^3}{EI}$

8–97,98. $\dfrac{417 \text{ k} \cdot \text{ft}^3}{EI}$

8–99. 49.1 mm

8–101,102. 1.70 in.

Chapter 9

9–1. $B_y = 3M/L$, $A_x = 0$, $A_y = 3M/L$, $M_A = 0.5M$

9–2. $A_y = 3M/2L$, $B_x = 0$, $B_y = 3M/2L$, $M_B = M/2$

9–3. $B_y = 900$ lb, $A_y = 1500$ lb, $A_x = 0$, $M_A = 3600$ lb·ft

9–5. $C_y = 3.68$ k, $A_x = 0$, $A_y = 36.3$ k, $M_A = 126$ k·ft

9–6. $B_y = 5.94$ k, $A_x = 0$, $A_y = 4.06$ k, $M_A = 10.9$ k·ft

9–7. $A_x = 0$, $A_y = 2\,wL/5$, $M_A = wL^2/15$

9–9. $B_y = 35.6$ kN, $C_y = 2.41$ kN, $A_y = 26.8$ kN, $B_x = 0$

9–10. $B_y = 68.8$ kN, $A_y = 15.6$ kN, $C_y = 15.6$ kN

9–11. $B_y = 6.66$ k, $C_y = 3.95$ k, $A_y = 1.02$ k, $A_x = 0$

9–13. $B_y = 160$ k, $A_y = C_y = 80$ k

9–14. $M_B = 58.1$ kN·m, $M_C = 200$ kN·m

9–15. $B_y = 27.5$ k, $C_y = 7.5$ k, $M_{max} = -25$ k·ft

9–17. $M_A = M_B = PL/8$, $A_y = B_y = P/2$

9–18. $M_A = M_B = -wL^2/12$

9–19. $A_y = 2.61$ k, $M_C = 6.26$ k·ft, $C_y = 3.39$ k, $C_x = 0$

9–21. $C_y = 28$ kN, $A_x = 2$ kN, $A_y = 44$ kN, $M_A = 78$ kN·m

9–22. $C_y = 27.0$ k, $A_y = 25.0$ k, $A_x = 0$, $M_A = 134$ k·ft

9–23. $A_y = 15.0$ k, $D_y = 15.0$ k, $D_x = 2$ k, $M_D = 19.5$ k·ft

9–25. 4.05 kN (C)

9–26. 2.56 kN (T)

9–27. $F_{CB} = 3.06$ k (C), $F_{AC} = 8.23$ k (C), $F_{DC} = 6.58$ k (T), $F_{DB} = 5.10$ k (T), $F_{AB} = 10.1$ k (C), $F_{DA} = 4.94$ k (T)

9–29. $F_{AD} = 8.54$ k (C), $F_{AB} = 6.04$ kN (T), $F_{AE} = 6.04$ k (T), $F_{EB} = 5.61$ kN (T), $F_{ED} = 3.96$ kN (C), $F_{BC} = 14.1$ kN (T), $F_{DC} = 10.0$ kN (C), $F_{DB} = 14.0$ kN (C)

9–30. 7.78 kN (T)

9–31. 2.95 kN (T)

9–33. 28.0 k

9–34. $T = 24.6$ kN, $A_x = 19.7$ kN, $A_y = 35.3$ kN, $M_A = 41.0$ kN·m

9–35. $F_{CD} = 64.7$ k (C), $F_{CB} = F_{AC} = 84.1$ k (T)

9–37. $F_{CE} = 43.1$ k (T), $F_{AC} = F_{EF} = 53.9$ k (T), $F_{BC} = F_{DE} = 32.3$ k (C)

9–38. $A_y = 15.8$ k, $C_y = 9.98$ k, $B_y = 49.2$ k

9–39. $A_y = 17.6$ kN, $C_y = 11.6$ kN, $B_y = 60.8$ kN

9–41. $A_y = 5.21$ k, $D_y = 12.6$ k, $B_y = 26.1$ k, $C_y = 40.1$ k

9–42. $A_y = 6.65$ k, $C_y = 16.3$ k, $B_y = 39.6$ k

9–43. (0, 0), (5, 0.425), (10, 1), (15, 0.275), (20, 0)

9–45. (0, 1), (5, 0.633), (10, 0.312), (15, 0.0859), (20, 0)

9–46. (0, 0), (5, 0.481), (10, 0.852), (15, 1), symmetric

9–47. 46.9 kN·m

9–49. 9.00 kN·m

Chapter 10

10–1. $M_{AB} = -230$ k·ft, $M_{BA} = 187$ k·ft, $M_{BC} = -187$ k·ft, $M_{CB} = 122$ k·ft

10–2. $M_{AB} = -11.6$ k·ft, $M_{BA} = 12.8$ k·ft, $M_{BC} = -12.8$ k·ft, $M_{CB} = 13.8$ k·ft

10–3. $M_{AB} = -34.8$ k·ft, $M_{BA} = 45.6$ k·ft, $M_{BC} = -45.6$ k·ft, $M_{CB} = 67.2$ k·ft

10–5. $M_{AB} = -2680$ k·ft, $M_{BA} = -1720$ k·ft, $M_{BC} = 1720$ k·ft, $M_{CB} = 0$

10–6. $M_{BA} = 38.7$ kN·m, $M_{BC} = -38.7$ kN·m

10–7. $M_{AB} = 146$ k·ft, $M_{DC} = -146$ k·ft, $A_x = 29.3$ k, $D_y = 96$ k, $A_y = 96$ k, $D_x = 29.3$ k

10–9. $M_{AB} = -51.9$ kN·m, $M_{BA} = 85.2$ kN·m, $M_{BC} = -85.2$ kN·m

10–10. $M_{AB} = 44.4$ k·ft, $M_{BA} = 88.8$ k·ft, $M_{BC} = -88.8$ k·ft, $M_{CB} = 72.4$ k·ft, $M_{CD} = -72.4$ k·ft, $M_{DC} = -36.2$ k·ft

10–11. $M_{CD} = -38.4$ k·ft, $M_{BA} = -9.60$ k·ft, $M_{BC} = 9.60$ k·ft, $M_{CB} = 38.4$ k·ft

10–13. $M_{BA} = 76.1$ k·ft, $M_{BC} = -97.9$ k·ft, $M_{CB} = 89.4$ k·ft, $M_{BD} = 21.8$ k·ft, $M_{CE} = -89.4$ k·ft

10–14. $M_{BA} = 181$ k·ft, $M_{BC} = -103$ k·ft, $M_{CB} = 38.3$ k·ft, $M_{BD} = -77.3$ k·ft

10–15. $M_{BA} = -41.1$ lb·ft, $M_{EF} = 41.1$ lb·ft, $M_{BC} = 41.1$ lb·ft, $M_{ED} = -41.1$ lb·ft, $M_{CB} = 214$ lb·ft, $M_{DE} = -214$ lb·ft, $M_{CD} = -214$ lb·ft, $M_{DC} = 214$ lb·ft

10–17. $M_{BC} = -19.9$ k·ft, $M_{BA} = 19.9$ k·ft, $M_{CB} = 22.4$ k·ft, $M_{CD} = -6.78$ k·ft, $M_{EC} = 1.18$ k·ft, $M_{CE} = -15.6$ k·ft

10–18. $M_{DB} = 0$, $M_{AB} = -330$ kN·m, $M_{CB} = 400$ kN·m

10–19. $M_{AB} = -24.8$ k·ft, $M_{BA} = 26.1$ k·ft, $M_{BC} = -26.1$ k·ft, $M_{CB} = 50.7$ k·ft, $M_{CD} = -50.7$ k·ft, $M_{DC} = -40.7$ k·ft

10–21. $M_{AB} = 128$ k·ft, $M_{BA} = 218$ k·ft,

$M_{BC} = -218\,\text{k} \cdot \text{ft}, M_{CB} = 175\,\text{k} \cdot \text{ft},$
$M_{CD} = -175\,\text{k} \cdot \text{ft}, M_{DC} = -55.7\,\text{k} \cdot \text{ft}$

10–22. $M_{BA} = -104\,\text{k} \cdot \text{ft}, M_{BC} = 104\,\text{k} \cdot \text{ft},$
$M_{CB} = 196\,\text{k} \cdot \text{ft}, M_{CD} = -196\,\text{k} \cdot \text{ft}$

10–23. $M_{AB} = 25.4\,\text{k} \cdot \text{ft}, M_{DC} = -56.7\,\text{k} \cdot \text{ft}$

Chapter 11

11–1. $M_{AB} = -47.5\,\text{k} \cdot \text{ft}, M_{BA} = 31.5\,\text{k} \cdot \text{ft},$
$M_{BC} = -31.5\,\text{k} \cdot \text{ft}, M_{CB} = 40.5\,\text{k} \cdot \text{ft}$

11–2. $M_{AB} = 0, M_{BA} = 84.0\,\text{k} \cdot \text{ft}, M_{BC} = -84.0\,\text{k} \cdot \text{ft},$
$M_{CB} = 84.0\,\text{k} \cdot \text{ft}, M_{CD} = -84.0\,\text{k} \cdot \text{ft}, M_{DC} = 0$

11–3. $M_{AB} = -30\text{kN} \cdot \text{m}, M_{BA} = 15\,\text{kN} \cdot \text{m},$
$M_{BC} = -15\text{kN} \cdot \text{m}, M_{CB} = 0, A_y = 33\,\text{kN},$
$B_y = 33\,\text{kN}, C_y = 6\,\text{kN}$

11–5. $M_{AB} = 10.4\,\text{k} \cdot \text{ft}, M_{BA} = 20.7\,\text{k} \cdot \text{ft},$
$M_{BC} = -20.7\,\text{k} \cdot \text{ft}, M_{CB} = 7.5\,\text{k} \cdot \text{ft},$
$M_{CD} = -7.5\,\text{k} \cdot \text{ft}$

11–6. $M_{AD} = 48\,\text{k} \cdot \text{ft}, M_{AB} = -48\,\text{k} \cdot \text{ft},$
$M_{BA} = -24\,\text{k} \cdot \text{ft}, M_{BC} = 24\,\text{k} \cdot \text{ft}, M_{CB} = 48\,\text{k} \cdot \text{ft},$
$M_{CE} = -48\,\text{k} \cdot \text{ft}$

11–7. $M_{AB} = 0, M_{BA} = 9\,\text{k} \cdot \text{ft}, M_{BC} = -9\,\text{k} \cdot \text{ft}, A_x = 0,$
$A_y = 11.25\,\text{k}, B_y = 12.75\,\text{k}, C_y = 12.75\,\text{k},$
$D_y = 11.25\,\text{k}$

11–9. $M_{BA} = 9\,\text{kN} \cdot \text{m}, M_{BC} = -9\,\text{kN} \cdot \text{m},$
$M_{CB} = -9.25\,\text{kN} \cdot \text{m}, M_{CD} = 9.25\,\text{kN} \cdot \text{m},$
$M_{DC} = 36\,\text{kN} \cdot \text{m}, M_{DE} = -36\,\text{kN} \cdot \text{m}$

11–10. $M_{AB} = -126\,\text{k} \cdot \text{ft}, M_{BA} = 72\,\text{k} \cdot \text{ft},$
$M_{BC} = -72\,\text{k} \cdot \text{ft}, M_{CB} = -36\,\text{k} \cdot \text{ft}$

11–11. $C_x = B_x = 4.55\,\text{k}, A_x = 2.95\,\text{k}, B_y = 0.903\,\text{k},$
$C_y = A_y = 4.10\,\text{k}$

11–13. $M_{AB} = 0, M_{BA} = 42.9\,\text{k} \cdot \text{ft}, M_{BC} = -22.1\,\text{k} \cdot \text{ft},$
$M_{BD} = 20.1\,\text{k} \cdot \text{ft}, M_{DB} = 64\,\text{k} \cdot \text{ft}, M_{DE} = -64$
$\text{k} \cdot \text{ft}, M_{CB} = -11.1\,\text{k} \cdot \text{ft}$

11–14. $M_{AB} = 0, M_{BA} = 8.78\,\text{k} \cdot \text{ft}, M_{BC} = -23.4\,\text{k} \cdot \text{ft},$
$M_{BD} = 14.6\,\text{k} \cdot \text{ft}, M_{DB} = 7.32\,\text{k} \cdot \text{ft}, M_{CB} = 0$

11–15. $M_{AB} = 20.6\,\text{k} \cdot \text{ft}, M_{BA} = 41.1\,\text{k} \cdot \text{ft},$
$M_{BC} = -41.1\,\text{k} \cdot \text{ft}, M_{CB} = 41.1\,\text{k} \cdot \text{ft},$
$M_{CD} = -41.1\,\text{k} \cdot \text{ft}, M_{DC} = -20.6\,\text{k} \cdot \text{ft}$

11–17. $M_{AB} = 3.93\,\text{k} \cdot \text{ft}, M_{BA} = 7.85\,\text{k} \cdot \text{ft},$
$M_{BC} = -7.85\,\text{k} \cdot \text{ft}, M_{CB} = 7.85\,\text{k} \cdot \text{ft},$
$M_{CD} = -7.85\,\text{k} \cdot \text{ft}, M_{DC} = -3.93\,\text{k} \cdot \text{ft}$

11–18. $M_{AB} = 0, M_{BA} = 9.68\,\text{k} \cdot \text{ft}, M_{BC} = -9.68\,\text{k} \cdot \text{ft},$
$M_{CB} = 25.2\,\text{k} \cdot \text{ft}, M_{CD} = -25.2\,\text{k} \cdot \text{ft}, M_{DC} =$
$9.68\,\text{k} \cdot \text{ft}, M_{DE} = -9.68, M_{ED} = 0$

11–19. $M_{AB} = 5.56\,\text{k} \cdot \text{ft}, M_{BA} = 11.1\,\text{k} \cdot \text{ft},$
$M_{BC} = -11.1\,\text{k} \cdot \text{ft}, M_{CB} = 24.4\,\text{k} \cdot \text{ft},$

$M_{CD} = -24.4\,\text{k} \cdot \text{ft}, M_{DC} = 11.1\,\text{k} \cdot \text{ft},$
$M_{DE} = -11.1\,\text{k} \cdot \text{ft}, M_{ED} = -5.56\,\text{k} \cdot \text{ft}$

11–21. $M_{AB} = -464\,\text{k} \cdot \text{ft}, M_{BA} = -110\,\text{k} \cdot \text{ft},$
$M_{BC} = 110\,\text{k} \cdot \text{ft}, M_{CB} = 155\,\text{k} \cdot \text{ft},$
$M_{CD} = -155\,\text{k} \cdot \text{ft}, M_{DC} = -243\,\text{k} \cdot \text{ft}$

11–22. $M_{AB} = -19.5\,\text{k} \cdot \text{ft}, M_{BA} = -15.0\,\text{k} \cdot \text{ft},$
$M_{BC} = 15.0\,\text{k} \cdot \text{ft}, M_{CB} = 20.1\,\text{k} \cdot \text{ft},$
$M_{CD} = -20.1\,\text{k} \cdot \text{ft}, M_{DC} = -36.9\,\text{k} \cdot \text{ft}$

11–23. $M_{AB} = -2251\,\text{k} \cdot \text{ft}, M_{BA} = -493\,\text{k} \cdot \text{ft},$
$M_{BC} = 493\,\text{k} \cdot \text{ft}, M_{CB} = 496\,\text{k} \cdot \text{ft},$
$M_{CD} = -496\,\text{k} \cdot \text{ft}$

11–25. $M_{AB} = 25.3\,\text{k} \cdot \text{ft}, M_{DC} = -56.7\,\text{k} \cdot \text{ft}$

Chapter 12

12–1. $C_{AB} = C_{BA} = 0.599, 24.1(10^3)\,\text{k} \cdot \text{ft}$

12–2. $C_{AB} = C_{BA} = 0.451, 7.21\,\text{MN} \cdot \text{m}$

12–3. $M_C = -26.4\,\text{k} \cdot \text{ft}, M_A = -26.9\,\text{k} \cdot \text{ft}$

12–5. $C_{BA} = 0.633, C_{AB} = 0.390, K_B = 104(10^3)\,\text{k} \cdot \text{ft},$
$K_A = 163(10^3)\,\text{k} \cdot \text{ft}$

12–6. $C_{AB} = 1.92, K_A = 1.97(10^3)\,\text{kN} \cdot \text{m}$

12–7. $\theta_B = 0.609(10^{-3})\,\text{rad}$

12–9. $M_C = 5.16\,\text{k} \cdot \text{ft}$

12–10. $C_{AB} = 0.707, C_{BA} = 0.315, K_A = 10.8(10^3)\,\text{k} \cdot \text{ft},$
$K_B = 24.1(10^3)\,\text{k} \cdot \text{ft}$

12–11. $M_{AB} = 113\,\text{k} \cdot \text{ft}, M_{BA} = 151\,\text{k} \cdot \text{ft}, M_{BC} =$
$-151\,\text{k} \cdot \text{ft}, M_{CB} = 461\,\text{k} \cdot \text{ft}$

12–13. $M_{AB} = 8.16\,\text{k} \cdot \text{ft}, M_{BA} = 16.3\,\text{k} \cdot \text{ft}, M_{BC} =$
$-16.3\,\text{k} \cdot \text{ft}, M_{CB} = 66.4\,\text{k} \cdot \text{ft}$

12–14. See Prob. 12–13

12–15. $M_{AB} = 1.75\,\text{k} \cdot \text{ft}, M_{BA} = 3.51\,\text{k} \cdot \text{ft}, M_{BC} =$
$-3.51\,\text{k} \cdot \text{ft}, M_{CB} = 3.51\,\text{k} \cdot \text{ft}, M_{CD} = -3.51\,\text{k} \cdot \text{ft},$
$M_{DC} = -1.75\,\text{k} \cdot \text{ft}$

12–17. $M_{AB} = 0, M_{BA} = -42.3\,\text{k} \cdot \text{ft}, M_{BF} = -44.7\,\text{k} \cdot \text{ft},$
$M_{BC} = 87.1\,\text{k} \cdot \text{ft}, M_{CB} = 185\,\text{k} \cdot \text{ft}, M_{CE} = 213$
$\text{k} \cdot \text{ft}, M_{CD} = -398\,\text{k} \cdot \text{ft}, M_{EC} = 106\,\text{k} \cdot \text{ft}, M_{FB} =$
$-22.4\,\text{k} \cdot \text{ft}, M_{DC} = 0$

12–18. See Prob. 12–17

12–19. $M_{AB} = 0, M_{BA} = 29.0\,\text{k} \cdot \text{ft}, M_{BC} = -29.0\,\text{k} \cdot \text{ft},$
$M_{CB} = 29.0\,\text{k} \cdot \text{ft}, M_{CD} = -29.0\,\text{k} \cdot \text{ft}, M_{DC} = 0$

Chapter 13

13–1. (a) $\begin{bmatrix} 8 & 1 & -3 \\ 8 & 1 & 13 \end{bmatrix}$; (b) $\begin{bmatrix} -4 & 4 & -3 \\ 2 & 1 & -11 \end{bmatrix}$

Answers

13–2. $\begin{bmatrix} 7 \\ 7 \\ 6 \end{bmatrix}$

13–3. $\begin{bmatrix} 6 & 3 \\ 3 & 14 \end{bmatrix}$

13–5. $\begin{bmatrix} -2 & 22 & 18 \\ 7 & -13 & 1 \\ 6 & -8 & 8 \end{bmatrix}$

13–6. $\begin{bmatrix} -21 & 4 & 2 \\ 37 & 2 & 0 \\ -35 & 22 & 12 \end{bmatrix}$

13–7. $\begin{bmatrix} 79 \\ -5 \end{bmatrix}$

13–9. $\begin{bmatrix} 72 \\ -3 \end{bmatrix}$

13–10. $\begin{bmatrix} 2 & 48 & 2 \\ 0 & 17 & 2 \\ -6 & -5 & 2 \end{bmatrix}$

13–11. $\begin{bmatrix} 29 & 17 \\ 17 & 10 \end{bmatrix}$

13–14. $-33, -18$

13–15. $\begin{bmatrix} \frac{2}{13} & \frac{1}{13} \\ \frac{3}{13} & -\frac{5}{13} \end{bmatrix}$

13–17. $\begin{bmatrix} 0.75 \\ 0.25 \\ 0.75 \end{bmatrix}$

13–18. $x_2 = 0.25, x_3 = 0.75, x_1 = 0.75$

13–19. $\begin{bmatrix} 1.0 \\ 1.0 \\ -1.0 \end{bmatrix}$

Chapter 14

14–1. $K_{11} = (510.7), K_{12} = 0, K_{13} = -201.4, K_{14} = 0,$
$K_{15} = -154.7, K_{16} = -116, K_{17} = -154.7,$
$K_{18} = 116$

14–2. 0.0230 in.

14–3. $q_1 = -3.33 \text{ k}, q_2 = 0, q_3 = 3.33 \text{ k}$

14–5. $D_2 = -330 \text{ ft/AE}, q_1 = -2.83 \text{ k}$

14–6. $K_{11} = 113.4, K_{12} = 28.8, K_{13} = -38.8,$
$K_{14} = -28.8, K_{15} = 0, K_{16} = 0, K_{17} = -75,$
$K_{18} = 0$

14–7. $q_6 = 729 \text{ N}$

14–9. $q_8 = -2.83 \text{ kN}, q_5 = 2.00 \text{ kN}$

14–10. $K_{11} = 0.08019, K_{12} = 3.806(10^{-3}),$
$K_{13} = -0.04419, K_{14} = 0.04419, K_{15} = 0,$
$K_{16} = 0, K_{17} = -0.036, K_{18} = -0.048$

14–11. $q_1 = 4.52 \text{ k}, q_2 = 1.20 \text{ k}, q_3 = -3.00 \text{ k}$

14–13. $D_2 = -6.89(10^{-3}) \text{ in.}, q_2 = 50.9 \text{ lb}$

14–14. $K_{11} = 0.15(10^6), K_{12} = 0, K_{13} = 0, K_{14} = 0,$
$K_{15} = -0.075(10^6), K_{16} = 0, K_{17} = -0.075(10^6),$
$K_{18} = 0$

14–15. $D_2 = -0.540(10^{-6}) \text{ m}, q_4 = 10.0 \text{ kN}$

Chapter 15

15–1. $M_1 = 18.5 \text{ kN} \cdot \text{m}$, $M_3 = -20.375 \text{ kN} \cdot \text{m}$

15–2. $M_1 = 2.25 \text{ kN} \cdot \text{m}$, $M_2 = 4.50 \text{ kN} \cdot \text{m}$

15–3. $M_1 = M_3 = 44.2 \text{ kN} \cdot \text{m}$

15–5. $F_1 = F_3 = 25.5 \text{ k}$, $F_2 = 21.0 \text{ k}$

15–6. $K_{11} = 130.9$, $K_{12} = 0$, $K_{13} = -7854.17$, $K_{14} = -130.9$, $K_{15} = 0$, $K_{16} = -7854.17$, $K_{17} = 0$, $K_{18} = 0$, $K_{19} = 0$

15–7. $D_1 = 0.761 \text{ in.}$, $D_2 = -0.423 \text{ in.}$, $D_3 = 6.84(10^{-3}) \text{ rad}$

15–9. $D_1 = -0.406(10^{-3}) \text{ in.}$, $D_3 = 0.149(10^{-3}) \text{ rad}$, $Q_6 = -0.818 \text{ k}$, $Q_7 = 3.05 \text{ k}$, $Q_8 = 0.818 \text{ k}$, $Q_9 = 1.95 \text{ k}$

15–10. $K_{11} = 6123.25$, $K_{12} = 0$, $K_{13} = 5873.84$, $K_{14} = 0$, $K_{15} = 5873.84$, $K_{16} = -6041.67$, $K_{17} = 0$, $K_{18} = -81.581$, $K_{19} = 0$

15–11. $Q_7 = 17.0 \text{ k}$, $Q_8 = -5.00 \text{ k}$, $Q_9 = 7.00 \text{ k}$

15–13. $Q_5 = 77.4 \text{ kN}$, $Q_6 = -60.5 \text{ kN}$, $Q_7 = -77.4 \text{ kN}$, $Q_8 = 60.5 \text{ kN}$, $Q_9 = 129 \text{ kN} \cdot \text{m}$

15–14. $K_{11} = 4806.58$, $K_{12} = 0$, $K_{13} = 13216.15$, $K_{14} = -4531.25$, $K_{15} = 0$, $K_{16} = 0$, $K_{17} = 0$, $K_{18} = 13216.15$, $K_{19} = 0$, $K_{110} = 0$, $K_{111} = -275.336$, $K_{112} = 0$

15–15. $Q_9 = -300 \text{ lb}$, $Q_{10} = 600 \text{ lb}$, $Q_{11} = -300 \text{ lb}$, $Q_{12} = -600 \text{ lb}$

15–17. $D_1 = 0.0783 \text{ in.}$, $Q_7 = -3.02 \text{ k}$, $Q_8 = -3.96 \text{ k}$, $Q_9 = 243 \text{ k} \cdot \text{in.}$, $Q_{10} = -2.98 \text{ k}$, $Q_{11} = 3.96 \text{ k}$, $Q_{12} = 240 \text{ k} \cdot \text{in.}$

Index

Geometric Properties of Areas

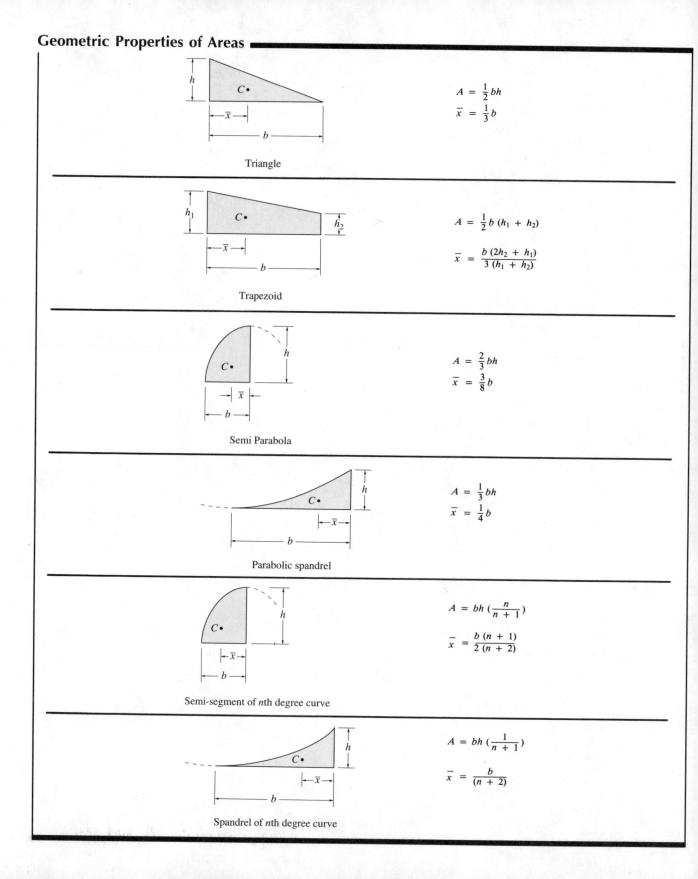

$$A = \tfrac{1}{2} bh$$
$$\bar{x} = \tfrac{1}{3} b$$

Triangle

$$A = \tfrac{1}{2} b \, (h_1 + h_2)$$

$$\bar{x} = \frac{b \, (2h_2 + h_1)}{3 \, (h_1 + h_2)}$$

Trapezoid

$$A = \tfrac{2}{3} bh$$
$$\bar{x} = \tfrac{3}{8} b$$

Semi Parabola

$$A = \tfrac{1}{3} bh$$
$$\bar{x} = \tfrac{1}{4} b$$

Parabolic spandrel

$$A = bh \left(\frac{n}{n+1} \right)$$

$$\bar{x} = \frac{b \, (n+1)}{2 \, (n+2)}$$

Semi-segment of nth degree curve

$$A = bh \left(\frac{1}{n+1} \right)$$

$$\bar{x} = \frac{b}{(n+2)}$$

Spandrel of nth degree curve